RAPID METHODS IN FOOD MICROBIOLOGY

RAPID METHODS IN FOOD MICROBIOLOGY

Edited by:

Chenggui Sun

Delve
PUBLISHING

www.delvepublishing.com

Rapid methods in Food Microbiology

Chenggui Sun

Delve Publishing
2010 Winston Park Drive,
2nd Floor
Oakville, ON L6H 5R7
Canada
www.delvepublishing.com
Tel: 001-289-291-7705
001-905-616-2116
Fax: 001-289-291-7601
Email: orders@arclereducation.com

ISBN: 978-1-77361-369-7 (Hardcover)

DECLARATION

ABOUT THE EDITOR

Dr. Chenggui Sun obtained his Ph.D. degree in Chemical Engineering from the University of Waterloo, a public research university with a main campus in Waterloo, Ontario, Canada. His academic background is in the area of nanotechnology and membrane technology, which is at the interface between Chemical Engineering and Environmental Engineering. Dr. Sun enjoys the challenge of combining technology, engineering, and business expertise to contribute to organizational development in the context of a more sustainable, healthier world. He has proven ability to understand client needs and perspectives in order to formulate and implement strategic plans, and known networking and collaboration skills with a commitment to fostering growth in others. Dr. Sun is also reputable for approaching tasks with creativity, integrity, and drive to inspire cooperation and exceed expectations.

TABLE OF CONTENTS

LIST OF CONTRIBUTORS

Eugenia Bezirtzoglou
Department of Microbiology

Vassiliki Maipa
Department of Hygiene, Medical School, University of Ioannina, Ioannina, Greece and

Chrissoula Voidarou
Department of Microbiology

Arsenis Tsiotsias
Department of Microbiology

Maria Papapetropoulou
Department of Environmental Microbiology, Medical School, University of Patra, Patra, Greece

J. Schlundt
World Health Organization, Headquarters, Geneva, Switzerland

Marion J.Healy
Food Standards Australia New Zealand, Barton, Australian Capital Territory, 2905, Australia

WeidaTong
National Center for Toxicological Research, US Food and Drug Administration, Jefferson, AR, 72079-9502, US

StephenOstroff
US Food and Drug Administration, Silver Spring, MD, 20993-0002, USA

Hans-GeorgEichler
European Medicines Agency, London, E14 5EU, United Kingdom

AlexPatak
European Commission, Joint Research Centre, Ispra, Varese, 21027, Italy

Margaret Neuspiel
Canadian Food Inspection Agency, Ottawa, Ontario, Canada

Hubert Deluyker
European Food Safety Authority, Parma, 43126, Italy

William SlikkerJr.
National Center for Toxicological Research, US Food and Drug Administration, Jefferson, AR, 72079-9502, US

Rebecca L. Bell
Center for Food Safety and Applied Nutrition, U.S. Food and Drug Administration, College Park, MD, USA

Karen G. Jarvis
Center for Food Safety and Applied Nutrition, U.S. Food and Drug Administration, Laurel, MD, USA

Andrea R. Ottesen
Center for Food Safety and Applied Nutrition, U.S. Food and Drug Administration, College Park, MD, USA

Melinda A. McFarland
Center for Food Safety and Applied Nutrition, U.S. Food and Drug Administration, College Park, MD, USA

Eric W. Brown
Center for Food Safety and Applied Nutrition, U.S. Food and Drug Administration, College Park, MD, USA

Thadiyam Puram Ramees
Division of Veterinary Public Health, ICAR-Indian Veterinary Research Institute (IVRI), Bareilly, India

Kuldeep Dhama
Division of Pathology, ICAR-Indian Veterinary Research Institute (IVRI), Bareilly, India

Kumaragurubaran Karthik
Central University Laboratory, Tamil Nadu Veterinary and Animal Sciences University, Chennai, India

Ramswaroop Singh Rathore
Division of Veterinary Public Health, ICAR-Indian Veterinary Research Institute (IVRI), Bareilly, India

Ashok Kumar
Division of Veterinary Public Health, ICAR-Indian Veterinary Research Institute (IVRI), Bareilly, India

Mani Saminathan
Division of Pathology, ICAR-Indian Veterinary Research Institute (IVRI), Bareilly, India

Ruchi Tiwari
Department of Veterinary Microbiology and Immunology, College of Veterinary Sciences, UP Pandit Deen Dayal Upadhayay Pashu Chikitsa Vigyan Vishwavidyalay Evum Go-Anusandhan Sansthan (DUVASU), Mathura, India

Yashpal Singh Malik
Division of Biological Standardization, ICAR-Indian Veterinary Research Institute (IVRI), Bareilly, India

Raj Kumar Sing
ICAR-Indian Veterinary Research Institute (IVRI), Bareilly, India

Ruud H.Deurenberg
Department of Medical Microbiology, University of Groningen, University Medical Center Groningen, The Netherlands

ErikBathoorn
Department of Medical Microbiology, University of Groningen, University Medical Center Groningen, The Netherlands

Monika A.Chlebowicz
Department of Medical Microbiology, University of Groningen, University Medical Center Groningen, The Netherlands

NatachaCouto
Department of Medical Microbiology, University of Groningen, University Medical Center Groningen, The Netherlands

MithilaFerdous
Department of Medical Microbiology, University of Groningen, University Medical Center Groningen, The Netherlands

SilviaGarcía-Cobos
Department of Medical Microbiology, University of Groningen, University Medical Center Groningen, The Netherlands

Anna M.D.Kooistra-Smid
Department of Medical Microbiology, University of Groningen, University Medical Center Groningen, The Netherlands
Certe, Department of Medical Microbiology, Groningen, The Netherlands

Erwin C.Raangs
Department of Medical Microbiology, University of Groningen, University Medical Center Groningen, The Netherlands

SigridRosema
Department of Medical Microbiology, University of Groningen, University Medical Center Groningen, The Netherlands

Alida C.M.Veloo
Department of Medical Microbiology, University of Groningen, University Medical Center Groningen, The Netherlands

KaiZhou
State Key Laboratory for Diagnosis and Treatment of Infectious Diseases, Collaborative Innovation Center for Diagnosis and Treatment of Infectious Diseases, The First Affiliated Hospital of Medicine School, Zhejiang University, Hangzhou, China

Alexander W.Friedrich
Department of Medical Microbiology, University of Groningen, University Medical Center Groningen, The Netherlands

John W.A.Rossen
Department of Medical Microbiology, University of Groningen, University Medical Center Groningen, The Netherlands

Kalliopi Rantsiou
Department of Agriculture, Forest and Food Sciences, University of Turin, Largo Paolo Braccini 2, 10095 Grugliasco, Italy

Sophia Kathariou
North Carolina State University, Department of Food, Bioprocessing and Nutrition Sciences, Raleigh, NC, USA

Annet Winkler
Mondelez International, Kraft Foods R&D, Unterbiberger St. 15, D-81737 Munich, Germany

Panos Skandamis
Laboratory of Food Quality Control & Hygiene, Department of Food Science & Technology, Agricultural University of Athens, Iera Odos 75, 118 55 Athens, Greece

Manuel Jimmy Saint-Cyr
SECALIM Unit UMR1014, Oniris, INRA, Université Bretagne Loire, Nantes, France

Katia Rouzeau-Szynalski
Food Safety Microbiology, Nestec Ltd, Nestlé Research Center, Lausanne, Switzerland

Alejandro Amézquita
Safety and Environmental Assurance Centre, Unilever, Colworth Science Park, Sharnbrook, Bedfordshire MK44 1LQ, UK

María-José Chapela
New Preservation Technologies Division, ANFACO-CECOPESCA, Campus Univ. 16, 36310 Vigo, Pontevedra, Spain

Alejandro Garrido-Maestu
Microbiology and Bioassays Laboratory, ANFACO-CECOPESCA, Campus Univ. 16, 36310 Vigo, Pontevedra, Spain.

Food Safety Division, ANFACO-CECOPESCA, Campus Univ. 16, 36310 Vigo, Pontevedra, Spain

Ana G. Cabado
Food Safety Division, ANFACO-CECOPESCA, Campus Univ. 16, 36310 Vigo, Pontevedra, Spain

Carolina Ripolles-Avila
Hygiene and Food Inspection Unit, Department of Food and Animal Science, Universitat Autònoma de Barcelona, Barcelona, Spain

Abel G. Ríos-Castillo
Hygiene and Food Inspection Unit, Department of Food and Animal Science, Universitat Autònoma de Barcelona, Barcelona, Spain

José J. Rodríguez-Jerez
Hygiene and Food Inspection Unit, Department of Food and Animal Science, Universitat Autònoma de Barcelona, Barcelona, Spain

Yingsha Liu
College of Food Engineering and Nutritional Science, Shaanxi Normal University, Xi' an, P. R. China

Jianke Li
College of Food Engineering and Nutritional Science, Shaanxi Normal University, Xi' an, P. R. China

Lin Zhang
College of Food Engineering and Nutritional Science, Shaanxi Normal University, Xi' an, P. R. China

Xiaoxia Wu
College of Food Engineering and Nutritional Science, Shaanxi Normal University, Xi' an, P. R. China

Feng Wang
College of Food Engineering and Nutritional Science, Shaanxi Normal University, Xi' an, P. R. China

Bingbing Song
College of Food Engineering and Nutritional Science, Shaanxi Normal University, Xi' an, P. R. China

Soad A Abdallah
General Science Unit, College of Nursing, Public Authority for Applied Education & Training (PAAET), Al-Shuwaikh B, 70466 Kuwait City, Kuwait. On tenured leave from Women's College for Arts, Science & Education, Botany Department, Ain Shams University. Cairo, Egypt

Laila A Al-Shatti
General Science Unit, College of Nursing, Public Authority for Applied Education & Training (PAAET), Al-Shuwaikh B, 70466 Kuwait City, Kuwait. On tenured leave from Women's College for Arts, Science & Education, Botany Department, Ain Shams University. Cairo, Egypt

Ali F Alhajraf
Biomedical Sciences Department, College of Nursing, PAAET, Kuwait City, Kuwait

Noura Al-Hammad
General Science Unit, College of Nursing, Public Authority for Applied Education & Training (PAAET), Al-Shuwaikh B, 70466 Kuwait City, Kuwait. On tenured leave from Women's College for Arts, Science & Education, Botany Department, Ain Shams University. Cairo, Egypt

Bashayer Al-Awadi
General Science Unit, College of Nursing, Public Authority for Applied Education & Training (PAAET), Al-Shuwaikh B, 70466 Kuwait City, Kuwait. On tenured leave from Women's College for Arts, Science & Education, Botany Department, Ain Shams University. Cairo, Egypt

Noordiana Nordin
Institute of Advanced Technology, Universiti Putra Malaysia (UPM), 43400 Serdang, Selangor, Malaysia
Food Safety Research Centre, Faculty of Food Science and Technology, Universiti Putra Malaysia, 43400 Serdang, Selangor, Malaysia

Nor Azah Yusof
Institute of Advanced Technology, Universiti Putra Malaysia (UPM), 43400 Serdang, Selangor, Malaysia
Department of Chemistry, Faculty of Science, Universiti Putra Malaysia (UPM), 43400 Serdang, Selangor, Malaysia

Jaafar Abdullah
Institute of Advanced Technology, Universiti Putra Malaysia (UPM), 43400 Serdang, Selangor, Malaysia
Department of Chemistry, Faculty of Science, Universiti Putra Malaysia (UPM), 43400 Serdang, Selangor, Malaysia

Son Radu
Food Safety Research Centre, Faculty of Food Science and Technology, Universiti Putra Malaysia, 43400 Serdang, Selangor, Malaysia

Roozbeh Hushiarian
La Trobe Institute for Molecular Science, La Trobe University, Melbourne, VIC 3086, Australia
Department of Life Sciences, Imperial College London, London SW7 2AZ, UK

Florian Matthias Freimoser
Agroscope, Institute for Plant Production Sciences IPS, Schloss 1, 8820 Wädenswil, Switzerland

Maja HilberBodmer
Agroscope, Institute for Plant Production Sciences IPS, Schloss 1, 8820 Wädenswil, Switzerland

René Brunisholz
Functional Genomics Center Zurich, ETH Zurich and University of Zurich, Winterthurerstr. 190, 8057 Zurich, Switzer land

David Drissner
Agroscope, Institute for Food Sciences IFS, Schloss 1, 8820 Wädenswil, Switzerland

Marwa M Saleh
Department of Biotechnology, Al-Farabi Kazakh National University, 71, Al-Farabi, ave, Almaty 050040, Kazakhstan

Botany Department, Faculty of Science, Suez Canal University, Ismaillia 41522, Egypt

Dmitry N Matorin
Faculty of Biology, Moscow State University, Vorobyevi Gory 119992, Moscow, Russia

Bolatkhan K Zayadan
Department of Biotechnology, Al-Farabi Kazakh National University, 71, Al-Farabi, ave, Almaty 050040, Kazakhstan

Daria A Todorenko
Faculty of Biology, Moscow State University, Vorobyevi Gory 119992, Moscow, Russia

Evgenii P Lukashov
Faculty of Biology, Moscow State University, Vorobyevi Gory 119992, Moscow, Russia

Mona M Gaballah
Botany Department, Faculty of Science, Suez Canal University, Ismaillia 41522, Egypt

LIST OF ABBREVIATIONS

AFA	Adaptive Focused Acoustics
AFLP	Amplified fragment length polymorphism analysis
ARG	Antibiotic resistance genes
ASB	Arcobacter selective enrichment broth
ACT	Artemis Comparison Tool
BIP	Border Inspection Post
CIEF	Capillary isoelectric focusing
CZE	Capillary zone electrophoresis
CDC	Centers for Disease Control and Prevention
CFU	Colony forming units
DNA	Deoxyribonucleic acid
DEM	Direct epifluorescence microscopy
ECL	Electrochemiluminescence
ECDC	European Center for Disease Prevention and Control
EPS	Extracellular polymeric substances
FBO	Food business operators
FSMS	Food safety management systems
FINS	Forensically informative nucleotide sequencing
GI	Gastro-intestinal
GMMs	Genetically modified microorganisms
GWAS	Genome–wide association studies
GCRSR	Global Coalition for Regulatory Science Research
GCRSR	Global Coalition on Regulatory Science Research
GSRS	Global Summit on Regulatory Science
GALT	Gut-associated immune system
GALT	Gut-associated lymphoid tissue

HACCP	Hazard analysis critical control point
ICS	Immunochromatographic strip
IMS	Immunomagnetic separation
IGS	Intergenic sequence
ITS	Internal transcribed spacer
IoT	Internet of Things
LT	Labile toxins
LOD	Limit of detection
LC-MS	Liquid chromatography mass spectrometry
MAQC	MicroArray Quality Control
MRA	Microbiological Risk Assessment
ICMSF	Microbiological Specifications for Foods
mCCDA	Modified charcoal cefoperazone deoxycholate
MLST	Multi-locus sequencing typing
NCBI	National Center for Biotechnology Information
NGS	Next Generation Sequencing
ETR	Noncyclic electron transport
PCR	Polymerase chain reaction
PI	Propidium iodide
PFGE	Pulsed-field gel electrophoresis
qPCR	quantitative polymerase chain reaction
QSARs	Quantitative Structure-Activity Relationships
RSD	Relative standard deviation
CARD	Resistance Database
RTPCR	Reverse-transcription polymerase chain reaction
rDNA	ribosomal deoxyribonucleic acid
SEM	Scanning electron microscopy
SRM	Selected reaction monitoring
SMRT	Single molecule real-time
SNPs	Single Nucleotide Polymorphisms
SNP	Single-nucleotide polymorphisms
SD	Sprague Dawley

TVC	Total viable counts
TFA	Trifluoroacetic acid
TAP	Tris-acetate-phosphate
TSB	Tryptic soy broth
UPB	Universal Pre-enrichment Broth
UMCG	University Medical Center Groningen
VFDB	Virulence Factor Database
WGS	Whole genome sequencing
WHO	World Health Organization
YACs	Yeast artificial chromosomes

PREFACE

Microorganisms have been very important part in food production, and foods are also ideal for microbial to survive and grow. Undesirable microorganism in foods can cause foodborne disease, which may lead to major public hearth problem. Compared to conventional methods, rapid methods are time-efficient way to detect food microbiology. So, they are very important for providing safe food supply, preventing and treating foodborne diseases effectively. Over years, a variety of rapid detection methods have been developed and applied. The major content of this book is to introduce the recent advances in rapid detection methods. The negative role of food microorganisms will be discussed and safety assessment and regulations about food productions will be reviewed as well. More details about the content of this book are as follows:

- Foodborne intestinal bacterial pathogens and safety assessment of food derived from genetically modified microorganisms are reviewed;
- "Regulatory bioinformatics", which refer to the efforts to develop policies, strategies, and approaches to analyses and interpret data generated by emerging life-science technologies for regulatory purpose, has been striving to develop and implement a standardized and transparent bioinformatic framework. Those efforts are reported in this book.
- Salmonella and Arcobacter is two foodborne pathogens responsible for serious illnesses. Rapid, reliable detection and identification of these pathogens is important for food safety. The approaches in the detection and identification of these pathogens are reviewed.
- Reviews about methods that can detect foodborne pathogen, including sequencing technologies and quantitative polymerase chain reaction (qPCR), are presented.
- Recent researches on the microbiology detection methods for food industries are reported. These researches include work on peroxide biodetector, colloidal gold-based immunochromatographic assay, electronic nose, electrochemical biosensor, mass spectrometry and the construction of stable laboratory control strains for food safety relevant Enterobacteriaceae.

CHAPTER 1

FOOD-BORNE INTESTINAL BACTERIAL PATHOGENS

Eugenia Bezirtzoglou[1], Vassiliki Maipa[2], Chrissoula Voidarou[1], Arsenis Tsiotsias[1] and Maria Papapetropoulou[3]

[1]Department of Microbiology,

[2]Department of Hygiene, Medical School, University of Ioannina, Ioannina, Greece and

[3]Department of Environmental Microbiology, Medical School, University of Patra, Patra, Greece

ABSTRACT

The foods we eat are rarely sterile. They often provide an ideal environment for microbial survival and growth. Microorganisms present in food can originate from the natural microflora of the raw material or can be introduced in the course of harvesting, slaughter, processing, storage and distribution.

CITATION: Bezirtzoglou, E., et al., "Food-Borne Intestinal Bacterial Pathogens", Microbial Ecology in Health and Disease (2000), https://doi.org/10.1080/089106000750060350.

Microbial growth in foods involves changes including intrinsic or extrinsic factors interacting with the microbial community over time. Undesirable microorganisms in foods can manifest their presence in two different ways: they can cause spoilage of food and/or they can cause food borne illness. On the other hand, some microorganisms can transform food in a beneficial way; this is called food fermentation.

INTRODUCTION

Foods are not sterile: they carry a natural and transient flora reflecting their environment. Foods carry microbial associations the composition of which varies. These microorganisms are introduced in the food during the course of harvesting or slaughter or they come from the natural micro flora of the raw material (1) (2). However, in most cases, food is consumed without objection and consequences and, in same cases, microorganisms manifest their presence by causing spoilage of food or are associated with food-borne illness in man. It is of interest to report here that some microorganisms have the ability to transform food in a beneficial way; this is called food fermentation. Fermented foods have the reputation to be beneficial to human health, especially lactic acid bacteria. Live cultures of lactic acid bacteria and Bifidobacteria in foods are termed «Probiotics» (3). Organisms enter the alimentary tract through foods, water and air. There is infection when there is presence of large numbers of pathogens in the vehicle (4). Unwashed hands, lesions on skin and mucous membranes of food handlers are reported to be important sources of food-borne illness (5) (6).

Human or animal faecal contamination of food or water is another possibility; untreated water supplies, flies and fingers constitute the vehicle in this last case. Non properly cooked, stored or refrigerated food may be heavily contaminated (7). Two primary types of food-related diseases are described:

FOOD-BORNEINFECTIONS FOOD-BORNEINTOXICATIONS

A food-borne infection involves the ingestion of the pathogen, followed by growth accompanied by tissue invasion and: or the release of toxins.

An intoxication is a true food poisoning; it is generally associated with consumption of food containing preformed toxins produced by microbial growth prior to ingestion. In this communication, we report the most common bacterial agents introduced by food responsible for foodborne diseases.

Aeromonas hydrophila

This bacterium has the ability to grow at low temperatures (-0.1°C) and its prevalence is increasing by consumption of chilled foods. The microorganism likes to grow in water (hydrophila¾friend of water in Greek). The microorganism is found in drinking water, where it can multiply in the piped water systems. Its principal reservoir is the aquatic environment and especially freshwater lakes, streams and wastewater systems, where it can be found in numbers as high as 10^8 CFU: ml.

In immunosuppressed persons, food and: or water can be a source of severe extra intestinal infection. Symptoms of gastroenteritis accompanied by profuse watery diarrhea occur in young children (\5 years old).

Aerolysin, a cytotoxin and 3 cytotonic enterotoxins acting like cholera toxin are produced by *Aeromonas*. *Aeromonas* can cause infections when the skin is punctured, sometimes as a result of a swimming accident. *A. hydrophila* has been also associated with acute diarrhoea of travellers in Asian countries. Fresh foods, salad vegetables, water, .sh, meat, poultry and raw milk are reported to be the most common vehicle for the microorganism (6).

Brucella

The genus *Brucella* includes 4 species that are human pathogens: *B. melitensis*, *B. abortus*, *B. suis* and *B. canis*. *Brucella* is mainly contracted by consumption of milk and milk products, or inadequately cooked meat from infected animals. Normal milk pasteurisation process kills the microorganism. It is also reported as an occupational disease in farmers and veterinarians, contracted from close association with infected animals. All *Brucella* species contains a mixture of two antigens: antigen A and antigen M, which are complexes of proteins and lipopolysaccharides. However, a quantitative difference in antigens is associated with the different *Brucella* spp.

Brucellosis has an incubation period of one to 6 weeks, followed by undulant fever with fatigue, sweats, headache, anorexia, pains in the limbs and back and weight loss. Skin, mucosa and conjunctiva constitute the main portal of entry for the microorganism. The microorganism spreads through the lymphatic route to ganglions and organs. Brucellosis can be asymptomatic in professionals being in close association with infected animals. Acute brucellosis is reported when the antibodies titre is superior to 1:1280. Recurrent brucellosis occurs sometimes years after a primary infection. Localised brucellosis, especially in untreated patients, is associated with the urogenital, central nervous and the cardiovascular system, as well as with lesions in the lung, bone and articulations (6).

Bacillus cereus and other Bacillus spp.

B. cereus shows a uniform distribution in the environment, soil, water, and vegetation. Twenty-nine serotypes of the organism are reported and are based in differences in the flagella (H) antigen. Epidemiological studies by serotyping are necessary in case of outbreaks. *B. cereus* is described in outbreaks involving important numbers of people (300–600 persons).

The usual food vehicle is reheated fried rice. In Norway, a vanilla sauce stored at room temperature before serving has been incriminated in an outbreak. Enterotoxins are formed in food or in the human gut from growth of *B.cereus*. Two different types of foodborne disease have been reported:

DIARRHOEAL SYNDROME AND COLITIS

The disease is associated with consumption of food including spores. Toxin production occurs *in vivo*. The onset of the disease is 8 to 16 hours after food consumption and lasts between 12 to 24 hours. The disease is caused by a heat-labile toxin which acts on adenylocyclase and this results in a profuse watery diarrhoea accompanied by abdominal pain and rectal tenesmus. The diarrhoeal syndrome is associated mainly with meat products, soups, vegetables, puddings and sauces.

EMETIC SYNDROME AND GASTRITIS

This type of disease shows a shorter incubation period than the diarrhoeal syndrome, usually 1 to 5 hours. Nausea and vomiting, main symptoms of the disease, last between 6 to 24 hours. The disease is due to preformed enterotoxin in food. This enterotoxin is heat resistant.

4The serotype-1 is associated with the emetic syndrome (64%). The emetic syndrome is also associated with consumption of starchy products such as rice and pastries. The disease is referred to as the Chinese restaurant syndrome. Rice prepared in bulk in advance contains spores surviving precooking which germinate and grow by producing an emetic toxin during storage. This could be prevented by chilling of the prepared rice.

Another *Bacillus*, *B. subtilis* is associated with products such as meat pies, pastries, meats, rices and bread. The «Ropey bread» phenomenon occurs when spores of *B. subtilis* surviving baking degrade the loaf's internal structure and produce a sticky slime.

A very rapid onset of the disease occurs (10 minutes after consumption of the bread) including symptoms such as nausea, vomiting, diarrhoea, headache and chills.

Campylobacter

The thermophilic (growth at 42°C) *Campylobacter*, *C. jejuni* and *C. coli* are a major cause of diarrhoeal illness in many countries (8). *C. laridis*, *C. hyointestinalis* and *C. concisus* have been also incriminated as diarrhoeal agents, and .nally *C. pylori* (reclassi.ed as *Helicobacter pylori*) has been associated with gastritis, stomach and duodenal ulcer. Only 500 cells of enteropathogenic *Campylobacter* can cause an acute enterocolitis. The infective dose is closely related to the health status of the individual. Young adults and children are particularly susceptible. The incubation period is from 1 to 11 days, followed by symptoms including malaise, fever, abdominal pain, and diarrhoea. The diarrhoea produces foul-smelling stools (106–109 cells:g of faeces) and can vary from profuse watery to bloody and dysenteric lasting a week. Excretion of microorganisms in the faeces continues for up 2 to 3 weeks. Complications includes arthritis and neurological disease referred to as the Guillain- Barre syndrome. *Campylobacter* penetrates the mucus which covers the epithelial surface of the intestine and adhere through adhesins to epithelial cells.

The different symptoms of the disease reflect different pathogenic mechanisms:

- Profuse watery diarrhoea is due to the elaboration of an enterotoxin similar to the cholera toxin. There is production of a heat labile, acid-sensitive enterotoxin which stimulates the adenylate cyclase activity.
- Dysenteric syndrome: Invasion and necrosis of the cells of the terminal ileum and colon occur producing blood and pus in stools. In this case, a heat-labile, trypsin-sensitive cytotoxin has been isolated in almost 70%of the cases.

Food is incriminated as the principal vehicle (9). The microorganism can be also contracted by infected animals (kitten, puppies) and water. Direct transmission from person- to-person occurs. *Campylobacter* cannot survive correct pasteurisation procedures and the majority of outbreaks have involved unpasteurized milk.

Clostridium botulinum

The disease «botulism» results from the ingestion of an exotoxin produced by *Clostridium botulinum* growing in the food (10). The disease derives its name from *botulus*, the Latin word for sausage, because the earliest recorded cases were traced to the consumption of contaminated sausage products. The microorganism was isolated by Van Ermengen (University of Gent, Belgium) in 1896 who gave it the name *Bacillus botulinus*. In 1923, it was reclassi.ed as *Clostridium botulinum*.

In Europe, most cases are still due to sausages and home-preserved meats, whereas in the U.S., they are mainly associated with home-canned fruits and vegetables. The food can be contaminated at the source or during processing with spores or vegetative cells of *C. botulinum*. Food consumed cold or after mild heat treatment are insuf.cient to inactivate the botulism toxin and .nally anaerobic conditions in the food favours *C. botulinum* growth.

Fish, smoked .sh, raw fermented .sh (sushi) or seal meat, vacuum packed hot-smoked .sh and mushrooms are associated with the disease, because salting and smoking treatment are often insuf.cient to eliminate *C. botulinum* or inhibit its growth during storage. A minimum salt concentration (3%) and a temperature of 63°C during smoking are recommended.

The most potent toxin even identi.ed is produced by *C. botulinum*. Eight serologically distinct toxins are recognised (A, B, C_1, C_2, D, E, F and G), and a single strain of *C. botulinum* will usually produce only one type. As little as 1 g of these potent toxins ingested can inhibit nerve function and produce fatal paralysis. The botulism toxins are neuneurotoxins, which act locally on the gut, after affecting primarily the cholinergic nerves of the peripheral nervous system.

The toxin is not inactivated by proteolytic enzymes in the gut, is absorbed in the upper part of the intestine and reaches the bloodstream via the lymphatics. It binds to the nerve endings of the nerve-muscle junction, blocking release of the acetyl choline responsible for transmission of stimulus, thus producing flaccid paralysis.

The incubation period of the disease varies from 8 h to 8 days, after consumption of the toxin-containing food. The toxin is produced when foods contaminated with *C. botulinum* endospores are kept under anaerobic conditions. Canned foods containing no air encourage spore germination and bacterial multiplication. Low-acid foods present a more favourable environment: *C. botulinum* is rarely found in high-acid products because a pH below 5.3 inhibits germination of its spores.

Symptoms of botulism appear between 12 to 36 hours after consumption of the toxin and include weakness, blurred or double vision, dysphagia, dysphonia, dry mouth, vomiting, constipation, neurological and gastrointestinal symptoms and, if untreated, respiratory or heart failure and paralysis. The disease is characterized as a neuroparalytic syndrome with gastrenteritis. Surviving patients may take as long as 8 months to recover fully. Diagnosis is con.rmed by immunological identi.cation of toxin in serum or faeces (11). The treatment includes stomach washing to remove any toxic food, mechanical respiratory support, polyvalent antitoxins to neutralize the toxin and neuromuscular blockade antagonists (aminopyridine). Since the disease is not an infection, antibiotics are of no therapeutic value.

The mortality rate is high (20–50%) and depends on the amount and type of food ingested, the extent of treatment and the type of toxin (Toxin A shows a higher mortality rate). In 1976 infant botulism resulting from ingestion of food containing spores of *C. botulinum* was .rst described in the U.S.. These ingested spores are transformed in vegetative forms in the large intestine and production of toxin occurs *in situ*. The disease is observed in newborns between 2 weeks and 6 months, when non milk foods are introduced containing viable spores, at a stage at which the infant's intestinal flora is not fully developed to exclude *C. botulinum*.

In the U.S., honey has been reported as the incriminated food; viable spores of *C. botulinum* have been isolated in samples of honey. It is therefore inadvisable to feed honey to children under one year of age. Infant botulism occurs as a mild disease with symptoms of muscular atonia. Death is rarely the outcome of the disease, and has been reported as the sudden infant death syndrome.

Clostridium perfringens

In 1892 *C. welchii* was reported as the agent of gas gangrene. Later, this microorganism was reclassi.ed as *C.perfringens* and was associated with outbreaks of gastroenteritis (food poisoning) (12).

The *C. perfringens* species are classi.ed in 5 types: A, B, C, D, E based on the production of 4 major toxins $\eta \cdot \beta \cdot \mu$,β and 8 minor ones ($\kappa$, λ, π, ϑ, σ, θ, ω, ς). *C. perfringens* type A is usually associated with diseases in humans ranging from gas gangrene, acute necrotising disease of the intestine infection and toxaemia to food poisoning. Large numbers of vegetative cells (10^8:g of food) of the microorganism introduced by food and surviving the acidity of the stomach could be responsible, after 8 to 24 hours, for the clinical symptoms including nausea, abdominal pain, diarrhoea, vomiting and enteritis. The vegetative cells pass to the small intestine where they grow, sporulate and release an enterotoxin (13). The microorganism is detected by the presence of large numbers (\$10^6$:g) of spores in patient faeces, large numbers of vegetative cells (\$10^6$:g) in food consumed, as well as by the presence of enterotoxin in the faeces.

C. perfringens food poisoning occurs when food is being prepared some time in advance of consumption. Meat and meat pies are the main food incriminated. Adequate refrigeration remains the key to its control.

Escherichia coli

E. coli is an inhabitant of the human and animal gut, and is the main indicator of faecal water contamination. Some strains, however are pathogenic and cause gastroenteritis. It can be an opportunistic pathogen (causing sepsis, urinary tract infections, and pneumonia in immunosuppressed patients or meningitis in neonates). Strains producing diarrhoea are classi.ed following their plasmid-encoded virulence properties in 4 major categories:

ENTEROTOXIGENIC *E. COLI* (ETEC)

The disease occurs generally after 12 to 36 hours after ingestion of the microorganisms. They adhere to the intestinal epithelium through .mbriae. Two toxin types are produced: the heat-stable toxins (ST) including ST_A and ST_B toxins without a known mechanism of action, and the heat-labile toxins (LT). The latter type of toxins are subdivided in LT I, which is a cholera-like toxin, and LT II, which is also a cholera-like toxin, but does not cross-react with antiserum to LT I.

The disease includes symptoms such as mild watery diarrhoea. In regions usually characterised by poor sanitation, the disease is reported as the «traveller's diarrhoea». The pathogens do not appear to pose a health problem to local populations who developed toxin-neutralising antibodies.

The disease remains a danger for travellers to these areas after consumption of faecally contaminated raw fruits, vegetables or water.

ENTEROINVASIVE *E. COLI* (EIEC)

The ingested microorganism invades and multiplies within the epithelial cells of the colon causing ulceration and inflammation. The symptoms of the disease includes fever, abdominal pains, malaise, watery diarrhoea and stools with blood, mucus and leukocytes.

ENTEROPATHOGENIC *E. COLI* (EPEC)

The disease occurs 12 to 36 hours after ingestion of the contaminated food. The EPEC strains have the ability to adhere closely to the enterocyte membrane causing cell loss from the villus tips. Malaise, vomiting and diarrhoea with stools containing mucus but rarely blood are the main symptoms of the illness, which is reported more severe in infants.

ENTEROHAEMORRHAGIC *E. COLI* (EHEC) OR VEROTOXIN-PRODUCING *E.COLI* (VTEC) (14)

The serotype mainly incriminated in this category of the illness is O157:H7. Adhesion by plasmid encoding .mbriae is an important factor of virulence for O157:H7 strains. Stomach cramps, acute bloody diarrhoea and haemorrhagic colitis, mainly during the summer period, could be life-threatening in the elderly.

Haemolytic uraemic syndrome, characterised by acute renal failure, haemolytic anaemia and thrombocytopaenia, occurring more often in children, is part of the spectrum of the disease. Finally, thrombocytopaenic purpura related to the above syndrome but also including neurological symptoms and fever have been reported as a severe complication.

Outbreaks with EHEC O157:H7 are associated with consumption of undercooked ground meat, raw milk or undercooked hamburgers. Food contaminated during processing or handling and faecal contaminated waters are involved in outbreaks of EPEC, EIEC and ETEC. Furthermore, mould-ripened soft cheeses, sushi, potato salad, vegetables and coffee have been incriminated in some cases of serious gastroenteritis by *E. coli*.

Plesiomonas shigelloides

This microorganism is found in stools of patients with diarrhoea and is recovered in 5.5% of human faeces in Thai adults. The infection is more common in warmer climates and in travellers returning from warmer climates (15).

The disease is reported as a mild watery diarrhea occurring 48 hours after consumption of the food. In immunosuppressed persons, the disease manifests itself more as a cholera-like syndrome.

Little is known for the pathogenesis of the disease, but the motility of the microorganism seems to be an important factor, as well as the production of an enterotoxin. Shell.sh (crab, shrimp, cuttle .sh, oysters) and .sh are the principle reservoirs of the illness.

Listeria monocytogenes

L. monocytogenes is an important human and animal pathogen. In animals the disease presents 2 forms: meningoencephalitis in adult ruminants, and a visceral form causing stillbirth, abortion and septicaemia in young ruminants and monogastrics. Food is the major source of infection in humans. It is ubiquitous in the environment although the incidence of infection remains low (16) (17).

L. monocytogenes is a facultative intracellular pathogen which can survive and multiply in cells of the monocytemacrophage system. The microorganisms attach to intestinal cells and induce endocytosis, processes promoted by the virulence factor p60. Once inside the phagocytic vacuole, it produces high levels of the enzymes superoxide dismutase and catalase to be protected from toxic superoxide anions. Finally, the microorganisms escape into the cytoplasm from the vacuole. Production of a haemolysin named listeriolysin O is essential in this process.

Invasive infection occurs only in susceptible individuals including immunocompromized persons, persons under prolonged antibiotic or corticosteroid therapy, pregnant women, cirrhosis, diabetic and cancer patients. The disease is strongly associated with suppression of T-cell mediated immunity. However, the illness is not very common in AIDS patients. The symptoms of the disease appears within 1 to 90 days ranging from a mild flu-like disease to meningitis or meningoencephalitis. In pregnant women the disease is usually associated with fever, headache and gastrointestinal symptoms.

Vegetables and cereals from listeria-contaminated area can cause disease in the afore-mentioned health states. Direct transmission from animal to man or from man to man or animal to animal has been described. Human or animal faeces of contaminated subjects spread the disease (18). A different disease is described in newborns, namely granulomatosis infantiseptica. This disease is characterized by a transplacental foetal infection which generally results in abortion, stillbirth or premature labour. Two types of disease can occur in newborns: (i) an early-onset syndrome characterized by rapid contamination at birth, possibly by aspiration of infected amniotic fluid from an *in utero* infection. Pneumonia, septicaemia and disseminated granulomas (abscesses) are the main symptoms of this type of disease, which carry an elevated mortality rate of 38%; (ii) a late-onset syndrome occurs when the newborn is infected by his mother during passage from the birth canal or sometimes is acquired after delivery. Meningitis is the main symptom of the disease in 93% of the cases presented. However, the mortality rate in this case remains lower (26%).

The microorganism possess the ability to grow on most non-acid foods. Coleslaw and cabbages as well as vacuumpacked ham and pate´ are involved in listeriosis of serotype 4b. Cabbages grown in .elds fertilized by fresh or composted manure from sheep and stored during winter, permit multiplication to high levels. Meat and meat products such as pork sausages, turkey frankurters, chicken, and chicken nuggets are incriminated in cases of serotype 1 disease (19) (20) (21). Dairy products such as raw or pasteurized milk and soft cheeses have been associated with the disease, when there is improper pasteurisation. Association with soft cheeses (Camembert, brie) is due to the processing of the soft cheeses through ripening. During the ripening process, microbial utilisation of lactate and release of amines increase the surface pH, allowing *Listeria* to multiply to dangerous levels.

Salmonella

Typhoid fever, a severe disease, is one of the most important causes of foodborne illness worldwide. Many serotypes of *Salmonella* have been described. The Kaufman-White serotyping scheme based on the presence of O, H and Vi antigen distinguished 100 different serotypes in the year 1941. Currently, it encompasses 2200 serotypes. *S. typhi*, *S.paratyphi* A, *S. paratyphi* B are found in humans, and *S. typhimurium*, *S.cholerae-suis*, *and S. abortusuis* in animals. Different serovar names are based on the geographical location where the disease occurs: *S.dublin*, *S. montevideo*,

S. minneapolis, *S. guilford*. *Salmonella* are responsible for different clinical syndromes:

ENTERITIS

Infection can range from asymptomatic carriage to severe diarrhoea. Human illness is associated with a limited number of serotypes and *S. enteritidis*, *S.typhimurium*, *S. virchow* are the most common. The incubation period varies from 6 to 48 hours. The main symptoms reported are mild fever, nausea, vomiting, abdominal pain, and diarrhoea. The disease can be more severe in susceptible groups such as the very young, old or ill persons. Ingested microorganisms surviving the stomach acid adhere to the epithelial cells of the ileum via mannose-resistant .mbriae. Then they are engulfed by the cells through endocytosis in a membrane-bound vacuole, where they multiply and then are released into the lamina propria via the basal cell membrane. This causes an influx of inflammatory cells leading to the release of prostaglandins, which activate adenylase cyclase-producing fluid secretion into the intestinal lumen. However, the cause seems to be multifactorial. The diarrhoegenic Salmonellas also produce enterotoxins.

The infectious dose of *Salmonella* is usually over 10^6, and factors predisposing are virulence of the serotype, susceptibility of the individual and food vehicle involved. An infective dose of 10–100 cells in susceptible individuals (children, elderly) can be fatal. An infective dose of less than 200 cells has been associated with waterborne outbreak.

The illness is associated with fatty foods such as, cheese, salami, and chocolate. High fat content in some foods affords the bacteria some protection from stomach acidity. After the end of symptoms, carriage of the microorganism and its passage in high numbers in the stools may occur for a few weeks or months.

SYSTEMIC DISEASE

Host-adapted serotypes are more invasive and tend to cause systemic disease in their host. In humans, *S. typhi*, *S. paratyphi* A, B are responsible for septicaemia and enteric fever. The incubation period varies from 3 to 56 days.

The disease present different stages:

During the .rst stage the invasive *Salmonella* penetrate the intestinal epithelium and are subsequently carried by the lymphatics to the mesenteric lymph nodes. After multiplication in the macrophages, they are released in the blood stream and are disseminated around the body (septicaemia). The microorganism can be cultured from the blood. The main symptoms are fever, headache, abdominal tenderness, constipation, and red spots on the skin. During the second stage, the microorganism reaches the gall bladder where it multiplies in the bile. The flow of infected bile reinfects the small intestine causing inflammation and ulceration. Symptoms such as fever persist, and 'pea soup' stools with large numbers of bacteria are produced.

The microorganism can also be found in the urine. There is haemorrhage due to ulceration and perforation of the intestine is observed leading to peritonitis. In milder cases, the ulcers heals and fever falls with recovery after 4–5 weeks.

The third stage includes remission and a carrier state for months or years as bacteria colonising the gall bladder are discharged with the bile into the faeces. This is most common in women and elderly persons. Some very famous *Salmonella* cases are reported in literature, e.g., Master Baker's Wife, the Folkestone Milker, and the famous Typhoid Mary, Mary Mellon who worked as a cook in households in the U.S. and who disappeared after an outbreak of typhoid fever. She was a carrier until her death.

Nowadays chronic carriers are treated with antibiotics and cholecystectomy (=surgical removal of the gall bladder) is performed in some cases. Non human-adapted serotypes such as *S. blegdam*, *S. bredeny*, *S.cholerae-suis*, *S. dublin*, *S. enteritidis*, *S.panama*, *S. typhimurium*, and *S. virchow* can be invasive in susceptible individuals, by causing less severe forms of enteric fever and septicaemia, and focal infections in the heart, appendix, gall bladder, peritoneum, lungs, urinary tract, brain, meninges, and spleen. Meningitis in infants, pneumonia in the elderly, osteomyelitis in patients with sicklecell anemia is also reported for some of the serotypes.

Salmonellosis is a zoonotic infection since the major source for human illness is infected animals. Transmission is through the faecal-oral route (intestinal contents from the infected animal are ingested with food or water). Meat, milk, poultry (22) (23), eggs are primary vehicles when undercooked or when cross-contamination occurs after cooking. Human carriers are less important than animals. Human transmission can occur if the faecally contaminated hands of an infected food handler touch food which is then inadequately cooked.

There is a direct person-to-person spread possible by the faecal-oral route in institutional outbreaks. Animals may acquire *Salmonella* infection on the farm from birds and rodents. An important factor for maintaining the cycle of *Salmonella* infection is the practice of using animal byproducts as animal feeds such as meat and bone meal. The heat process destroys *Salmonella*, but they are subject to post-process contamination.

Poultry and poultry products (24) are reported to be the major source of *Salmonella* infection. *S. enteritidis* is the commonest serotype. Eggs are the most frequent vehicle. Outbreaks are associated with raw eggs (25) in products such as mayonnaise (4), ice-cream, or 'body-building' drink. Contamination of eggs with *Salmonella* is due to contamination of the eggshell with faecal material in the hen's cloaca or after laying in the nest. The shell can then contaminate the contents when the egg is broken. Contamination of the yolk of intact hen's eggs occurs when microorganisms infect the bird's ovaries and oviduct and thereby contaminate the egg contents (26). Mild cooking procedures of eggs is insuf.cient to kill the microorganism and the fat content of the yolk may protect the microorganism from gastric acidity. Raw milk is incriminated, when inadequately pasteurized or post-process contamination. *S. ealing* has been isolated from dried baby-milk when the milk was reconstituted in water. Fish, .sh products, and shell.sh from polluted water (prawns) are also involved. Cocoa beans for chocolate production have been associated with *S. eastbourne*, *S. napoli*, and *S.typhimurium*.

Shigella

Classical bacillary dysentery, also called shigellosis, is associated with infections with four species of *Shigella* (*S. dysentariae*, *S. flexneri*, *S. boydii* and *S.sonnei*). *S. dysentariae* causes a more severe disease than the other *Shigella* species. Unlike most enteric infections, shigellosis can be initiated with a low infectious dose (10–100 microorganisms). The incubation period of the illness varies from 7 hours to 7 days and the person-to-person spread is the main mode of transmission. The illness is more prevalent in child-rich communities and in persons who do not respect hygiene practices. The disease ranges from asymptomatic cases to life-threatening dysentery. The microorganisms penetrate into and multiply in the super.cial epithelium of the colon where they are believed to release an endotoxin (27). The endotoxin induces infl- ammation and local damage and, when absorbed into the blood, results in symptoms of fever. An exotoxin responsible for the diarrhoeal disease is also produced. The stools contain blood, mucus, and pus.

S. sonnei is associated with a watery diarrhoea and recovery confers serotype-speci.c immunity. The source of the microorganism is often a human carrier involved in the chain of food preparation. In areas where sewage disposal is inadequate, the microorganism could be transferred from human faeces by flies. Uncooked food having been subjected to extensive handling, such as prawn, tuna, salad, or buffet meals are involved in cases of the disease.

Staphylococcus aureus

The microorganism derives its name from the Greek words staphyle (=bunch of grapes) and coccus (=grain), following its appearance under the microscope. 27 different species and 7 subspecies of the genus *Staphylococcus* have been reported. The principal habitat of the *Staphylococci* is the skin, skin glands and the mucous membranes of warm blooded animals.

Several species are associated with particular hosts, e.g., *S. gallinarum* with chicken and *S. hyicus* with pigs. *S. aureus* occurs most frequently on the skin of humans and animals. On the skin, *S. aureus* can cause skin abscesses, acting as an opportunistic pathogen when the skin barrier is breached or the host resistance is low. It is also found in the nasal passages of healthy individuals (20–50%). It has been isolated from faeces and from environmental sites (soil, marine and fresh water, plant surfaces, dust, air). *S. aureus* produces 7 exotoxins (A, B, C_1, C_2, C_3, D, E). Exotoxin Type A is involved in 52%of outbreaks and D in 6%.These toxins are resistant to gut proteases and heat stable, being inactivated only by prolonged boiling. *S. aureus* toxins are strict neurotoxins. They elicit the emetic response by acting on receptors in the gut, which stimulate the vomiting centre in the brain via the vagus and sympathetic nerves.

The incubation period of the food-borne disease is very short (2–4 h). Symptoms includes nausea, vomiting, stomach cramps, and diarrhoea. In severe cases, dehydratation, marked pallor and collapse require treatment by intravenous infusion (28).

The disease occurs after ingestion of a pre-formed toxin in food. The presence of small numbers of *S. aureus* in food is not uncommon and they can be eliminated by cooking or pasteurisation. Outbreaks due to dried and chocolate milk have been described, through bacterial growth and enterotoxin production. Contamination by food handlers in view of the high rate of human carriage is the most common mode of transmission. Contamination from the nose and throat and skin or by coughing and sneezing is another common mode of transmission. Numbers superior to 10^6: g are required for the production of enough toxin to cause illness.

The most common food vehicles are poultry products, cold cooked meats (29), salted meats, ham, corned beef, buffet meals and canned foods. Hard cheeses, cold sweets, cream-.lled bakery products, rice balls (Japan), and icecream (Hungary) have been implicated. When the food prepared in advance and stored at ambient temperature or inadequately chilled, it can be responsible for the disease.

Vibrio cholerae

In India more than 20 million of people have been killed this century by cholera. *Vibrio cholerae* was discovered by Koch in 1886 and thereafter has been many times involved in serious outbreaks. The incubation period of the disease varies from 1 to 3 days and the symptoms range from mild, self-limiting diarrhoea to a severe, life-threatening disorder. The infectious dose for initiation of the disease remains large (1010 cells).

Hypochlorohydric persons are more susceptible to acquire cholera. Cholera is a non-invasive infection; the microorganism colonises the intestinal lumen and produces a potent enterotoxin which activates adenylate cyclase and this leads to an accumulation of cAMP which inhibits absorption of Na^+ and Cl^-, HCO_3^- and Na^+ ions (30). To maintain an osmotic balance, the transfer is accompanied by a massive outflow of water into the intestinal lumen. This results in watery diarrhoea containing flakes of mucus described as rice water and containing 10^{10} vibrios: ml.

Other symptoms include vomiting, nausea, fever, fall in blood volume and pressure, and increase in blood viscosity, renal failure and circulatory collapse. Massive losses of fluid and electrolytes are seen. The death rate is high, ranging from 30–50%if the patient is left untreated and these numbers are reduced to less than 1%if rehydration using an electrolyte solution is administered.

Vibrio Parahaemolyticus

V. parahaemolyticus is a halophile most commonly found in marine environments (31). The bacterium causes gastroenteritis after ingestion of contaminated seafood. The incubation period varies from some hours (9–25 h) to several days (2–4 days) and the symptoms include profuse watery diarrhoea with mucus (10%), blood (5%), abdominal pain, vomiting and fever.

V. parahaemolyticus is more enteroinvasive than *V.*

cholerae and penetrates the intestinal epithelium to reach the lamina propria (32). Kanagawa reaction in Wagatsuma's agar is used to observe that the haemolysin can lyse fresh human or rabbit blood cells. 96.5%of strains causing foodborne disease are Ka (+); environmental strains are Ka (-) (99%).

The disease is associated with eating contaminated raw or inadequately cooked seafood. During summer season, the pathogen found in coastal waters may infect shrimp, crabs, oysters and Fishes.

Yersinia enterocolitica

Y. enterocolitica can occasionally cause gastroenteritis in humans following ingestion of faecally contaminated food or water or by direct contact with contaminated faeces. Yersiniosis is most common in the cooler climates of northern Europe and North America. There is seasonal variation with a peak in the autumn and winter. The disease is most common in children under 7 years. The incubation period of the disease varies from 1 to 11 days and includes symptoms such as mild fever, vomiting, arthritis, and inflammation of the mesenteric lymph nodes. Pigs are chronic carriers of *Y. enterocolitica*. The serotypes (03, 05, 27, 08, 09) are associated with human disease. Erythema nodosum is usually associated with serotypes 03 and 09. The microorganism is found in the tongue, tonsils, gut, and caecum of healthy animals, contaminated milk, and water. Control of infection by proper transport and slaughter practices, evisceration, excision of the tongue and tonsils is imperative (33).

Ingested microorganisms survive passage through the stomach acid and adhere to the mucosal cells of Peyer's patches through bacterial outer membrane proteins. There is endocytosis and microorganisms are released into the lamina propria where they invade phagocytic cells and multiply extracellularly producing a local inflammatory response. Damage in epithelial surface results in malabsorption and an osmotic fluid loss is characterized by diarrhoea (34)

CONCLUSIONS

Food can be preserved by a variety of methods. It is important to eliminate populations able to cause spoilage and disease and to maintain the microbiological quality of a food. To best manage the potential sources of contamination from production to consumption, the hazard analysis critical control point (HACCP) approach is used. Until now, we have been mostly concerned with the negative role of microorganisms in foods. We should not forget other roles for microorganisms: fermentation, where microbial activity is essential for the production of food, and other bene.cial effects (6).

ACKNOWLEDGEMENTS

This review has been carried out with .nancial support from the Commission of the European Communities, Agriculture and Fisheries (FAIR) speci.c RTD programme PL98 4230 «Intestinal Flora: Colonisation Resistance and Other Effects». It does not necessarily reflect its views and in no way anticipates the commission's future policy in this area.

REFERENCES

1. Brock TD, Smith DW, Madigan MT. Biology of Microorgan-isms. Englewood Cliffs, N.J.: Prentice-Hall, 1984.
2. Christian JHB. Water activity and the growth of microorgan-isms. In: Recent Advances in Food Science, Vol. 3, ed. Leitch JM, Rhodes DN, London, Butterworths 1963; 248–55.
3. Bezirtzoglou E. Human intestinal microflora. Ed. University of Ioannina, 1996.
4. Appleman MD, Hess EP, Rittenberg SC. An investigation of a mayonnaise spoilage. Food Technol 1949; 3: 201–3.
5. Florian MLE, Trussell PC. Bacterial spoilage of shell eggs. IV. Identification of spoilage organisms. Food Technol 1957; 11: 56–60.
6. Watt BK, Merrill AL. Composition of foods-raw, processed, prepared. Agricultural Handbook No 8, Washington, D.C.: USA; 1950.
7. Beumer RR, Cruysen JJM, Birtantie IRK. The occurrence of Campylobacter jejuni in raw cow's milk. J Appl Bacteriol 1988; 65: 93–6.
8. Griffiths PL, Park RWA. Campylobacters associated with human diarrhoeal disease. J Appl Bacteriol 1990; 69: 281–301.
9. Lynt RK, Kautter DA, Read RB Jr. Botulism in commer-cially canned foods. J Milk Food Technol 1975; 38: 546–50.
10. Midura TF, Arnon SS. Infant botulism: identification of Clostridium botulinurn and its toxins in faeces. Lancet 1976; 2: 934–6.
11. Saito M. Production of enterotoxin by Clostridium perfringens derived from humans, animals, foods, and the natural envi-ronment in Japan. J Food Protect 1990; 53: 115–8.
12. Shimizu T, Okabe A, Minami J, Hayashi H. An upstream regulatory sequence stimulates expression of the per-fringolysin 0 gene of Clostridium perfringens. Infect Immun 1991; 59: 137–42.
13. Smith HB, Cheasty T, Roberts D, Thomas A, Rowe B. Examination of retail chickens and sausages in Britain for vero cytotoxin-producing Escherichia coli. Appl Environ Mi-crobiol 1991; 57: 2091–3.
14. Paul R, Siitonen A, Karkkainen P. Plesiomonas shigelloides bacteremia in a healthy girl with mild gastroenteritis. J Clin Microbiol 1990; 28: 1445–6.
15. Pearson LJ, Marth EH. Listeria monocytogenes-threat to a safe food

supply: a review. J Dairy Sci 1990; 73: 912–28.

16. Wesley IV, Ashton F. Restriction enzyme analysis of Listeria monocytogenes strains associated with food-borne epidemics. Appl Environ Microbiol 1991; 57: 969–75.
17. Petran RL, Zottola EA, Gravani RB. Incidence of Listeria monocytogenes in market samples of fresh and frozen vegeta-bles. J Food Sci 1988; 53: 1238–40.
18. Pini PN, Gilbert RJ. The occurrence in the U.K. of Listeria species in raw chickens and soft cheeses. Int .1 Food Microbiol 1988; 6: 317–26.
19. Shelef LA. Survival of Listeria monocytogenes in ground beef or liver during storage at 40° and 250°C. J Food Protect 1989; 52: 379–83.
20. Zaika LL, Palumbo SA, Smith JL, Del Corral F, Bhaduri S, Jones CO, Kim AH. Destruction of Listeria monocytogenes during frankfurter processing. J Food Protect 1990; 53: 18–21.
21. Hinton M, Threlfall El, Rowe B. The invasive potential of Salmonella enteritidis phage types for young chickens. Lett Appl Microbiol 1990; 10: 237–9.
22. Machardo J, Bernardo F. Prevalence of Salmonella in chicken carcasses in Portugal. J Appl Bacteriol 1990; 69: 477–80.
23. Mead GC. Microbiology of poultry and game birds. In: Meat Microbiology, ed. Brown MH, London: Applied Science, 1982; 67–101.
24. Mayes FJ, Takeballi MA. Microbial contamination of the hen's egg: a review. J Food Protect 1983; 46: 1092–8.
25. Mead GC, Barrow PA. Salmonella control in poultry by competitive exclusion or immunization. Lett Appl Microbiol 1990; 10: 221–7.
26. Zaika LL, Engel LS, Kim AH, Palumbo SA. Effect of sodium chloride, pH and temperature on growth of Shigella flex neri. J Food Protect 1989; 52: 356–9.
27. Minor TE, Marth EH. Staphylococci and their Significance in Foods. New York: Elsevier, 1976.
28. Sokari TG, Anozie SO. Occurrence of enterotoxin producing strains of Staphylococcus aureus in meat and related samples from traditional markets in Nigeria. J Food Protect 1990; 53: 1069–70.
29. Ogawa A, Kato JI, Watanabe H, Nair BG, Takeda T. Cloning and nucleotide sequence of a heat-stable enterotoxin gene from Vibrio

cholerae non-01 isolated from a patient with traveler's diarrhea. Infect Immun 1990; 58: 3325–9.

30. Chan KY, Woo ML, Lam LY, French GL. Vibrio para-haemolyticus and other halophilic vibrios associated with seafood in Hong Kong. J Appl Bacteriol 1989; 66: 57–64.
31. Nakasone N, Iwanaga M. Pili of Vibrio parahaemolyticus strain as a possible colonization factor. Infect Immun 1990; 58: 61–9.
32. Hanski C, Naumann M, Grutzkau A, Pluschke G, Friedrich B, Hahn H, Riecken EO. Humoral and cellular defense against intestinal murine infection with Yetsinia enterocolitica. Infect Immun 1991; 59: 1106–11.
33. Walker SJ, Gilmour A. Production of enterotoxin by Yersinia species isolated from milk. J. Food Protect 1990; 53: 751–4.

CHAPTER 2

SAFETY ASSESSMENT OF FOODS DERIVED FROM GENETICALLY MODIFIED MICROORGANISMS

J. Schlundt
World Health Organization, Headquarters, Geneva, Switzerland

ABSTRACT

Microorganisms have a long history of use in food production, e.g. in the production of sausages, cheeses, etc. Roughly one quarter of all food products rely on microbiological processes, and the safe use of microorganisms for food production is essential. The transfer of novel traits to food microorganisms through recombinant gene technology will result in new potential food safety issues. This requires the elaboration of criteria for safety assessment of foods derived from genetic microorganisms. This

---

CITATION: Schlundt, J., "Safety Assessment of Foods Derived from Genetically Modified Microorganisms", Microbial Ecology in Health and Disease (2001), https://doi.org/10.1080/08910600175334 1273.

joint FAO WHO Expert Consultation was convened to consider criteria for the safety and risk assessment of food and food ingredients produced with the aid of, or containing, viable or non-viable genetically modified microorganisms (GMMs). The outcome is meant to contribute to safety and regulatory considerations by Member States and to support the development of international safety assessment guidelines for food produced with the aid of GMMs (GMM food) by the Codex ad hoc Intergovernmental Task Force on Foods Derived from Biotechnology. The Consultation concluded that the safety and risk assessment of GMM foods should proceed on a case-by-case basis considering both safety and nutritional aspects. The assessment should use the concept of substantial equivalence as a starting point, identifying differences and similarities between GMM foods and their conventional counterpart. While these differences guide the assessment, a number of additional issues require special consideration. Examples are the potential persistence of the GMM in the host gut and its interaction with the endogenous microflora, including the transfer of new genes to the gut microflora. Likewise the interaction of new genetic material with human cells in the gut, including the immune system, should be considered. The assessment of GMM foods take into consideration the history of safe use of microorganisms used in the construction. The genetic stability and pathogenic potential of the strains, as well as possible secondary effects of the genetic modification on metabolism, including the concentration of critical nutrients, antinutrients, etc. are important factors to consider. Antibiotic resistance marker genes should be avoided and their absence from the final GMM food ensured. In order to provide the basis for improved safety and risk assessment of GMM food, the Consultation recommends strengthening the capability to analyze the complex ecosystem in the human gastrointestinal tract, including work on colonization, persistence, gene transfer and host interactions. New technologies, such as molecular profiling, are developing rapidly with a potential to enhance the assessment of GMMs. The Consultation encourages further development and validation of such methods. Fermentation provides a simple technology used globally to produce food of high nutritional and hygienic quality; a technology, which is especially important in developing countries. The Consultation recommends capacity-building efforts to support the needs of developing countries in improving and evaluating this technology and its potential also in relation to GMM use. To enable efficient interaction and transparency in the safety and risk assessment of GMM food, FAO and WHO is urged to coordinate efforts to enhance the communication of safety and risk assessment principles of GMM food to all stakeholders including the public.

INTRODUCTION

A Joint FAO WHO Expert Consultation on Foods Derived from Biotechnology—Safety Assessment of Foods Derived from Genetically Modified Microorganisms (GMMs), was held at the Headquarters of the World Health Organization (WHO) in Geneva from 24 to 28 September 2001. A total of 27 experts, including authors of discussion papers, participated in the Consultation. The complete list of participants is given in Appendix 1. Ms Ann Kern, Executive Director, Cluster of Sustainable Development and Healthy Environments, opened the Consultation on behalf of Director-General of WHO and the Food and Agriculture Organization of the United Nations (FAO). Ms Kern stated that WHO and FAO have been organizing Consultations of this kind since 1990 to provide scientific and technical guidance to Member States and to the Codex Alimentarius Commission. Ms Kern also expressed the appreciation of the two Organizations to the Government of Japan for its generosity in supplying additional funding for this Consultation. She acknowledged the interest of Member States in these sometimes hotly debated issues, and the need for sound scientific advice developed and formulated by the Expert Consultations upon which Governments can base their discussions. Clear assessment and communication of scientific data is becoming increasingly important so that the scientific risk assessment process is accurately reflected in the risk management process. Ms Kern suggested that the issues of safety and nutritional assessment of foods derived from biotechnology would be even more important in the near future with the rapid development of new foods with potential benefits related to health.

The Consultation elected Dr Ian Munro as Chairperson and Dr Bodil Lund Jacobsen, Professor Ingolf Nes, Dr Ruud Valyasevi and Dr Christopher Viljoen as Rapporteurs. The Consultation also decided to ask all participants to assist Rapporteurs by drafting each of the sections of the report of the Consultation. Dr Thomas Cebula and Dr James Maryanski (USA) and Dr William Yan (Canada) took part in the discussions through teleconference from their countries. All participants completed a Declaration of Interest as defined by FAO and WHO.

BACKGROUND

The FAO and WHO have embarked on an initiative to organize a series of scientific Expert Consultations to provide scientific and technical advice to their Member States. The scientific advice derived from the Joint FAO WHO Expert Consultations can be used by the Member States of the FAO and WHO directly. It will also serve as the scientific foundation for the work of the Codex Alimentarius Commission in their deliberation on safety assessment guidelines for foods derived from biotechnology presently being developed by the Codex *ad hoc* Intergovernmental Task Force on Foods Derived from Biotechnology. The FAO and WHO have to date organized two Expert Consultations. The first Consultation held in Geneva in June–July 2000 addressed the overall aspects of safety assessment of genetically modified foods of plant origin and responded to five specific questions raised by the First Session of the Task Force (1). The second Consultation held in Rome in January 2001 specifically addressed the allergenicity of foods derived from biotechnology (2).

At the 24th Session of the Codex Alimentarius Commission held in July 2001, it was decided that the Task Force expand its work in starting the work on drafting a Proposed Draft Guideline for the Conduct of Food Safety Assessment of Modified Microorganisms in Food and established a new Working Group to prepare a draft document on this. The FAO and WHO at that time announced that they would jointly convene a third joint Consultation to consider the scientific aspects of the safety assessment of genetically modified microorganisms in foods in support of this new work.

The FAO and WHO convened this Consultation to evaluate the build on experience gained since the previous two FAO WHO Consultations and to assess whether currently available approaches for assessing the safety of foods and food ingredients derived from genetically modified plants could be applied to GMMs. The Consultation examines unique aspects to be considered in the safety assessment of foods produced with the aid of GMMs. The safety and risk assessment of foods, including genetically modified foods, are generally considered within a framework of risk analysis. Within this framework, reference can be made to the use of precaution in risk management and risk assessment. Ongoing discussions within the Codex system will help guide these considerations further.

SCOPE

The Consultation was convened to consider criteria for the safety assessment of food derived from microorganisms that have been genetically modified using recombinant DNA techniques. Specifically, the Consultation was requested to provide the FAO, WHO and their Member Countries with scientific advice in relation to the safety assessment of GMMs in food.

The Consultation agreed to the following definitions for its purposes. "Genetically modified microorganisms" (GMMs) means:

Bacteria, yeasts or filamentous fungi in which the genetic material has been changed through modern biotechnology in a way that does not occur naturally by multiplication and or natural recombination.

"Modern biotechnology"* means the application of:

- *In vitro* nucleic acid techniques†, including recombinant deoxyribonucleic acid (DNA) and direct injection of nucleic acid into cells or organelles, or
- Fusion of cells‡ beyond the taxonomic family that overcome natural physiological, reproductive or recombination barriers and that are not techniques used in traditional breeding and selection.

The Consultation agreed to confine its discussion to food produced with the aid of GMMs, namely:

- Foods and food ingredients consisting of or containing viable GMMs.
- Foods and food ingredients consisting of or containing non-viable GMMs.
- Foods and food ingredients produced by fermentation using GMMs and from which the GMM has been removed.

The Consultation did not consider highly purified products such as food additives, enzymes, polysaccharides, flavors etc. derived from GMMs. The Consultation noted that such purified food additives have been produced and permitted in a number of countries for over a decade (e.g. chymosin) and the safety of many has been assessed by the Joint FAO WHO Expert Committee on Food Additives (3). However, the Consultation agreed that if viable GMMs and or microbial components are present in such products, the general principles for the safety assessment of foods derived from GMMs outlined in this report could be applied.

The Consultation agreed that, although beyond its scope, the concepts and principles described in this report are equally applicable to all foods produced with the aid of microorganisms.

The report does not specifically address the presence of GMMs used for agricultural purposes such as microbial plant protection agents, feed additives, bio-fertilizers etc. that may find their way into food. However, the Consultation agreed that the general principles for the safety assessment of foods produced with the aid of GMMs could be applied in such cases.

The Consultation also recognized safety issues that exist in relation to the exposure of workers (occupational health) involved in the production of the food using GMMs but found that this issue was not within the scope of this Consultation.

The Consultation did not consider environmental safety issues related to the release of GMMs into the environment as these were outside its defined scope. Similarly, it did not consider socioeconomics, risk management and public perception issues.

As a background to its discussions, the Consultation received papers and or presentations on:

1. General consideration of the safety assessment of GMMs in foods, including:

- Use of living microorganisms in food and perspectives on the application of genetic modification in such microorganisms.
- Methodologies available for the safety assessment of the GMMs.
- Conventional safety assessment methodologies used for foods derived from biotechnology including foods of plant origin.
- Issues specific to foods produced with the aid of GMMs.

2. Other specific topics arising in relation to the safety assessment of GMMs in food, including:

- Gene transfer.
- Genetic familiarity and stability of microorganisms.
- Colonization and persistence of microorganisms in the gut and their impact on the gut ecosystem with special reference to effects on human health.
- Genetic basis of pathogenicity of microorganisms and its possible change by genetic modification.
- Nutritional and toxicological aspects of bacterial or fungal

metabolites and their possible alteration by genetic modification.

- The allergenicity of microorganisms and its potential for induction or change by genetic modification.
- Survival propagation of microorganisms, including
- GMMs, in the environment and the implications to public health.

A list of agenda items and documents is given in Appendix 2.

SAFETY ASSESSMENT

General approach to safety evaluation

Roughly one-quarter of food production involves microbial fermentation processes. This includes foods such as bread, sour dough, sour milk and cream, yogurt, cheese, sour vegetables, fermented meat, vinegar, wine and beer. Fermentation provides a simple technique to produce food of high nutritional and hygienic quality. This technology has a long history of food use and is especially important in developing countries. The influence of microorganisms is thus of great importance for the safety and nutritional status of food. The introduction of foods produced with the aid of GMMs into the food supply brings potentially new issues of food safety. This section outlines general principles for the safety assessment of GMMs in food with reference to the established principles applied to the safety assessment of genetically modified plants and food safety issues specific to the nature and use of microorganisms in food.

Several international organizations have already addressed the issues associated with the safety assessment of novel foods including genetically modified plants and microorganisms (4–9). It is generally agreed that such an assessment requires an integrated and stepwise, case-by case approach using the concept of *substantial equivalence* (see 'Application of the concept of *substantial equivalence* to GMMs') that is directed by the results of a comparison between the genetically modified plant or microorganism and its conventional counterpart. The FAO WHO Expert Consultation on Safety Aspects of Genetically Modified Foods of Plant Origin (1) provides important recommendations on the use of *substantial equivalence* as a concept to guide the further safety assessment process. Some authorities have developed decision trees to assist in determining the extent of testing required in specific cases (9–11).

The following general principles are important to consider in the safety evaluation of food produced from GMMs:

- The safety assessment should address the health aspects for the whole population, including immuno-compromised individuals, infants and the elderly.
- The safety assessment should be based on sound scientific data and should use the most appropriate safety assessment methods. If new information becomes available, the safety assessment may need to be revised.
- The safety assessment should involve the characterization of the genetic modification, including deletion or insertion of DNA sequences, characterization of recipient microorganisms, the ultimate donor organism, the vector(s) used in the construction of the GMM, the construct and the GMM itself.

Other important considerations in the safety assessment of food produced with the aid of GMMs include:

- Information on the direct and indirect exposure§ of humans to the food or to the GMM itself. This should also consider the potential level of intake and dietary impact of the GMM.
- Possible secondary effects from expression of the inserted gene or the disruption of host DNA or metabolic pathways, including composition of critical macro- and micronutrients, anti-nutrients, endogenous toxicants, allergens, and physiologically active substances.
- The inherent differences between microbes and plants and the influence of the food matrix on the GMM should be taken into consideration when applying the concept of *substantial equivalence* (discussed in 'Application of the concept of *substantial equivalence* to GMMS').

In addition, the following elements should be taken into account in the safety assessment. This list is not exhaustive and, in some cases, not all elements mentioned are relevant.

- Techniques used for genetic modification.
- Strain identification and characterization (recipient, donor [if appropriate] and the GMM itself).
- Natural habitat.
- History of use.

- Gene transfer.
- Genetic stability.
- Pathogenic potential.
- Characterization and verification of the expected protein expression product of the novel DNA.
- Composition of the food containing the GMM.
- Safety and nutritional assessment (including potential toxicity and nutritional aspects).
- Interactions between the GMM, the gastrointestinal flora and the mammalian host.
- Impact on the immune system.
- Effects of processing, cooking and storage.

Aspects specific to microorganisms

The genetic modification of microorganisms involves similar recombinant DNA techniques to those used in the production of GM plants. However, there are distinct genetic characteristics of microorganisms that require specific aspects of their safety to be addressed. Microorganisms that are used in food production include gram-positive and gram-negative bacteria, yeasts and filamentous fungi. The genome structures and the available genetic technologies differ for bacteria, yeasts and filamentous fungi, although some common techniques are used.

The ready exploitation of homologous recombination in bacteria is a major advantage that facilities good control over genetic modification procedures. Integration sites can be used by design and unwanted DNA can be removed with relative ease. Systems for the selection and maintenance of introduced DNA can be designed using homologous genes and selection methods developed that are compatible with safe food use.

Microbial genomes are relatively small and several bacterial genomes and some yeast genomes have been sequenced including that of *Saccharomyces cerevisiae*. The acquisition of the complete genome sequence for a particular bacterial species is now a realizable scientific objective. The availability of such genome sequence data greatly enhances the knowledge base that is available to support safety evaluation. Post-genomic analytical techniques provide a valuable opportunity to analyze gene expression at the level of the entire genome. DNA microarray technology involves the use of nucleic acid probes for all the genes in the genome. This can be used to compare the presence of individual genes in different strains and gene expression in different strains and different environments. Proteomics separates proteins isolated from the whole cell using two-dimensional gel electrophoresis and allows comparisons to be made between strains and different environments. Individual proteins can be identified using mass spectrometry to relate separated protein spots to specific genes.

Microorganisms used in food may remain viable during food production and following their consumption. For this reason, they have a potential to interact with the consumer both directly and indirectly. It is important to ensure that the recipient microorganisms are not pathogenic, toxigenic or allergenic and that the genetic modification does nothing to compromise this status. Also, the fate of consumed GMM and its impact on both the gastro-intestinal (GI) tract and its micro flora need to be considered.

Application of the concept of Substantial Equivalence to GMMs

The concept of *substantial equivalence* was developed by the WHO, OECD and FAO following recognition of the limitations of conventional toxicology for the safety evaluation of novel foods. Quoting the OECD publication (1993), the concept 'embodies the idea that existing organisms used as food or as a source of food can be used as the basis for comparison when assessing the safety of human consumption of a food or food component that has been modified or is new'.

The FAO WHO report on 'Safety aspects of genetically modified food of plant origin' (2000) addressed criticism of the application of the concept of *substantial equivalence* and reaffirmed its usefulness. It emphasized that the determination of *substantial equivalence* is not in itself an end-point but rather the starting point for safety evaluation.

The concept of *substantial equivalence* is also of value for the safety assessment of GMMs. Microbial genomes have natural genetic plasticity and this may become apparent in foods or during food processing and where complex communities are involved. In addition, the gene expression of a microorganism is expected to vary according to its environment. This is especially pertinent for safety assessment where a variety of data might be gathered under *in vitro* laboratory conditions, in a food matrix, or in the GI tract following consumption. These constraints suggest that the concept of *substantial equivalence* should be applied both to the GMM itself and food produced with its aid. In applying this concept, it is important to remember that minor differences can separate pathogenic strains from non-pathogenic strains of microorganisms.

The FAO WHO report (1) recognized the need to keep abreast of developments in genetic modification technology and noted that new methodologies, such as molecular profiling techniques, may provide a more detailed analytical comparison. The concept of *substantial equivalence* involves a targeted analysis of the composition and phenotype of the GMM compared with its conventional counterpart. Molecular profiling provides a non-targeted and more holistic approach to this analysis. Microorganisms are especially amenable to the use of DNA microarrays and proteomics. Metabolic profiling has advanced, using a range of analytical techniques, and it may be of special value in the assessment of GMMs where metabolic rerouting is the intended outcome. The major limitation of profiling is the need to accommodate the background of normal variation and the need to interpret the significance of any differences that are detected. Several steps must be taken before the full potential of these techniques can be realized in routine safety assessments. First, the methodologies must be validated to ensure their reproducibility and robustness. Then, agreement must be reached in assessing their performance. That is, what is the range of differences in a given array or profile that will be considered as "normal variation". Any profile differences considered not to be within this normal variation must be evaluated from a safety perspective.

SPECIFIC FOOD SAFETY ISSUES

Introduction

This section deals with specific issues that are relevant to the safety of foods produced with the aid of GMMs. These issues include the potential of gene transfer between the GMM and other organisms found in food and the GI tract. Also, the safety of genetic markers used for selection (such as antimicrobial resistance genes), and the potential of GMM interaction with the intestinal micro flora and the immune system are evaluated. The discussion that follows provides an appraisal of existing knowledge on these topics and suggests scientific approaches that may be used to assess possible health risks.

Techniques for genetic modification

In general, the techniques for genetic modification of microorganisms are better understood than in plants. For example, in bacteria, the inserted gene recombinant constructs can be integrated at specific sites on the chromosome or on plasmids. However, the techniques for genetic modification applied to GMMs remain an important safety consideration. The factors to be considered are discussed below.

Bacteria

Host. The host microorganism should have a history of safe consumption either as a food or as a food component or its safety must otherwise be established. *Inserted gene.* The inserted gene(s) may be sourced from the same microbial species or from evolutionarily more distant organisms. The gene product or any inserted gene(s) should have a history of safe use in food or its safety should otherwise be established. The food safety evaluation will be facilitated by reducing to a minimum any extraneous DNA sequences.

Vector and construct. The DNA sequence of the whole vector should be characterized, including replicons, promoters, selective markers, linkers, and any foreign DNA, in case the vector is part of the GMM. It is recommended that vectors consisting only of nucleotide sequences from microorganisms with a history of safe use in food should be used. Any selectable markers should be carefully chosen and based on safe use, information on the sequence similarities and the protein function of the selective marker should be available to assess any safety hazard. In particular, antimicrobial resistance markers should be avoided and should not be present in the final GMM. Several techniques are available to remove selectable markers in GMMs, such as sequence specific recombination.

Methods of DNA transfer to the host. Methods of transfer of DNA into microorganisms can involve physical, chemical or biological approaches. DNA transformation methods that minimize major genetic rearrangement in the host genome should be used. When the inserted gene is integrated into the genome, the nucleotide sequence of the flanking regions at the integration site of the chromosome should be characterized.

Yeast and filamentous fungi. Most methods of genetic manipulation for bacterial systems can also be applied to yeast and filamentous fungi used in food production, thus similar safety considerations as in bacteria should be applied. Specific cloning vectors (e.g. centromeric plasmids, yeast artificial chromosomes, plasmids based on killer factor determinants etc.) have also been constructed. Reliable methods are available in certain species for directed integration of *in vitro* modified or composed gene constructs into specific chromosomal sites and for deletion of genes. Transgenic constructs made by these methods are highly stable during vegetative propagation of cells. They can, however, recombine at mating with related strains of the native myco-flora. In species with less developed genetics, there is insufficient information available on the recombinant processes to precisely predict the mechanism and the site of integration of novel genes. For this reason, methods currently used for genetic manipulation of yeast and filamentous fungi may allow integration at variable sites, resulting in transgenic strains with variable biotechnological performance and genetic stability.

Strain Identification and Characterization

First and foremost, the host microorganism should be safe, well characterized and stable. The origin of the strain should be known. Proper state-of-art taxonomy should be applied to describe the strain. Although many genotypic and phenotypic methods exist, each with its own merits and limitations, the Consultation recognized the need for the host microbial strain to be adequately characterized from a scientific, manufacturing, and safety perspective. This currently includes DNA DNA hybridization and 16S rRNA sequence the determination, which provide crucial information on taxonomic classification of the organism. Standard physiological biochemical methods for phenotypic characterization are commercially available and are commonly used. Information on pathogenic traits within the host genus can also provide important guidance for the characterization. Secondly, the GMM strain produced should be as safe as the host organism. The novel status of the GMM strain, including its phenotypic and genotypic characteristics, should be characterized in order to assess its safety. Existing molecular techniques provide precise tools for such characterization and for the comparison of microorganisms at species and strain level. Comparative assessment of the GMM strain with its host strain can be undertaken using approaches such as restriction analysis, random amplified polymorphic DNA analysis (RAPD-PCR), amplified fragment length polymorphism analysis (AFLP), protein profiling, etc. Further analysis can extend to genome sequencing.

Additionally, the effects of the genetic modification on the properties of the host organism, the desired stability of the genetic system, and the desired functional properties of the gene construct are important factors that should be considered in the GMM.

Gene Transfer

Bacteria. Prokaryotic microorganisms have developed a variety of mechanisms by which they can transfer DNA to other cells. This can result in the transfer of heritable traits. These transfer mechanisms allow bacteria to respond to environmental changes by acquiring new genetic information that might provide a selective advantage under changed environmental conditions. An example would be the worldwide spread of antimicrobial resistance genes among microorganisms since the introduction of antimicrobial agents in human health care, veterinary medicine and agriculture. One mechanism of gene transfer, termed *conjugation*, relies on a plasmid (autonomously replicating DNA molecule) in the donor cell or the presence of conjugative transposons in the chromosome. These genetic elements direct cell-to-cell contact during which a copy of the plasmid or the transposon is transferred to the recipient cell. Various types of plasmids have been identified in bacteria and some of them can elicit the transfer of other plasmids that do not have capability for their own transfer (12).

In nature, bacterial populations and communities often contain considerable fractions of cells with plasmids, and several different plasmids may be present in the same cell. Plasmids and transposons may confer new properties to the cells. Conjugative gene transfer from bacteria to eukaryotic cells, including plant cells, yeast, filamentous fungi and animal cells, has been observed (in nature or in experimental systems).

Natural transformation is another gene transfer process that involves the active uptake of extracellular DNA by bacteria into their cytoplasm. This process has so far been identified in a limited number of bacteria belonging to major trophic and taxonomic groups (13). DNA uptake occurs mostly during a specific growth phase of cells (*competence*). Both chromosomal DNA fragments and plasmids can be taken up. Under particular physical or chemical conditions, DNA may enter bacteria that do not actively take up DNA (a type of transformation often used in gene technology) Finally, transfer of bacterial genes can also occur by *transduction*, wherein transfer is mediated by bacterial viruses that have incidentally packaged DNA of the last host cell instead of viral DNA (14).

Conjugative processes, transformations and transduction may occur within members of a species but also between members of different species and genera. Extensive studies, including whole genome sequence analyses, have indicated that horizontal gene transfers have strongly contributed to the genomic structure of bacterial species. Other studies provide evidence that various gene transfer mechanisms are active in natural habitats of bacteria including soil, sediment, river epiliton, rhizosphere, phylloplane, foodstuffs, intestine, mammalian oral cavity (15–18).

DNA transferred to a new host cell may establish by genomic integration (e.g. homologous recombination; (19)) or by plasmid formation (e.g. when an origin of replication is present). Establishment may be inhibited, for example, by lack of nucleotide sequence homology or the presence of restriction endonucleases (17, 20). When the new genetic information provides a selective advantage to the cell, the trait may become fixed in the population when selection pressure prevails long enough. Gene transfer processes must be considered as part of the nature of prokaryotic organisms (21). Clearly, the spread of a gene or a gene assembly, as well as the formation of new gene assemblies, in the microbial community is mainly driven by the selection pressure in the habitat (22).

In order to reduce the potential for the spread of a recombinant construct, its chromosomal integration may be preferable to localization on a plasmid. To limit the selection of bacterial strains coexisting with the GMM into which the recombinant construct has been unintentionally transferred, genes in the construct that could provide a selective advantage under certain conditions should be avoided (e.g. antimicrobial resistance determinants). Finally, to limit the chance of unintended integration into other genomes, any sequences that would stimulate such integration should be avoided in the construct. *Yeast and filamentous fungi*. Due to the more complex structure of the eukaryotic microorganisms, the processes for gene transfer in yeast and filamentous fungi are different than bacteria. Natural cell-hybridization and genetic recombination occur frequently in species that have sexual (mating, meiosis and sporulation) or parasexual (anastomosis, nuclear fusion and haploidization by gradual loss of chromosomes) life cycles. In certain genera, interspecific hybridization can also occur between closely related species.

Synthetic gene transfer from yeast to mammalian cells has also been successfully demonstrated using yeast artificial chromosomes (YACs) which have significant potential as gene therapy vectors (23, 24).

Genetic Stability

The chromosomes of microorganisms are generally much more fluid than the chromosomes of higher eukaryotes. They grow faster and are required to adapt rapidly to changing environments. They are consequently more prone to genetic change than higher organisms. In bacteria, various mechanisms of horizontal gene transfer have been identified and are frequently seen. Further, mobile genetic elements are actively involved in reorganization of the genetic material of bacteria, and this may result in new phenotypic properties, gene inactivation, destabilization of the integrity of the genetic material, and gene loss. Mobile DNA elements include insertion sequences (IS), transposons, plasmids and prophages. Many bacteria have large numbers and different types of IS elements and some of these are very active in transposition. Such genetic changes often happen in a non-random way and may include participation of specific DNA sequences.

Genomes of certain yeasts and filamentous fungi are also prone to undergoing rearrangements. These changes take place during cultivation and are probably due to spontaneous transpositions of mobile elements (e.g. Ty retrotransposons) and chromosomal segments (manifested in chromosomal length polymorphism).

The genetic plasticity of microorganisms (25–27) may influence the fate of the recombinant DNA in GMM and, therefore, this has to be taken into consideration when the stability of a GMM is evaluated. The genetic stability of the recombinant DNA molecule is also dependent on the localization of the cloned gene(s), whether it is on the chromosome or on a plasmid. Plasmids can be lost by segregation or integrated into the chromosome or other plasmids. If the vector system is chosen carefully, one will expect that in either of the two cases—high copy number vector and chromosomally integrated insert DNA—the stability of the new genetic information will follow basic biological mechanisms and the stability of the transferred genes should be comparable with that of host genes. If high stability is required, one should not include DNA sequences that may represent a risk for stability, sites for know transposons and IS element insertions, or attachment sites for temperate phages.

Pathogenic Potential

Microorganisms used for the production of fermented food (e.g. acetic, propionic, and lactic acid bacteria, yeasts and certain filamentous fungi) have a long history of safe use. Although some enteric lactic acid bacteria have, in rare instances, been identified as the cause of bacteremia or endocarditis in patients with severe underlying disease, they can be no means be regarded as food-borne pathogens (28).

Food-borne pathogens are either invasive and or toxinogenic in the food or in the human intestine. Opportunistic pathogens in food may not be hazardous for the healthy consumer, but may pose a threat to some health-compromised persons. The genomes of many of the important food-borne pathogens and of some opportunistic pathogens have been fully sequenced, and the genes responsible for pathogenicity traits have been identified (29, 30). This opens the way to identify similar genetic information in the genomes of microorganisms used in food fermentation. Of the several genomes of microorganisms used in food fermentation that have been completely sequenced, two examples (*S. cerevisiae* and *Lactococcus lactis*) have been published and reported to be free of known pathogenicity traits. If strains of species known to carry potential toxin genes are subjected to genetic modification, they must not carry such genetic information.

The long history of safe use and the available genetic evidence suggest that the genetic background of the majority of microorganisms used for food fermentation is free from pathogenicity islands and other pathogenicity determinants.

In addition to this, the following need to be considered:

- The genetic modification could produce a metabolic imbalance that may enhance the level of common metabolites that are normally not toxicologically significant in food to levels that are unacceptable (e.g. formic acid, acetaldehyde, biogenic amines in lactic acid bacteria or yeast; cyclopiazonic acid or roquefortin in *Penicillium camemberti roqueforti*).
- The genetic modification could switch on genes coding for normally unexpressed toxins in the microorganism.
- The genetic modification (i.e. the expression of new protein(s)) could change the 'cross-talk' between the microbe and the intestinal immune system of the consumer leading, for example, to an undesirable immune reaction or undesirable reactions with other cells (e.g. enterocytes) of the GI tract.

The first and second of these points may be addressed with *in vitro* studies. The third point may need to be addressed by *in vivo* studies in suitable animal models or human volunteers. If required, such studies will need to be carried out in accordance with good practice guidelines and ethical standards.

Safety and Nutritional Assessment

As highlighted in 'Application of the concept of *substantial equivalence* to GMMs', the assessment of the safety and nutritional aspects of GMMs should take into consideration the outcome of comparisons according to the concept of *substantial equivalence*. When deemed appropriate, animal studies may be used to evaluate the safety of GMMs where the recipient organism, donor organism or gene or gene product do not have a history of safe use in food. However, as previously noted, animal studies have both strengths and weaknesses (1). While animal studies may be useful for identifying potential hazards in food produced with the aid of GMMs, and in certain cases to establish dose –response relationships, the major limitation is that the response in the animal has to be correlated with that in humans. Often, differences in the anatomy and physiology of animals and humans lead to substantial differences in dose relationships between the two and this needs to be taken into consideration in the interpretation and assessment of the relevance of animal studies on food produced with the aid of GMMs. In general, the need to use experimental animals should be carefully evaluated, taking into consideration the potential to obtain meaningful information, as well as data from other relevant test method alternatives.

The safety of GMMs intended for use in food should be evaluated in terms of the food matrix in which they are consumed and this raises issues regarding the safety assessment of whole foods using animal safety studies. Guidance regarding the role of animal studies in assessing the safety of whole food products has been provided by previous consultations (1, 4). Further advice concerning the role of animal studies in the safety assessment of whole foods has been given by the Scientific Committee for Food (9) and the UK ACNFP (10, 31).

In the use of animal studies, it is important that the experimental design addresses the specific safety issue under investigation.

The following considerations (*inter alia*) should be taken into account when designing animal feeding studies:

- The use of appropriate and relevant control groups.
- Dose to be administered.
- Endpoints to be measured (for example the impact on the micro

flora and the GI tract).

- Statistical power.
- Duration of the test.
- Rigorous control of confounding factors.

Previous recommendations with regard to animal testing are not specific for GMMs and it is, therefore, recommended that guidelines be established to assess when and what kinds of animal studies are appropriate for testing GMMs.

Interactions between the GMM, the Intestinal Flora and the Mammalian Host

Throughout adult life, the human gastrointestinal (GT) tract is populated by a huge number (up to 1014) of live microorganisms, outnumbering the number of somatic cells by 1–2 orders of magnitude. The exact composition of the flora, estimated to consist of at least 400 species, is poorly known, as the majority of species cannot be routinely analyzed by commonly used cultivation methods or molecular biology techniques. The flora fluctuates qualitatively and quantitatively from the oral cavity (dominated by lactic acid bacteria, streptococci and certain anaerobic species), to the stomach (transient acid-tolerant organisms), to the small intestine (with a transition to colonlike flora), to the colon where bacterial densities up to 1012 per g dry weight are reached. The colonic micro flora is dominated by anaerobes like *Bi. dobacterium*, *Eubac - terium*, and *Clostridium*, while the counts of microaerophilic microorganisms and facultative anaerobes such as Lactobacilli, Enterococci and coliforms usually are 3–4 orders of magnitude lower (32, 33). Throughout the GI tract, the endogenous micro flora represents the main barrier against exogenous microbes (colonization resistance (34)).

At any time after birth, the qualitative and quantitative composition of GI micro flora depends on previous and ongoing exposure to and interactions with environmental factors (diet, antimicrobial therapy, disinfectants, food additives, occupation, climate, etc.), factors associated with the mammalian host (age, gender, intestinal motility, transit time, pH, bile acids defensing, etc.), and factors associated with flora itself (competition for nutrients, oxygen, H» acceptors receptors, production of antimicrobials, organic acids, NH3, H2S, etc.). The relative importance of these factors in the formation of GI ecosystems (s) has not yet been elucidated.

The GI flora may interfere with various mammalian host-associated structures and functions in a compartmentalized way and at an organ, cell, and molecular level (35–37). Some important interactions include:

- Role of enterocytic mitosis.
- Intestinal motility.
- Development of gut-associated immune system (GALT).
- Interference with enterohepatic circulation.
- Production of organic acids, nucleotides etc.
- Prokaryotic–eukaryotic cross-talk at the cellular level.

The relative importance of these interactions may vary according to the age and health status of the individual. The ability of an exogenous microbe (including a GMM) to survive in the GI tract depends on its ability to tolerate and adapt to the various micro flora and host-associated factors mentioned above (survival being defined as the detection of a microorganism for a limited period to time where the growth rate is lower than the rate of elimination (38)). In relation to colonization resistance, other strain-specific properties necessary for survival are not very well understood (39). The strain may even leave the gut lumen and end up elsewhere by mechanisms usually termed translocation. Consequently, simple *in vitro* trials are not sufficient to predict the intestinal survival and suitable animal models, simulated human GI systems, and human trials may be required. Reliable strain identification is also essential for these studies.

A GMM that survives ingestion may only be a transient passenger or may establish itself for a variable time in the GI system. Colonization has been defined as the detection of a microorganism for a relevant period of time at a constant level (38).

While a permanent, life-long colonization of an adult by an exogenous microorganism is apparently a rare event, experiences with certain probiotic strains have shown that strains can be recovered in the feces and colonic mucosa for weeks after discontinuation of oral administration (40). The term 'persistence' has been introduced to describe the survival of a microorganism in the GI tract for longer than two intestinal transit times (11). Whether the GMM is established in the GI tract or not, the possibility remains that it might influence the micro flora or the mammalian host. The effects on the flora might partly depend on the functions expressed by the GMM (phenotypic expression), and potentially on horizontal gene transfer. The influences on the host can be direct or indirect. A direct effect can take place on all structures and functions listed previously, and indirect effects can be mediated through interference with the active part of endogenous flora. Even non-viable microorganisms are known to retain functional properties (i.e. cell adhesion, binding of chemicals, fiimmunomodulating activities), which can have direct or indirect on both micro flora- and host-associated functions (reviewed by (41)). Additional liberation of biologically active compounds (toxins, enzymes, etc.) may take place.

Conjugative transfer between microorganisms in the gut is known to occur. It is reasonable to assume that its probability depends both on the relatedness of the GMM to the intestinal micro flora and on its residence time in the GI tract, this being more likely with a persistent or colonizing strain than with a transient strain. The possibility of transfer of DNA from lyzed GMMs should not be overlooked, since bacterial transformation in the human oral cavity has been demonstrated (18). Furthermore, the work of Schubbert et al. (42) indicated measurable persistence of DNA in the intestinal tract, and studies on the fate of DNA from food in the mammalian GI tract showed that plant and recombinant DNA could enter the blood stream, tissue cells and even nuclei (42 –44).

Exposure

The degree of intake of food GMMs needs to be considered in the pre-market safety evaluation and for monitoring any impact in the food chain.

The following factors should be considered when assessing exposure to such food GMMs:

- The food matrices in which the GMM or components thereof are consumed.
- Whether the exposure is to viable or non-viable GMMs, genes or gene products.
- The dose and duration of exposure (in assessing the potential hazard).
- Exposure should be assessed within the entire population, taking into consideration special population groups (e.g. immune-deficient individuals, children and the elderly) and different geographical regions.
- Exposure to GMMs and their components during their production and preparation in the food chain.

As GMMs gain use in the food supply, consideration needs to be given to methods that measure the potential effects of exposure on the general population. The issues relating to conducting health surveillance for genetically modified foods have been dealt with by a previous Consultation who found that: "The change in nutrient levels in a particular crop plant may impact on overall dietary intake. In such cases, it is important to determine alterations in nutrient content and bio-availability, and their stability with time, processing and storage, as well as to monitor changes in dietary patterns as a result of the introduction of the genetically modified food and evaluate its potential effect on nutritional and health status of consumers. However, an assessment of the impact on nutritional status of consumers is important for all significant dietary changes and not specific to the introduction of genetically modified foods" (1).

As noted by a previous Consultation (2), very little is known about the potential long-term health effect of any food and this situation is compounded by the wide genetic variability in human populations. Given the complexity of monitoring the effects of human exposure to food produced with the aids of GMMs, it is recognized that it would be dif. cult to identify any effects against the background of conventional foods, unless studies were designed to answer very specific questions. Nevertheless, it is also recognized that it is important to develop methods to monitor (trace) exposure to GMMs.

Impact on the immune system

To assess the immune-modulating potential of a transgene in a GMM, a case-by-case consideration is recommended. A separate Consultation dealing with allergenicity in terms of genetically modified organisms has already made several recommendations in this regard (2). It must be noted that interactions of gut micro flora with the immune system occur. Gut-associated lymphoid tissue (GALT) has important interactions with the immune system and it is well established that microbial stimuli are the main antigenic forces in the development and maintenance of GALT and acquired immunity. It should be emphasized that GMMs, in contrast to food derived from GM plants, may establish themselves within the GI tract, thereby prolonging potential immune-modulating effects.

CONCLUSIONS

(i) The Consultation agreed that the safety assessment of GMMs should proceed on a case-by-case basis aided by a series of well-defined questions. The Consultation confirmed that a comparative approach, using the concept of *substantial equivalence*, provided a practical means of identifying similarities and differences between food produced with the aid of GMMs and their appropriate comparators. These differences would then be the focus of the safety evaluation.

(ii) The Consultation noted that there are intrinsic properties of microorganisms that require special consideration in the application of the concept of *substantial equivalence*. In particular, it noted that the food matrix in which GMMs may be consumed could influence its safety, thus the impact of the food matrix needs to be considered. It may, therefore, be necessary to apply the concept of *substantial equivalence* both to the GMM and to the food produced with the aids of GMMs. In doing this, it may be necessary to examine additional parameters such as pathogenicity and persistence in the mammalian host GI tract.

(iii) The Consultation noted that the potential range of GMMs in food included viable and non-viable microorganisms that may be consumed as such or may be integral components of foods. The Consultation noted that because of the wide range of products involved, the safety assessment needed to take into consideration the specific uses and exposures to the GMM being considered.

(iv) The Consultation noted that the use of microorganisms in food production is of great importance to the nutritional quality and safety of the food supply. Therefore, the evaluation of GMMs should encompass both safety and nutritional aspects.

(v) The Consultation noted that microorganisms in the GI tract exert important effects on the immune system. While previous recommendations (2) relating to the allergenicity of new proteins expressed in GM plants can be used in the safety assessment of foods produced with the aids of GMMs, it must be noted that the possible effect of GMMs or their components on the immune system in the mammalian host requires additional consideration.

(vi) The Consultation noted that genetic material from food has the potential to transfer to gut micro flora and to cells of the mammalian host *in vivo*. The safety concerns of such gene transfer need careful evaluation based on the properties of the GMM and its components.

vii) The Consultation concluded that in developing a GMM for use in food production, vectors should be used which consist only of nucleotide sequences from microorganisms with a history of safe use in food. Any selectable markers should be carefully chosen and based on safe use. In particular, antimicrobial resistance marker genes should be avoided and not be present in the final GMM.

RECOMMENDATIONS

(i) It is recognized that the complex ecosystem of the human gastrointestinal tract is subject to increasing and successful analysis. It is recommended that this analysis should be enhanced, including examination of the ecological components of the GI tract, the prevailing selective conditions, and the effect of nutritional conditions and host factors on the interactions. These studies would provide the basis for improved risk assessment.

(ii) While a permanent life-long colonization of an adult by an exogenous microbe is apparently a rare event, strains can be recovered from the intestinal tract for weeks after the discontinuation of exposure through food. Whether a GMM is established in the GI tract or not, the possibility remains that it might influence the micro flora and or the mammalian host. The effects on the flora might relate to functions expressed or to horizontal gene transfer. There is a need to improve the methods for evaluating the function of microorganisms in the GI tract.

(iii) Fermentation provides a simple technique to produce food of high nutritional and hygienic quality. This technology is used globally and is especially important in developing countries. The continued improvement of such technologies could involve the use of GMMs. The Consultation recommended that the FAO WHO promote capacity-building efforts to support the needs of developing countries in improving and evaluating this technology.

(iv) The Consultation recognized the need for efficient communication of issues related to the development and safety evaluation of GMM food. Specific communication of the principles guiding the safety assessment to the public etc. would enable efficient interaction and transparency in the evaluation process. The Consultation recommended that the FAO WHO coordinate the efforts to achieve this.

(v) The Consultation noted that specific methodologies are available which enhance the safety of GMMs through an improved understanding of the biology of microorganisms. New technologies are developing rapidly with a potential to enhance the safety evaluation of GMMs, especially bacteria. These are described within the report and include molecular profiling. The Consultation encourages further development and validation of such methods.

(vi) The Consultation identified a number of aspects that it recommends should be taken into account in assessing the safety of food produced with the aid of GMMs.

Details of these are given in the body of the report and include:

- application of the concept of *substantial equivalence* to the GMM and foods produced from the GMM;
- consideration of the techniques used for the development of the GMM—especially the history of safe use of host microorganisms and also of the microorganisms from which inserted gene(s) and vector(s) are derived—and avoidance of the use of antimicrobial resistance marker genes;
- strain identification and characterization; transfer of genetic material from the food to the gut micro flora and mammalian host cells;
- genetic stability of the GMM;
- pathogenic potential of the GMM;
- impact of the GMM on the human immune system; and
- human exposure to the GMM and the effects of food processing, production and storage.

REFERENCES

1. FAO/WHO. Safety aspects of genetically modified of plant origin. Geneva: World Health Organization, 2000.
2. FAO/WHO. Evaluation of the allergenicity of genetically modified foods. Rome: Food and Agriculture Organization of the United Nations, 2001.
3. JECFA, Joint FAO/WHO Expert Committee on Food Additives. FAO Food and Nutrition Papers 52 Addendums 1-8. 1999.
4. FAO/WHO. Strategies for assessing the safety of foods produced by biotechnology. Geneva: World Health Organization, 1991.
5. OECD. Safety evaluation of foods derived by modem biotechnology: concepts and principles. Paris: Organization for Economic Cooperation and Development, 1993.
6. WHO. Application of the principle of substantial equivalence to the safety evaluation of foods for food components from plants derived by modern biotechnology. Geneva: World Health Organization, 1995.
7. FAO/WHO. Biotechnology and food safety. FAO Food and Nutrition Paper 61. Rome: Food and Agriculture Organization of the United Nations, 1996.
8. ILSI. The safety assessment of novel foods. Brussels: International Life Sciences Institute, Europe, 1995.
9. UK ACNFP. Structured approach to safety assessment of novel foods and processes. Advisory Committee on Novel Foods and Processes, Annual Report for 1994. London: Ministry of Agriculture Fisheries and Food, 1995.
10. ILSI. The safety assessment of viable genetically modified microorganism used as food. Brussels: International Life Sciences Institute, Europe, 1999.
11. Clewell DB. Bacterial Conjugation. New York: Plenum Publishing Press, 1993.
12. Lorenz MG, Wackernagel W. Bacterial gene transfer by natural genetic transformation in the environment. Microbiol Rev 1994; 58: 563-602.
13. Masters M. Generalised transduction. In: Neidhart FC, ed. Escherichia coll and Salmonella. Washington, DC: ASM Press, 1996:2421-41.
14. Lorenz MG, Wackernagel W. Bacterial gene transfer in the environment. In: Wöhrmann W, Tomiuk J, eds. Transgenic Organisms, Risk

Assessment of Deliberate Release. Basel, Boston, Berlin: Birkhäuser, 1993: 43-64.

15. Bräutigam M, Hertel C, Hammes WP. Evidence for natural transformation of Bacillus subtillis in foodstuffs. FEMS Microbiol Lett 1997; 155: 93-8.
16. Davison J. Genetic exchange between bacteria in the environment. Plasmid 1999; 42: 73-91.
17. Mercer D, Scott K, Melville C, Glover A, Flint H. Transformation of an oral bacterium via chromosomal integration of free DNA in the presence of human saliva. FEMS Microbiol Lett 2001; 200: 163-7.
18. de Vries J, Meier P, Wackernagel W. The natural transformation of the soil bacteria Pseudomonas stutzen and Acinetobacter sp. by transgenic plant DNA strictly depends on homologous sequences in recipient cells. FEMS Microbiol Lett 2001; 195:211-5.
19. Majewski J. Sexual isolation in bacteria. FEMS Microbiol Lett 2001; 199: 161-9.
20. Syvanen M, Kado CI. Horizontal Gene Transfer. London: Chapman and Hall, 1998.
21. Lawrence J, Roth J. Roles of horizontal gene transfer in bacterial evolution. In: Syvanen M, Kado CI, eds. Horizontal gene transfer. London: Chapman and Hall, 1998: 208-25.
22. Giraldo P, Montoliu L. Size matters: use of YACs, BACs, and PACs in transgenic animals. Trans Res 2001; 10: 83-103.
23. Fabb SA, Ragoussis J. Yeast artificial chromosome vectors. MoI Cell Biol (Hum Dis Ser) 1995; 5: 104-24.
24. Stibitz S, Yang MS. Genomic plasticity in natural populations of Bordetella pertussis. J Bacteriol 1999; 181: 5512-5.
25. Brunder W, Karch H. Genome plasticity in Enterobacteriaceae. Int J Med Microbiol 2000; 290 (2): 153-65.
26. Le Bourgeois P, Daveran-Mingot ML, Ritzenthaler P. Genome plasticity among related Lactococcus strains: identification of genetic events associated with macrorestriction polymorphisms. J Bacteriol 2000; 182: 2481-91.
27. Gasser F. Safety of lactic acid bacteria and their occurrence in human clinical infection. Bull Inst Pasteur 1994; 92: 45-67.
28. Finlay BB, Cossart P. Exploitation of mammalian host cell functions

by bacterial pathogens. Science 1997; 276: 718-25.

29. Morschhauser J, Kohler G, Ziebuhr W, Blum-Oehler G, Dobrindt U, Hacker J. Evolution of microbial pathogens. Phil Trans R Soc London 2000; 355: 695-704.

30. UK ACNFP. The role of lexicological studies in assessing the safety of novel foods for humans. Advisory Committee on Novel Foods and Processes, Annual Report for1998. London: Ministry of Agriculture, Fisheries and Food,1999.

31. Mikkelsaar M, Mändar R, Sepp E. Lactic acid microflora in the human microbial ecosystem and its development In: Salminen S, von Wright A, eds. Lactic Acid Bacteria: Microbiology and Functional Aspects. New York: Marcel Dekker Inc. 1997: 279-342.

32. Willis CL, Gibson GR. The natural microflora of humans. In: Robinson RK, Batt CA, Patel PD, eds. Encyclopaedia of Food Microbiology. London: Academic Press, 2000: 1351-5.

33. Araneo BA, Cebra JJ, Beuth J, Fuller R, Heidt PJ, Midtvedt T, Nord CE, Nieuwenhuis P, Manson WL, Pulverer G, Rusch VC, Tanala R, van der Waaij D, Walker RI, Wells CL. Problems and priorities for controlling opportunistic pathogens with new antimicrobial strategies: an overview of current literature. Zentralblat für Bakteriologie 1996; 283: 431-65.

34. Falk PG, Hooper LV, Midtvedt T, Gordon JI. Creating and maintaining the gastrointestinal ecosystem: what we know and need to know from gnotobiology. Microbiol MoI Biol Rev 1998; 62: 1157-70.

35. Midtvedt T. Microbial functional activities. In: Hansson LAA, Yolken RH, eds. Probiotics, Other Nutritional Factors, and Intestinal Microflora. Philadelphia, PA: Lippincott-Raven Publisher, 1999: 79-96.

36. Moreau MC, Gaboriau-Ruthiau V. Influence of resident intestinal microflora on the development and functions of the gut-associated lymphoid tissue. Microb Ecol Health Dis 2001; 13: 65-86.

37. Benbadis L, Cohen PS, Contrepois M, Conway P, Duvallflah Y, Gasson MJ, Blix, Gundersen W, Hentges DJ, Jacobsen BL, Klemm P, Krivan H, Krogfeldt KA, Lintermanns P, Nielsen M011er E, Norin E, Schlundt J, Saadbye P, Vuust J. Microbial colonisation of the body. Microbial Ecol Health Dis 1995; 8 (Suppl 1):S33-S34.

38. Cesena C, Morelli L, Alander M, Siljander T, Tuomola E, Salminen S,

Mattila-Sandholm T, Vilpponen-Salmela T, von Wright A. Lactobacillus crispatus and its nonaggregating mutant in human colonisation trials. J Dairy Sci 2001; 84: 1001-10.

39. von Wright A, Salminen S. Probiotics: established facts and open questions. Eur J Gastroenterol Hepatol 1999; 11: 1195-8.
40. Ouwehand A, Salminen S. The health effects of cultured milk products with viable and non-viable bacteria. Int Dairy J 1998; 8: 749-58.
41. Schubbert R, Renz D, Schmidt B, Dorfler W. Foreign (M 13) DNA ingested by mice reaches peripheral leukocytes, spleen and liver via the intestinal wall mucosa and can be covalently linked to mouse DNA. Proc Natl Acad Sei USA 1997; 94: 961-6.
42. Einspanier R, Klotz A, Kraft J, Aulrich K, Poser R, Schwägele F, Jahreis G, Flachowsky G. The fate of forage plant DNA in farm animals: a collaborative case-study investigating cattle and chicken fed recombinant plant material. Eur Food Res Technd 2001; 212: 129-34.
43. Hohlweg U, Doerfler W. On the fate of plant or other foreign genes upon the uptake of food or after intramuscular injection in mice. Mol Genet Genomics 2001; 265: 225-33.

CHAPTER 3

REGULATORY BIOINFORMATICS FOR FOOD AND DRUG SAFETY

Marion J.Healy[a], WeidaTong[b], StephenOstroff[c] Hans-GeorgEichler[d], AlexPatak[e], Margaret Neuspiel[f], Hubert Deluyker[g], William SlikkerJr.[b]

[a]Food Standards Australia New Zealand, Barton, Australian Capital Territory, 2905, Australia

[b]National Center for Toxicological Research, US Food and Drug Administration, Jefferson, AR, 72079-9502, US

[c]US Food and Drug Administration, Silver Spring, MD, 20993-0002, USA

[d]European Medicines Agency, London, E14 5EU, United Kingdom

[e]European Commission, Joint Research Centre, Ispra, Varese, 21027, Italy

[f]Canadian Food Inspection Agency, Ottawa, Ontario, Canada

[g]European Food Safety Authority, Parma, 43126, Italy

CITATION: Healy, M.J., et al., "Regulatory bioinformatics for food and drug safety", Regulatory Toxicology and Pharmacology (2016), https://doi.org/10.1016/j.yrtph.2016.05.021

ABSTRACT

"Regulatory Bioinformatics" strives to develop and implement a standardized and transparent bioinformatic framework to support the implementation of existing and emerging technologies in regulatory decision-making. It has great potential to improve public health through the development and use of clinically important medical products and tools to manage the safety of the food supply. However, the application of regulatory bioinformatics also poses new challenges and requires new knowledge and skill sets. In the latest Global Coalition on Regulatory Science Research (GCRSR) governed conference, Global Summit on Regulatory Science (GSRS2015), regulatory bioinformatics principles were presented with respect to global trends, initiatives and case studies. The discussion revealed that datasets, analytical tools, skills and expertise are rapidly developing, in many cases via large international collaborative consortia. It also revealed that significant research is still required to realize the potential applications of regulatory bioinformatics. While there is significant excitement in the possibilities offered by precision medicine to enhance treatments of serious and/or complex diseases, there is a clear need for further development of mechanisms to securely store, curate and share data, integrate databases, and standardized quality control and data analysis procedures. A greater understanding of the biological significance of the data is also required to fully exploit vast datasets that are becoming available. The application of bioinformatics in the microbiological risk analysis paradigm is delivering clear benefits both for the investigation of food borne pathogens and for decision making on clinically important treatments. It is recognized that regulatory bioinformatics will have many beneficial applications by ensuring high quality data, validated tools and standardized processes, which will help inform the regulatory science community of the requirements necessary to ensure the safe introduction and effective use of these applications.

Keywords: Bioinformatics, Regulatory bioinformatics, Drug safety, Food safety, Regulatory science, GCRSR, GSRS, Next-generation sequencing, Microbiome, Genomics

INTRODUCTION

Emerging life-science technologies offer major opportunities to improve public health through the development and use of innovative medical products and by improving the safety of the food supply. However, these technologies often utilize and generate large amounts of data. In order to effectively use these data sets for regulatory purposes, policies, strategies and approaches to curate, analyse and interpret the data must be development. Such efforts, which can be thought of as "regulatory bioinformatics" can strengthen the scientific evidence base and produce tools to support, improve, and enhance the efficiency of regulatory decision making. Initiatives to develop regulatory bioinformatics capabilities are underway globally that aim to maximize the health benefits presented by the emerging technologies of today and tomorrow.

Bioinformatics is considered a priority area within the regulatory science agenda (http://www.fda.gov/ScienceResearch/SpecialTopics/RegulatoryScience/ucm228131.htm). To move this agenda forward requires enhancing global scientific capacity and cooperation among academia, regulatory agencies and industry. One mechanism to accomplish this goal was the establishment in 2013 of the Global Coalition for Regulatory Science Research (GCRSR), an international collaboration that aims to build knowledge, to promote the development of regulatory science, and to identify research needs to support regulatory decision making. A major focus is establishing best practices to understand and interpret data from innovative technologies and facilitate the translation of basic science innovation into regulatory application (Tong et al., 2015).

The GCRSR consists of representative governmental agencies that regulate food and medical products in many countries (Tong et al., 2015). Its mission is to communicate, among the global researchers/regulators, current advances in science and their potential to impact the global regulatory process. Even before the coalition was formally established in 2013, GCRSR began holding annual conferences, named Global Summit on Regulatory Science (GSRS) to facilitate scientific exchanges and enhanced understanding about emerging technologies and their potential application in regulatory science (Tong et al., 2015, Howard et al., 2014, Miller et al., 2013, Slikker et al., 2012). Each of the five conferences since 2011 has focused on a specific topic in the context of regulatory research. The two most recent conferences focused on nanotechnology in 2013 (Howard et al., 2014) and on regulatory genomics in 2014 (Tong et al., 2015). At both of these conferences, the crucial role of bioinformatics for successful regulation of nanotechnology and genomics was readily apparent. Therefore, it was decided that the theme of the 2015 conference (GSRS2015) would be "regulatory bioinformatics." This meeting was held on October 12–13, 2015 at the headquarters of the European Food Safety Authority in Parma, Italy (http://www.fda.gov/

AboutFDA/CentersOffices/OC/OfficeofScientificandMedicalPrograms/NCTR/WhatWeDo/).

"Regulatory bioinformatics" strives to develop and implement a standardized and transparent bioinformatic framework to support the implementation of existing and emerging technologies in regulatory decision-making. As shown in Fig. 1, GSRS2015 explored a range of applications of regulatory bioinformatics in the development and use of medical products and as tools to manage the safety of the food supply. These examples illustrate the potential benefits of the technologies, highlight areas of research and development, and the novel approaches required to fully realize these benefits. These examples also reveal the variable state of development of applications. In the drug development arena, next generation sequencing (NGS) has proven to be an excellent discovery tool, but improvements are needed in bioinformatics to use the enormous amount of data generated by NGS for beneficial clinical applications. In contrast, several applications using different and large datasets within the microbiological risk analysis framework in the drug and food safety fields are delivering clear benefits. In the following sections, we summarize key discussions from GSRS2015 on these topics depicted in Fig. 1.

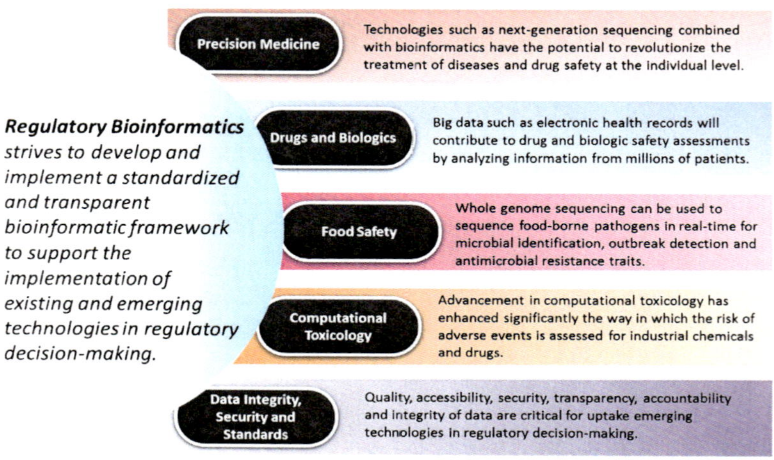

Figure 1. Overview of regulatory bioinformatics and its associated impact on regulatory decision-making.

PRECISION MEDICINE, NEXT-GENERATION SEQUENCING AND REGULATORY APPLICATION

Technologies such as NGS, when combined with bioinformatics analysis, have the potential to revolutionize the treatment of diseases and the evaluation of drug safety. NGS enables human genomes to be sequenced rapidly and cost effectively, and tools are now available for analysing the consequent large datasets, providing a mechanism for treating patients individually to enhance the health outcome as well as improving the success rate in the development of new therapeutics. Individualized treatment of patients (personalized medicine) takes into account their genetic variation, environment and lifestyle. Several countries/regions are implementing programs to support the application of precision medicine (Winship, 2015). The United States is implementing the multifaceted Precision Medicine Initiative to develop high quality, curated databases and interoperable standards and requirements that address privacy and enable secure exchange of data across systems; to determine genomic drivers of diseases such as cancer and identify appropriate biomarkers and develop more effective treatments; and improve understanding of the environmental, genetic, biochemical and other factors predictive of disease risk, response to therapy, and disease outcomes (Collins and Varmus, 2015).

The new approaches required to identify effective treatments at the individual level, classify diseases and the analytical tools needed to effectively deliver personalized medicine are illustrated by the transition from therapies that target individual mutations to applied systems that address the complexity of diseases such as cancers (Klauschen et al., 2014). Routine mutational profiling in cancer diagnostics in concert with an increasing availability of drugs targeting specific genetic alterations in tumors have started to transform oncology. However, current targeted approaches that are mostly monotherapies against single actionable mutations show variable clinical response in most patients even with the identical mutations and resistance develops frequently during therapy.

This is likely due to the mutational complexity observed in most cancers that may induce multiple oncogenic mechanisms (Heim et al., 2014). There is currently a lack of information about the clinical relevance of the many mutations found within tumors. Moreover, actionable mutations are normally present in only a minor fraction of the tumors, which makes the recruitment of a sufficient number of suitable patients exceedingly difficult for drug combinations because of the high combinatorial complexity. Therefore current clinical trial design strategies need to be reconsidered to solve the combinatorial complexity problem.

Novel experimental approaches that integrate genomic with dynamic proteomicprofiling and computational modeling facilitate the analysis of the functional effects of the observed mutational profiles. This allows for a substantial reduction of the mutational complexity and identification of driver mutations that propose combination therapies for individual patients. This approach may serve as the basis for future novel clinical design strategies that no longer only test specific drugs for a particular molecular cancer type but also evaluate the capability of systems biological approaches to predict optimal targeted (combination) therapies for individual patients.

One of the key issues in applying NGS/bioinformatics in regulatory decision making is ensuring the accuracy and quality of information. To gain a full understanding of this issue, the United States has created a community research and development portal that allows for testing, piloting, and validating existing and new bioinformatics approaches to NGS processing (PrecisionFDA, https://precision.fda.gov). PrecisionFDA is focused on software to evaluate and characterize NGS applications within an open-source cloud infrastructure and is considering secure storage of human genomes, storage and sharing software for NGS, reference genomes for use in identifying variants and how to use ethnically diverse and admixed genomes in validating annotation tools. China is also implementing an initiative to speed the development of precision medicine. The China initiative will utilize information from a range of 'omics techniques (genomics, transcriptomics, proteomics, and metabolomics); consider phenotypic information from sources such as electronic health records, clinical information, survival data and the Human Phenomics Project; and develop big data tools that address data quality, standardization, data mining and standard operating procedures. China's approach to precision medicine includes diagnosis, prognosis, therapy, monitoring and prevention and is focused on ten types of cancers, type 2 diabetes and various rare diseases.

Many of the current issues with NGS resemble challenges faced a decade ago when microarray technologies were the principle tool for gene expression assessment. At that time, a number of FDA efforts were initiated to utilize and evaluate these technologies within inter-center and cross-community collaborations. Examples include the FDA's Voluntary eXploratory Data Submission program to develop standards and bioinformatics solutions to receive and evaluate these new data in the review process (Goodsaid et al., 2010), and the FDA led community wide MicroArray Quality Control (MAQC) consortium to address the quality control and proper use of these new genomic methodologies (e.g. SEQC/MAQC SEQC/MAQC-III Consortium, 2014). Global partnerships and cooperation among the regulatory agencies of different countries are essential for building consensus and best practices to expedite the translation from the exploratory nature of NGS to regulatory application.

DRUGS AND BIOLOGICS

A novel approach to the post market monitoring of the safety of drugs and biologics, making use of large datasets of patient information relating to insurance claims, diagnostic and clinical procedures and patient information held by insurance companies, hospitals and government has been explored by Japan, amongst other countries. The incorporation of electronic health records into drug safety assessments can overcome some of the weaknesses of conventional approaches (under reporting of adverse events, difficulty in attributing adverse events to drug effects) by allowing assessment of information from millions of patients (https://www.pmda.go.jp/english/safety/surveillance-analysis/0001.html).

The pharmacoepidemiological methodology of using electronic health information for regulatory decision making on drug safety has been conducted in a number of pilot studies that tested well characterized safety issues (e.g., the concomitant use of sitagliptin and sulfonylurea). The approach is being progressively implemented with the aim to use it in regulatory post-marketing safety assessments and re-examination of approved drugs. Successful implementation requires a medical information database with data standardization and quality assurance procedures to integrate vast quantities of patient records from different clinical settings with real time synchronization of information. To reduce the risk of biased information, a robust sample of hospitals need to participate, and currently there is no link between hospital information for individual patients. Community acceptance of the use of these electronic records necessitates transparent rules for secondary use of the records, with these rules being understood by the community.

Electronic health records have also been integrated in the classical infectious disease risk assessment paradigm, which has been adapted to take advantage of big data sets and bioinformatic analytical tools. This approach has been demonstrated through analysis of records of US

Medicare beneficiaries to determine the relative effectiveness of different influenza vaccines (Izurieta et al., 2015). The successful study of a large population group demonstrated differential effectiveness of the vaccines in older individuals covered by Medicare, and identified for the first time that influenza related hospital admissions were significantly reduced for specific vaccines. A second example is examination of adverse events in infants in response to vaccination with the second-generation rotavirus vaccine (Yih et al., 2014). The large number of infants in the study (approx. 1.2 million) provided sufficient statistical power to determine a low frequency adverse event that had not been detected in earlier relatively large studies involving about 60,000 infants.

These examples demonstrate successful application of bioinformatics to provide new information for regulatory decision makers. However, studies to date emphasize the importance of capturing metadata in a structured way and adequate cross linking of the metadata to the experimental data, such as adverse events, medical records, and outcomes of outbreak investigations.

FOOD SAFETY AND MICROBIAL GENOMICS

Leading edge genomics technologies are opening new regulatory possibilities for comprehensive analyses of a variety of microbes, including those recovered from food inspection samples. Whole genome sequencing (WGS) is now making it feasible to sequence food-borne isolates in real-time, such as during food-borne illness outbreak investigations (Elkins et al., 2013). Sequencing pathogenic bacteria, whether in the context of outbreak investigations or information gathering in the course of research, can yield an unprecedented quality of information regarding the relatedness of isolates and the presence of virulence and other marker genes of relevance to pathogen identification and risk characterization. Such advances are resulting in the integration of WGS approaches into food microbiology regulatory decision making, although effective implementation requires a mechanism to rapidly collate, analyse and report information from distributed sources (Lambert et al., 2015).

One example, the Center for Genomic Epidemiology (Lyngby, Demark, www.genomicepidemiology.org) focuses on WGS of bacterial pathogens and particularly how to assemble process and handle the large amount of data generated in a standardized way that will make the information useful, especially for diagnostic and surveillance purposes. The Center employs a range of analytical tools to identify certain phenotypes, such as acquired antimicrobial resistance genes, virulence genes, various typing functions and the development of phylogenetic trees using different data types. It establishes processes for use of WGS in outbreak investigations and clinical settings, thereby demonstrating how the data can be practically applied, as well as contributing to the development of a global system for microbiological identification and epidemiological surveillance.

Another example is related to the application genomic and bioinformatics analyses to the variable global incidence of food borne disease associated with contamination by *E. coli* 0157, a major food borne pathogen. The phylogenetic linkages of a collection of isolates from geographically distinct regions, namely Argentina, the United States and Australia, have been further refined based on molecular typing and the relationship between Shiga toxin genes (*stx*) subtypes and total Shiga toxin production. Examination of isolates revealed strong geographical segregation of Single Nucleotide Polymorphisms (SNPs) lineages, and several new lineages were identified that are regionally distinct. Furthermore, there are strong associations between specific SNP lineages, *stx* subtype and *stx* production (Mellor et al., 2015). The regional distribution of these lineages are correlated with regional differences in *E. coli* 0157 disease incidence, suggesting that these subtype differences influence the patterns of disease seen in each location.

Studies of pathogenic *E. coli* have also highlighted some of the challenges in metagenomic analyses. Different enrichment procedures and analytical tools can result in significant differences in determining phylogenetic diversity, intraspecies diversity and marker abundance, highlighting the need for caution in the interpretation and the conduct of multiple independent methods. Metagenomics has emerged as an important area to understand the interactions of the human microbiome and host to unravel the complex relationship between the composition of the microbiome and human health. The human microbiome, estimated to consist of 100 trillion bacterial cells, varies in its composition at different sites of the body. The gut microbiota is considered the largest and most complex, with many aspects of the host (e.g. age, diet, genetics) influencing the microbial ecology. Clearly, the application of genomics and bioinformatics to determining the phylogeny of gut microflora and high resolution microbiota profiling and linking changes to health status is the focus of ever more attention, and will undoubtedly lead to new therapeutic approaches.

However, these techniques are also being applied to food safety risk assessment questions such as the role of residues of antimicrobial veterinary drugs in food products contributing to resistance to clinically important antibiotics. The impact of residual levels of antimicrobials in food on the bacteria and selection of resistant bacteria in the gastrointestinal tract has not been fully understood, although is now being elucidated through genomic and bioinformatic analysis of the microbiome in individuals treated with varying concentrations of antimicrobials. These studies are demonstrating in greater clarity that both the relative abundance of species in the gut and expression of resistance genes change in response to antimicrobial use.

5. Computational toxicology and its role in regulatory decision-making

Recent development in the tools and applications of predictive toxicology is also significantly changing the way in which the risk of adverse events is assessed for chemicals used as drugs, added to food and used in other contexts. One of the methodological approaches, Quantitative Structure-Activity Relationships (QSARs) is a well-established multi-disciplinary methodology that integrates biological, chemical and computational models to provide a new paradigm for chemical risk assessment (Sturla et al., 2014). QSARs therefore provide models for considering the standards for use of predictive toxicology in regulatory decision making.

Barriers to the acceptance of QSARs in regulatory settings range from scientific concerns (e.g., relevance, reliability and interpretability), practical considerations (e.g., availability and ease of use, portability/interoperability), to policy change that permits incorporation of QSARs in a modern toxicological framework with comprehensive assessment and credible outcomes. These barriers have been addressed through the initiatives of regulatory agencies and multilateral bodies. Scientific concerns have been addressed for the use of QSARs through the development of internationally accepted reporting formats for models and predictions and guidance on how to interpret the prediction results. For example, guidance on use of QSARs in pesticide assessments, the grouping of chemicals and the reporting of read-across and chemical categories (e.g., OECD, 2014) have been established. The practical use of QSAR is supported by freely available software tools and training, such as the ChemoTyper application, which allows for searching and highlighting chemical chemotypes (chemical substructures or subgraphs) in datasets of molecules, and the Joint Research Centre model database. Validation of different software packages, however, remains an issue. Finally the positioning of QSARs as part of an integrated framework to testing and assessment should enable QSAR to be used in concert with other strategies.

DATA INTEGRITY, SECURITY AND STANDARDS

Data Integrity and Security

The data sources to underpin personalized medicine, such as the genetic basis of disease and the variation in pertinent genetic factors, health information from electronic records held by hospitals and medical practitioners and health related information generated through mobile personal devices, are becoming available. Many of these sources of information pose new, confronting and unresolved challenges that have not been appropriately addressed; they include data integrity, traceability, accuracy and security. A clear example is the hackabilty of medical devices (http://go.bloomberg.com/tech-blog/2012-02-29-hacker-shows-off-lethal-attack-by-controlling-wireless-medical-device/http://go.bloomberg.com/tech-blog/2012-02-29-hacker-shows-off-lethal-attack-by-controlling-wireless-medical-device/). For example, the growing popularity of the Internet of Things (IoT) crosses multiple industries and disciplines. (The Internet of Things is the network of physical objects that contain embedded technology to communicate and sense or interact with their internal states or the external environment.) However, the Quantified Self movement and mobile medical care raise similar big data concerns relating to hacking, data reliability, privacy anonymity, data standardization, usability, reusability, and interoperability of the data.

The Quantified Self movement seeks to use proprietary wearables, smartphones, and other sensor technologies to collect, analyze and quantify basic aspects of daily lives. Applications range from collection of health related information by consumer products to the academic effort to catalogue all available personal 'omic data' (e.g., the Snyderome). Often consumer-facing apps, falling outside regulatory oversight, collect data using variable brands of sensors running on hundreds of different types of smartphones with thousands of different operating system configurations. This variability may result in unreliable and noisy data that are of questionable utility in monitoring and treating chronic disease.

Similarly, remote medical care, including telemedicine, and particularly chronic disease management, intends to employ smartphones, sensors, and third party devices to track and assess the health of patients. Often this software is configured to report back to health care professionals, especially when emergency medical intervention is needed. Without strong encryption, sensor data in transit can be hijacked, and unencrypted Wi-Fi enabled medical devices can be maliciously manipulated.

One possible solution is implementation of middleware, a third party gateway between the consumer/patient and the end-point data analysis. The gateway could provide encryption and calibration services, and confirm that data are standardized. The middleware would reject the data if key conditions are not fulfilled. Moreover, a consumer could track their personal information to assess to whom and where their data is transferred. Such an approach allows for self-regulation to provide technological options to assure that the data are safely transferred and are usable, reusable, transferable, and accessible.

Automated data gathering technologies, such as those used with imaging techniques, are resulting in huge amounts of data being gathered and integrated to inform clinical research and medical practice. One of the major challenges in developing beneficial applications from this knowledge is access to validated tools to assist analysis of information, access to expertise to interrogate and integrate multiple data sets, mechanisms to share information, and mechanisms to classify and archive the information. The required expertise, tools and computing power are largely being housed within major institutions such as the European Molecular Biology Laboratory – European Bioinformatics Institute (EMBL-EBI, http://www.ebi.ac.uk/). The role of these institutions is continuing to evolve with new types of data sets being added, for example NGS data on human gene variation; images; and medical data.

However, there are ongoing challenges to facilitate effective use of the data, particularly because of the diverse range of user communities, need to integrate different types of data; the requirement for establishing mechanisms and tools for data sharing; and for providing secure environments for data analysis. The importance of these institutions in facilitating the translation of data to information for health care application is illustrated by the EMBL-EBI. EMBL-EBI as an international and data sharing organization has a rich history of successful partnerships with clinical researchers and occasionally healthcare professionals. It has developed a coherent strategy for medicine to focus on data that are open, research-based, health-care-system agnostic and relevant to human health. EMBL-EBI is also a major driver for the pan European research infrastructure for biological information, ELIXIR, and contributes to the goal of accelerating the potential of genomic medicine to advance human health through its membership in the Global Alliance for Global Health (https://genomicsandhealth.org/).

Data standards and ontology

The large quantities of data generated by technologies such as NGS and predictive toxicology are often still not managed effectively, despite significant investment. While it is recognized that data should be findable, accessible, interoperable and reusable, in practice, data are usually difficult to find and understand (e.g., Sansone and Rocca-Serra, 2012). A diverse range of organizations are involved in developing content standards that describe data better, but the multiple programs have resulted in a plethora of standards. One tool that assists in navigating and selecting content standards is a web-based, curated and searchable registry that ensures standards and databases are registered, informative and discoverable. It also monitors development and evolution of standards, their use in databases and adoption of both in data policies (https://biosharing.org/pages/about/).

However, there remain significant inadequacies in the development of content standards, which are now being addressed through major initiatives, such as the European Innovative Medicines Initiative, the pan-European infrastructure for biological information (Elixir programme) and the National Institutes of Health Big Data to Knowledge initiative (http://www.imi.europa.eu; https://www.elixir-europe.org/about/elixir-programme-2014-2018; https://datascience.nih.gov/bd2k). These initiatives focus on best practice data storage, analysis and use that ultimately should improve accessibility and usability of the data being generated.

DISCUSSION

Personalized and precision medicine are the wave of the future, where bioinformaticsplays a significant role. Significant resources are being devoted to realizing the regulatory potential of these approaches that involve cross sector collaborations and multi-disciplinary approaches. A range of data sources are being exploited in the development of these approaches, such as electronic health records, patient data, administrative, genomic sequences, meta-data, and predictive toxicology. Similarly, regulatory bioinformatics is modernizing approaches to food safety. Its use can drastically decrease the time taken to identify a pathogen and increase the efficiency and precision of risk characterizations. WGS and bioinformatics can be integrated into food safety regulatory decision making in a variety of ways, including species identification, virulence profiling, serotyping and molecular typing, signature sequences analysis, risk and phylogenetic profiling, comprehensive information analysis and antimicrobial resistance profiling. In a number of these applications, proof of concept has been demonstrated and there is sufficient confidence in the reproducibility and quality assurance procedures for use in regulatory applications.

There are many different types of big data sets and the power of analysis and data mining comes from the analysts ability to integrate many different layers of datasets for the synthesis of new information. Big data can include both structured and unstructured data sets and these can be records with narratives and descriptors and databases of reports of adverse effects and each database typically develops its own lexicons and rules. In order to support good regulatory decision making, datasets must be of good quality, robust, verified and have sufficient metadata.

Big data pose new regulatory challenges and requires new skill sets and new knowledge for successful interpretation. In addition to computer and data analysis expertise, key skill sets include a basic understanding of risk analysis integrated with statistical knowledge. Big data analysis can be used for the individual risk assessments elements; dose response using databases (GenBank, EMBL, Electronic Health Records, Administrative Records); exposure (sources, contamination, potential for growth and the internet of things); and risk characterization (distribution curves, millions of iterations and simulations, visualization tools).

To fully reap the benefits of the emerging use of bioinformatics across a broad range of data sets and provide a robust basis for regulatory decision making, a number of challenges must continue to be addressed. There is an urgent need for standards and validation processes that are applicable internationally. Data must be appropriately stored and be shared and accessible. The challenges of integrating big data sets that have different degrees of data integrity and standardization and application of different software and computational methods must be resolved. These issues highlight the important role of broad based international collaborations and projects in achieving international consensus.

The use of certain data sets for health and medical purposes raise significant privacy and ethical issues, for example electronic health records. Patient consent may be required in some cases and there is a risk that biases occur in the datasets if subsets of patients decline to allow their personal information to be used. As individual patient data may have implications for family members, ethical issues also arise.

Finally, big data repositories are growing exponentially and in some cases faster than knowledge of the basic biology underpinning the science and biological relevance of the data. The challenge remains in our ability to interpret the meaning of the data and synthesize relevant information and knowledge so it can be effectively applied to deliver improved health outcomes and sound regulatory decisions for our communities.

In summary, regulatory bioinformatics offers exciting possibilities for developing targeted medicine and therapeutics as well as post market monitoring of the safety of medical products and food. There have been a limited number of applications of these technologies, which have demonstrated the enormous potential health benefits of the future. However, there remain challenges for regulatory bioinformatics with respect to quality, accessibility, security, transparency, accountability and integrity. The feasibility of developing a standard for both bioinformatics analysis and reporting of new data stream to assure the reliability of processes used in deriving critical regulatory information is being discussed in GSRS2015. At the time of this writing, the technical Working Group on Bioinformatics governed by GCRSR has undertaken two key bioinformatics gap analyses for regulatory uptake of genomic information including comparison of existing standards and best practice elements (e.g., fitness for purpose, traceability, auditability and documentation). This practice will further our knowledge to develop robust regulatory bioinformatics framework for enhanced application of emerging technologies in regulatory decision-making.

DISCLAIMER

The opinions expressed by the authors do not reflect the opinions or policies of their respective institutions. Any statements in this article should not be considered present or future policy of any regulatory agency.

ACKNOWLEDGEMENTS

The authors would like to recognize contributions to this topic from speakers at the Global Summit on Regulatory Science 2015 – Regulatory Bioinformatics: Russ Altman, Andrew Worth, Leming Shi, Susanna-Assunta Sansone, Dov Greenbaum, Frederick Klauschen, Rolf Apweiler, Rolf Sommer Kaas, Carl Cerniglia, Burton Blais, Atsuchi Noguchi, Scott Chandry and Mark Walderhaug. We also thank the European Food Safety Authority, which hosted this discussion.

REFERENCES

1. Collins, Francis S., Varmus, Harold, 2015. A new initiative on precision medicine. N. Engl. J. Med. 372, 793e795.
2. Elkins, C.A., Kolewicz, M.L., Jackson, S.A., Lacher, D.W., Abu-Ali, G.S., Patel, I.R., 2013. Food Addit. Contam., Part A 30, 1422e1436.
3. Goodsaid, F.M., Amur, S., Aubrecht, J., Burczynski, M.E., Catalona, J., Carl, K., Charlab, R., Cornu-Artis, C., Essioux, L., Fornace Jr., A.J., Hinman, L., Hong, H., Hunt, I., JacobsonKram, D., Jawaid, A., Li, H.H., Close, S., Laurie, D., Lesko, L., Lindpainter, K., Mayne, J., Morrow, P., Papaluca-Amati, M., Robison, T.W., Roth, J., Schuppe-Koistinen, I., Shi, L., Spleiss, O., Tong, W., Truter, S.L., Vonderscher, J., Westelinck, A., Zhang, L., Zineh, I., 2010. Voluntary exploratory data submissions to the US FDA and the EMA: experience and impact. Nat. Rev. Drug Discov. 9, 435e445.
4. Heim, D., Budczies, J., Stenzinger, A., Treue, D., Hufnagi, P., Denkert, C., Dietel, M., Klauschen, F., 2014. Cancer beyond organ and tissue specific: next generation sequencing gene mutation data reveal complex genetic imilarities across major cancers. Int. J. Cancer 135, 2362e2369.
5. Howard, P.C., et al., 2014. Global Summit on regulatory science 2014. Regul. Toxicol. Pharmacol. 70, 728e732.
6. Izurieta, Hector S., Thadani, Nicole, Shay, David K., Lu, Yun, Maurer, Aaron, Foppa, Ivo M., Franks, Riley, Pratt, Douglas, Forshee, Richard A., MaCurdy, Thomas, Worrall, Chris, Howery, Andrew, Kelman, Jeffrey, 2015.
7. Comparative effectiveness of high dose versus standards dose influenza vaccines in US residents aged 65 years and older from 2012 to 2013 using Medicare data: a retrospective cohort analysis. Lancet Infect. Dis. 15, 293e300.
8. Klauschen, Frederick, Andreeff, Michael, Keilholz, Ulrich, Dietel, Manfred, Stenzinger, Albrecht, 2014. The combinatorial complexity of cancer precision medicine. Oncoscience 1, 504e509.
9. Lambert, Dominic, Carrillo, Catherine D., Koziol, Adam G., Manninger, Paul, Blais, Burton W., 2015. GeneSipper: a rapid whole-genome approach for the identification and characterization of foodborne pathogens such as priority shiga toxigenic Escherichia coli. PLoS One.

http://dx.doi.org/10.1371/journal.- pone.0122928. Published: April 10, 2015.

10. Mellor, Glen E., Fegan, Narelle, Gobius, Kari S., Smith, Helen V., Jennison, Amy V., D'Astek, Beatriz A., Rivas, Marta, Shringi, Smitri, Baker, Katherine, N.K., Besser, Thomas, 2015. Geographically distinct Escherichia coli 0517 isolates differ by lineage, shiga toxin genotype, and total shiga toxin production. J. Clin. Microbiol. 53, 579e586.
11. Miller, M.A., et al., 2013. 2012 global Summit on regulatory science (GSRS-2012)d Modernizing toxicology. Toxicol. Sci. 131, 9e12.
12. OECD, 2014. Guidance on Grouping of Chemicals, second ed. OECD Environment, Health and Safety Publications Series on Testing and Assessment No. 194.
13. Sansone, Susanna-Assunta, Rocca-Serra, Philippe, 2012. On the evolving portfolio of community standards and data sharing policies: turning challenges into new opportunities. GigaScience 1, 10.
14. SEQC/MAQC-III Consortium, 2014. A comprehensive assessment of RNA-seq accuracy, reproducibility and information content by the sequence quality control consortium. Nat. Biotechnol. 32, 903e914.
15. Slikker Jr., W., et al., 2012. Advancing global health through regulatory science research: summary of the global Summit on regulatory science research and innovation. Regul. Toxicol. Pharmacol. 62, 471e473.
16. Sturla, Shana, J., Boobis, Alan R., FitzGerald, Rex, E., Hoeng, Julia, Kavlock, Robert, J., Schirmer, Kristin, Whelan, Maurice, Wilks, Martin F., Peitsch, Manuel C., 2014.
17. Systems toxicology: from basic research to risk assessment. Chem. Res. Toxicol. 27, 314e329.
18. Tong, Weida, Ostroff, Stephen, Blais, Burton, Silva, Prima, Dubuc, Martine, Healy, Marion, Slikker, William, 2015. Genomics in the land of regulatory science. Regul. Toxicol. Pharmacol. 72, 102e106.
19. Winship, Ingrid, 2015. Precision medicine: are we there yet? Med. J. Aust. 203,132e133. Yih, Katherine W., Lieu, Tracy A., Kulldorff, M., Martin, David, McMahillWalravem, Cheryl N., Platt, Richard, Selvam, Nandini, Selvan, Mano, Lee, Grace, Nguyen, Michael, 2014. Intussusception risk after rotavirus vaccination in U.S. infants. N. Engl. J. Med. 370, 503e512.

CHAPTER 4

RECENT AND EMERGING INNOVATIONS IN SALMONELLA DETECTION: A FOOD AND ENVIRONMENTAL PERSPECTIVE

Rebecca L. Bell[1], Karen G. Jarvis[2], Andrea R. Ottesen[1], Melinda A. McFarland[1] and Eric W. Brown[1]

[1]Center for Food Safety and Applied Nutrition, U.S. Food and Drug Administration, College Park, MD, USA.

[2]Center for Food Safety and Applied Nutrition, U.S. Food and Drug Administration, Laurel, MD, USA.

CITATION: Bell, R.L., et al., "Recent and emerging innovations in Salmonella detection: a food and environmental perspective", Microbial Biotechnology (2016), https://doi.org/10.1111/1751-7915.12359.

SUMMARY

Salmonella is a diverse genus of Gram-negative bacilli and a major foodborne pathogen responsible for more than a million illnesses annually in the United States alone. Rapid, reliable detection and identification of this pathogen in food and environmental sources is key to safeguarding the food supply. Traditional microbiological culture techniques have been the 'gold standard' for State and Federal regulators. Unfortunately, the time to result is too long to effectively monitor foodstuffs, especially those with very short shelf lives. Advances in traditional microbiology and molecular biology over the past 25 years have greatly improved the speed at which this pathogen is detected. Nonetheless, food and environmental samples possess a distinctive set of challenges for these newer, more rapid methodologies. Furthermore, more detailed identification and subtyping strategies still rely heavily on the availability of a pure isolate. However, major shifts in DNA sequencing technologies are meeting this challenge by advancing the detection, identification and subtyping of Salmonella towards a culture-independent diagnostic framework. This review will focus on current approaches and state-of-the-art next-generation advances in the detection, identification and subtyping of Salmonella from food and environmental sources.

INTRODUCTION

Members of the genus Salmonella, belonging to the family Enterobacteriaceae, are Gram-negative, non-spore forming, predominantly motile, facultative anaerobic bacilli. The genus is composed of two species, enterica and bongori, with six subspecies of S. enterica (McQuiston et al., 2008). The genus is further subdivided into serotypes based on the presence of specific surface molecules, namely O-antigen (O-Ag), present in lipopolysaccharide, and H-antigen (H-Ag), typically the major protein of the flagellar complex, flagellin. Collectively, there are over 2500 serotypes of salmonellae, all which are capable of causing disease in humans (Grimont and Weill, 2007).

The majority of salmonellae cause gastroenteritis, however, a few serotypes, such as S. Typhi, S. Paratyphi A, S. Paratyphi B and S. Paratyphi C (typhoidal salmonellae), are capable of causing enteric fever, a severe illness characterized by the onset of high fever with abdominal pain and general malaise but not typically with diarrhoea or vomiting. This is in contrast to salmonellosis, a gastroenteritis caused by non-typhoidal salmonellae, which is characterized by fever, vomiting and severe diarrhoea. In a majority of the cases, salmonellosis is self-limiting, resolving in about a week. Occasionally, however, the infection becomes systemic, a much more severe disease requiring antibiotic interventions. Yearly in the United States, it is estimated that Salmonella is responsible for over a million illnesses, 19 000 hospitalizations and almost 400 deaths (Scallan et al., 2011). Studies on the infectious dose of Salmonella in humans indicate a wide range for the number of cells required to cause disease. Clinical studies conducted using human volunteers indicate a range of 105 to 1010 cells. In contrast, enumeration of food commodities indicate much lower numbers of organisms, as low as 10 cells, were present to cause illness (Blaser and Newman, 1982). Infection typically occurs after the ingestion of contaminated food or water. It is estimated that 95% of Salmonella infections are due to the consumption of contaminated foodstuffs (Fatica and Schneider, 2011). These data suggest that salmonellae may be present at very low levels in food and still be able to cause a significant number of infections.

Until the 1990s, most illnesses were due to the consumption of animal products: poultry, poultry products, meat and dairy. Recently, an increasing number of illnesses have been associated with the consumption of raw, fresh, ready-to-eat produce, such as tomatoes, melons, sprouts, leafy greens and berries (Painter et al., 2013). In most instances, contamination levels in these commodities are quite low and sporadic as demonstrated by several prevalence studies conducted in the United States (Sapers and Doyle, 2014; Bell et al., 2015). As such, rapid, reliable and sensitive detection of Salmonella in food and environmental sources is essential to safeguard the food supply effectively and subsequently ensure public health.

There has been a thrust in the past 25 years to develop much faster methods to detect, identify and subtype Salmonella specifically in food and environmental samples. This review will focus on the current culture-dependent methods while highlighting some promising innovative culture-independent methods for the rapid, accurate detection, identification and subtyping of salmonellae in food and environmental samples.

CULTURE-DEPENDENT METHODS

Currently used Method to Detect, Identify and Subtype

Current testing of food and environmental samples for the presence of Salmonella can be divided into three stages: (i) detection of the pathogen;

(ii) identification of the isolate as Salmonella and its specific serovar designation; and (iii) subtyping of the isolate for association with any clinical cases of salmonellosis. Detection methods rely on traditional bacterial culture procedures that employ the use of serial enrichments with increasing selectivity culminating in the isolation of Salmonella on selective-differential agar plates (Fig. 1) (Andrews et al., 2011; USDA FSIS, 2014). The process takes up to 5 days to gather a presumptive positive isolate. Confirmation relies on traditional biochemical testing of sugar and nutrient utilization media, which can take days to complete. Even with newer automated technologies that permit simultaneous testing of multiple analytes at least 24 h is needed for a confirmation of Salmonella. DNA finger printing techniques, such as pulsed-field gel electrophoresis (PFGE), ribotyping and intergenic sequence (IGS) ribotyping have all been used to subtype Salmonella isolates. All these techniques are based on a similar idea of examining DNA size differences on an agarose gel. For ribotyping, genomic DNA is digested, separated on an agarose gel and then hybridized to rRNA operons to visualize the banding pattern. After comparison to a database of fingerprints species, serovar and occasionally strain identifications can be made (Bailey et al., 2002). More discriminatory power maybe available with IGS ribotyping, where the size differences found within the intergenic spacer regions between 16S and 23S rRNA regions are examined. Here, the regions are amplified by polymerase chain reaction (PCR) before gel electrophoresis is done. The banding patterns allow for differentiation between strains of Salmonella within a serovar (Brown, 2001). Neither of these techniques has been widely adopted. Federal and State agencies within the United States and many other countries around the world rely on PFGE to subtype Salmonella. For this technique, genomic DNA is digested by the restriction endonuclease XbaI. The DNA fragments are separated on an agarose gel subjected to a pulsed electric field. DNA is visualized by ethidium bromide staining and fingerprints are analysed using specific software available in BioNumberics (Applied Maths) (Ribot et al., 2006). The power of PFGE is the ability to compare the resultant fingerprint patterns to a large national database housed and maintained by the Centers for Disease Control and Prevention (CDC), aka the PulseNet reference library (Swaminathan et al., 2001). The use of PFGE and the implementation of PulseNet have greatly increased the United State's ability of track and trace back illness clusters and outbreaks. Unfortunately, PFGE still requires a pure isolate and a minimum of 3 days to complete (Ribot et al., 2006).

While this analytical schema is the 'gold standard' of regulatory agencies due to its sensitivity, a detection limit assumed to be 1 cfu per portion of food tested; and its ability to provide a pure culture of Salmonella, many pitfalls remain associated with this approach (Fig. 1). First, as mentioned, the time to result is quite long, taking at least a week to get a confirmed isolate and longer for serotyping and subtyping. For many food commodities, especially fresh produce, this time frame is far too long to effectively test food vehicles before they are consumed or to hold in warehouses while awaiting test results before they spoil. Second, at each step, the amount of media used to find Salmonella increases, resulting in numerous plates required for each sample. The process is very labour intensive and necessitates large areas of space, particularly if many samples are being tested. Finally, food samples, especially fresh produce and spices, can be notably difficult for traditional microbiological methods due to the high numbers of indigenous microbiota and the presence of antimicrobials found within the food commodity (Jameson, 1962; Arora and Kaur, 1999; Singer et al., 2009; Kim et al., 2011; Bell et al., 2012; Gorski, 2012; Pettengill et al., 2012)

Advances in Detection: PCR and Real-time PCR

The largest advance towards faster detection of Salmonella has been in the realm of molecular biology, where PCR and real-time, quantitative PCR (qPCR) are predominantly being applied as the methods of choice for the detection stage of this process. Many different protocols targeting different genes or gene regions specific to Salmonella have been published (Cohen et al., 1996; Guo et al., 2000; Malorny et al., 2003, 2004; Chen et al., 2010; Postollec et al., 2011; Gonzalez-Escalona et al., 2012; Cheng et al., 2015). Additionally, many targets have been investigated for the specific and sensitive detection of all salmonellae in food and environmental samples. By far the most popular gene target is invA, with some assays multiplexed to improve sensitivity and specificity (Malorny et al., 2003; Postollec et al., 2011; Gonzalez-Escalona et al., 2012; Cheng et al., 2015). While the obvious advantages of PCR/ qPCR assays are the rapid time to result (Table 1), and sensitivity and specificity of detection, there are many disadvantages as well. These include the need for expensive equipment and trained personnel, the use of extensive DNA clean-up chemistries before addition to the PCR/qPCR reaction, the need to culture the samples to meet the limit of detection threshold, and a lack of distinction between live and dead cells.

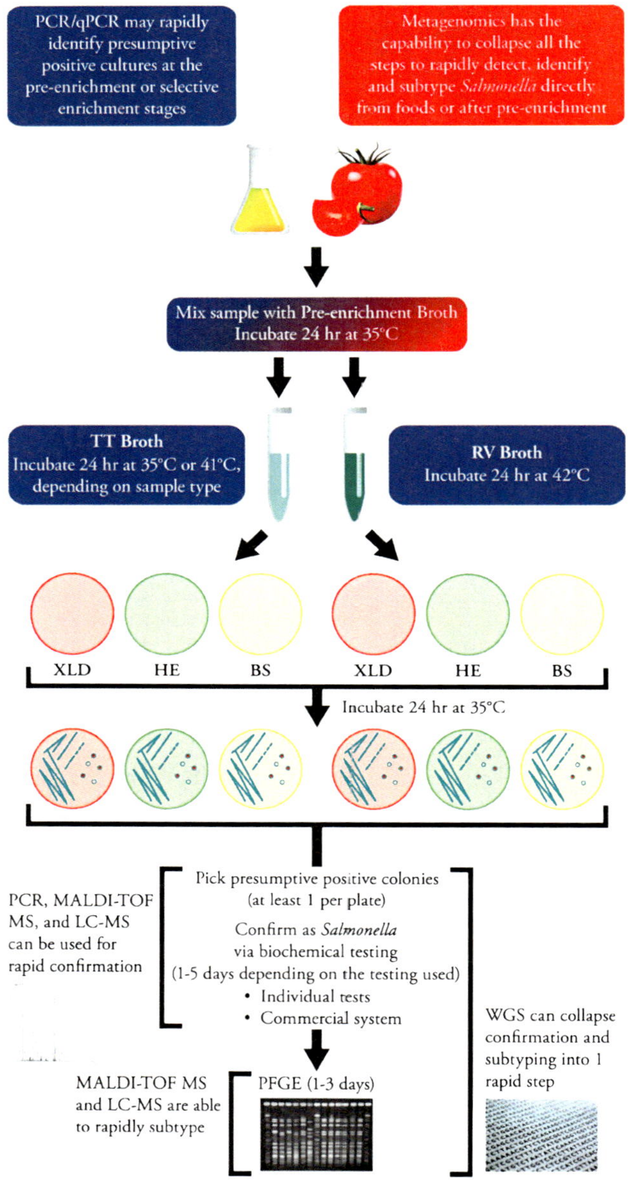

Figure 1. Overview of the U.S. Food and Drug Administration Bacterial Analytical Manual (FDA-BAM) workflow for the detection, isolation and subtyping of Salmonella (Andrews et al., 2011). Detection and isolation of Salmonella requires 5 days. Subsequent confirmation and subtyping may take up to a week

longer. Various, newer molecular methods such as PCR/qPCR, MS, WGS and metagenomics, may shorten the time to result and may be incorporated into the workflow at the indicated steps.

Food and environmental samples pose a unique set of challenges for PCR/qPCR reactions. The detection limit of qPCR is approximately 10^2 cells/reaction. Naturally contaminated food typically will not contain high enough numbers of salmonellae to reach this detection limit in order to attempt direct detection from such food items.

This is also true for certain environmental samples such as soils, sediments and waters. As such, there is a need to culture these sample types to enrich for Salmonella at levels more easily detected by PCR/qPCR. However, several difficulties may be encountered during enrichment. First, enrichment bias may occur where Salmonella cells are outcompeted by the natural microbiota found in the food/environmental sample or the salmonellae are outright inhibited by specialized metabolites, such as antibiotics, produced by these same organisms (Jameson, 1962; Singer et al., 2009; Bell et al., 2012; Gorski, 2012; Ottesen et al., 2013a; Allard et al., 2014). Subsequently, the number of Salmonella cells may never reach the required threshold for detection. As a result of this bias, the exact time necessary for the enrichment step is largely unknown with suggested ranges of 4– 24 h depending on the sample type, the starting concentration of Salmonella in the sample and the health of those salmonellae (Myint et al., 2006; Josefsen et al., 2007; Tatavarthy et al., 2009; Cheng et al., 2015). Some have even suggested the need to use a secondary, selective enrichment to enhance the chances of detecting Salmonella in challenging samples of this nature (Myint et al., 2006). Consequently, this greatly lengthens the time to result and reduces the usefulness of PCR/ qPCR for faster detection of Salmonella in contaminated food and environmental samples. Culturing samples to increase the salmonellae population also increases the background microbiota populations. Unfortunately, the sensitivity of qPCR reactions has been found to decrease in the presence of high concentration of nonspecific DNA suggesting that even higher levels of Salmonella must be present in order for detection to occur (Jyoti et al., 2011). Compounding this problem, the food or environmental sample may contain PCR/qPCR inhibitors that are not removed during the culture enrichment step. These inhibitors include high concentrations of fats and proteins in meat and dairy as well as polysaccharides and polyphenols in fruits and vegetables (Rossen et al., 1992; Wilson, 1997). Humic and fulvic acids may be present in environmental samples, especially soils and sewage, which will prevent

PCR/qPCR reactions from occurring (Wilson, 1997). Likewise, the media used for enrichment may contain inhibitors, which also will decrease the sensitivity of the PCR/qPCR reaction (Gorski and Liang, 2010). Protocol adjustments to alleviate the impact of inhibitors inherent to the sample include variation in DNA extraction and clean-up methods and the addition of facilitators to the PCR/qPCR reaction (Wilson, 1997; Ma and Michailides, 2007; Chua and Bhagwat, 2009). Moreover, internal amplification controls to identify PCR inhibition have become the standard to confirm the efficacy of the sample preparation and clean-up steps (Malorny et al., 2003). Finally, as PCR/qPCR assays are quite sensitive in the detection of DNA molecules, there is a concern over the detection of live versus dead cells because DNA may linger for prolonged periods after the death of the cell. To address this concern, the use of reverse-transcriptase-qPCR assays have been developed to detect RNA which will only be produced in living cells (Gonzalez-Escalona et al., 2009). Also the use of propidium monoazide has been proposed as a way to differentiate living from dead cells in a qPCR assay (Li and Chen, 2013).

Table 1. Comparison of traditional methods to molecular methods for Salmonella detection, identification and subtyping.

			Technical competency		
Method	Resolving power	Accuracy	Performance	Analysis	Time to result
PFGE	Sub-serotype	100%[d]	Highly trained/Must be certified	Highly trained/Must be certified	1–3 days
Traditional serology	Serotype	Approx. 80%[e]	Highly trained/Must be certified	Highly trained/Must be certified	Up to 3 days
Phage typing[a]	Sub-serotype	Approx. 80%[f]	Highly trained/Must be certified	Highly trained/Must be certified	1–2 days
PCR/qPCR	Genus to serotype[b]	Varies with protocol and matrix[g]	Moderately trained	Moderately trained	4–6 h
MALDI-TOF MS	Species	>98% at species level	Easy for clinical workflow	Easy for clinical workflow	<5 min
LC-MS	Serotype to sub-serotype level	98% at serotype level[h]	Moderately trained	Highly trained	<1 day
WGS	Strain	100%	Easy to perform	Highly trained	3–4 days[j]
Metagenomics	Genus to strain[c]	Approx. 11%[i]	Easy to perform	Highly trained	3–4 days[j]

a. Only used for S. Typhi, S. Paratyphi A, S. Typhimurium and S. Enteritidis.

b. Depends on matrix and primer sets used, some only detect genus, some will test for specific serotypes.

c. Depends on sequencing depth, analysis pipeline and available database to query sequences against.

d. If performed within PulseNet certification standards.

e. Due to rough, mucoid and non-motile strains.

f. Due to ambiguous lysis reactions.

g. Must have a minimum of 102 genomes in the reaction in order to get a positive detection, see text for discussions on other limitations.

h. Based on single lab evaluation studies.

i. Based on current pipelines and databases in naturally contaminated cilantro (Jarvis et al., 2015).

j. Depending on analysis time.

Advances in sample preparation to concentrate Salmonella

Of all the shortcomings mentioned above, the most problematic is enriching the Salmonella population to detectable levels in the limited sample volumes used in PCR/ qPCR. There has been a thrust in up-front processing steps to help selectively separate the Salmonella population from the background microorganisms. The most widely used method employs the use of anti-Salmonella antibodies bound to paramagnetic beads. This process, known as immunomagnetic separation (IMS), allows for the specific separation of Salmonella from other organisms within the food or environmental sample by first mixing the anti-Salmonella magnetic bead with a portion of the pre-enrichment culture and then separating them out with the use of a magnet (Cudjoe et al., 1994; Shaw et al., 1998). This not only decreases the number of background microorganisms but it also concentrates the Salmonella to the levels required for detection. This may also allow for shorter pre-enrichment periods, such as 6 h as opposed to the full 24 h typically used, which would allow detection to occur in 1 day, much earlier than traditional culture methods (Tatavarthy et al., 2009). Unfortunately, the sensitivity of IMS is limited by the binding specificity of the antibody to all salmonellae cells

- This is critical to prevent false-negative results as all salmonellae have the ability to cause disease in humans
- Leaving holes in this method if it is used to screen the food supply (Life Technologies 2012).

Along with antibodies, other biomolecules have been explored as possible means to selectively capture and concentrate Salmonella from cultures. Aptamers are single-stranded oligonucleotides, DNA or RNA that take on unique 3-D structures based on their primary nucleotide sequence, rendering them capable of binding to specific ligands similar to an antibody interacting with an antigen. Aptamers offer some advantages over antibodies in that they are relatively inexpensive to synthesize, and they provide more batch-to-batch consistency (Bruno, 2014). Limited studies have demonstrated their specific use in concentrating Salmonella Typhimurium from river water and faecal samples (Joshi et al., 2009; Jyoti et al., 2011; Singh et al., 2012). Bacteriophages have also been explored as a means to capture Salmonella cells (Bennett et al., 1997; Laube et al., 2014). Phages may offer some advantages over antibodies given their inherent specificity for host cells, their ease of production in bacteria versus animals or eukaryotic cell culture and their relative stability in harsh conditions such as pH and temperature extremes (Laube et al., 2014)

In a move towards culture-independent detection others have focused on methods to concentrate all cells within the sample before the pre-enrichment step. This concentration of the Salmonella population allows for direct detection from food and environmental samples. These methods focus mainly on filtering liquids, rinsates, or mechanically disintegrated (i.e. blended or stomached) samples. While this approach has been widely used to test large volumes of water, the testing of food samples was problematic due to clogging of filter membranes by large food particles (Li et al., 2013; Vibbert et al., 2015). To overcome this problem, endopeptidases have been added to the stomached food samples. These degrade the small, soluble proteins and peptides so that they are unable to clog the filter and pass through with the permeate (Vibbert et al., 2015). Additional filtration advances include the use of hollow-fibre filter membrane cartridges, which have higher surface-tovolume ratios than flat-sheet membranes, as well as cross-flow, or tangential flow, where the sample flows across the membranes to further aid in the reduction of membrane fouling (Cho et al., 2014). This method has recently been awarded the grant prize in the U.S. Food and Drug Administration's first ever Food Safety Challenge competition (http://www.foodsafetychallenge.com), signifying its potential to greatly enhance the detection of Salmonella directly from foods.

Advances in Identification and Subtyping: Mass Spectrometry

Identification of bacteria by mass spectrometry (MS) has been an active research area for decades (Anhalt and Fenselau, 1975). Most MS-based bacterial identification methods rely on the reproducible patterns generated from measurement of masses of proteins and/or lipids from intact bacterial cells or cell extracts. MS-based methods are intriguing because they provide direct detection of the presence of expressed bacterial proteins. Tens to hundreds of proteins are measured in a single experiment. The result is a reproducible mass fingerprint with features that are unique to a given genus or species. The significant advantage of MS methods over phenotypic and molecular methods for bacterial identification is that no advanced knowledge of the microorganism is necessary and there is no need for assay selection. The mass spectral fingerprint contains a snap shot of reproducible observable proteins, which makes this an inherently multiplexed screening method.

Matrix-associated laser desorption ionization–time of flight mass spectrometry (MALDI-TOF MS) is the most common technique used for bacterial analysis by MS (Claydon et al., 1996; Holland et al., 1996; Krishnamurthy and Ross, 1996). It is easy to operate, per sample cost of analysis is low, excluding the initial cost of the mass spectrometer, and analysis is as fast as 10 min from colony selection to identification (Table 1). Early work focused on proof of principle of genus and species level differentiation and method standardization (Fenselau and Demirev, 2001; Lay, 2001; Williams et al., 2003; Wunschel et al., 2005). As multiple laboratories began developing libraries of reference spectra representing species of interest, it became clear that lab-tolab reproducibility was going to require standardization of spectral libraries. In 2008, multiple commercial MALDI-TOF MS instruments were released that included reference spectra libraries (Martiny et al., 2012). Two systems received FDA approval for clinical use in 2013. The research use versions of these systems also include methods for creating libraries customized to the user laboratory. A thorough review of bacterial identification by MALDI-TOF MS for clinical use can be found in Clark et al. (2013).

The critical determinant for accurate identification is the presence of the genus or species of interest in the spectral library. A transferrable spectral library database containing sufficient microbial representation is required for the deployment of an MS-based assay to multiple laboratories. Most commercially available MALDI-TOF MS bacterial identification instruments contain databases populated with composite reference spectra from clinically relevant isolates. Consequently, MALDI-TOF MS is an effective and rapid tool for genus and some species level identifications. Unfortunately, commercially available libraries currently lack the breadth and specificity that is ultimately needed to analyse complex matrices such as food. In addition, commercial identification algorithms struggle or fail to identify mixtures of bacteria. Therefore, a single colony or pure culture is generally required. However, the speed at which even a genus level identification can be made makes MALDI-TOF MS a potentially important screening tool.

Species level identification of Salmonella by MS is relatively straight forward when analysing a pure colony. Comparatively few studies have been published on MSbased subspecies detection of Salmonella (Lynn et al., 1999; Leuschner et al., 2004; Dieckmann et al., 2008; Kuhns et al., 2012; Sparbier et al., 2012). Ribosomal proteins whose masses are sufficiently different at the genus level dominate MALDI-TOF MS spectra of bacteria. However, Salmonella subspecies and serovar-level identification continues to be a challenge by MALDI-TOF MS. MALDI-TOF MS spectra of complex mixtures suffer from poor spot to spot mass accuracy and peak reproducibility. The current method for MALDI-TOF MS spectral library creation minimizes these limitations by creating reference spectra that are composites of spectra built from different isolates and different culture conditions while also matching peaks with a wide mass accuracy tolerance. Although this serves to increase reproducibility, it is possible that spectral compositing lowers specificity by masking the ability to detect mass differences created by single amino acid differences. Some specificity can be gained back, however, by adding the appropriate reference spectra to the reference library. Nonetheless, reference libraries populated with too many serovar-specific reference spectra may challenge commercial matching algorithms because of the high homology across serovars.

Impressive work by Dieckmann and Malorny (2011) has shown that detection of serovar-specific combinations of proteins is possible by MALDI-TOF MS. This was accomplished by extending the upper mass range of detection from 20 000 Da up to 40 000 Da. The authors were able to identify several non-ribosomal serovar-specific combinations of proteins as markers, but no single marker was serovar-specific. Combinations of marker masses will be required to achieve such specificity. As an additional caveat, a MALDI-TOF MS method capable of extending above 20 000 Da is not currently supported by commercial vendors making it difficult to reproduce in other laboratories. As an aside, the authors also identified a large number of the masses that were detected. Such knowledge of the identity of marker proteins being used in reference libraries would go a long way in standardizing across laboratories and instruments.

An alternative to MALDI-TOF MS is separation and detection of bacterial proteins by high-performance liquid chromatography mass spectrometry (LC-MS). LC-MS of intact bacterial lysates has shown great promise for differentiating closely related species within the Enterobacteriaceae family, thermophilic versus non-thermophilic groups of Enterobacter sakazaki and identifying bacterial biomarker proteins for the development of PCR probes (Krishnamurthy et al., 1999; Ho and Hsu, 2002; Williams et al., 2004, 2005; Mott et al., 2010). More recently, LCMS of intact lysates has been used for Salmonella serovar-level identification (Callahan et al., 2009). This method chromatographically separates the intact bacterial proteins prior to detection by MS. Consequently, many more proteins and, by extension, more serovarspecific marker proteins are detected. Recent work by McFarland et al. (2014) shows that a powerful advantage of this method is that it can be performed on mass spectrometers with improved mass accuracy that are capable of dissociating the proteins so they can be identified. Once the detected proteins are identified they can be used for further analysis such as in silico validation of the markers, extension to unknown isolates and crossreferencing to genomic-based assays. Because the proteins are measured intact, this method can recognize changes to detected proteins even if the specific substitution is unknown. This work demonstrates that serovarlevel identification of Salmonella is possible by LC-MS. One disadvantage is that it is slower than MALDI-TOF MS analysis (Table 1). Nonetheless its robustness at serovar identification and its ability to identify pathogens without an extensive reference library makes this a promising method for further analysis of isolates that are positive for Salmonella by MALDI-TOF MS screening.

The role of Next-generation Sequencing

Next-generation sequencing (NGS) or whole genome sequencing (WGS) refers to highly automated and parallelized genome sequencers used to sequence the entire genome(s) of bacterial pathogens in a matter of hours. When coupled with analytical bioinformatic pipelines such as the one established at CFSAN-FDA (https:// github.com/FDA/open. fda.gov), accurate and stable genetic changes can be identified that can distinguish foodborne outbreak strains of Salmonella down to the source level including specific farms, food types and geographic regions. Also, the rapid increase in Salmonella genome sequences in the NCBI database and other public domains, such as GenomTrakr (http:// www.fda.gov/Food/ FoodScienceResearch/WholeGenom eSequencingProgramWGS/default. htm), are changing the way that public health laboratories establish linkages between Salmonella, the environment and illness. Moreover, WGS has begun to streamline laboratory testing of salmonellae into a single microbiological workflow, supplanting phenotypic, serologic and other less robust genotypic typing schemes. For example, WGS outputs can now be readily shunted into bioinformatic pipelines that accurately predict antimicrobial susceptibility patterns, determine serotype and provide multiple virulence profiles for a single Salmonella strain (Zankari et al., 2013; Tyson et al., 2015; Zhang et al., 2015). In the United States, the National Antimicrobial Resistance Monitoring System has evaluated the use of WGS data to determine the presence of antimicrobial resistance genes. In an initial study, focused on Escherichia coli, comparison of WGS data to a database containing over 2500 resistant genes and gene variants was able to predict antimicrobial resistant profiles that highly correlated with phenotypic profiles (Tyson et al., 2015). In work conducted in Denmark, WGS data funnelled into the ResFinder web server (www.genomicepidemiology.org) showed 100% concordance in predicting S. Typhimurium resistance profiles when compared with phenotypic data (Zankari et al., 2013). The use of WGS data in this way should greatly enhance antimicrobial monitoring systems allowing for better tracking of antimicrobial resistant isolates within the food supply and augmenting epidemiological tracebacks during outbreak situations. Additionally, Salmonella serotype may also be predicted from WGS data with the use of the SEQSERO program, recently developed by Deng and colleagues. This simple computer dashboard program quickly extracts both O- and H-antigen types from genomic data quickly predicting serotype with high accuracy when compared with traditional serotyping data (Zhang et al., 2015).

Whole genome sequencing is changing the laboratory landscape for foodborne outbreak investigations of Salmonella contamination as well. WGS of Salmonella strains is supplanting conventional approaches, such as PFGE, which often fall short in ample characterization of a Salmonella strain as well as delimiting the scope or tracking the source of contamination potentially associated with such a strain (Allard et al., 2012). In certain cases of foodborne contamination, highly clonal strains of Salmonella have been known to confound outbreak investigations because conventional subtyping approaches often lack the resolution for differentiating certain closely related isolates (Table 1). As an example, investigations of S. Enteritidis outbreaks are taxing because nearly 40% of all egg isolates have the same PFGE pattern, making epidemiological investigations difficult due to a common fingerprint pattern found in a widely distributed food (Allard et al., 2013). In response to such challenges, Federal public health and food safety laboratories are exploiting NGS to define complex outbreak scenarios involving Salmonella. For example, WGS analysis of S. Bareilly isolates responsible for the 2012 tuna scrape outbreak established a clear link between outbreak isolates and a specific manufacturing facility in India (Hoffmann et al., 2015). WGS analysis of S. Enteritidis isolates from an egg outbreak in the United Kingdom revealed a clear linkage between human, egg and environmental S. Enteritidis isolates specific to the outbreak (Inns et al., 2015). Additionally, WGS has identified outbreak clusters of Salmonella in outbreaks associated with black pepper (S. Montivideo), tomato (S. Newport), cucumber (S. Newport), watermelon (S. Newport) and peanut butter (S. Tennessee) (Bakker et al., 2011; Cao et al., 2013; Byrne et al., 2014; Angelo et al., 2015; Bell et al., 2015, M. R. Wilson, unpublished). These genomic studies are providing vital information, retrospectively and in real-time, for foodborne outbreak investigations. Deployments of WGS have been undertaken using the technology at the U.S. FDA, the CDC, Public Health Canada, Harvard and Cornell Universities, The Sanger-Wellcome Trust, The Danish Technical University and Danish Food Institutes, numerous public health institutes in Germany, and various industry colleagues engaged in this technology development.

Culture-independent Approache

While highly effective for providing important and unique identifiers among individual salmonellae, it is notable that WGS technology, in its most effective form, is requisite on a pure culture for the generation of a complete genome sequence. This caveat makes any current applications for pathogen surveillance or detection directly from complex food matrices or pre-enriched foods a challenging task. Metagenomic approaches, however, are now beginning to provide a path forward in the use of WGS technology for detection of Salmonella in situ in food and environmental backgrounds (Ottesen et al., 2013a). Multiple proof-of-concept studies for the use of metagenomic methods for pathogen detection have been undertaken in a wide range of fields including the detection of causal agents in human matrices such as plasma, sputum, saliva and faeces (Finkbeiner et al., 2008; Rogers et al., 2009; Cho and Blaser, 2012; Loman et al., 2013; Doughty et al., 2014); identifying the causal agents of crop diseases, bee colony collapse; investigation of fermentation processes; and spoilage (Jung et al., 2011;

Nieminen et al., 2012; Delhalle et al., 2013; Ma et al., 2014; Wanglin and Longhao, 2015). Also the ecologies of poultry, meats, fresh produce and a variety of processed food have all been interrogated using targetbased and/or shotgun metagenomic approaches (Lusk et al., 2012; Ottesen et al., 2013b; Jackson et al., 2015; Jarvis et al., 2015).

Metagenomics is defined as the analysis of genomic DNA from a whole community; this distinguishes it from genomics, which is the analysis of genomic DNA from an individual organism or cell (Chen and Pachter, 2005; Gilbert and Dupont, 2011). Definitions vary to include any study that analyses a whole community, for example, from directed studies of 16S rRNA gene diversity within an environment, to isolation and analysis of total DNA from environmental samples without prior cultivation (Chen and Pachter, 2005; Gilbert and Dupont, 2011). A recent publication by Marchesi and Ravel (2015) defines a metagenome as the collection of genomes and genes from the members of a microbiota obtained through shotgun sequencing of DNA extracted from a sample. The microbiota consists of all the microorganisms present in a defined environment as described by amplifying and sequencing specific marker genes, such as 16S rRNA and 18S rRNA (Marchesi and Ravel, 2015). The term metataxomics is proposed for the high-throughput process used to characterize the entire microbiota by sequencing specific marker genes that can be used to infer taxonomies (Marchesi and Ravel, 2015).

Target-based bacterial metagenomic studies have focused on the use of 16S rRNA gene sequences. 16S rRNA gene sequencing takes advantage of the ability to use the conserved regions of the 16S rRNA gene to identify bacterial taxa. PCR primers specific to variable regions of the 16S rRNA gene are capable of amplifying all of the 16S rRNA genes present in a sample, followed by high-throughput sequencing. Samples can be barcoded and multiplexed at various levels depending on the depth of sequencing desired. The 16S rRNA gene sequencing data reveal the bacterial taxonomy of a sample as operational taxonomic units, which can be clustered into similar sequences using software programs such as QIIME, MOTHUR and Resphera Insights (http://www.respherabio.com/Resphera_Insight_v2.2.pdf) (Schloss et al., 2009; Caporaso et al., 2010). It is now notable the number of known 16S rRNA gene sequences has surpassed the number of cultured organisms and therefore this analysis reveals the total microbial population in the sample. Further analysis of the taxonomic breakdown of a sample can reveal the proportional abundances of each taxon. For example, taxonomic composition can also be classified according to alpha (within sample) and beta (between sample) diversity algorithms to establish similarities and differences between sample populations (Lozupone and Knight, 2005; Caporaso et al., 2010). In contrast to 16S rRNA studies, shotgun metagenomic studies are valuable for looking deeper into a sample and may allow for genome assembly depending on the abundance of the organism(s) of interest. Additionally, because shotgun metagenomic studies are not target specific, the data reveal information about all domains present in a sample. Uniformly sheared genomic DNA is barcoded and sequenced. Multiplexing considerations are more critical for shotgun metagenomic sequencing as no amplification step is necessary. Subsequently, sequencing depth becomes more important, especially with high diversity samples.

This leads to a discussion of the many details that must be considered before metagenomic approaches become reliable applications for pathogen, specifically Salmonella, detection. First, the question of how much sequence data are needed to reliably detect a pathogen, such as Salmonella, should be addressed. This is contingent upon species richness and abundance in the food or environmental matrix, the size of all genomes present, the number of salmonellae present in the sample, and the number and length of all the sequences that have been generated. This also depends on the selection of gene targets when using targeted metagenomic approaches. For example, fragments of 16S rRNA genes, such as the popular V4-V6 variable region and nucleotides 533–1492 at the 30 end of the molecule, have shown to be useful for inferring taxonomy of bacterial genera and sometimes species; however, there are numerous situations where the taxonomic resolution achievable with these regions is not specific enough to identify Salmonella from other genera within Enterobacteriaceae (Lane et al., 1985; Weisburg et al., 1991; Fox et al., 1992; DeSantis et al., 2006; Caporaso et al., 2012). Conversely, recent work has demonstrated that the V1-V3 region, located at the beginning of the 16S rRNA gene, was able to assign a call of Salmonella to the genus level with 99% confidence (Jarvis et al., 2015). Next, the question of database quality needs to be examined. Genomic and metagenomic data sets are deposited in national and international repositories. The utility of these data sets is dependent on the quality of the sequence data and whether or not detailed and accurate metadata has been submitted. To identify a sequence as Salmonella with high confidence, highly curated genetic signatures and genomes of the target pathogen and other coexisting organisms must be available. Studies that demonstrated the functionality of metagenomic methods to recover pathogens of interest knew the target ahead of time and had the full genome as a reference and/or other diagnostic genetic signatures available (Loman et al., 2013; Leonard et al., 2015). Finally, the choice of bioinformatic pipelines, which are used to annotate metagenomic data sets, needs to be considered. These pipelines tend to be computationally intense and ultimately rely on well-curated reference databases and pathogen genomes in order to provide reliable taxonomic descriptions and detections. Each pipeline and reference database must be tailored for each target pathogen and the respective environment from which it is being detected.

Culture-independent analyses for public health investigations and surveillance are transforming the way samples are screened for genetic content. The ability to sequence all of the genomic material in a sample is changing the way microbial ecologists, clinicians and food scientists analyse and identify organisms of interest.

Recently published studies that used 16S rRNA gene sequencing to detect Salmonella in tomato and cilantro, also investigated microbiota shifts during the culturebased FDA-BAM method for the detection of Salmonella in produce. These studies concluded that the bacterial phyllosphere changed from a predominance of Proteobacteria, which includes Salmonella, to a predominance of Firmicutes after a 24-h non-selective preenrichment (Pettengill et al., 2012; Ottesen et al., 2013a; Jarvis et al., 2015). The diversity of the uncultured samples was higher than the cultured samples, and there were inherent beta-diversity differences between replicate samples for each commodity tested. An interesting finding in the tomato study was that even though RTPCR and FDA-BAM culture methods were negative for Salmonella, the bioinformatic analysis of 16S rRNA gene sequences showed putative hits to Salmonella (Ottesen et al., 2013a). Conversely, in the cilantro study, only one of six Salmonella culture-positive samples contained Salmonella-specific 16S rRNA gene sequences (Jarvis et al., 2015). Further characterization of the culture-positive cilantro metagenomes showed a clear difference in levels of Salmonella genomic DNA, suggesting that contamination levels varied among the six culture-positive cilantro samples tested (Jarvis et al., 2015). Both the tomato and the cilantro culture-independent studies demonstrate the importance of considering the impact of competing microorganisms on our ability to detect and identify pathogens such as Salmonella in the complex environments such as food, despite the specific genetic or genomic method being employed.

Metagenomic studies also help to establish microbial baselines for foodstuffs, where a 'healthy' microbiome for fresh and processed foods may be defined. These culture-independent data sets provide valuable starting points to expand current understandings of food ecologies and how they respond to intentional and unintentional perturbations, such as agricultural management practices and contamination events. For example, Ottesen et al. (2013b) performed a baseline survey of the tomato microbiome to characterize the tomato phyllosphere. In this study, metataxonomic (16S and 18S rRNA gene sequencing) and metagenomic sequencing were performed on tomato plant parts including roots, bottom leaves, stems, tomatoes, flowers and top leaves. The authors observed a gradient of diversity from low to high going from the top to the bottom of the plant (Ottesen et al., 2013b). The tomato and cilantro studies revealed the core microbiomes of these food plants and lay the groundwork for understanding how pathogens such as Salmonella survive and thrive in the environment and in foods (Ottesen et al., 2013b; Jarvis et al., 2015).

Along with pathogen detection, metagenomic studies may be useful to identify co-culturing microbes. These include native communities that may one day be used as indicator organisms or signature consortia that correlate with risk of pathogen presence, persistence, or virulence. The principle that isolation of a microbe must involve separating it from its surrounding community has formed the basis of culture-based microbiology. However, a paradigm shift is underway where the recognition that no organism is an island and pathogens exits within complex microbial communities and may rely on population density or other external conditions to express certain virulence traits (Rogers et al., 2009).

CONCLUSION

In summary, the various methods reviewed here each have utility for Salmonella testing in the food safety sector. It is important to recall, however, that not every method will be recommended or even suited for every situation in testing all food varieties for this pathogen. For instance, application to specific food samples will be dictated by method performance. As noted previously, method performance is dependent on several things including matrix driven effects, general sensitivity and specificity of the method, and its technical complexity. Other extrinsic factors include user skill set and technical prowess, cost of the equipment and cost per sample. Ultimately, systematic validation of a method will ascertain its specific utility and application across the food supply.

In considering the current spectrum of rapid molecular methods for detecting Salmonella, it is clear that several approaches have emerged including PCR-based, mass spectrometric and others encompassing those stemming from the current genomics era (see Table 1 for comparison to traditional methods). All of these methods are moving towards greater automation, network integration, with the ultimate goal of a fully automated laboratory test workflow. It is important to recall that outputs from one approach will serve to strengthen other approaches with many of these methods being either directly or tangentially related. For example, the explosion of WGS data is contributing not only to better characterizations of salmonellae from foods but also contributes directly to enhanced PCR-based screening tools for this diverse genus of foodborne pathogen. Ultimately, it seems that a suite of tools is emerging for the food safety microbiologist, each with its specific strengths and weaknesses but all with the ability to rapidly and accurately detect Salmonella with greater specificity and early in its contamination of the human and veterinary food supply

REFERENCES

1. Allard, M.W., Luo, Y., Strain, E., Li, C., Keys, C.E., Son, I., et al. (2012) High resolution clustering of Salmonella
2. enterica serovar Montevideo strains using a next-generation sequencing approach. BMC Genom 13: 32.
3. Allard, M.W., Luo, Y., Strain, E., Pettengill, J., Timme, R., Wang, C., et al. (2013) On the evolutionary history, population genetics and diversity among isolates of Salmonella Snteritidis PFGE pattern JEGX01.0004. PLoS ONE 8: e55254.
4. Allard, S., Enurah, A., Strain, E., Millner, P., Rideout, S.L., Brown, E.W., and Zheng, J. (2014) In situ evaluation of Paenibacillus alvei in reducing carriage of Salmonella enterica serovar Newport on whole tomato plants. Appl Environ Microbiol 80: 3842–3849.
5. Andrews, W.H., Jacobson, A., and Hammack, T.S. (2011) Salmonella. In Bacteriological Analytical Manual. U.S. Food and Drug Administration. http://www.fda.gov/Food/Food ScienceResearch/LaboratoryMethods/ucm070149.htm
6. Angelo, K.M., Chu, A., Anand, M., Nguyen, T.A., Bottichio, L., Wise, M., et al. (2015) Outbreak of Salmonella Newport infections linked to cucumbers - United States, 2014.
7. MMWR Morb Mortal Wkly Rep 64: 144–147. Anhalt, J.P., and Fenselau, C. (1975) Identification of bacteria using mass-spectrometry. Anal Chem 47: 219–225.
8. Arora, D.S., and Kaur, J. (1999) Antimicrobial activity of spices. Int J Antimicrob Agents 12: 257–262.
9. Bailey, J.S., Fedorka-Cray, P.J., Stern, N.J., Craven, S.E., Cox, N.A., and Cosby, D.E. (2002) Serotyping and ribotyping of Salmonella using restriction enzyme PvuII. J Food Prot 65: 1005–1007.
10. Bakker, H.C., Switt, A.I., Cummings, C.A., Hoelzer, K., Degoricija, L., Rodriguez-Rivera, L.D., et al. (2011) A whole-genome single nucleotide polymorphism-based approach to trace and identify outbreaks linked to a common Salmonella enterica subsp. enterica serovar Montevideo pulsed-field gel electrophoresis type. Appl Environ Microbiol 77: 8648–8655.
11. Bell, R.L., Cao, G., Allard, M.W., Keys, C., Hill, T., Ottesen, A., and Brown, E.W. (2012) Salmonella Newport contamination of produce:

ecological, genetic, and epidemiological aspects. In Salmonella: Classification, Genetics and Disease. Monte, A.S., and De Santos, P.E. (eds). Hauppauge, NY: Nova Publishers, pp. 155–173.

12. Bell, R.L., Zheng, J., Burrows, E., Allard, S., Wang, C.Y., Keys, C.E., et al. (2015) Ecological prevalence, genetic diversity, and epidemiological aspects of Salmonella isolated from tomato agricultural regions of the Virginia eastern shore. Front Microbiol 6: 415.
13. Bennett, A.R., Davids, F.G., Vlahodimou, S., Banks, J.G., and Betts, R.P. (1997) The use of bacteriophage-based systems for the separation and concentration of Salmonella. J Appl Microbiol 83: 259–265.
14. Blaser, M.J., and Newman, L.S. (1982) A review of human salmonellosis: I. Infective dose. Rev Infect Dis 4: 1096– 1106.
15. Brown, E.W. (2001) Molecular differentiation of bacterial strains. In Molecular Epidemiology. A Practical Approach. Carrington, M. and Hoelzel, A.R. (eds). London: Oxford University Press, pp. 11–41.
16. Bruno, J.G. (2014) Application of DNA aptamers and quantum dots to lateral flow test strips for detection of foodborne pathogens with improved sensitivity versus colloidal gold. Pathogens 3: 341–355.
17. Byrne, L., Fisher, I., Peters, T., Mather, A., Thomson, N., Rosner, B., et al. (2014) A multi-country outbreak of
18. Salmonella Newport gastroenteritis in Europe associated with watermelon from Brazil, confirmed by whole genome sequencing: October 2011 to January 2012. Euro Surveill 19: 6–13.
19. Callahan, J.H., McFarland, M.A., Musser, S.M., Bell, R., and Andrzejewski, D. (2009) Anyl 282-intact protein liquid chromatography mass spectrometry for bacterial identification and differentiation. Abstracts of Papers of the American Chemical Society. http://oasys2.confex.com/ acs/238nm/techprogram. Cao, G., Meng, J., Strain, E., Stones, R., Pettengill, J., Zhao, S., et al. (2013) Phylogenetics and differentiation of Salmonella Newport lineages by whole genome sequencing. PLoS ONE 8: e55687.
20. Caporaso, J.G., Kuczynski, J., Stombaugh, J., Bittinger, K., Bushman, F.D., Costello, E.K., et al. (2010) Qiime allows analysis of high-throughput community sequencing data. Nat Methods 7: 335–336.
21. Caporaso, J.G., Lauber, C.L., Walters, W.A., Berg-Lyons, D., Huntley, J., Fierer, N., et al. (2012) Ultra-high-throughput microbial community analysis on the Illumina Hiseq and Miseq platforms. ISME J 6: 1621–

1624.
22. Chen, K., and Pachter, L. (2005) Bioinformatics for wholegenome shotgun sequencing of microbial communities.
23. PLoS Comput Biol 1: 106–112.
24. Chen, J., Zhang, L., Paoli, G.C., Shi, C., Tu, S.I., and Shi, X. (2010) A real-time PCR method for the detection of
25. Salmonella enterica from food using a target sequence identified by comparative genomic analysis. Int J Food Microbiol 137: 168–174.
26. Cheng, C.M., Doran, T., Lin, W., Chen, K.S., Williams-Hill, D., and Pamboukian, R. (2015) Interlaboratory validation for a real-time PCR Salmonella detection method using the ABI 7500 fast real-time PCR system. J Food Prot 78: 1119–1124.
27. Cho, I., and Blaser, M.J. (2012) The human microbiome: at the interface of health and disease. Nat Rev Genet 13: 260–270.
28. Cho, I.H., Radadia, A.D., Farrokhzad, K., Ximenes, E., Bae, E., Singh, A.K., et al. (2014) Nano/micro and spectroscopic approaches to food pathogen detection. Annu Rev Anal Chem (Palo Alto Calif) 7: 65–88.
29. Chua, T., and Bhagwat, A.A. (2009) A rapid and simple DNA extraction procedure to detect Salmonella spp. and
30. Listeria monocytogenes from fresh produce using realtime PCR. Food Anal Methods 2: 96–101.
31. Clark, A.E., Kaleta, E.J., Arora, A., and Wolk, D.M. (2013) Matrix-assisted laser desorption ionization-time of flight mass spectrometry: a fundamental shift in the routine practice of clinical microbiology. Clin Microbiol Rev 26: 547–603.
32. Claydon, M.A., Davey, S.N., Edwards-Jones, V., and Gordon, D.B. (1996) The rapid identification of intact microorganisms using mass spectrometry. Nat Biotechnol 14: 1584–1586.
33. Cohen, H.J., Mechanda, S.M., and Lin, W. (1996) PCR amplification of the fimA gene sequence of Salmonella
34. typhimurium, a specific method for detection of Salmonella spp. Appl Environ Microbiol 62: 4303–4308.
35. Cudjoe, K.S., Krona, R., and Olsen, E. (1994) IMS: a new selective enrichment technique for detection of Salmonella in foods. Int J Food Microbiol 23: 159–165.
36. Delhalle, L., Taminiau, B., Nezer, C. and Daube, G. (2013) Metagenomic

analysis targeted on the 16S ribosomal

37. DNA to study the quality of meat: a example with raw minced beef meat. In Microbial Spoilers in Food. Quimper,
38. France. DeSantis, T.Z., Hugenholtz, P., Larsen, N., Rojas, M., Brodie, E.L., Keller, K., et al. (2006) Greengenes, a chimerachecked 16S rRNA gene database and workbench compatible with arb. Appl Environ Microbiol 72: 5069–5072.
39. Dieckmann, R., and Malorny, B. (2011) Rapid screening of epidemiologically important Salmonella enterica subsp. enterica serovars by whole-cell matrix-assisted laser desorption ionization-time of flight mass spectrometry. Appl Environ Microbiol 77: 4136–4146.
40. Dieckmann, R., Helmuth, R., Erhard, M., and Malorny, B. (2008) Rapid classification and identification of salmonellae at the species and subspecies levels by whole-cell matrixassisted laser desorption ionization-time of flight mass spectrometry. Appl Environ Microbiol 74: 7767–7778.
41. Doughty, E.L., Sergeant, M.J., Adetifa, I., Antonio, M. and Pallen, M.J. (2014) Culture-independent detection and
42. characterisation of Mycobacterium tuberculosis and M. africanum in sputum samples using shotgun metagenomics on a benchtop sequencer. PeerJ 2, e585.
43. Fatica, M.K., and Schneider, K.R. (2011) Salmonella and produce: survival in the plant environment and implications in food safety. Virulence 2: 573–579.
44. Fenselau, C., and Demirev, P.A. (2001) Characterization of intact microorganisms by MALDI mass spectrometry.
45. Mass Spectrom Rev 20: 157–171.
46. Finkbeiner, S.R., Allred, A.F., Tarr, P.I., Klein, E.J., Kirkwood, C.D., and Wang, D. (2008) Metagenomic analysis of human diarrhea: viral detection and discovery. PLoS Pathog 4: e1000011.
47. Fox, G.E., Wisotzkey, J.D., and Jurtshuk, P. Jr (1992) How close is close: 16S rRNA sequence identity may not be sufficient to guarantee species identity. Int J Syst Bacteriol 42: 166–170.
48. Gilbert, J.A., and Dupont, C.L. (2011) Microbial metagenomics: beyond the genome. Ann Rev Mar Sci 3: 347–371.
49. Gonzalez-Escalona, N., Hammack, T.S., Russell, M., Jacobson, A.P., De Jesus, A.J., Brown, E.W., and Lampel, K.A. (2009) Detection of

live Salmonella sp. cells in produce by a taqman-based quantitative reverse transcriptase real-time PCR targeting invA mRNA. Appl Environ Microbiol 75: 3714–3720.

50. Gonzalez-Escalona, N., Brown, E.W., and Zhang, G. (2012) Development and evaluation of a multiplex real-time PCR (qPCR) assay targeting ttrRSBCA locus and invA gene for accurate detection of Salmonella spp. in fresh produce and eggs. Food Res Int 48: 202–208.
51. Gorski, L. (2012) Selective enrichment media bias the types of Salmonella enterica strains isolated from mixed strain cultures and complex enrichment broths. PLoS ONE 7: e34722.
52. Gorski, L., and Liang, A.S. (2010) Effect of enrichment medium on real-time detection of Salmonella enterica from lettuce and tomato enrichment cultures. J Food Prot 73: 1047–1056.
53. Grimont, P.A.D., and Weill, F.X. (2007) Antigenic Formulae of the Salmonella Serovars. Paris, France: WHO Collaborating Centre for Reference and Research on Salmonella. Guo, X., Chen, J., Beuchat, L.R., and Brackett, R.E. (2000) PCR detection of Salmonella enterica serotype Montevideo in and on raw tomatoes using primers derived from hilA. Appl Environ Microbiol 66: 5248–5252.
54. Ho, Y.P., and Hsu, P.H. (2002) Investigating the effects of protein patterns on microorganism identification by highperformance liquid chromatography-mass spectrometry and protein database searches. J Chromatogr A 976: 103–111.
55. Hoffmann, M., Luo, Y., Monday, S.R., Gonzalez-Escalona, N., Ottesen, A.R., Muruvanda, T., et al. (2015) Tracing origins of the Salmonella Bareilly strain causing a food-borne outbreak in the United States. J Infect Dis 213: 502–508.
56. Holland, R.D., Wilkes, J.G., Rafii, F., Sutherland, J.B., Persons, C.C., Voorhees, K.J., and Lay, J.O. Jr (1996) Rapid identification of intact whole bacteria based on spectral patterns using matrix-assisted laser desorption/ionization with time-of-flight mass spectrometry. Rapid Commun Mass Spectrom 10: 1227–1232.
57. Inns, T., Lane, C., Peters, T., Dallman, T., Chatt, C., McFarland, N., et al. (2015) A multi-country Salmonella Enteritidis phage type 14b outbreak associated with eggs from a German producer: 'Near real-time' application of whole genome sequencing and food chain investigations,

United Kingdom, May to September 2014. Eurosurveillance 20: 15–22.

58. Jackson, C.R., Stone, B.W., and Tyler, H.L. (2015) Emerging perspectives on the natural microbiome of fresh produce vegetables. Agriculture 5: 170–187.
59. Jameson, J.E. (1962) A discussion of the dynamics of Salmonella enrichment. J Hyg (Lond) 60: 193–207.
60. Jarvis, K.G., White, J.R., Grim, C.J., Ewing, L., Ottesen, A.R., Beaubrun, J.J., et al. (2015) Cilantro microbiome before and after nonselective pre-enrichment for Salmonella using 16S rRNA and metagenomic sequencing. BMC Microbiol 15: 160.
61. Josefsen, M.H., Krause, M., Hansen, F., and Hoorfar, J. (2007) Optimization of a 12-hour taqman PCR-based method for detection of Salmonella bacteria in meat. Appl Environ Microbiol 73: 3040–3048.
62. Joshi, R., Janagama, H., Dwivedi, H.P., Senthil Kumar, T.M., Jaykus, L.A., Schefers, J., and Sreevatsan, S. (2009) Selection, characterization, and application of DNA aptamers for the capture and detection of Salmonella enterica serovars. Mol Cell Probes 23: 20–28.
63. Jung, J.Y., Lee, S.H., Kim, J.M., Park, M.S., Bae, J.W., Hahn, Y., et al. (2011) Metagenomic analysis of kimchi, a traditional korean fermented food. Appl Environ Microbiol 77: 2264–2274.
64. Jyoti, A., Vajpayee, P., Singh, G., Patel, C.B., Gupta, K.C., and Shanker, R. (2011) Identification of environmental reservoirs of nontyphoidal salmonellosis: aptamer-assisted bioconcentration and subsequent detection of Salmonella Typhimurium by quantitative polymerase chain reaction. Environ Sci Technol 45: 8996–9002.
65. Kim, S.Y., Kang, D.H., Kim, J.K., Ha, Y.G., Hwang, J.Y., Kim, T., and Lee, S.H. (2011) Antimicrobial activity of plant extracts against Salmonella Typhimurium, Escherichia coli O157:H7, and Listeria monocytogenes on fresh lettuce. J Food Sci 76: M41–M46.
66. Krishnamurthy, T., and Ross, P.L. (1996) Rapid identification of bacteria by direct matrix-assisted laser desorption/ ionization mass spectrometric analysis of whole cells. Rapid Commun Mass Spectrom 10: 1992–1996.
67. Krishnamurthy, T., Davis, M.T., Stahl, D.C., and Lee, T.D. (1999) Liquid chromatography/microspray mass spectrometry for bacterial investigations. Rapid Commun Mass Spectrom 13: 39–49.

68. Kuhns, M., Zautner, A.E., Rabsch, W., Zimmermann, O., Weig, M., Bader, O., and Gross, U. (2012) Rapid discrimination of Salmonella enterica serovar Typhi from other serovars by MALDI-TOF mass spectrometry. PLoS ONE 7: e40004.
69. Lane, D.J., Pace, B., Olsen, G.J., Stahl, D.A., Sogin, M.L., and Pace, N.R. (1985) Rapid determination of 16S ribosomal RNA sequences for phylogenetic analyses. Proc Natl Acad Sci U S A 82: 6955–6959.
70. Laube, T., Cortes, P., Llagostera, M., Alegret, S., and Pividori, M.I. (2014) Phagomagnetic immunoassay for the rapid detection of Salmonella. Appl Microbiol Biotechnol 98: 1795–1805.
71. Lay, J.O. Jr (2001) MALDI-TOF mass spectrometry of bacteria. Mass Spectrom Rev 20: 172–194.
72. Leonard, S.R., Mammel, M.K., Lacher, D.W., and Elkins, C.A. (2015) Application of metagenomic sequencing to food safety: detection of shiga toxin-producing Escherichia coli on fresh bagged spinach. Appl Environ Microbiol 81: 8183–8191.
73. Leuschner, R.G., Beresford-Jones, N., and Robinson, C. (2004) Difference and consensus of whole cell Salmonella enterica subsp. enterica serovars matrix-assisted laser desorption/ionization time-of-flight mass spectrometry spectra. Lett Appl Microbiol 38: 24–31.
74. Li, B., and Chen, J.Q. (2013) Development of a sensitive and specific qPCR assay in conjunction with propidium
75. monoazide for enhanced detection of live Salmonella spp. in food. BMC Microbiol 13: 273.
76. Li, X., Ximenes, E., Amalaradjou, M.A., Vibbert, H.B., Foster, K., Jones, J., et al. (2013) Rapid sample processing for detection of food-borne pathogens via cross-flow microfiltration. Appl Environ Microbiol 79: 7048–7054.
77. Life Technologies (2012) Dynabeads anti-Salmonella. In Product Manual. Life Technologies: CA, USA.
78. Loman, N.J., Constantinidou, C., Christner, M., Rohde, H., Chan, J.Z., Quick, J., et al. (2013) A culture-independent sequence-based metagenomics approach to the investigation of an outbreak of shiga-toxigenic Escherichia coli O104:H4. JAMA 309: 1502–1510.
79. Lozupone, C., and Knight, R. (2005) Unifrac: a new phylogenetic method for comparing microbial communities. Appl Environ Microbiol 71: 8228–8235.

80. Lusk, T.S., Ottesen, A.R., White, J.R., Allard, M.W., Brown, E.W., and Kase, J.A. (2012) Characterization of microflora in latin-style cheeses by next-generation sequencing technology. BMC Microbiol 12: 254.

81. Lynn, E.C., Chung, M.C., Tsai, W.C., and Han, C.C. (1999) Identification of Enterobacteriaceae bacteria by direct matrix-assisted laser desorptiom/ionization mass spectrometric analysis of whole cells. Rapid Commun Mass Spectrom 13: 2022–2027.

82. Ma, Z.H., and Michailides, T.J. (2007) Approaches for eliminating PCR inhibitors and designing PCR primers for the detection of phytopathogenic fungi. Crop Prot 26: 145–161.

83. Ma, L.M., Fletcher, J. and Zhang, G. (2014) Detection of human pathogens on plants. In Detection and Diagnostics of Plant Pathogens. Gullino, M.L. and Bonants, P.J.M. (eds). Dordrecht, the Netherlands: Springer Netherlands, pp. 87–102.

84. Malorny, B., Hoorfar, J., Bunge, C., and Helmuth, R. (2003) Multicenter validation of the analytical accuracy of Salmonella PCR: towards an international standard. Appl Environ Microbiol 69: 290–296.

85. Malorny, B., Paccassoni, E., Fach, P., Bunge, C., Martin, A., and Helmuth, R. (2004) Diagnostic real-time PCR for detection of Salmonella in food. Appl Environ Microbiol 70: 7046–7052.

86. Marchesi, J.R., and Ravel, J. (2015) The vocabulary of microbiome research: a proposal. Microbiome 3: 31.

87. Martiny, D., Busson, L., Wybo, I., El Haj, R.A., Dediste, A., and Vandenberg, O. (2012) Comparison of the Microflex LT and Vitek MS systems for routine identification of bacteria by matrix-assisted laser desorption ionization-time of flight mass spectrometry. J Clin Microbiol 50: 1313–1325.

88. McFarland, M.A., Andrzejewski, D., Musser, S.M., and Callahan, J.H. (2014) Platform for identification of Salmonella serovar differentiating bacterial proteins by top-down mass spectrometry: S. Typhimurium vs S. Heidelberg. Anal Chem 86: 6879–6886.

89. McQuiston, J.R., Herrera-Leon, S., Wertheim, B.C., Doyle, J., Fields, P.I., Tauxe, R.V., and Logsdon, J.M. Jr (2008) Molecular phylogeny of the salmonellae: relationships among Salmonella species and subspecies determined from four housekeeping genes and evidence of lateral gene transfer events. J Bacteriol 190: 7060–7067.

90. Mott, T.M., Everley, R.A., Wyatt, S.A., Toney, D.M., and Croley, T.R.

(2010) Comparison of MALDI-TOF/MS and LC-QTOF/MS methods for the identification of enteric bacteria. Int J Mass Spectrom 291: 24–32.

91. Myint, M.S., Johnson, Y.J., Tablante, N.L., and Heckert, R.A. (2006) The effect of pre-enrichment protocol on the sensitivity and specificity of PCR for detection of naturally contaminated Salmonella in raw poultry compared to conventional culture. Food Microbiol 23: 599–604.
92. Nieminen, T.T., Koskinen, K., Laine, P., Hultman, J., Sade, E., Paulin, L., et al. (2012) Comparison of microbial communities in marinated and unmarinated broiler meat by metagenomics. Int J Food Microbiol 157: 142–149.
93. Ottesen, A.R., Gonzalez, A., Bell, R., Arce, C., Rideout, S., Allard, M., et al. (2013a) Co-enriching microflora associated with culture based methods to detect Salmonella from tomato phyllosphere. PLoS ONE 8: e73079.
94. Ottesen, A.R., Gonzalez Pena, A., White, J.R., Pettengill, J.B., Li, C., Allard, S., et al. (2013b) Baseline survey of the anatomical microbial ecology of an important food plant: Solanum lycopersicum (tomato). BMC Microbiol 13: 114.
95. Painter, J.A., Hoekstra, R.M., Ayers, T., Tauxe, R.V., Braden, C.R., Angulo, F.J., and Griffin, P.M. (2013) Attribution of foodborne illnesses, hospitalizations, and deaths to food commodities by using outbreak data, United States, 1998-2008. Emerg Infect Dis 19: 407–415.
96. Pettengill, J.B., McAvoy, E., White, J.R., Allard, M., Brown, E., and Ottesen, A. (2012) Using metagenomic analyses to estimate the consequences of enrichment bias for pathogen detection. BMC Res Notes 5: 378.
97. Postollec, F., Falentin, H., Pavan, S., Combrisson, J., and Sohier, D. (2011) Recent advances in quantitative PCR (qPCR) applications in food microbiology. Food Microbiol 28: 848–861.
98. Ribot, E.M., Fair, M.A., Gautom, R., Cameron, D.N., Hunter, S.B., Swaminathan, B., and Barrett, T.J. (2006)
99. Standardization of pulsed-field gel electrophoresis protocols for the subtyping of Escherichia coli O157:H7, Salmonella, and Shigella for PulseNet. Foodborne Pathog Dis 3: 59–67.
100. Rogers, G.B., Carroll, M.P., and Bruce, K.D. (2009) Studying bacterial

infections through culture-independent approaches. J Med Microbiol 58: 1401–1418.

101. Rossen, L., Norskov, P., Holmstrom, K., and Rasmussen, O.F. (1992) Inhibition of PCR by components of food
102. samples, microbial diagnostic assays and DNA-extraction solutions. Int J Food Microbiol 17: 37–45.
103. Sapers, G.M., and Doyle, M.P. (2014) Scope of the produce contamination problem. In The Produce Contamination Problem. Matthews, K.R., Sapers, G.M., and Gerba, C.P. (eds). San Diego, CA: Elsevier, pp. 3–20.
104. Scallan, E., Hoekstra, R.M., Angulo, F.J., Tauxe, R.V., Widdowson, M.A., Roy, S.L., et al. (2011) Foodborne illness acquired in the United States-major pathogens. Emerg Infect Dis 17: 7–15.
105. Schloss, P.D., Westcott, S.L., Ryabin, T., Hall, J.R., Hartmann, M., Hollister, E.B., et al. (2009) Introducing mothur: open-source, platform-independent, community-supported software for describing and comparing microbial communities. Appl Environ Microbiol 75: 7537–7541.
106. Shaw, S.J., Blais, B.W., and Nundy, D.C. (1998) Performance of the dynabeads anti-Salmonella system in the detection of Salmonella species in foods, animal feeds, and environmental samples. J Food Prot 61: 1507–1510.
107. Singer, R.S., Mayer, A.E., Hanson, T.E., and Isaacson, R.E. (2009) Do microbial interactions and cultivation media decrease the accuracy of Salmonella surveillance systems and outbreak investigations? J Food Prot 72: 707–713.
108. Singh, G., Vajpayee, P., Rani, N., Jyoti, A., Gupta, K.C., and Shanker, R. (2012) Bio-capture of S. Typhimurium from surface water by aptamer for culture-free quantification. Ecotoxicol Environ Saf 78: 320–326.
109. Sparbier, K., Weller, U., Boogen, C., and Kostrzewa, M. (2012) Rapid detection of Salmonella sp. by means of a combination of selective enrichment broth and MALDITOF MS. Eur J Clin Microbiol Infect Dis 31: 767–773.
110. Swaminathan, B., Barrett, T.J., Hunter, S.B., Tauxe, R.V., and Force, C.D.C.P.T. (2001) Pulsenet: the molecular subtyping network for foodborne bacterial disease surveillance, United States. Emerg Infect Dis 7: 382–389.

111. Tatavarthy, A., Peak, K., Veguilla, W., Cutting, T., Harwood, V.J., Roberts, J., et al. (2009) An accelerated method for isolation of Salmonella enterica serotype Typhimurium from artificially contaminated foods, using a short preenrichment, immunomagnetic separation, and xylose-lysine-desoxycholate agar (6IX method). J Food Prot 72: 583–590.

112. Tyson, G.H., McDermott, P.F., Li, C., Chen, Y., Tadesse, D.A., Mukherjee, S., et al. (2015) WGS accurately predicts antimicrobial resistance in Escherichia coli. J Antimicrob Chemother 70: 2763–2769.

113. USDA FSIS (2014) Isolation and identification of Salmonella from meat, poultry, pasteurized egg, and catfish products and carcass and environmental sponges. In MLG 4.08

114. United States Department of Agriculture, Food Safety and Inspection Service. Vibbert, H.B., Ku, S., Li, X., Liu, X., Ximenes, E., Kreke, T., et al. (2015) Accelerating sample preparation through enzyme-assisted microfiltration of Salmonella in chicken extract. Biotechnol Prog 31: 1551–1562.

115. Wanglin, H. and Longhao, Y. (2015) Advances on Shelf-life Prediction of Low-temperature Meat Products Based on Spoilage Microorganisms. 27: 5156.

116. Weisburg, W.G., Barns, S.M., Pelletier, D.A., and Lane, D.J. (1991) 16S ribosomal DNA amplification for phylogenetic study. J Bacteriol 173: 697–703.

117. Williams, T.L., Andrzejewski, D., Lay, J.O., and Musser, S.M. (2003) Experimental factors affecting the quality and reproducibility of MALDI TOF mass spectra obtained from whole bacteria cells. J Am Soc Mass Spectrom 14: 342–351.

118. Williams, T.L., Musser, S.M., Nordstrom, J.L., DePaola, A., and Monday, S.R. (2004) Identification of a protein biomarker unique to the pandemic O3:K6 clone of Vibrio parahaemolyticus. J Clin Microbiol 42: 1657–1665.

119. Williams, T.L., Monday, S.R., Edelson-Mammel, S., Buchanan, R., and Musser, S.M. (2005) A top-down proteomics approach for differentiating thermal resistant strains of Enterobacter sakazakii. Proteomics 5: 4161–4169.

120. Wilson, I.G. (1997) Inhibition and facilitation of nucleic acidamplification. Appl Environ Microbiol 63: 3741–3751.

121. Wunschel, S.C., Jarman, K.H., Petersen, C.E., Valentine, N.B., Wahl, K.L., Schauki, D., et al. (2005) Bacterial analysis by MALDI-TOF mass spectrometry: an inter-laboratory comparison. J Am Soc Mass Spectrom 16: 456–462.
122. Zankari, E., Hasman, H., Kaas, R.S., Seyfarth, A.M., Agerso, Y., Lund, O., et al. (2013) Genotyping using whole-genome sequencing is a realistic alternative to surveillance based on phenotypic antimicrobial susceptibility testing. J Antimicrob Chemother 68: 771–777.
123. Zhang, S., Yin, Y., Jones, M.B., Zhang, Z., Deatherage Kaiser, B.L., Dinsmore, B.A., et al. (2015) Salmonella serotype determination utilizing high-throughput genome sequencing data. J Clin Microbiol 53: 1685–1692.

CHAPTER 5

ARCOBACTER: AN EMERGING FOOD-BORNE ZOONOTIC PATHOGEN, ITS PUBLIC HEALTH CONCERNS AND ADVANCES IN DIAGNOSIS AND CONTROL – A COMPREHENSIVE REVIEW

Thadiyam Puram Ramees[a], Kuldeep Dhama[b] , Kumaraguruba-ran Karthik[c] , Ramswaroop Singh Rathore[a], Ashok Kumar [a], Mani Saminathan[b], Ruchi Tiwari [d], Yashpal Singh Malik[e] and Raj Kumar Sing[f]

[a] Division of Veterinary Public Health, ICAR-Indian Veterinary Research Institute (IVRI), Bareilly, India;

CITATION: Ramees, T.P., et al., "Arcobacter: an emerging food-borne zoonotic pathogen, its public health concerns and advances in diagnosis and control – a comprehensive review". VETERINARY QUARTERLY (2017). https://doi.org/10.1080/01652176.2017.1323355.

[b] Division of Pathology, ICAR-Indian Veterinary Research Institute (IVRI), Bareilly, India;

[c] Central University Laboratory, Tamil Nadu Veterinary and Animal Sciences University, Chennai, India;

[d] Department of Veterinary Microbiology and Immunology, College of Veterinary Sciences, UP Pandit Deen Dayal Upadhayay Pashu Chikitsa Vigyan Vishwavidyalay Evum Go-Anusandhan Sansthan (DUVASU), Mathura, India;

[e] Division of Biological Standardization, ICAR-Indian Veterinary Research Institute (IVRI), Bareilly, India;

[f] ICAR-Indian Veterinary Research Institute (IVRI), Bareilly, India

ABSTRACT

Arcobacter has emerged as an important food-borne zoonotic pathogen, causing sometimes serious infections in humans and animals. Newer species of *Arcobacter* are being incessantly emerging (presently 25 species have been identified) with novel information on the evolutionary mechanisms and genetic diversity among different *Arcobacter* species. These have been reported from chickens, domestic animals (cattle, pigs, sheep, horses, dogs), reptiles (lizards, snakes and chelonians), meat (poultry, pork, goat, lamb, beef, rabbit), vegetables and from humans in different countries. *Arcobacters*are implicated as causative agents of diarrhea, mastitis and abortion in animals, while causing bacteremia, endocarditis, peritonitis, gastroenteritis and diarrhea in humans. Three species including *A. butzleri*, *A. cryaerophilus* and *A. skirrowii* are predominantly associated with clinical conditions. *Arcobacters* are primarily transmitted through contaminated food and water sources. Identification of *Arcobacter* by biochemical tests is difficult and isolation remains the gold standard method. Current diagnostic advances have provided various molecular methods for efficient detection and differentiation of the *Arcobacters* at genus and species level. To overcome the emerging antibiotic resistance problem there is an essential need to explore the potential of novel and alternative therapies. Strengthening of the diagnostic aspects is also suggested as in most cases *Arcobacters* goes unnoticed and hence the exact epidemiological status remains uncertain. This review updates the current knowledge and many aspects of this important food-borne pathogen, namely etiology, evolution and emergence, genetic diversity, epidemiology, the disease in animals and humans, public health concerns, and advances in its diagnosis, prevention and control.

Keywords: Arcobacter; epidemiology; emergence; animals; food- borne; zoonosis; public health; diagnosis control

INTRODUCTION

Food-borne infections are gaining importance due to alarming increase in the number of outbreaks in the recent years throughout the world (Scallan et al. 2011Scallan E, Griffin PM, Angulo FJ, Tauxe RV, Hoekstra RM. 2011. Foodborne illness acquired in the United States – unspecified agents. Emerg Infect Dis. 17:16. Dhama et al. 2013aDhama K, Rajagunalan S, Chakraborty S, Verma AK, Kumar A, Tiwari R, Kapoor S. 2013a. Food-borne pathogens of animal origin-diagnosis, prevention, control and their zoonotic significance: a review. Pak J Biol Sci. 16:1076–1085 ; Hu et al. 2014Hu Q, Lyu D, Shi X, Jiang Y, Lin Y, Li Y, Qiu Y, He L, Zhang R, Li Q. 2014. A modified molecular beacons-based multiplex real-time pcr assay for simultaneous detection of eight foodborne pathogens in a single reaction and its application. Foodborne Pathog Dis. 11:207–214. Lehmann et al. 2015Lehmann D, Alter T, Lehmann L, Uherkova S, Seidler T, Gölz G. 2015. Prevalence, virulence gene distribution and genetic diversity of *Arcobacter* in food samples in Germany. Berl Munch Tierarztl Wochenschr. 128:163–168). Important food-borne bacterial pathogens include *Escherichia coli*, *Salmonella*, *Campylobacter*, *Yersinia enterocolitica*, *Staphylococcus*, *Bacillus cereus*, *Clostridium perfringens*, *Cl. botulinum* and *Arcobacter*(Collado & Figueras 2011Collado L, Figueras MJ. 2011. Taxonomy, epidemiology and clinical relevance of the genus *Arcobacter*. Clin Microbiol Rev. 24:174–192. Calvo et al. 2013Calvo G, Arias ML, Fernández H. 2013. *Arcobacter*: a foodborne emerging pathogen. Arch Latinoam Nutr. 63:164–172 ; Dhama et al. 2013aDhama K, Rajagunalan S, Chakraborty S, Verma AK, Kumar A, Tiwari R, Kapoor S. 2013a. Food-borne pathogens of animal origin-diagnosis, prevention, control and their zoonotic significance: a review. Pak J Biol Sci. 16:1076–1085 ; Hsu & Lee 2015Hsu TT, Lee J. 2015. Global distribution and prevalence of arcobacter in food and water. Zoonoses Public Health. 62:579–589. Hänel et al. 2016Hänel I, Tomaso H, Neubauer H. 2016. Arcobacter an underestimated zoonotic pathogen? Bundesgesundheitsblatt Gesundheitsforschung Gesundheitsschutz. 59:789–794 ,). While the virulence and pathogenesis of certain pathogens has been dissected thoroughly, pathogens like *Arcobacter* are still in the dark zone needing immediate attention. During recent years, *Arcobacter* spp. has been identified as an emerging food-borne zoonotic pathogen worldwide and the International Commission on Microbiological Specifications for Foods (ICMSF) has classified *Arcobacter* as a serious hazard to human health. *Arcobacter* was very first isolated and described from aborted bovine fetal tissues (Ellis et al. 1977Ellis WA, Neill SD, O' Brien JJ, Ferguson, HW, Hanna J. 1977. Isolation of *Spirillum*/*Vibrio*-like organisms from bovine fetuses. Vet Record. 100:451–452 ,) and later from porcine fetuses (Ellis et al. 1977Ellis WA, Neill SD, O' Brien JJ, Ferguson, HW, Hanna J. 1977. Isolation of *Spirillum*/*Vibrio*-like organisms from bovine fetuses. Vet Record. 100:451–452 , , 1978Ellis WA, Neill SD, O' Brien JJ, Hannna J. 1978. Isolation of *Spirullum*-like organisms from pig fetuses. Vet Record. 102:106. Aydin et al. 2007Aydin F, Gumussoy KS, Atabay HI, Ica T, Abay S. 2007. Prevalence and distribution of *Arcobacter* species in various sources in Turkey and molecular analysis of isolated strains by ERIC-PCR. J Appl Microbiol. 103:27–35. Rasmussen et al. 2013Rasmussen LH, Jette K, Jens PC, Hanne I. 2013. Multilocus sequence typing and biocide tolerance of *Arcobacter butzleri* from Danish broiler carcasses. BMC Res Notes. 6:322). *Arcobacters* are implicated as causative agents of diarrhea, mastitis and abortion in animals, while causing bacteremia, endocarditis, peritonitis, gastroenteritis and diarrhea in humans (Jiang et al. 2010Jiang ZD, DuPont HL, Brown EL, Nandy RK, Ramamurthy T, Sinha A, Ghosh, S, Guin S, Gurleen K, Rodrigues S, et al. 2010. Microbial etiology of travelers' diarrhea in Mexico, Guatemala, and India: importance of enterotoxigenic *Bacteroides fragilis* and *Arcobacter*species. J Clin Microbiol. 48:1417–1419. Figueras et al. 2014Figueras MJ, Levican A, Pujol I, Ballester F, Rabada, Quilez MJ, Gomez-Bertomeu F. 2014. A severe case of persistent diarrhoea associated with *Arcobacter cryaerophilus* but attributed to *Campylobacter* sp. and a review of the clinical incidence of *Arcobacter*spp. New Microbes New Infect. 2:31–37 ; Ferreira et al. 2016Ferreira S, Queiroz JA, Oleastro M, Domingues FC. 2016. Insights in the pathogenesis and resistance of *Arcobacter*: a review. Crit Rev Microbiol. 42:364–383. Their presence in blood samples of humans with cirrhosis and gangrenous appendicitis (Lau et al. 2002Lau SKP, Woo PCY, Teng J, Leung KW, Yuen KY. 2002. Identification by 16S ribosomal RNA gene sequencing of *Arcobacter*bacteraemia in a patient with acute gangrenous appendicitis. Mol Pathol. 55:182–185) furthermore complicates the situation. Three species of *Arcobacter* namely *A. butzleri*, *A. cryaerophilus*and *A. skirrowii* are more commonly associated with clinical conditions (Gill 1983Gill KP.1983. Aerotolerant *campylobacter* strain isolated from a bovine preputial sheath washing. Vet Rec. 112:459. Kiehlbauch et al. 1991Kiehlbauch JA,

Brenner DJ, Nicholson MA, Baker CN, Patton CM, Steigerwalt AG, Wachsmuth IK. 1991. *Campylobacter butzleri* sp. nov. isolated from humans and animals with diarrheal illness. J Clin Microbiol. 29:376–385 ; de Oliveira et al. 1997de Oliveira SJ, Wesley I, Harmon KM. 1997. Classification of *Arcobacter* species isolated from aborted pig fetuses and sows with reproductive problems in Brazil. Vet Microbiol. 57:347–354. Wesley 1997Wesley IV. 1997. *Helicobacter and Arcobacter*: potential human foodborne pathogens? Trends Food Sci Technol. 8:293–299., ; Wesley et al. 2000Wesley IV, Wells SJ, Harmon KM, Green A, Schroeder-Tucker L, Glover M, Siddique I. 2000. Fecal shedding of *Campylobacter* and *Arcobacter* spp. in dairy cattle. Appl Environ Microbiol. 66:1994–2000. On et al. 2002On SLW, Jensen TK, Bille-Hansen V, Jorsal SE, Vandamme P. 2002. Prevalence and diversity of *Arcobacter* spp. isolated from the internal organs of spontaneous porcine abortions in Denmark. Vet Microbiol. 85:159–167. Collado & Figueras 2011Collado L, Figueras MJ. 2011. Taxonomy, epidemiology and clinical relevance of the genus *Arcobacter*. Clin Microbiol Rev. 24:174–192. Patyal et al. 2011Patyal A, Rathore RS, Mohan HV, Dhama K, Kumar A. 2011. Prevalence of *Arcobacter* spp. in humans, animals and foods of animal origin including sea food from India. Transbound Emerg Dis. 58:402–410. Bagalakote et al. 2014Bagalakote PS, Rathore RS, Ramees TP, Mohan HV, Sumankumar M, Agarwal RK, Kumar A, Dhama K. 2014. Molecular characterization of *Arcobacter* isolates using randomly amplified polymorphic DNA- polymerase chain reaction (RAPD-PCR). Asian J Anim Vet Adv. 9:543–555., ; Ramees et al. 2014aRamees TP, Rathore RS, Bagalakote PS, Mohan HV, Kumar A, Dhama K. 2014a. Multiplex PCR detection of*Arcobacter butzleri* and*Arcobacter cryaerophilus* in skin of poultry. J Pure Appl Microbiol. 8:1755–1758. , 2014bRamees TP, Rathore RS, Bagalkot PS, Mohan HV, Kumar A, Dhama K. 2014b. Detection of *Arcobacter butzleri* and *Arcobacter cryaerophilus* in clinical samples of humans and foods of animal origin by cultural and multiplex PCR based methods. Asian J Anim Vet Adv. 9:243–252 2014cRamees TP, Rathore RS, Bagalkot PS, Ravi Kumar GVPPS, Mohan HV, Anoopraj R, Kumar A, Dhama K. 2014c. Real-time PCR detection of *Arcobacter butzleri* and *Arcobacter cryaerophilus* in chicken meat samples. J Pure Appl Microbiol. 8:3165–3169.). As the *Arcobacter* organisms are commonly isolated from fecal samples of healthy animals and humans and are unable to satisfy the Koch's postulate, their pathology and pathogenicity is under dispute (Shah et al. 2011Shah AH, Saleha AA, Zunita Z, Murugaiyah M. 2011. *Arcobacter*– an emerging threat to animals and animal origin food products? Tr Food Sci

Technol. 22:225–236., ; Ongor et al. 2004Ongor H, Cetinkaya B, Açik MN, Atabay HI. 2004. Investigation of Arcobacters in meat and fecal samples of clinically healthy cattle in Turkey. Lett Appl Microbiol. 38:339–344. Houf & Stephan 2007Houf K, Stephan R. 2007. Isolation and characterization of the emerging food-borne pathogen *Arcobacter* from human stool. J Microbiol Methods. 68:408–413. Figueras et al. 2014Figueras MJ, Levican A, Pujol I, Ballester F, Rabada, Quilez MJ, Gomez-Bertomeu F. 2014. A severe case of persistent diarrhoea associated with *Arcobacter cryaerophilus* but attributed to *Campylobacter* sp. and a review of the clinical incidence of *Arcobacter*spp. New Microbes New Infect. 2:31–37).

Arcobacters have been isolated from healthy cattle, sheep and pigs (Ho et al. 2006aHo HT, Lipman LJ, Gaastra W. 2006a. *Arcobacter*, what is known and unknown about a potential food-borne zoonotic agent.Vet Microbiol. 115:1–13 , , 2006bHo HT, Lipman LJA, Van der Graaf-van Bloois L, Van Bergen M, Gaastra W. 2006b. Potential routes of acquisition of *Arcobacter* species by piglets. Vet Microbiol. 114:122–133., ; De Smet et al. 2011De Smet S, De Zutter L, Houf K. 2011. Small ruminants as carriers of the emerging food-borne pathogen *Arcobacter* on small and medium farms. Small Rumin Res. 97:124–129., ; Patyal et al. 2011Patyal A, Rathore RS, Mohan HV, Dhama K, Kumar A. 2011. Prevalence of *Arcobacter* spp. in humans, animals and foods of animal origin including sea food from India. Transbound Emerg Dis. 58:402–410. Ferreira et al. 2016Ferreira S, Queiroz JA, Oleastro M, Domingues FC. 2016. Insights in the pathogenesis and resistance of *Arcobacter*: a review. Crit Rev Microbiol. 42:364–383). Poultry act as an important reservoir of *Arcobacters* and as a major source of infection spread (de Boer et al. 1996de Boer E, Tilburg JJHC, Woodward DL, Loir H, Johnson WM. 1996. A selective medium for the isolation of *Arcobacter* from meats. Lett Appl Microbiol. 23:64–66. Lammerding et al. 1996Lammerding AM, Harris JE, Lior H, Woodward D, Cole L, Muckle CA. 1996. Isolation method for recovery of *Arcobacter butzleri* from fresh poultry and poultry products. In Newell DG, Ketley J, Feldman RA, editors. 8th International Workshop on '*Campylobacters*, *Helicobacters*and Related Organisms'. New York (NY): Plenum Press; p. 329–334., ; Atabay et al. 1998Atabay HI, Corry JEL, On SLW. 1998. Diversity and prevalence of Arcobacter spp. in broiler chickens. J Appl Microbiol. 84:1007–1016. Manke et al. 1998Manke TR, Wesley IV, Dickson JS, Harmon KM. 1998. Prevalence and genetic variability of *Arcobacter*species in mechanically separated turkey. J Food Prot. 61:1623–1628. Johnson & Murano 1999aJohnson LG, Murano EA. 1999a. Development of a new medium for the isolation of *Arcobacter* spp. J Food Prot. 62:456–462. Wesley & Baetz 1999Wesley IV, Baetz AL. 1999. Natural and experimental infections of *Arcobacter* in poultry. Poult Sci. 78:536–545. González et al. 2000González I, Garcia T, Antonlin A, Hernandez PE, Martin R. 2000. Development of a combined PCR-culture technique for the rapid detection of *Arcobacter* spp. in chicken meat. Lett Appl Microbiol. 30:207–212. Houf et al. 2000Houf K, Tutenel A, De Zutter L, Van Hoof J, Vandamme P. 2000. Development of a multiplex PCR assay for the simultaneous detection and identification of *Arcobacter butzleri, Arcobacter cryaerophilus* and *Arcobacter skirrowii*. FEMS Immunol Med Microbiol. 49:337–345. ; Atabay & Aydin 2001Atabay HI, Aydin F. 2001. Susceptibility of *Arcobacter butzleri*isolates to 23 antimicrobial agents. Lett Appl Microbiol. 33:430–433. Atabay et al. 2008Atabay HI, Unver A, Sahin M, Otlu S, Elmali M, Yaman H. 2008. Isolation of various *Arcobacter* species from domestic geese (Anser anser). Vet Microbiol. 128:400–405. Collado & Figueras 2011Collado L, Figueras MJ. 2011. Taxonomy, epidemiology and clinical relevance of the genus *Arcobacter*. Clin Microbiol Rev. 24:174–192. Rahimi 2014Rahimi E. 2014. Prevalence and antimicrobial resistance of *Arcobacter* species isolated from poultry meat in Iran. Br Poult Sci. 55:174–180). Recently, the presence of *Arcobacter* has been reported from a wild bird, namely collared doves (*Streptopelia decaocto*) (Di Francesco et al. 2014Di Francesco A, Delogu M, Giacometti F, Stancampiano L, Grilli E, Guarniero I, Serraino A. 2014. First detection of Arcobacter sp. in Eurasian collared doves (Streptopelia decaocto). Vet Ital. 50:313–315). Poultry intestines have been proposed to harbor *Arcobacters* and contaminate the slaughter houses during processing of the carcasses thereby enhancing the chances of further contamination (Ho et al. 2008aHo HT, Lipman LJ, Gaastra W. 2008a. The introduction of *Arcobacter* spp. in poultry slaughterhouses. Int J Food Microbiol. 125:223–229 ,). *Arcobacters* have also been recovered from a variety of foods of animal origin, namely beef, pork, milk, and sea foods, and also from water and vegetables (Son et al. 2006Son I, Englen MD, Berrang ME, Fedorka-Cray PJ, Harrison MA. 2006. Genetic diversity of *Arcobacter* and *Campylobacter* on broiler carcasses during processing. J Food Prot. 69:1028–1033. Collado et al. 2010Collado L, Kasimir G, Perez U, Bosch A, Pinto R, Saucedo G, Huguet JM, Figueras MJ. 2010. Occurrence and diversity of *Arcobacter* spp. along the Llobregat river catchment, at sewage effluents and in a drinking water treatment plant. Water Res. 44:3696–3702. Hausdorf et al. 2011Hausdorf L, Fröhling A, Schlüter O, Klocke M. 2011. Analyses of the bacterial community within carrot wash water. Can J

Microbiol. 57:447–452. Patyal et al. 2011Patyal A, Rathore RS, Mohan HV, Dhama K, Kumar A. 2011. Prevalence of *Arcobacter* spp. in humans, animals and foods of animal origin including sea food from India. Transbound Emerg Dis. 58:402–410. Shah et al. 2012aShah AH, Saleha AA, Murugaiyah M, Zunita Z, Memon AA. 2012a. Prevalence and distribution of *Arcobacter* spp. in raw milk and retail raw beef. J Food Prot. 75:1474–1478. Hausdorf et al. 2013Hausdorf L, Neumann M, Bergmann I, Sobiella K, Mundt K, Fröhling A, Schlüter O, Klocke M. 2013. Occurrence and genetic diversity of *Arcobacter* spp. in a spinach-processing plant and evaluation of two *Arcobacter*-specific quantitative PCR assay. Syst Appl Microbiol. 36:235–243. Lee & Choi 2013Lee MH, Choi, C. 2013. Survival of *Arcobacter butzleri* in apple and pear purees. J Food Safety. 33:333–339., ; Rahimi et al. 2014Rahimi E. 2014. Prevalence and antimicrobial resistance of *Arcobacter* species isolated from poultry meat in Iran. Br Poult Sci. 55:174–180 ; Ramees et al. 2014aRamees TP, Rathore RS, Bagalakote PS, Mohan HV, Kumar A, Dhama K. 2014a. Multiplex PCR detection of *Arcobacter butzleri* and *Arcobacter cryaerophilus* in skin of poultry. J Pure Appl Microbiol. 8:1755–1758. ; Lehmann et al. 2015Lehmann D, Alter T, Lehmann L, Uherkova S, Seidler T, Gölz G. 2015. Prevalence, virulence gene distribution and genetic diversity of *Arcobacter* in food samples in Germany. Berl Munch Tierarztl Wochenschr. 128:163–168 ; Ferreira et al. 2016Ferreira S, Queiroz JA, Oleastro M, Domingues FC. 2016. Insights in the pathogenesis and resistance of *Arcobacter*: a review. Crit Rev Microbiol. 42:364–383). Contaminated water and meat play an important role in its transmission (Snelling et al. 2006Snelling WJ, Matsuda M, Moore JE, Dooley JS. 2006. Under the microscope: *Arcobacter*. Lett Appl Microbiol. 42:7–14. Collado et al. 2010Collado L, Kasimir G, Perez U, Bosch A, Pinto R, Saucedo G, Huguet JM, Figueras MJ. 2010. Occurrence and diversity of *Arcobacter* spp. along the Llobregat river catchment, at sewage effluents and in a drinking water treatment plant. Water Res. 44:3696–3702. González & Ferrus 2011González AM, Ferrus MA. 2011. Study of *Arcobacter* spp. contamination in fresh lettuces detected by different cultural and molecular methods. Int J Food Microbiol. 145:311–314. Rahimi 2014Rahimi E. 2014. Prevalence and antimicrobial resistance of *Arcobacter* species isolated from poultry meat in Iran. Br Poult Sci. 55:174–180). In abattoir, *Arcobacters* have been frequently reported from various sources and their persistence has been recorded even after follow up of disinfection procedures, therefore abattoirs act as a potential source of spread, but such full epidemiological information to draw conclusive remark is lacking (Collado et al. 2010Collado

L, Kasimir G, Perez U, Bosch A, Pinto R, Saucedo G, Huguet JM, Figueras MJ. 2010. Occurrence and diversity of *Arcobacter* spp. along the Llobregat river catchment, at sewage effluents and in a drinking water treatment plant. Water Res. 44:3696–3702 ,). Therefore, contamination of food samples and changing trends in human food habits lead to increasing infections of this food-borne pathogen in the human community (Wesley et al. 2000Wesley IV, Wells SJ, Harmon KM, Green A, Schroeder-Tucker L, Glover M, Siddique I. 2000. Fecal shedding of *Campylobacter* and *Arcobacter* spp. in dairy cattle. Appl Environ Microbiol. 66:1994–2000. On et al. 2002On SLW, Jensen TK, Bille-Hansen V, Jorsal SE, Vandamme P. 2002. Prevalence and diversity of *Arcobacter* spp. isolated from the internal organs of spontaneous porcine abortions in Denmark. Vet Microbiol. 85:159–167. Patyal et al. 2011Patyal A, Rathore RS, Mohan HV, Dhama K, Kumar A. 2011. Prevalence of *Arcobacter* spp. in humans, animals and foods of animal origin including sea food from India. Transbound Emerg Dis. 58:402–410. Dhama et al. 2013aDhama K, Rajagunalan S, Chakraborty S, Verma AK, Kumar A, Tiwari R, Kapoor S. 2013a. Food-borne pathogens of animal origin-diagnosis, prevention, control and their zoonotic significance: a review. Pak J Biol Sci. 16:1076–1085 ; Rahimi 2014Rahimi E. 2014. Prevalence and antimicrobial resistance of *Arcobacter* species isolated from poultry meat in Iran. Br Poult Sci. 55:174–180 ; Ramees et al. 2014aRamees TP, Rathore RS, Bagalakote PS, Mohan HV, Kumar A, Dhama K. 2014a. Multiplex PCR detection of *Arcobacter butzleri* and *Arcobacter cryaerophilus* in skin of poultry. J Pure Appl Microbiol. 8:1755–1758. , 2014bRamees TP, Rathore RS, Bagalkot PS, Mohan HV, Kumar A, Dhama K. 2014b. Detection of *Arcobacter butzleri* and *Arcobacter cryaerophilus* in clinical samples of humans and foods of animal origin by cultural and multiplex PCR based methods. Asian J Anim Vet Adv. 9:243–252 2014cRamees TP, Rathore RS, Bagalkot PS, Ravi Kumar GVPPS, Mohan HV, Anoopraj R, Kumar A, Dhama K. 2014c. Real-time PCR detection of *Arcobacter butzleri* and *Arcobacter cryaerophilus* in chicken meat samples. J Pure Appl Microbiol. 8:3165–3169.). A higher amount of genetic diversity and increase in antibiotic resistance has been reported among *Arcobacter* isolates recovered from different parts of the world (Bagalakote et al. 2014Bagalakote PS, Rathore RS, Ramees TP, Mohan HV, Sumankumar M, Agarwal RK, Kumar A, Dhama K. 2014. Molecular characterization of *Arcobacter* isolates using randomly amplified polymorphic DNA- polymerase chain reaction (RAPD-PCR). Asian J Anim Vet Adv. 9:543–555., ; Mohan et al. 2014Mohan HV, Rathore RS, Dhama K, Ramees TP, Patyal A, Bagalkot PS, Wani MY,

Bhilegaonkar KN, Kumar A. 2014. Prevalence of *Arcobacter*spp. in humans, animals and foods of animal origin in India based on cultural isolation, antibiogram, PCR and multiplex PCR detection. Asian J Anim Vet Adv. 9:452–466., ; Rahimi 2014Rahimi E. 2014. Prevalence and antimicrobial resistance of *Arcobacter* species isolated from poultry meat in Iran. Br Poult Sci. 55:174–180 ; Ferreira et al. 2016Ferreira S, Queiroz JA, Oleastro M, Domingues FC. 2016. Insights in the pathogenesis and resistance of *Arcobacter*: a review. Crit Rev Microbiol. 42:364–383).

Isolation of *Arcobacter* is considered as gold standard method for diagnostic purposes (Atabay & Corry 1998Atabay HI, Corry JEL. 1998. Evaluation of a new *Arcobacter*enrichment medium and comparison with new media developed for enrichment for *Campylobacter spp*. Int J Food Microbiol. 41:53–58. Aydin et al. 2007Aydin F, Gumussoy KS, Atabay HI, Ica T, Abay S. 2007. Prevalence and distribution of *Arcobacter* species in various sources in Turkey and molecular analysis of isolated strains by ERIC-PCR. J Appl Microbiol. 103:27–35. Rahimi 2014Rahimi E. 2014. Prevalence and antimicrobial resistance of *Arcobacter* species isolated from poultry meat in Iran. Br Poult Sci. 55:174–180 ; Ramees et al. 2014bRamees TP, Rathore RS, Bagalkot PS, Mohan HV, Kumar A, Dhama K. 2014b. Detection of *Arcobacter butzleri* and *Arcobacter cryaerophilus* in clinical samples of humans and foods of animal origin by cultural and multiplex PCR based methods. Asian J Anim Vet Adv. 9:243–252.,). Various types of growth media have been employed for cultural isolation of *Arcobacter*species (de Boer et al. 1996de Boer E, Tilburg JJHC, Woodward DL, Loir H, Johnson WM. 1996. A selective medium for the isolation of *Arcobacter* from meats. Lett Appl Microbiol. 23:64–66. Atabay & Corry 1998Atabay HI, Corry JEL. 1998. Evaluation of a new *Arcobacter*enrichment medium and comparison with new media developed for enrichment for *Campylobacter spp*. Int J Food Microbiol. 41:53–58. Johnson & Murano 1999bJohnson LG, Murano EA. 1999b. Comparison of three protocols for the isolation of *Arcobacter* from poultry. J Food Prot. 62:610–614. Ohlendorf & Murano 2002Ohlendorf DS, Murano EA. 2002. Prevalence of *Arcobacter* spp. in raw ground pork from several geographical regions according to various isolation methods. J Food Protect. 65:1700–1705. Scullion et al. 2004Scullion R, Harrington CS, Madden RH. 2004. A comparison of three methods for the isolation of *Arcobacter* spp. from retail raw poultry in Northern Ireland. J Food Prot. 67:799–804. Collado & Figueras 2011Collado L, Figueras MJ. 2011. Taxonomy, epidemiology and clinical relevance of the genus *Arcobacter*. Clin Microbiol Rev. 24:174–192. Rahimi 2014Rahimi E. 2014. Prevalence and antimicrobial resistance of *Arcobacter* species isolated from poultry meat in Iran. Br Poult Sci. 55:174–180 ; Ramees et al. 2014bRamees TP, Rathore RS, Bagalkot PS, Mohan HV, Kumar A, Dhama K. 2014b. Detection of *Arcobacter butzleri* and *Arcobacter cryaerophilus* in clinical samples of humans and foods of animal origin by cultural and multiplex PCR based methods. Asian J Anim Vet Adv. 9:243–252.,); among these most commonly used method is the one which is developed by Houf et al. (2001Houf K, Devriese LA, De zutter L, Van Hoof J, Vandamme P. 2001. Susceptibility of *Arcobacter butzleri, Arcobacter cryaerophilus*, and *Arcobacter skirrowii* to antimicrobial agents used in selective media. J Clin Microbiol. 39:1654–1656 ,). Use of membrane filter is considered superior as compared to the use of antibiotics for isolation, since *Arcobacters* are more sensitive to growth inhibition by antibiotics used in selective media (Atabay & Corry 1998Atabay HI, Corry JEL. 1998. Evaluation of a new *Arcobacter*enrichment medium and comparison with new media developed for enrichment for *Campylobacter spp*. Int J Food Microbiol. 41:53–58. Engberg et al. 2000Engberg J, On SLW, Harrington CS, Gerner-Smidt P. 2000. Prevalence of *Campylobacter*, *Arcobacter*, *Helicobacter* and *Sutterella* spp. in human fecal samples as estimated by a reevaluation of isolation methods for campylobacters. J Clin Microbiol. 38:286–291). Despite the availability of numerous isolation techniques, there is no recommended standard method for isolation of *Arcobacters* (Shah et al. 2011Shah AH, Saleha AA, Zunita Z, Murugaiyah M. 2011. *Arcobacter*– an emerging threat to animals and animal origin food products? Tr Food Sci Technol. 22:225–236.,). Due to this limiting factor, many of the important cases may go undetected, resulting in underestimation of the prevalence and epidemiological status of *Arcobacters* (Harrab et al. 1998Harrab B, Schwarz S, Wenzel S. 1998. Identi¢cation and characterization of *Arcobacter*isolates from broilers by biochemical tests, antimicrobial resistance patterns and plasmid analysis. J Vet Med. 45:87–94., ; Calvo et al. 2013Calvo G, Arias ML, Fernández H. 2013. *Arcobacter*: a foodborne emerging pathogen. Arch Latinoam Nutr. 63:164–172), and this also complicates the result interpretation as reported in various earlier studies (Brann 2001Brann OS. 2001. Infectious complications of cirrhosis. Curr Gastroenterol Rep. 3:285–292 ; Prouzet-Mauleon et al. 2006Prouzet-Mauleon V, Labadi L, Bouges N, Menard A, Megraud F. 2006. *Arcobacter butzleri*: underestimated enteropathogen. Emerg Infect Dis. 12:307–309 ,).

Arcobacter species are fastidious in nature and are biochemically inert similar to *Campylobacter* species, which complicates their phenotypic differentiation. Hence, polymerase chain reaction (PCR)-based methods are more commonly used for specific detection and identification purposes (Brightwel et al. 2007Brightwell G, Mowat E, Clemens R, Boerema J, Pulford DJ, On SL. 2007. Development of a multiplex and real time PCR assay for the specific detection of *Arcobacter butzleri*and *Arcobacter cryaerophilus*. J Microbiol Methods. 68:318–325. Douidah et al. 2010Douidah L, Zutter LD, Vandamme P, Houf K. 2010. Identification of five human and mammal associated *Arcobacter* species by a novel multiplex-PCR assay. J Microbiol Methods. 80:281–286 ,), especially the multiplex PCR-based methods targeting genus- and species-specific sequences (Douidah et al. 2010Douidah L, Zutter LD, Vandamme P, Houf K. 2010. Identification of five human and mammal associated *Arcobacter* species by a novel multiplex-PCR assay. J Microbiol Methods. 80:281–286. Houf et al. 2000Houf K, Tutenel A, De Zutter L, Van Hoof J, Vandamme P. 2000. Development of a multiplex PCR assay for the simultaneous detection and identification of *Arcobacter butzleri, Arcobacter cryaerophilus* and *Arcobacter skirrowii*. FEMS Immunol Med Microbiol. 49:337–345. ; Ramees et al. 2014aRamees TP, Rathore RS, Bagalakote PS, Mohan HV, Kumar A, Dhama K. 2014a. Multiplex PCR detection of *Arcobacter butzleri* and *Arcobacter cryaerophilus* in skin of poultry. J Pure Appl Microbiol. 8:1755–1758.) along with other advanced diagnostics (Suarez et al. 1997Suarez DL, Wesley IV, Larson DJ. 1997. Detection of *Arcobacter*species in gastric samples from swine. Vet Microbiol. 57:325–336. Antolin et al. 2001Antolin A, González I, Garcia T, Hernandez PE, Martin R. 2001. *Arcobacter* spp. enumeration in poultry meat using a combined PCR-ELISA assay. Meat Sci. 59:169–174. Moreno et al. 2003Moreno Y, Botella S, Alonso JL, Ferrus MA, Hernandez M, Hernandez J. 2003. Specific detection of *Arcobacter* and *Campylobacter* strains in water and sewage by PCR and fluorescent *in situ* hybridization. Appl Environ Microbiol. 69:1181–1186. González et al. 2006González A, Moreno Y, González R, Herna´ndez J, Ferrus MA. 2006. Development of a simple and rapid method based on polymerase chain reaction-based restriction fragment length polymorphism analysis to differentiate *Helicobacter*, *Campylobacter* and *Arcobacter* species. Curr Microbiol. 53:416–421. Ramees et al. 2014aRamees TP, Rathore RS, Bagalakote PS, Mohan HV, Kumar A, Dhama K. 2014a. Multiplex PCR detection of *Arcobacter butzleri* and *Arcobacter cryaerophilus* in skin of poultry. J Pure Appl Microbiol. 8:1755–1758.). Perhaps, conventional PCR is evolving as a more rapid tool for detection of *Arcobacter* from food and other samples (González et al. 2010González A, Suski J, Ferrus MA. 2010. Rapid and accurate detection of *Arcobacter* contamination in commercial chicken products and waste water samples by real-time polymerase chain reaction. Foodborne Pathog Dis. 7:327–338. de Boer et al. 2013de Boer RF, Ott A, Güren P, van Zanten E, van Belkum A, Kooistra-Smid AM. 2013. Detection of *Campylobacter* species and *Arcobacter butzleri* in stool samples by use of real-time multiplex PCR. J Clin Microbiol. 51:253–259 ,). More studies are essential in the field of diagnosis of *Arcobacters* to improve the sensitivity and specificity of the diagnostic methods that are currently in practice.

In the present review, all the salient aspects on this important food-borne pathogen with main focus on etiological details, evolution, emergence and genetic diversity of different *Arcobacter* species, prevalence and epidemiology in animals, the disease in general in animals and humans, public health significance, and trends and advances in diagnosis, prevention and control have been emphasized.

ETIOLOGY

The genus *Arcobacter* was proposed by Vandamme et al. (1991Vandamme P, Falsen E, Rossau R, Hoste B, Segers P, Tytgat R, De Ley J. 1991. Revision of *Campylobacter*, *H elicobacter* and *Wolinella*taxonomy: emendation of generic descriptions and proposal of *Arcobacter* genera. Int J Syst Bacteriol. 41:88–103) and was previously known as 'aerotolerant campylobacters' due to their resemblance to the organisms of the genus *Campylobacter* (Neill et al. 1979Neill SD, Ellis WA, O'brien JJ. 1979. Designation of aerotolerant *Campylobacter* like organisms from porcine and bovine abortions to the genus *Campylobacter*. Res Vet Sci. 27:180–186 ; Vandamme et al. 1991Vandamme P, Falsen E, Rossau R, Hoste B, Segers P, Tytgat R, De Ley J. 1991. Revision of *Campylobacter*, *H elicobacter* and *Wolinella* taxonomy: emendation of generic descriptions and proposal of *Arcobacter* genera. Int J Syst Bacteriol. 41:88–103 ; Aydin et al. 2007Aydin F, Gumussoy KS, Atabay HI, Ica T, Abay S. 2007. Prevalence and distribution of *Arcobacter* species in various sources in Turkey and molecular analysis of isolated strains by ERIC-PCR. J Appl Microbiol. 103:27–35. Collado et al. 2010Collado L, Kasimir G, Perez U, Bosch A, Pinto R, Saucedo G, Huguet JM, Figueras MJ. 2010. Occurrence and diversity of *Arcobacter* spp. along the Llobregat river catchment, at sewage effluents and in a drinking water treatment plant. Water Res. 44:3696–3702. Shah et al. 2012aShah AH, Saleha AA, Murugaiyah M, Zunita Z, Memon AA. 2012a. Prevalence and distribution of *Arcobacter* spp. in raw milk and retail raw beef. J Food Prot.

75:1474–1478. Levican & Figueras 2013Levican A, Figueras MJ. 2013. Performance of five molecular methods for monitoring *Arcobacter*spp. BMC Microbiol. 13:220 ,). Later, it has been revealed that these organisms can be distinguished from *Campylobacter* by their ability to grow between 15 °C and 30 °C temperature aerobically and need of microaerophilic condition for primary isolation (Collado & Figueras 2011Collado L, Figueras MJ. 2011. Taxonomy, epidemiology and clinical relevance of the genus *Arcobacter*. Clin Microbiol Rev. 24:174–192. Ferreira et al. 2016Ferreira S, Queiroz JA, Oleastro M, Domingues FC. 2016. Insights in the pathogenesis and resistance of *Arcobacter*: a review. Crit Rev Microbiol. 42:364–383). Furthermore, based on extensive immunotyping, SDS-PAGE analysis of cellular proteins, fatty acid analysis, DNA–rRNA and DNA hybridization studies, a new genus within the family of *Campylobacteriaceae* was named as *Arcobacter* by Vandamme and De Ley (1991Vandamme P, De Ley J. 1991. Proposal for a new family *Campylobacteraceae*. Int J Syst Bacteriol. 41:451–455.,). Vandamme et al. (1992bVandamme P, Vanvanneyt M, Pot B, Mels L, Hoste B, Dewettinck D, Vlaes L, Van Den Borre C, Higgins R, Hommez J, et al. 1992b. Polyphasic taxonomic study of the emended genus *Arcobacter* with *Arcobacter butzleri* comb. Nov. and *Arcobacter skirrowii* sp. Nov., and aerotolerant bacterium isolated from veterinary specimens. Int J Syst Bacteriol. 42:344–356) formally transferred *Campylobacter butzleri* to the genus *Arcobacter* as *Arcobacter butzleri* and also identified five major groups using DNA–DNA hybridization data, these included *A. cryaerophilus* (two distinct electrophoretic subgroups), *A. butzleri*, *A. nitrophilus* and a new species *A. skirrowii*. The species *A. butzleri* was named in the honor of Jean-Paul Butzler, a Belgian clinician and microbiologist. The organism has been identified from a wide variety of habitats and hosts which makes the genus atypical within the epsilon subdivision of the *Proteobacteria*(Debruyne et al. 2008Debruyne L, Gevers D, Vandamme P. 2008. Taxonomy of the family *Campylobacteraceae*. In: Nachamkin I, Szymanski C, Blaser M, editors. *Campylobacter*. Vol. 3. Washington (DC): ASM press; p. 3–25.,).

Arcobacter in Latin means 'arc-shaped bacterium', and is included in *Campylobacteraceae*family consisting of Gram-negative, helical, non-spore forming rods. Together with the genus *Helicobacter*, *Arcobacters* form a phylogenetically distinct group referred to either as rRNA superfamily VI or as the epsilon division of the *Proteobacteria* (Vandamme et al. 1991Vandamme P, Falsen E, Rossau R, Hoste B, Segers P, Tytgat R, De Ley J. 1991. Revision of *Campylobacter*, *H elicobacter* and *Wolinella*taxonomy: emendation of generic descriptions and proposal of *Arcobacter* genera.

Int J Syst Bacteriol. 41:88–103 ; Lehner et al. 2005Lehner A, Tasara T, Stephan R. 2005. Relevant aspects of *Arcobacter* spp. as potential food-borne pathogens. Int J Food Microbiol. 102:127–135. Collado & Figueras 2011Collado L, Figueras MJ. 2011. Taxonomy, epidemiology and clinical relevance of the genus *Arcobacter*. Clin Microbiol Rev. 24:174–192 ,). These bacteria are usually helical, curved or 'S'-shaped when viewed by conventional light microscopy. *Arcobacters*are generally 0.2–0.9 μm wide and 1–3 μm long (Wesley 1994Wesley IV. 1994. *Arcobacter* infections. In: Beran GW, Steele JH, editors. In handbook of zoonoses. CRC Press; p. 181–190.). These organisms are motile by means of single unsheathed polar flagellum and exhibit a darting corkscrew movement which enables them to pass through membrane filters that retain almost all other bacteria, a feature that is used for their isolation. Biochemically, the majority of *Arcobacter* species show positive reaction to oxidase, catalase, indoxyl acetate hydrolysis test and nitrate reduction test whereas hippurate hydrolysis test, urease test, and H_2S production from triple sugar iron agar (TSI) are negative. The G+C content of their DNA is 27.2–28.2 mol% (Mesbah et al. 1989Mesbah M, Premachandran U, Whitman WB. 1989. Precise measurement of the G+C content of deoxyribonucleicacid by high-performance liquid chromatography. Int J Syst Bacteriol. 39:159–167., ; Vandamme & De Ley 1991Vandamme P, De Ley J. 1991. Proposal for a new family *Campylobacteraceae*. Int J Syst Bacteriol. 41:451–455.,).

GROWTH PARAMETERS

Arcobacter spp. differ physiologically from the *Campylobacter* spp. as they can tolerate oxygen and can grow at atmospheric temperature below 30 °C (Kjeldgaard et al. 2009Kjeldgaard J, Jorgensen K, Ingmer H. 2009. Growth and survival at chiller temperature of *Arcobacter butzleri*. Int J Food Microbiol. 131:256–259 ,). *A. butzleri* strain can grow within a range from 15 °C to 37 °C and with optimal growth at 30 °C (Hilton et al. 2001Hilton CL, Mackey BM, Hargreaves AJ, Forsythe SJ. 2001. The recovery of *Arcobacteri* NCTC 12481 from various temperature treatments. J Appl Microbiol. 91:929–932 ,). For primary isolation, microaerobic condition is necessary (Vandamme et al. 1991Vandamme P, Falsen E, Rossau R, Hoste B, Segers P, Tytgat R, De Ley J. 1991. Revision of *Campylobacter*, *H elicobacter* and *Wolinella*taxonomy: emendation of generic descriptions and proposal of *Arcobacter* genera. Int J Syst Bacteriol. 41:88–103) and no detectable growth is usually seen at 40 °C (Vandamme et al.

1992bVandamme P, Vanvanneyt M, Pot B, Mels L, Hoste B, Dewettinck D, Vlaes L, Van Den Borre C, Higgins R, Hommez J, et al. 1992b. Polyphasic taxonomic study of the emended genus *Arcobacter* with *Arcobacter butzleri* comb. Nov. and *Arcobacter skirrowii* sp. Nov., and aerotolerant bacterium isolated from veterinary specimens. Int J Syst Bacteriol. 42:344–356). It can withstand freezing for up to 6 months at −20 °C and for up to 24 months at −70 °C, but at 55 °C and above the organism gets rapidly inactivated (D'Sa & Harrison 2005D'Sa EM, Harrison MA. 2005. Effect of pH, NaCl content, and temperature on growth and survival of *Arcobacter* spp. J Food Protect. 68:18–25 ,). *Arcobacter* strains can tolerate a wide range of pH from 5.5 to 9.5 but optimal growth occurs between pH 6.8–8.0 (Neill et al. 1979Neill SD, Ellis WA, O'brien JJ. 1979. Designation of aerotolerant *Campylobacter* like organisms from porcine and bovine abortions to the genus *Campylobacter*. Res Vet Sci. 27:180–186). *Arcobacters* have a respiratory type of metabolism and use organic and amino acids as carbon sources. Carbohydrates are neither fermented nor oxidized (Ellis et al. 1977Ellis WA, Neill SD, O' Brien JJ, Ferguson, HW, Hanna J. 1977. Isolation of *Spirillum/Vibrio*-like organisms from bovine fetuses. Vet Record. 100:451–452 ,). Hitherto studies reported that *A. butzlerii* was able to grow at 10 °C, the minimum detectable growth temperature for this microorganism (D'Sa & Harrison 2005D'Sa EM, Harrison MA. 2005. Effect of pH, NaCl content, and temperature on growth and survival of *Arcobacter* spp. J Food Protect. 68:18–25. Kjeldgaard et al. 2009Kjeldgaard J, Jorgensen K, Ingmer H. 2009. Growth and survival at chiller temperature of *Arcobacter butzleri*. Int J Food Microbiol. 131:256–259 ,) and there was no growth at 40 °C (Hilton et al. 2001Hilton CL, Mackey BM, Hargreaves AJ, Forsythe SJ. 2001. The recovery of *Arcobacteri* NCTC 12481 from various temperature treatments. J Appl Microbiol. 91:929–932 ,). Confirmation regarding its tolerance to pH, temperature treatments and salt concentrations has been documented recently (Cervenka 2007Cervenka L. 2007. Survival and inactivation of *Arcobacter* spp., a current status and future prospect. Crit Rev Microbiol. 33:101–108 ; Ferreira et al. 2016Ferreira S, Queiroz JA, Oleastro M, Domingues FC. 2016. Insights in the pathogenesis and resistance of *Arcobacter*: a review. Crit Rev Microbiol. 42:364–383 ; Salas-Massó et al. 2016Salas-Massó N, Andree KB, Furones MD, Figueras MJ. 2016. Enhanced recovery of Arcobacter spp. using NaCl in culture media and re-assessment of the traits of *Arcobacter marinus* and *Arcobacter halophilus* isolated from marine water and shellfish. Sci Total Environ. 566–567:1355–1361 ,).

EVOLUTION AND EMERGENCE OF *ARCOBACTERS*

To date, 25 species of *Arcobacters* have been reported with a significant genetic diversity among and within the species (Collado & Figueras 2011Collado L, Figueras MJ. 2011. Taxonomy, epidemiology and clinical relevance of the genus *Arcobacter*. Clin Microbiol Rev. 24:174–192. Figueras et al. 2012Figueras MJ, Levican A, Collado L. 2012. Updated 16S rRNA-RFLP method for the identification of all currently characterised *Arcobacter*spp. BMC Microbiol. 12:292. Levican & Figueras 2013Levican A, Figueras MJ. 2013. Performance of five molecular methods for monitoring *Arcobacter*spp. BMC Microbiol. 13:220. Levican et al. 2013aLevican A, Collado L, Figueras MJ. 2013a. *Arcobacter cloacae* sp. nov. and *Arcobacter suis* sp. nov., two new species isolated from food and sewage. Syst Appl Microbiol. 36:22–27. Sasi Jyothsna et al. 2013Sasi Jyothsna TS, Rahul K, Ramaprasad EV, Sasikala Ch, Ramana ChV. 2013. Arcobacter anaerophilus sp. nov., isolated from an estuarine sediment and emended description of the genus *Arcobacter*. Int J Syst Evol Microbiol. 63:4619–4625. Figueras et al. 2014Figueras MJ, Levican A, Pujol I, Ballester F, Rabada, Quilez MJ, Gomez-Bertomeu F. 2014. A severe case of persistent diarrhoea associated with *Arcobacter cryaerophilus* but attributed to *Campylobacter* sp. and a review of the clinical incidence of *Arcobacter*spp. New Microbes New Infect. 2:31–37 ; Bagalakote et al. 2014Bagalakote PS, Rathore RS, Ramees TP, Mohan HV, Sumankumar M, Agarwal RK, Kumar A, Dhama K. 2014. Molecular characterization of *Arcobacter* isolates using randomly amplified polymorphic DNA- polymerase chain reaction (RAPD-PCR). Asian J Anim Vet Adv. 9:543–555., ; Levican et al. 2015Levican A, Rubio-Arcos S, Martinez-Murcia A, Collado L, José Figueras M. 2015. *Arcobacter ebronensis* sp. nov. and *Arcobacter aquimarinus*sp. nov., two new species isolated from marine environment. Syst Appl Microbiol. 38:30–35. Whiteduck-Le veillée et al. 2015Whiteduck-Léveillée K, Whiteduck-Léveillée J, Cloutier M, Tambong JT, Xu R, Topp E, Arts MT, Chao J, Adam Z, Lévesque CA, et al. 2015. *Arcobacter lanthieri* sp. nov., isolated from pig and dairy cattle manure. Int J Syst Evol Microbiol. 65:2709–2716. Park et al. 2016Park S, Jung YT, Kim S, Yoon JH. 2016. Arcobacter acticola sp. nov., isolated from seawater on the East Sea in South Korea. J Microbiol. 54:655–659. Zhang et al. 2016Zhang Z, Yu C, Wang X, Yu S, Zhang XH. 2016. *Arcobacter pacificus* sp. nov., isolated from seawater of the South Pacific Gyre. Int J Syst Evol Microbiol. 66:542–547). List of *Arcobacter* spp. and their origin has been presented in Table 1. Laishram et al. (2016Laishram M, Rathlavath S, Lekshmi M, Kumar S, Nayak BB. 2016. Isolation and characterization of Arcobacter spp. from fresh seafood and the aquatic environment. Int J Food Microbiol. 232:87–89 ,) for the first time reported the isolation of *Arcobacter mytili* from India. They also reported that *A. butzleri* isolated from seafood harbored putative virulence marker genes which are found in human pathogenic strains (Laishram et al. 2016Laishram M, Rathlavath S, Lekshmi M, Kumar S, Nayak BB. 2016. Isolation and characterization of Arcobacter spp. from fresh seafood and the aquatic environment. Int J Food Microbiol. 232:87–89 ,). Whiteduck-Léveillée et al. (2016bWhiteduck-Léveillée K, Whiteduck-Léveillée J, Cloutier M, Tambong JT, Xu R, Topp E, Arts MT, Chao J, Adam Z, Lévesque CA, et al. 2016b. Identification, characterization and description of Arcobacter faecis sp. nov., isolated from a human waste septic tank. Syst Appl Microbiol. 39:93–99 ,) identified and characterized the *Arcobacter faecis* (strain AF1078^{T}), from a human waste septic tank near Ottawa, Ontario, Canada.

Table 1. List of *Arcobacter* spp. and their origin.

S. no.	***Arcobacter* species**	**First time isolated from**	**References**
(1)	*A. nitrofigilis*	Roots of *Spartina alterniflora*	McClung et al. (1983McClung CR, Patriquin DG, Davis RE. 1983. Campylobacter nitrofigilis sp. nov., a nitrogen-fixing bacterium associated with roots of Spartina alterniflora. Int J Syst Bacteriol. 33:605–612.
(2)	*A. cryaerophilus*	Bovine abortus fetus	Neill et al. (1985Neill SD, Cambell JN, O'brien JJ, Weatherup STC, Ellis WA. 1985. Taxonomic position of *Campylobacter cryaerophila* sp. Nov. Int J Systemat Bacteriol. 35:342–356.
(3)	*A. butzleri*	Human feces	Kiehlbauch et al. (1991Kiehlbauch JA, Brenner DJ, Nicholson MA, Baker CN, Patton CM, Steigerwalt AG, Wachsmuth IK. 1991. *Campylobacter butzleri* sp. nov. isolated from humans and animals with diarrheal illness. J Clin Microbiol. 29:376–385.
(4)	*A. skirrowii*	Sheep feces	Vandamme et al. (1992bVandamme P, Vanvanneyt M, Pot B, Mels L, Hoste B, Dewettinck D, Vlaes L, Van Den Borre C, Higgins R, Hommez J, et al. 1992b. Polyphasic taxonomic study of the emended genus *Arcobacter* with *Arcobacter butzleri* comb. Nov. and *Arcobacter skirrowii* sp. Nov., and aerotolerant bacterium isolated from veterinary specimens. Int J Syst Bacteriol. 42:344–356.)
(5)	*Candidatus Arcobacter sulfidicus*	Coastal seawater	Wirsen et al. (2002Wirsen CO, Sievert SM, Cavanaugh CM, Molyneaux SJ, Ahmad A, Taylor LT, DeLong EF, Taylor CD. 2002. Characterization of an autotrophic sulfide-oxidizing marine Arcobacter sp. that produces filamentous sulfur. Appl Environ Microbiol. 68:316–325.)

(6)	*A. cibarius*	Chicken meat	Houf et al. (2003Houf K, Devriese LA, De Zutter L, Verbeke B, Van Hoof J, Vandamme P. 2003. Molecular characterization of *Arcobacter*isolates collected in a poultry slaughterhouse. J Food Prot. 66:364–369.)
(7)	*A. halophilus*	Hypersaline lagoon	Donachie et al. (2005Donachie SP, Bowman JP, On SLW, Alam M. 2005. *Arcobacter halophilus* sp. nov., the first obligate halophile in the genus Arcobacter. Int J Syst Evol Microbiol. 55:1271–1277.)
(8)	*A. mytili* sp. nov.	Mussels	Collado et al. (2009bCollado L, Cleenwerck I, Van Trappen S, De Vos P, Figueras MJ. 2009b. *Arcobacter mytili* sp. nov., an indoxyl acetate-hydrolysis-negative bacterium isolated from mussels. Int J Syst Evol Microbiol. 59:1391–1396.)
(9)	*A. thereius* sp. nov.	Porcine abortion	Houf et al. (2009Houf K, On SL, Coenye T, Debruyne L, De Smet S, Vandamme P. 2009. *Arcobacter thereius* sp. nov., isolated from pigs and ducks. Int J Syst Evol Microbiol. 59:2599–2604.)
(10)	*A. marinus* sp. nov.	Seawater, seaweed and a starfish	Kim et al. (2010Kim HM, Hwang CY, Cho BC. 2010. *Arcobacter marinus* sp. nov. Int J Syst Evol Microbiol. 60:531–536.)
(11)	*A. trophiarium*	Pig feces	De Smet et al. (2011De Smet S, De Zutter L, Houf K. 2011. Small ruminants as carriers of the emerging foodborne pathogen *Arcobacte*r on small and medium farms. Small Rumin Res. 97:124–129.
(12)	*A. defluvii* sp. Nov	Sewage samples	Collado et al. (2011Collado L, Levican A, Perez J, Figueras MJ. 2011. *Arcobacter defluvii* sp. nov., isolated from sewage samples. Int J Syst Evol Microbiol. 61:2155–2161.)
(13)	*A.molluscorum* sp. nov.	Shellfish	Figueras et al. (2011aFigueras MJ, Collado L, Levican A, Perez J, Solsona MJ, Yustes C. 2011a. *Arcobacter molluscorum*sp. nov., a new species isolated from shellfish. Syst Appl Microbiol. 34:105–109., 2011bFigueras MJ, Levican A, Collado L, Inza MI, Yustes C. 2011b. *Arcobacter ellisii* sp. nov., isolated from mussels. Syst Appl Microbiol. 34:414–418.)
(14)	*A. ellisii* sp. nov.	Shellfish	Figueras et al. (2011aFigueras MJ, Collado L, Levican A, Perez J, Solsona MJ, Yustes C. 2011a. *Arcobacter molluscorum*sp. nov., a new species isolated from shellfish. Syst Appl Microbiol. 34:105–109., 2011bFigueras MJ, Levican A, Collado L, Inza MI, Yustes C. 2011b. *Arcobacter ellisii* sp. nov., isolated from mussels. Syst Appl Microbiol. 34:414–418.)
(15)	*A. venerupis*sp. nov.	Shellfish	Levican et al. (2012Levican A, Collado L, Aguilar C, Yustes C, Dieguez AL, Romalde JL, Figueras MJ. 2012. *Arcobacter bivalviorum* sp. nov. and *Arcobacter venerupis* sp. nov., new species isolated from shellfish. Syst Appl Microbiol. 35:133–138.)

(16)	*A. bivalviorum* sp. nov.	Shellfish	Levican et al. (2012Levican A, Collado L, Aguilar C, Yustes C, Dieguez AL, Romalde JL, Figueras MJ. 2012. *Arcobacter bivalviorum* sp. nov. and *Arcobacter venerupis* sp. nov., new species isolated from shellfish. Syst Appl Microbiol. 35:133–138.)
(17)	*A. cloacae* sp. nov.	Sewage	Levican et al. (2013aLevican A, Collado L, Figueras MJ. 2013a. *Arcobacter cloacae*sp. nov. and *Arcobacter suis* sp. nov., two new species isolated from food and sewage. Syst Appl Microbiol. 36:22–27.)
(18)	*A. suis* sp. nov.	Pork meat	Levican et al. (2013aLevican A, Collado L, Figueras MJ. 2013a. *Arcobacter cloacae*sp. nov. and *Arcobacter suis* sp. nov., two new species isolated from food and sewage. Syst Appl Microbiol. 36:22–27.)
(19)	*A. anaerophilus*sp. nov.	Estuarine sediment	Sasi Jyothsna et al. (2013Sasi Jyothsna TS, Rahul K, Ramaprasad EV, Sasikala Ch, Ramana ChV. 2013. Arcobacter anaerophilus sp. nov., isolated from an estuarine sediment and emended description of the genus *Arcobacter*. Int J Syst Evol Microbiol. 63:4619–4625.)
(20)	*A. ebronensis* sp. nov.	Mussels	Levican et al. (2015Levican A, Rubio-Arcos S, Martinez-Murcia A, Collado L, José Figueras M. 2015. *Arcobacter ebronensis* sp. nov. and *Arcobacter aquimarinus* sp. nov., two new species isolated from marine environment. Syst Appl Microbiol. 38:30–35.)
(21)	*A. aquimarinus* sp. nov.	Sea water	Levican et al. (2015Levican A, Rubio-Arcos S, Martinez-Murcia A, Collado L, José Figueras M. 2015. *Arcobacter ebronensis* sp. nov. and *Arcobacter aquimarinus* sp. nov., two new species isolated from marine environment. Syst Appl Microbiol. 38:30–35.)
(22)	*A. lanthieri* sp. nov.	Cattle and pig manure	Whiteduck-Léveillée et al. (2015Whiteduck-Léveillée K, Whiteduck-Léveillée J, Cloutier M, Tambong JT, Xu R, Topp E, Arts MT, Chao J, Adam Z, Lévesque CA, et al. 2015. *Arcobacter lanthieri* sp. nov., isolated from pig and dairy cattle manure. Int J Syst Evol Microbiol. 65:2709–2716.)
(23)	*A. pacificus*	Seawater	Zhang et al. (2016Zhang Z, Yu C, Wang X, Yu S, Zhang XH. 2016. *Arcobacter pacificus* sp. nov., isolated from seawater of the South Pacific Gyre. Int J Syst Evol Microbiol. 66:542–547.)
(24)	*A. acticola* sp. nov.	Seawater	Park et al. (2016Park S, Jung YT, Kim S, Yoon JH. 2016. Arcobacter acticola sp. nov., isolated from seawater on the East Sea in South Korea. J Microbiol. 54:655–659.)
(25)	*Arcobacter lekithochrous* sp. nov.	Great scallop (*Pecten maximus*) larvae and tank seawater	Diéguez et al. (2017Diéguez AL, Balboa S, Magnesen T, Romalde JL. 2017. Arcobacter lekithochrous sp. nov., a new species isolated from a molluscan hatchery in Norway. Int J Syst Evol Microbiol. doi:10.1099/ijsem.0.001809.)

Among these species, *A. butzleri, A. cryaerophilus, A. skirrowii* and *A. cibarius* are considered as emerging food-borne pathogens (Ho et al. 2006aHo HT, Lipman LJ, Gaastra W. 2006a. *Arcobacter*, what is known and unknown about a potential food-borne zoonotic agent.Vet Microbiol.

115:1–13. Prouzet-Mauleon et al. 2006Prouzet-Mauleon V, Labadi L, Bouges N, Menard A, Megraud F. 2006. *Arcobacter butzleri*: underestimated enteropathogen. Emerg Infect Dis. 12:307–309. Snelling et al. 2006Snelling WJ, Matsuda M, Moore JE, Dooley JS. 2006. Under the microscope: *Arcobacter*. Lett Appl Microbiol. 42:7–14. Houf & Stephan 2007Houf K, Stephan R. 2007. Isolation and characterization of the emerging food-borne pathogen *Arcobacter* from human stool. J Microbiol Methods. 68:408–413. Collado et al. 2009aCollado L, Guarro J, Figueras M. 2009a. Prevalence of *Arcobacter* in meat and shellfish. J Food Protect. 72:1102–1110. Amare et al. 2011Amare LB, Sahela AA, Zunita Z, Jalila A, Hassan L. 2011. Prevalence of *Arcobacter* spp. on chicken meat at retail markets and in farm chickens in Selengor, Malaysia. Food Control. 22:732–736., ; McGrego & Wright 2015McGregor AC, Wright SG. 2015. Gastrointestinal symptoms in travellers. Clin Med. 15:93–95 ,). During the recent years, *Arcobacter* spp. have gained attention of worldwide researchers as an important emerging food-borne enteropathogen. Among all *Arcobacters*, *A. butzleri* is the most common species and has been associated with human disease, such as enteritis, severe diarrhea, bacteremia and septicemia (Engberg et al. 2000Engberg J, On SLW, Harrington CS, Gerner-Smidt P. 2000. Prevalence of *Campylobacter*, *Arcobacter*, *Helicobacter* and *Sutterella* spp. in human fecal samples as estimated by a reevaluation of isolation methods for campylobacters. J Clin Microbiol. 38:286–291 ; Lau et al. 2002Lau SKP, Woo PCY, Teng J, Leung KW, Yuen KY. 2002. Identification by 16S ribosomal RNA gene sequencing of *Arcobacter*bacteraemia in a patient with acute gangrenous appendicitis. Mol Pathol. 55:182–185 ; Collado & Figueras 2011Collado L, Figueras MJ. 2011. Taxonomy, epidemiology and clinical relevance of the genus *Arcobacter*. Clin Microbiol Rev. 24:174–192. Collado et al. 2014Collado L, Jara R, Vásquez N, Telsaint C. 2014. Antimicrobial resistance and virulence genes of *Arcobacter* isolates recovered from edible bivalve mollusks. Food Control. 46:508–512., ; Figueras et al. 2014Figueras MJ, Levican A, Pujol I, Ballester F, Rabada, Quilez MJ, Gomez-Bertomeu F. 2014. A severe case of persistent diarrhoea associated with *Arcobacter cryaerophilus* but attributed to *Campylobacter* sp. and a review of the clinical incidence of *Arcobacter*spp. New Microbes New Infect. 2:31–37 ; Van den Abeele et al. 2014Van den Abeele AM, Vogelaers D, Van Hende J, Houf K. 2014. Prevalence of *Arcobacter* species among humans, Belgium, 2008–2013. Emerg Infect Dis. 20:1731–1734. Webb et al. 2016Webb AL, Boras VF, Kruczkiewicz P, Selinger LB, Taboada EN, Inglis GD. 2016. Comparative detection and quantification of *Arcobacter butzleri* in stools from diarrheic and nondiarrheic people in Southwestern Alberta, Canada. J Clin Microbiol. 54:1082–1088 ,). Knowledge gap in the field of emergence and evolution of *Arcobacters* is very wide as several themes and questions remained unsolved. The genus *Arcobacter* is evolving rapidly from the day it has been defined. Recently, a symbiotic relationship of *Arcobacter*with *Acanthamoeba castellanii* has been identified showing its potential to survive and escape phagocytosis by amoeba (Medina et al. 2014Medina G, Flores-Martin S, Fonseca B, Otth C, Fernandez H. 2014. Mechanisms associated with phagocytosis of *Arcobacter butzleri*by *Acanthamoeba castellanii*. Parasitol Res. 113:1933–1942 ,). Evolution of this emerging food-borne pathogen needs to be elucidated in a better way so as to find out the epidemiology and virulence factors of this pathogen. Plasmids are usually present in many prokaryotes and they play a major role in evolution of these species (Ricci & Hernandez 2000Ricci JCD, Hernandez ME. 2000. Plasmid effects on *Escherichia coli*metabolism. Crit Rev Biotechnol. 20:79–108). Recently, several small and a large plasmid were isolated from *Arcobacter butzleri*. Complete sequence of the large plasmid revealed that putative protein peg21, peg22, peg24, peg25 Peg26 and Peg 29 were showing homology with type IV secretion which plays a role in transfer of DNA material and toxins. This large plasmid can play a role in transfer of genes within *Arcobacter* spp. leading to evolution of new species (Douidah et al. 2014aDouidah L, De Zutter L, Van Nieuwerburgh F, Deforce D, Ingmer H, Vandenberg O, Van den Abeele AM, Houf K. 2014a. Presence and analysis of plasmids in human and animal associated *Arcobacter*species. PloS One. 9:e85487 ,).

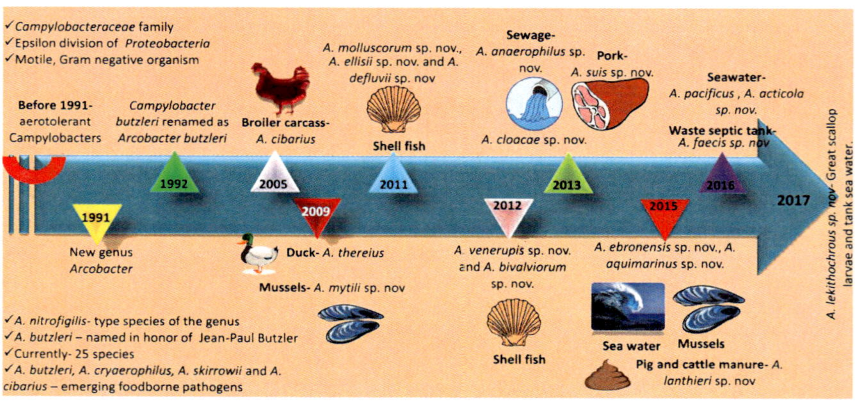

Figure 1. Evolution and emergence of *Arcobacter*. New genus was established

in the year 1991. New additions were incorporated over the timeline and currently 25 species are recognized. Some of the members of *Arcobacter* along with the source and year of isolation are depicted in the figure.

An evolutionary overview of emergence of *Arcobacter* species with new species additions over the years is presented in Figure 1.

VIRULENCE FACTORS

There is scare research information regarding the virulence genes/factors in *Arcobacters*and from available studies it is understood that adhesion, invasion of the pathogen, secretion of toxin and pro-inflammatory cytokine (IL-8) plays a major role in establishing infection in the host (Collado & Figueras 2011Collado L, Figueras MJ. 2011. Taxonomy, epidemiology and clinical relevance of the genus *Arcobacter*. Clin Microbiol Rev. 24:174–192 ,). *A. cryaerophilus* and *A. butzleri* are the two main species of *Arcobacter* studied so far for its adhesive and invasive properties. Studies with rat ileal loop assay showed that former species could cause accumulation of electrolytes and fluids (Ferna´ndez et al. 1995Ferna´ndez H, Eller G, Paillacar J, Gajardo T, Riquelme A. 1995. Toxigenic and invasive capacities: possible pathogenic mechanisms in *Arcobacter cryaerophilus*. Mem Inst Oswaldo. 90:633–634 ,). However, the latter one causes intestinal colonization in piglets (Ho et al. 2006bHo HT, Lipman LJA, Van der Graaf-van Bloois L, Van Bergen M, Gaastra W. 2006b. Potential routes of acquisition of *Arcobacter* species by piglets. Vet Microbiol. 114:122–133.,). Some studies also reported that Beltsville White turkey was the better animal model for studying the diarrheal infection caused by *Arcobacter* (Wesley & Baetz 1999Wesley IV, Baetz AL. 1999. Natural and experimental infections of *Arcobacter* in poultry. Poult Sci. 78:536–545 ,). Hemagglutinin property has also been reported to be associated with *A. butzleri* which interacted with the RBCs, hence helps in adhesion of the bacterium (Tsang et al. 1996Tsang RS, Luk JM, Woodward DL, Johnson WM. 1996. Immunochemical characterization of a haemagglutinating antigen of *Arcobacter* spp. FEMS Microbiol Lett. 136:209–213 ,). Impairment of epithelial function was noticed in human HT-29/B6 colonic epithelial monolayers due to *A. butzleri* infection. Claudin protein family is most important to the barrier function and claudin-1, 5 and 8 have barrier sealing properties (Bucker et al. 2009Bucker R, Troeger H, Kleer J, Fromm M, Schulzke JD. 2009. *Arcobacter butzleri* induces barrier dysfunction in intestinal HT-29/B6 cells. J Infect Dis. 200:756–764 ,). Tight junction proteins namely claudin-1, -5 and -8 have not been found to be expressed properly during *A. butzleri*infection in human colonic epithelial cells, which leads to epithelial barrier dysfunction (Bucker et al. 2009Bucker R, Troeger H, Kleer J, Fromm M, Schulzke JD. 2009. *Arcobacter butzleri* induces barrier dysfunction in intestinal HT-29/B6 cells. J Infect Dis. 200:756–764 ,). Similar effect of claudin-1 dysfunction was noticed in enteropathogenic *E. coli*-infected human intestinal T84 monolayers (Bucker et al. 2009Bucker R, Troeger H, Kleer J, Fromm M, Schulzke JD. 2009. *Arcobacter butzleri* induces barrier dysfunction in intestinal HT-29/B6 cells. J Infect Dis. 200:756–764 ,). Like *Campylobacter* and *Helicobacter,* the release of IL-8 in *Arcobacter* infection may play a role in causing diarrhea (Ho et al. 2007Ho HT, Lipman LJ, Hendriks HG, Tooten PC, Ultee T, Gaastra W. 2007. Interaction of *Arcobacter* spp. with human and porcine intestinal epithelial cells. FEMS Immunol Med Microbiol. 50:51–58 ; Ferreira et al. 2014bFerreira S, Queiroz JA, Oleastro M, Domingues FC. 2014b. Genotypic and phenotypic features of *Arcobacter butzleri* pathogenicity. Microb Pathog. 76:19–25 ,). Release of pro-inflammatory cytokines is a major virulence factor for *H. pylori* and *Campylobacter,*similarly *Arcobacter* leads to release of pro-inflammatory cytokines (Collado & Figueras 2011Collado L, Figueras MJ. 2011. Taxonomy, epidemiology and clinical relevance of the genus *Arcobacter*. Clin Microbiol Rev. 24:174–192 ,). Zur Bruegge et al. (2014Zur Bruegge J, Hanisch C, Einspanier R, Alter T, Gölz G, Sharbati S. 2014. *Arcobacter butzleri* induces a pro-inflammatory response in THP-1 derived macrophages and has limited ability for intracellular survival. Int J Med Microbiol. 304:1209–1217 ,) reported that *Arcobacter* has less potential for intracellular survival and also documented strain specific differences in invasion and survival capacities of the organism inside the host. *Arcobacter* isolates when cultured in Vero cells produced cell elongation, which showed their ability to produce enterotoxin and some isolates along with this ability produced vacuoles in cells showing that these produce vacuolizing toxin (Villarruel-López et al. 2003Villarruel-López A, Márquez-González M, Garay-Martínez LE, Zepeda H, Castillo A, Mota de la Garza L, Murano EA, Torres-Vitela R. 2003. Isolation of *Arcobacter* spp. from retail meats and cytotoxic effects of isolates against vero cells. J Food Prot. 66:1374–1378 ,). Putative role played by lipopolysaccharide (LPS) and the composition of LPS have not been elucidated properly in *Arcobacter* infection. Lipooligosaccharide (LOS) has been characterized in a marine *Arcobacter* (*A. halophilus*) and LPS has been reported to play a role in the survival of this organism in marine environment (Silipo et al. 2010Silipo A, Sturiale L, Perino V, Garozzo D, Lanzetta R, Parrilli M, Molinaro A. 2010.

The structure of the carbohydrate backbone of the lipooligosaccharide from the halophilic bacterium *Arcobacter halophilus*. Carbohydr Res. 345:850–853 ,). As *Arcobacter* is motile like *Campylobacter*, their flagella may also have a role in invasion of cells as reported in other bacteria, wherein flagella helps in motility and also in colonization of the pathogen (Ho et al. 2008bHo HT, Lipman LJ, Wosten MM, van Asten AJ, Gaastra W. 2008b. *Arcobacter* spp. possess two very short flagellins of which FlaA is essential for motility. FEMS Immunol Med Microbiol. 53:85–95). Studies regarding the role played by the host immune mechanisms need to be focused for exploring virulence mechanisms of *Arcobacters* as host can also participate during infection and some of the host immune mechanisms can even further aggravate diarrhea (Collado & Figueras 2011Collado L, Figueras MJ. 2011. Taxonomy, epidemiology and clinical relevance of the genus *Arcobacter*. Clin Microbiol Rev. 24:174–192 ,).

Ten putative genes namely *cadF, ciaB, cj1349, hecA, iroE, hecB, irgA, mviN, pldA* and *tlyA*have been recognized in *A. butzleri* ATCC 49616 genome (Miller et al. 2007Miller WG, Parker CT, Rubenfield M, Mendz GL, Wösten MM, Ussery DW, Stolz JF, Binnewies TT, Hallin PF, Wang G, et al. 2007. The complete genome sequence and analysis of the epsilonproteobacterium *Arcobacter butzleri*. PLoS One. 2:1358. Douidah et al. 2012Douidah L, de Zutter L, Baré J, De Vos P, Vandamme P, Vandenberg O, Van den Abeele AM, Houf K. 2012. Occurrence of putative virulence genes in *Arcobacter* species isolated from humans and animals. J Clin Microbiol. 50:735–741. Ferreira et al. 2014bFerreira S, Queiroz JA, Oleastro M, Domingues FC. 2014b. Genotypic and phenotypic features of *Arcobacter butzleri* pathogenicity. Microb Pathog. 76:19–25 ,). Functions of the different genes include: *cadF* gene and *cj1349* gene code for outer membrane proteins facilitating intestinal epithelial cell to cell contact by adhering to fibronectin; ciaB gene is involved in invasion of host cell, pldA gene codes for outer membrane phospholipase A which hydrolyzes acyl ester bonds, tlyA gene is a hemolysin gene, irgA gene codes for an outer membrane receptor for enterobactin, *hecA* gene is a member of filamentous hemagglutinin family *and hecB* gene codes for hemolysin activation protein (Douidah et al. 2012Douidah L, de Zutter L, Baré J, De Vos P, Vandamme P, Vandenberg O, Van den Abeele AM, Houf K. 2012. Occurrence of putative virulence genes in *Arcobacter* species isolated from humans and animals. J Clin Microbiol. 50:735–741 ,). A study conducted by Tabatabaei et al. (2014Tabatabaei M, Aski HS, Shayegh H, Khoshbakht R. 2014. Occurrence of six virulence-associated genes in *Arcobacter* species isolated from various sources in

Shiraz, Southern Iran. Microb Pathog. 66:1–4 ,) in Southern Iran revealed six virulence genes (*cadF, ciaB, cj1349, mviN, pldA and tlyA*) in *Arcobacter* spp. These six genes were present in all the *A. butzleri* isolates. The *cadF, ciaB, cj1349, mviNin, pldA* and *tlyA* genes were present in 55% and 53%, 98% and 87%, 45% and 60%, 90% and 80%, 33% and 13%, and 38% and 40% of *A. cryaerophilus* and *A. skirrowii*, respectively. Identification of virulence genes revealed the pathogenic nature of the isolated species of *Arcobacter*. It was well proved that various virulence genes were responsible for adhesion namely *CadF*, *HecA*and *Cj1349*, invasion *CiaB*, lysis of erythrocytes *HecB*, *TlyA* and *PldA*, iron acquisition and maintaining of infection *IrgA* and *IroE*, and peptidoglycan biosynthesis *MviN* in other bacteria (Ruiz 2008Ruiz N. 2008. Bioinformatics identification of MurJ (MviN) as the peptidoglycan lipid II flippase in Escherichia coli. Proc Natl Acad Sci USA. 105:15553–15557. Flanagan et al. 2009Flanagan RC, Neal-McKinney JM, Dhillon AS, Miller WG, Konkel ME. 2009. Examination of *Campylobacter jejuni* putative adhesins leads to the identification of a new protein, designated FlpA, required for chicken colonization. Infect Immun. 77:2399–2407 ,). However, it is unknown whether these putative virulence genes show similar actions in *Arcobacter*. Even though *A. butzleri* has been shown to adhere and invade into various cell lines *in vitro*, no correlation have been shown between *A. butzleri* virulence genes and adhesive and invasive phenotypes (Karadas et al. 2013Karadas G, Sharbati S, Hänel I, Messelhäußer U, Glocker E, Alter T, Gölz G. 2013. Presence of virulence genes, adhesion and invasion of *Arcobacter butzleri*. J Appl Microbiol. 115:583–590. Levican et al. 2013bLevican A, Alkeskas A, Günter C, Forsythe SJ, Figueras MJ. 2013b. Adherence to and invasion of human intestinal cells by *Arcobacter* species and their virulence genotypes. Appl Environ Microbiol. 79:4951–4957 ,). Similarly, presence of six virulence genes in the *Arcobacter* spp. isolates was also reported by Karadas et al. (2013Karadas G, Sharbati S, Hänel I, Messelhäußer U, Glocker E, Alter T, Gölz G. 2013. Presence of virulence genes, adhesion and invasion of *Arcobacter butzleri*. J Appl Microbiol. 115:583–590 ,) and Lehmann et al. (2015Lehmann D, Alter T, Lehmann L, Uherkova S, Seidler T, Gölz G. 2015. Prevalence, virulence gene distribution and genetic diversity of *Arcobacter* in food samples in Germany. Berl Munch Tierarztl Wochenschr. 128:163–168). Likewise, five virulence genes (*ciaB*, *cadF*, *cj1349*, *irgA* and *hecA*) were detected from *Arcobacter* isolated from different sources namely meat, shellfish, sewage, feces from pigs, sheep and chickens, environmental sources, piggery effluent, seawater roots of *Spartina alterni* flora and porcine aborted fetus from Nottingham, United Kingdom (Levican et al. 2013bLevican A, Alkeskas A, Günter C, Forsythe SJ, Figueras MJ. 2013b. Adherence to and invasion of human intestinal cells by *Arcobacter* species and their virulence genotypes. Appl Environ Microbiol. 79:4951–4957 ,). Draft genome of *Arcobacter* from various sources like pig, dairy cattle manure origin, broiler carcasses and human feces have been recently documented which can help in exploring the various genes and their roles in virulence and pathogenicity (Adam et al. 2014aAdam Z, Whiteduck-Leveillee K, Cloutier M, Chen W, Lewis CT, Lévesque CA, Topp E, Lapen DR, Tambong JT, Talbot G, Khan IU. 2014a. Draft genome sequence of *Arcobacter cibarius* Strain LMG21996T, isolated from Broiler Carcasses. Genome Announc. 2:e00034–14 2014bAdam Z, Whiteduck-Leveillee K, Cloutier M, Tambong JT, Chen W, Lewis CT, Lévesque CA, Topp E, Lapen DR, Talbot G, Khan IU. 2014b. Draft genome sequences of three *Arcobacter* strains of pig and dairy cattle manure origin. Genome Announc. 2:e00377–14 , 2014cAdam Z, Whiteduck-Léveillée K, Cloutier M, Chen W, Lewis CT, Lévesque CA, Topp E, Lapen DR, Tambong JT, Talbot G, Khan IU. 2014c. Draft genome sequences of two *Arcobacter* strains isolated from human feces. Genome Announc. 2:e00113–14.,). Whiteduck-Léveillée et al. (2016aWhiteduck-Léveillée J, Cloutier M, Topp E, Lapen DR, Talbot G, Villemur R, Khan IU. 2016a. Development and evaluation of multiplex PCR assays for rapid detection of virulence-associated genes in *Arcobacter* species. J Microbiol Methods. 121:59–65 ,) reported the prevalence of virulence-associated genes (VAGs) namely ciaB (90%), mviN (70%), tlyA (50%) and pldA (45%) genes were significantly higher when compared to hecA (16%), hecB (10%) and each of irgA and cj1349 (6%) genes, respectively in *A. butzleri*, *A. cryaerophilus* and *A. skirrowii*.

An overview of pathogenesis, mechanisms of virulence, transmission routes and ways/means of release of *Arcobacters* entering the human food chain is presented in Figure 2.

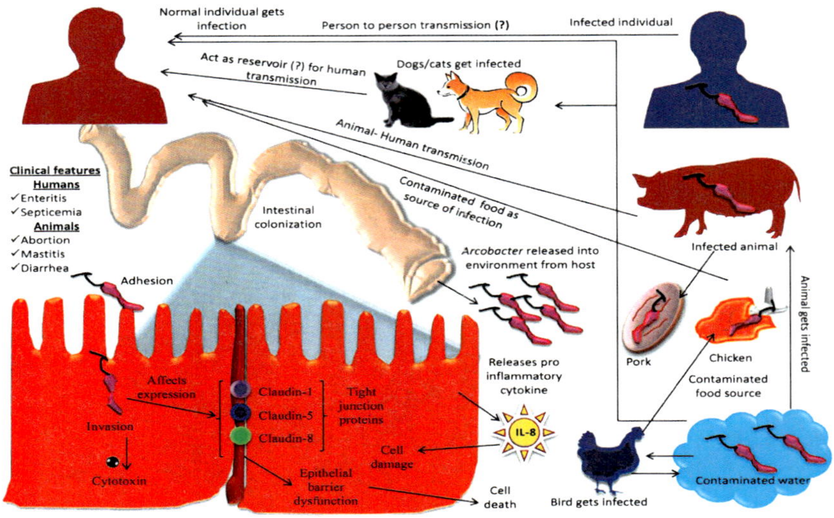

Figure 2. Pathogenesis and transmission of *Arcobacter*. *Arcobacter* after entry through the digestive tract, colonize the intestine. Several mechanisms like production of cytotoxin, affecting expression of tight junction proteins, release of pro-inflammatory cytokines and epithelial barrier dysfunction lead to cell death. The organism is released into the environment where it contaminates water and food sources thereby entering the human food chain.

EPIDEMIOLOGYANDTRANSMISSIONINANIMALS

Arcobacters have been reported worldwide from chickens, domestic animals (cattle, pigs, sheep, horses, dogs), reptiles (lizards, snakes and chelonians), meat (poultry, pork, goat, lamb, beef, rabbit), vegetables and from humans in different countries like Belgium, United States of America, Denmark, Brazil, Australia, Italy, the Netherlands, Malaysia, Japan, Spain, Czech Republic, Korea, Egypt and India (Lehner et al. 2005Lehner A, Tasara T, Stephan R. 2005. Relevant aspects of *Arcobacter* spp. as potential food-borne pathogens. Int J Food Microbiol. 102:127–135. Atabay et al. 2006Atabay HI, Wain M, Madsen M. 2006. Detection and diversity of various *Arcobacter* species in Danish poultry. Int J Food Microbiol. 9:139–145., ; Ho et al. 2006aHo HT, Lipman LJ, Gaastra W. 2006a. *Arcobacter*, what is known and unknown about a potential food-borne zoonotic agent. Vet Microbiol. 115:1–13. Snelling et al. 2006Snelling WJ, Matsuda M, Moore JE, Dooley JS. 2006. Under the microscope: *Arcobacter*. Lett Appl

Microbiol. 42:7–14. Pejchalova et al. 2008Pejchalova M, Dostali´kova, E. Slamova M, Brozkova I, Vytrasova J. 2008. Prevalence and diversity of *Arcobacter* spp. in the Czech Republic. J Food Protect. 71:719–727. Collado et al. 2010Collado L, Kasimir G, Perez U, Bosch A, Pinto R, Saucedo G, Huguet JM, Figueras MJ. 2010. Occurrence and diversity of *Arcobacter* spp. along the Llobregat river catchment, at sewage effluents and in a drinking water treatment plant. Water Res. 44:3696–3702. Lee et al. 2010Lee MH, Cheon DS, Choi S, Lee BH, Jung JY, Choi C. 2010. Prevalence of *Arcobacter* species isolated from retail meats in Korea. J Food Protect. 73:1313–1316. Vilar et al. 2010Vilar MJ, Pena FJ, Pe´rez I, Die´guez FJ, Sanjuan ML, Rodri´guez-Otero JL, Yus E. 2010. Presence of *Listeria*, *Arcobacter*, and *Campylobacter*spp. in dairy farms in Spain. Berl Munch Tierarztl Wochenschr. 23:58–62. ; Amare et al. 2011Amare LB, Sahela AA, Zunita Z, Jalila A, Hassan L. 2011. Prevalence of *Arcobacter* spp. on chicken meat at retail markets and in farm chickens in Selengor, Malaysia. Food Control. 22:732–736., ; González & Ferrús 2011González AM, Ferrus MA. 2011. Study of *Arcobacter* spp. contamination in fresh lettuces detected by different cultural and molecular methods. Int J Food Microbiol. 145:311–314. Patyal 2011Patyal A, Rathore RS, Mohan HV, Dhama K, Kumar A. 2011. Prevalence of *Arcobacter* spp. in humans, animals and foods of animal origin including sea food from India. Transbound Emerg Dis. 58:402–410. Suelam 2012Suelam II A. 2012. Isolation and identification of A*rcobacter* species recovered from rabbits in Zagazig, Egypt. Int J of Micr Res. 3:87–92. ; Ramees et al. 2014aRamees TP, Rathore RS, Bagalakote PS, Mohan HV, Kumar A, Dhama K. 2014a. Multiplex PCR detection of *Arcobacter butzleri* and *Arcobacter cryaerophilus* in skin of poultry. J Pure Appl Microbiol. 8:1755–1758. , 2014bRamees TP, Rathore RS, Bagalkot PS, Mohan HV, Kumar A, Dhama K. 2014b. Detection of *Arcobacter butzleri* and *Arcobacter cryaerophilus* in clinical samples of humans and foods of animal origin by cultural and multiplex PCR based methods. Asian J Anim Vet Adv. 9:243–252 2014cRamees TP, Rathore RS, Bagalkot PS, Ravi Kumar GVPPS, Mohan HV, Anoopraj R, Kumar A, Dhama K. 2014c. Real-time PCR detection of *Arcobacter butzleri* and *Arcobacter cryaerophilus* in chicken meat samples. J Pure Appl Microbiol. 8:3165–3169. ; Mohan et al. 2014Mohan HV, Rathore RS, Dhama K, Ramees TP, Patyal A, Bagalkot PS, Wani MY, Bhilegaonkar KN, Kumar A. 2014. Prevalence of *Arcobacter*spp. in humans, animals and foods of animal origin in India based on cultural isolation, antibiogram, PCR and multiplex PCR detection. Asian J Anim Vet Adv. 9:452–466., ; Gilbert et al. 2014Gilbert MJ, Kik M, Timmerman AJ,

Severs TT, Kusters JG, Duim B, Wagenaar JA. 2014. Occurrence, diversity, and host association of intestinal *Campylobacter*, *Arcobacter* and *Helicobacter* in Reptiles. PloS One. 9:e101599 ,). *Arcobacter* is a potential food- and water-borne pathogen, and thus pose serious public health concerns (González et al. 2007bGonzález A, Ferrús MA, González R, Hernández J. 2007b. Molecular fingerprinting of *Campylobacter*and *Arcobacter* isolated from chicken and water. Int Microbiol. 10:85–90 ; Gugliandolo et al. 2008Gugliandolo C, Irrera GP, Lentini V, Maugeri TL. 2008. Pathogenic *Vibrio*, *Aeromonas* and *Arcobacter*spp. associated with copepods in the straits of Messina (Italy). Mar Pollut Bull. 56:600–606. Miller et al. 2009Miller WG, Wesley IV, On SLW, Houf K, Megraud F, Wang G, Yee E, Srijan A, Mason CJ. 2009. First multi-locus sequence typing scheme for *Arcobacter* spp. BMC Microbiol. 14:196., ; Ferreira et al. 2016Ferreira S, Queiroz JA, Oleastro M, Domingues FC. 2016. Insights in the pathogenesis and resistance of *Arcobacter*: a review. Crit Rev Microbiol. 42:364–383). Among the 25 recognized species till date, *A. butzleri*, *A. cryaerophilus* and *A. skirrowii* are the species of veterinary importance. Different *Arcobacter* species have been isolated from a range of animals and their discharges such as saliva, feces, and from the vagina (Kabeya et al. 2003aKabeya H, Maruyama S, Morita Y, Kubo M, Yamamoto K, Arai S, Izumi T, Kobayashi Y, Katsube Y, Mikami T. 2003a. Distribution of *Arcobacter*species among livestock in Japan. Vet Microbiol. 93:153–158. Ho et al. 2006aHo HT, Lipman LJ, Gaastra W. 2006a. *Arcobacter*, what is known and unknown about a potential food-borne zoonotic agent.Vet Microbiol. 115:1–13 ,). Isolation of *Arcobacter* from feces of healthy pigs and also from dogs has been documented. In dogs, *A. cryaerophilus* was more commonly recovered (Ho et al. 2006bHo HT, Lipman LJA, Van der Graaf-van Bloois L, Van Bergen M, Gaastra W. 2006b. Potential routes of acquisition of *Arcobacter* species by piglets. Vet Microbiol. 114:122–133., ; Houf et al. 2008Houf K, Sarah DS, Julie B, Sylvie D. 2008. Dogs as carriers of the emerging pathogen; *Arcobacter*. J Vet Microbiol. 130:208–213.,). Various recovery rates have been documented from fecal samples of different species. From India, Patyal et al. (2011Patyal A, Rathore RS, Mohan HV, Dhama K, Kumar A. 2011. Prevalence of *Arcobacter* spp. in humans, animals and foods of animal origin including sea food from India. Transbound Emerg Dis. 58:402–410 ,) recorded a prevalence rate of 21% in pigs, 15% in poultry droppings and 3% in human diarrheal samples; Mohan et al. (2014Mohan HV, Rathore RS, Dhama K, Ramees TP, Patyal A, Bagalkot PS, Wani MY, Bhilegaonkar KN, Kumar A. 2014. Prevalence of

*Arcobacter*spp. in humans, animals and foods of animal origin in India based on cultural isolation, antibiogram, PCR and multiplex PCR detection. Asian J Anim Vet Adv. 9:452–466.,) recorded isolation rate of 8% in poultry droppings, 10% from bovine fecal samples, 12% from pigs and 2% in human diarrheal samples; while, Ramees et al. (2014aRamees TP, Rathore RS, Bagalakote PS, Mohan HV, Kumar A, Dhama K. 2014a. Multiplex PCR detection of *Arcobacter butzleri* and *Arcobacter cryaerophilus* in skin of poultry. J Pure Appl Microbiol. 8:1755–1758.) recorded an isolation rate of 2% from human diarrheal samples. From Belgium, Van Driessche et al. (2003Van Driessche E, Houf K, Van Hoof J, De Zutter L, Vandamme P. 2003. Isolation of *Arcobacter* species from animal feces. FEMS Microbiol. 229:243–248.,) recorded *Arcobacter*isolation of 44% (pigs), 39% (bovine), 16% (ovine) and 15% (equine) in fecal samples collected at slaughterhouses. *A. thereius* has also been isolated from liver and kidneys of pigs aborted spontaneously but its exact pathogenic role could not be elucidated (Houf et al. 2009Houf K, On SL, Coenye T, Debruyne L, De Smet S, Vandamme P. 2009. *Arcobacter thereius* sp. nov., isolated from pigs and ducks. Int J Syst Evol Microbiol. 59:2599–2604 ,). Feco-oral route of transmission play an important role in establishing infection to cattle (Shah et al. 2013Shah AH, Saleha AA, Zunita Z, Murugaiyah M, Aliyu AB, Jafri N. 2013. Prevalence, distribution and antibiotic resistance of emergent Arcobacter spp. from clinically healthy cattle and goats. Transbound Emerg Dis. 60:9–16. Scarano et al. 2014Scarano C, Giacometti F, Manfreda G, Lucchi A, Pes E, Spanu C, De Santis EP, Serraino A. 2014. *Arcobacter butzleri* in sheep ricotta cheese at retail and related sources of contamination in an industrial dairy plant. Appl Environ Microbiol. 80:7036–7041 ,). Shirzad Aski et al. (2016Shirzad Aski H, Tabatabaei M, Khoshbakht R, Raeisi M. 2016. Occurrence and antimicrobial resistance of emergent Arcobacter spp. isolated from cattle and sheep in Iran. Comp Immunol Microbiol Infect Dis. 44:37–40 ,) reported the occurrence of *Arcobacter spp.* in cattle and sheep from Southern Iran as 9% (27/308) by cultural isolation and 14% (44/308) by PCR. The predominant species was *A. butzleri* in both cattle (58%) and sheep (55%). These results suggested that cattle and sheep are significant intestinal carriers for *Arcobacter* spp. Very recently, water contamination of *Arcobacter* (80%) in a farm was documented with *A. skirrowii* (13%), *A. butzleri* (40%) and *A. cryaerophilus* (63%). Same *Arcobacter* species were isolated after two months from the water troughs which indicates the capacity to form biofilms or as a secondary bioformer (Giacometti et al. 2015Giacometti F, Lucchi A, Di Francesco A, Delogu M, Grilli E, Guarniero I, Stancampiano L, Manfreda

G, Merialdi G, Serraino A. 2015. *Arcobacter butzleri, Arcobacter cryaerophilus* and *Arcobacter skirrowii* circulation in a dairy farm and sources of milk contamination. Appl Environ Microbiol. 81:5055–5063 ,). This study also reported the presence of *A. butzleri* in oral cavity of cats. Similar reports regarding the presence of two different *Arcobacter* spp. (*A. butzleri* and *A. cryaerophilus*) from the oral cavity of cats have been reported earlier with 79% positive cases of *Arcobacter* by using PCR as a diagnostic tool (Fera et al. 2009Fera MT, La Camera E, Carbone H, Malara D, Pennisi MG. 2009. Pet cats as carriers of *Arcobacter* spp. in Southern Italy. J Appl Microbiol. 106:1661–1666 ,). Hence, pet animals may act as a carrier and can play a part in transmission of *Arcobacter* to humans.

Prevalence rate of *Arcobacter* spp. from healthy cattle has been reported to vary in different parts of the world indicating its presence without any clinical manifestation and suggest that cattle can act as a reservoir. Table 2 shows the prevalence of *Arcobacter* spp. in healthy cattle from different countries of the world. *A. butzleri* was the most common organism isolated in dairy farms when compared to other *Arcobacter* species which may be attributed to the ability of this organism to thrive in different environmental conditions (Golla et al. 2002Golla SC, Murano EA, Johnson LG, Tipton NC, Cureington EA, Savell JW. 2002. Determination of the occurrence of *Arcobacter butzleri* in beef and dairy cattle from Texas by various isolation methods. J Food Prot. 65:1849–1853. Giacometti et al. 2015Giacometti F, Lucchi A, Di Francesco A, Delogu M, Grilli E, Guarniero I, Stancampiano L, Manfreda G, Merialdi G, Serraino A. 2015. *Arcobacter butzleri, Arcobacter cryaerophilus* and *Arcobacter skirrowii* circulation in a dairy farm and sources of milk contamination. Appl Environ Microbiol. 81:5055–5063 ,). Merga et al. (2011Merga JY, Leatherbarrow AJH, Winstanley C, Bennett M, Hart CA, Miller WG, Williams NJ. 2011. Comparison of *Arcobacter* isolation methods, and diversity of *Arcobacter* spp. in Cheshire, United Kingdom. Appl Environ Microbiol. 77:1646–1650 , , 2013Merga JY, Williams NJ, William G. Miller WG, Leatherbarrow AJH, Bennett M, Hall N, Ashelford KE, Winstanley C. 2013. Exploring the diversity of *Arcobacter butzleri*from cattle in the UK using MLST and whole genome sequencing. PLoS One. 8:e55240 ,) reported prevalence of 43% from feces of cattle and 40% from feces of sheep in United Kingdom. A study in Belgium reported prevalence of 59% to 85% *Arcobacter* from fecal samples of pigs (Van Driessche et al. 2004Van Driessche E, Houf K, Vangroenweghe F, Nollet N, De Zutter L, Vandamme P, Van Hoof J. 2004. Occurrence and strain diversity of *Arcobacter* species isolated from healthy Belgian pigs. Res Microbiol.

155:662–666 ,). *Arcobacter* can colonize the gall bladder which shows its potential to cause liver and bile duct infections (Aydin et al. 2007Aydin F, Gumussoy KS, Atabay HI, Ica T, Abay S. 2007. Prevalence and distribution of *Arcobacter* species in various sources in Turkey and molecular analysis of isolated strains by ERIC-PCR. J Appl Microbiol. 103:27–35 ,). Both vertical and horizontal transmissions of *Arcobacters* have been documented in animals. Their transplacental transmission has also been documented in pigs (Ho et al. 2006bHo HT, Lipman LJA, Van der Graaf-van Bloois L, Van Bergen M, Gaastra W. 2006b. Potential routes of acquisition of *Arcobacter* species by piglets. Vet Microbiol. 114:122–133.,).

Table 2. Prevalence of *Arcobacter* spp. in healthy cattle from different countries of the world.

S . no.	Country	Prevalence	References
(1)	Turkey	6.9%	Aydin et al. (2007Aydin F, Gumussoy KS, Atabay HI, Ica T, Abay S. 2007. Prevalence and distribution of *Arcobacter* species in various sources in Turkey and molecular analysis of isolated strains by ERIC-PCR. J Appl Microbiol. 103:27–35.
(2)	Turkey	9.5%	Ongör et al. (2004Ongor H, Cetinkaya B, Açik MN, Atabay HI. 2004. Investigation of Arcobacters in meat and fecal samples of clinically healthy cattle in Turkey. Lett Appl Microbiol. 38:339–344.
(3)	Japan	3.6%	Kabeya et al. (2003aKabeya H, Maruyama S, Morita Y, Kubo M, Yamamoto K, Arai S, Izumi T, Kobayashi Y, Katsube Y, Mikami T. 2003a. Distribution of *Arcobacter* species among livestock in Japan. Vet Microbiol. 93:153–158.
(4)	Belgium	5.9%–11%	Van Driessche et al. (2005Van Driessche E, Houf K, Vangroenweghe F, De Zutter L, Van Hoof J. 2005. Prevalence, enumeration and strain variation of *Arcobacter* species in the faeces of healthy cattle in Belgium. Vet Microbiol. 105:149–154.
(5)	Belgium	39.2%	Van Driessche et al. (2003Van Driessche E, Houf K, Van Hoof J, De Zutter L, Vandamme P. 2003. Isolation of *Arcobacter* species from animal feces. FEMS Microbiol. 229:243–248.

(6)	Unites States of America	14.3%	Wesley et al. (2000Wesley IV, Wells SJ, Harmon KM, Green A, Schroeder-Tucker L, Glover M, Siddique I. 2000. Fecal shedding of *Campylobacter* and *Arcobacter* spp. in dairy cattle. Appl Environ Microbiol. 66:1994–2000
(7)	Italy	22.6%	Giacometti et al. (2015Giacometti F, Lucchi A, Di Francesco A, Delogu M, Grilli E, Guarniero I, Stancampiano L, Manfreda G, Merialdi G, Serraino A. 2015. *Arcobacter butzleri, Arcobacter cryaerophilus* and *Arcobacter skirrowii* circulation in a dairy farm and sources of milk contamination. Appl Environ Microbiol. 81:5055–5063.

Arcobacter has also been reported from monkey's feces (14%) in Canada (Higgins et al. 1999Higgins R, Messier S, Daignault D, Lorange M. 1999. *Arcobacter butzleri* isolated from a diarrhoeic non-human primate. Lab Anim. 33:87–90 ; Shah et al. 2011Shah AH, Saleha AA, Zunita Z, Murugaiyah M. 2011. *Arcobacter*– an emerging threat to animals and animal origin food products? Tr Food Sci Technol. 22:225–236.,). *A. butzleri* has been the only *Arcobacter* spp. isolated from both diarrheic as well as healthy non-human primates (Wesley et al. 2003Wesley IV, Schroeder-Tucker L, Franklin SL. 2003. Recovery of *Arcobacter* species from exotic animal species. Int J Med Microbiol. 293:57. ; Stirling et al. 2008Stirling J, Griffith M, Blair I, Cormican M, Dooley JS, Goldsmith CE, Glover SG, Loughrey A, Lowery CJ, Matsuda M, et al. 2008. Prevalence of gastrointestinal bacterial pathogens in a population of zoo animals. Zoonoses Public Health. 55:166–172 ,). Microscopic examination of the intestines from these non-human primates revealed chronic active colitis (Anderson et al. 1993Anderson KF, Kiehlbauch JA, Anderson DC. 1993. *Arcobacter*(*Campylobacter*) *butzleri*-associated diarrheal illness in a nonhuman primate population. Infect Immun. 61:2220–2223). Reports of *Arcobacter* spp. from raccoons (*Procyon lotor*) was documented for the first time in the year 2004, an issue of public health importance as raccoon share living environment with humans in several parts of the world (Hamir et al. 2004Hamir AN, Sonn RJ, Franklin S, Wesley IV. 2004. *Campylobacter*jejuni and *Arcobacter* species associated with intussusceptions in a raccoon (Procyon lotor). Vet Record. 155:338–340 ,). Other species like Galapagos turtle, white Rhinoceros, Gazelle, Rhea and Alpaca are all in the list of animals infected by *Arcobacter*spp. (Shah et al. 2011Shah AH, Saleha AA, Zunita Z, Murugaiyah M. 2011. *Arcobacter*– an emerging threat to animals and animal origin food products? Tr Food Sci Technol. 22:225–236.,).

Collado et al. (2014Collado L, Jara R, Vásquez N, Telsaint C. 2014. Antimicrobial resistance and virulence genes of *Arcobacter* isolates recovered from edible bivalve mollusks. Food Control. 46:508–512.,) reported the prevalence of *Arcobacter* from different marine species like surf clams (88%), where isolation was highest, followed by razor clams (65%), mussels (33%), clams (24%), scallops (18%) and oysters (15%). A moderate rate of prevalence has been reported from sea food (21%) from India (Patyal et al. 2011Patyal A, Rathore RS, Mohan HV, Dhama K, Kumar A. 2011. Prevalence of *Arcobacter* spp. in humans, animals and foods of animal origin including sea food from India. Transbound Emerg Dis. 58:402–410 ,). *A. butzleri* strains had the ability to survive inside the amoebae known as *Acanthamoeba castellanii* for the period of 240 h, which suggest that *A. butzleri* was resistant to amoebic digestion. Moreover, *A. castellanii* could act as an environmental reservoir and transmission vehicle for *Arcobacter* (Villanueva et al. 2016Villanueva MP, Medina G, Fernández H. 2016. *Arcobacter butzleri* survives within trophozoite of *Acanthamoeba castellanii*. Rev Argent Microbiol. 48:105–109).

Poultry act as potential reservoirs for *Arcobacters* and these organisms have been recovered from live poultry skin, droppings and meat (Corry & Atabay 2001Corry JE, Atabay HI. 2001. Poultry as a source of *Campylobacter* and related organisms. J Appl Microbiol. 90:96S–114S. Kabeya et al. 2004Kabeya H, Maruyama S, Morita Y, Ohsuga T, Ozawa S, Kobayashi Y, Abe M, Katsube Y Mikami T. 2004. Prevalence of *Arcobacter* species in retail meats and antimicrobial susceptibility of the isolates in Japan. Int J Food Microbiol. 90:303–308. Lehner et al. 2005Lehner A, Tasara T, Stephan R. 2005. Relevant aspects of *Arcobacter* spp. as potential food-borne pathogens. Int J Food Microbiol. 102:127–135. Ho et al. 2008bHo HT, Lipman LJ, Wosten MM, van Asten AJ, Gaastra W. 2008b. *Arcobacter* spp. possess two very short flagellins of which FlaA is essential for motility. FEMS Immunol Med Microbiol. 53:85–95 ; Collado et al. 2009aCollado L, Guarro J, Figueras M. 2009a. Prevalence of *Arcobacter* in meat and shellfish. J Food Protect. 72:1102–1110. Amare et al. 2011Amare LB, Sahela AA, Zunita Z, Jalila A, Hassan L. 2011. Prevalence of *Arcobacter* spp. on chicken meat at retail markets and in farm chickens in Selengor, Malaysia. Food Control. 22:732–736., ; Collado & Figueras 2011Collado L, Figueras MJ. 2011. Taxonomy, epidemiology and clinical relevance of the genus *Arcobacter*. Clin Microbiol Rev. 24:174–192. Ramees et al. 2014aRamees TP, Rathore RS, Bagalakote PS, Mohan HV, Kumar A, Dhama K. 2014a. Multiplex PCR detection of *Arcobacter butzleri* and *Arcobacter cryaerophilus* in skin of poultry. J Pure Appl Microbiol. 8:1755–1758. ; Rahimi et al. 2014). A study by Giacometti et al. (2015Giacometti F, Lucchi A, Di Francesco A, Delogu M, Grilli E, Guarniero I, Stancampiano L, Manfreda G, Merialdi G, Serraino A. 2015. *Arcobacter butzleri, Arcobacter cryaerophilus* and *Arcobacter skirrowii* circulation in a dairy farm and sources of milk contamination. Appl Environ Microbiol. 81:5055–5063 ,) indicated absence of *Arcobacter* spp. in pigeon gut samples (47 numbers tested) and that these birds do not have any role in transmission of *Arcobacter* to cattle. Various studies have documented the isolation of *Arcobacters* from feces of broilers with different percentages like 12% in Costa Rica, 14% in Japan and 1.3% from Nigeria (Kabeya et al. 2003aKabeya H, Maruyama S, Morita Y, Kubo M, Yamamoto K, Arai S, Izumi T, Kobayashi Y, Katsube Y, Mikami T. 2003a. Distribution of *Arcobacter*species among livestock in Japan. Vet Microbiol. 93:153–158. Bogantes et al. 2015Bogantes EV, Fallas-Padilla KL, Rodríguez-Rodríguez CE, Jaramillo HF, Echandi ML. 2015. Zoonotic species of the genus arcobacter in poultry from different regions of Costa Rica. J Food Prot. 78:808–811 ,). Bogantes et al. (2015Bogantes EV, Fallas-Padilla KL, Rodríguez-Rodríguez CE, Jaramillo HF, Echandi ML. 2015. Zoonotic species of the genus arcobacter in poultry from different regions of Costa Rica. J Food Prot. 78:808–811 ,) also reported 20%, 36% and 12% isolation of *Arcobacter* from feces of geese, duck and layers. Atabay et al. (2008Atabay HI, Unver A, Sahin M, Otlu S, Elmali M, Yaman H. 2008. Isolation of various *Arcobacter* species from domestic geese (Anser anser). Vet Microbiol. 128:400–405 ,) reported 18% prevalence of *Arcobacter* from geese in Turkey.

PATHOGENESIS OF *ARCOBACTER* SPP.

Presently, the pathogenic potential of *Arcobacter* remains unclear. Only few studies have attempted to explore the pathogenic mechanisms associated with *Arcobacter*. *In vitro* cell culture studies of animal and human origin revealed that *Arcobacter* spp. can adhere to eukaryotic cells, invade and produce toxins resulting in damage to host cells. Adhesion and invasion of bacteria to host cells are essential for successful colonization and establishment of infection (Kolling et al. 2012Kolling G, Wu M, Guerrant RL. 2012. Enteric pathogens through life stages. Front Cell Infect Microbiol. 2:114 ,), which might be responsible for potential pathogenic nature of various *Arcobacter* spp. The mechanism responsible for *A. butzleri* induced diarrhea was explored by Bucker et al. (2009Bucker R, Troeger H, Kleer J, Fromm M, Schulzke JD.

2009. *Arcobacter butzleri* induces barrier dysfunction in intestinal HT-29/B6 cells. J Infect Dis. 200:756–764 ,), wherein it was shown to be related to functional impairment of the epithelial barrier in human colonic carcinoma cells (HT-29/B6), by enhancing the macromolecular permeability through paracellular pathway and decreasing the epithelial resistance. Alteration in the tight junction protein composition and distribution was reported during *Arcobacter* infection, with decreased expression of sealing-associated proteins, namely claudin-1, claudin-5 and claudin-8, which might be responsible for diarrhea induced by *A. butzleri* infection through leak flux mechanism.

DISEASE IN ANIMALS CAUSED BY *ARCOBACTERS*

Most of the infections caused by *Arcobacter* remain asymptomatic in nature and only few set up clinical condition. *Arcobacter* in animals cause abortions, diarrhea and mastitis (Logan et al. 1982Logan EF, Neill SD, Mackie DP. 1982. Mastitis in dairy cows associated with an aerotolerant *Campylobacter*. Vet Res. 110:229–230. ; Neill et al. 1985Neill SD, Cambell JN, O'brien JJ, Weatherup STC, Ellis WA. 1985. Taxonomic position of *Campylobacter cryaerophila* sp. Nov. Int J Systemat Bacteriol. 35:342–356., ; Vandamme et al. 1992aVandamme P, Pugina P, Benzi G, Van Etterijck R, Vlaes L, Kersters K, Butzler JP, Lior H, Lauwers S. 1992a. Outbreak of recurrent abdominal cramps associated with *Arcobacter butzleri* in an Italian school. J Clin Microbiol. 30:2335–2337 ; On et al. 2002On SLW, Jensen TK, Bille-Hansen V, Jorsal SE, Vandamme P. 2002. Prevalence and diversity of *Arcobacter* spp. isolated from the internal organs of spontaneous porcine abortions in Denmark. Vet Microbiol. 85:159–167. Aydin et al. 2007Aydin F, Gumussoy KS, Atabay HI, Ica T, Abay S. 2007. Prevalence and distribution of *Arcobacter* species in various sources in Turkey and molecular analysis of isolated strains by ERIC-PCR. J Appl Microbiol. 103:27–35. Collado & Figueras 2011Collado L, Figueras MJ. 2011. Taxonomy, epidemiology and clinical relevance of the genus *Arcobacter*. Clin Microbiol Rev. 24:174–192 ,). In pigs, cattle, and horses, *A. butzleri*cause enteritis and/or diarrhea, while in sheep and cattle *A. skirrowii* cause diarrhea and/or hemorrhagic colitis (Ho et al. 2006bHo HT, Lipman LJA, Van der Graaf-van Bloois L, Van Bergen M, Gaastra W. 2006b. Potential routes of acquisition of *Arcobacter* species by piglets. Vet Microbiol. 114:122–133 Vandamme et al. 1992bVandamme P, Vanvanneyt M, Pot B, Mels L, Hoste B, Dewettinck D, Vlaes L, Van Den Borre C, Higgins R, Hommez J, et al. 1992b. Polyphasic taxonomic study of the emended genus *Arcobacter* with *Arcobacter butzleri* comb. Nov. and *Arcobacter skirrowii* sp. Nov., and aerotolerant bacterium isolated from veterinary specimens. Int J Syst Bacteriol. 42:344–356). *A. cryophillus*is the predominant species isolated from aborted animals, but it has also been isolated from preputial sheath washings of healthy bovine (Gill 1983Gill KP.1983. Aerotolerant *campylobacter* strain isolated from a bovine preputial sheath washing. Vet Rec. 112:459 ,) and from vaginal swabs of cows with no clinical disease (Kabeya et al. 2003aKabeya H, Maruyama S, Morita Y, Kubo M, Yamamoto K, Arai S, Izumi T, Kobayashi Y, Katsube Y, Mikami T. 2003a. Distribution of *Arcobacter*species among livestock in Japan. Vet Microbiol. 93:153–158 ,). It has also been isolated from feces of healthy animals such as pigs and cattle, thus raising doubt on its pathogenicity (Kabeya et al. 2003aKabeya H, Maruyama S, Morita Y, Kubo M, Yamamoto K, Arai S, Izumi T, Kobayashi Y, Katsube Y, Mikami T. 2003a. Distribution of *Arcobacter*species among livestock in Japan. Vet Microbiol. 93:153–158. Van Driessche et al. 2003Van Driessche E, Houf K, Van Hoof J, De Zutter L, Vandamme P. 2003. Isolation of *Arcobacter* species from animal feces. FEMS Microbiol. 229:243–248 2004Van Driessche E, Houf K, Vangroenweghe F, Nollet N, De Zutter L, Vandamme P, Van Hoof J. 2004. Occurrence and strain diversity of *Arcobacter* species isolated from healthy Belgian pigs. Res Microbiol. 155:662–666 , , 2005Van Driessche E, Houf K, Vangroenweghe F, De Zutter L, Van Hoof J. 2005. Prevalence, enumeration and strain variation of *Arcobacter* species in the faeces of healthy cattle in Belgium. Vet Microbiol. 105:149–154. Ongor et al. 2004Ongor H, Cetinkaya B, Açik MN, Atabay HI. 2004. Investigation of Arcobacters in meat and fecal samples of clinically healthy cattle in Turkey. Lett Appl Microbiol. 38:339–344. Aydin et al. 2007Aydin F, Gumussoy KS, Atabay HI, Ica T, Abay S. 2007. Prevalence and distribution of *Arcobacter* species in various sources in Turkey and molecular analysis of isolated strains by ERIC-PCR. J Appl Microbiol. 103:27–35 ,). To substantiate, *A. cibarius, A. butzleri* and *A. cryaerophilus* were recovered from piggery effluent and also from soil being irrigated by the effluent (Chinivasagam et al. 2007Chinivasagam HN, Corney BG, Wright LL, Diallo IS, Blackall PJ. 2007. Detection of *Arcobacter* spp. in piggery effluent and effluent-irrigated soils in southeast Queensland. J Appl Microbiol. 103:418–426 ,). *A. cryaerophilus* has been associated with porcine abortion, sows with reproductive problems, and preputial fluid of boars and fattening pigs (Neill et al. 1985Neill SD, Cambell JN, O'brien JJ, Weatherup STC, Ellis WA. 1985. Taxonomic position of *Campylobacter cryaerophila* sp. Nov. Int J

Systemat Bacteriol. 35:342–356., ; de Oliveira et al. 1997de Oliveira SJ, Wesley I, Harmon KM. 1997. Classification of *Arcobacter* species isolated from aborted pig fetuses and sows with reproductive problems in Brazil. Vet Microbiol. 57:347–354 ,). *A. butzleri* and *A. skirrowii* have been less frequently reported from aborted animals (de Oliveira et al. 1997de Oliveira SJ, Wesley I, Harmon KM. 1997. Classification of *Arcobacter* species isolated from aborted pig fetuses and sows with reproductive problems in Brazil. Vet Microbiol. 57:347–354. Collado & Figueras 2011Collado L, Figueras MJ. 2011. Taxonomy, epidemiology and clinical relevance of the genus *Arcobacter*. Clin Microbiol Rev. 24:174–192 ,). Recently, reports have been documented in South Africa regarding the involvement of *A. skirrowii*induced abortions in ewes. This report further highlighted the need to get insight into the ability of this pathogen to cause abortion in other animals (Bath et al. 2013Bath GF, Leask R, Pettey KP, Coetzee DJ. 2013. Abortions in sheep associated with *Arcobacter skirrowii* infection. J South Afr Vet Assoc. 84:4. Available from:https://doi.org/10.4102/jsava.v84i1.952,). However, pathology and pathogenesis of *Arcobacter* infection in animals need to be elucidated properly by unrevealing the pathogenesis of these organisms in different animal models and clinical experiments involving both *in vitro* and *in vivo* studies, the better understanding of which could pave the way for developing suitable control strategies and counter this important emerging pathogen.

Although reports from animals are restricted to terrestrial animals, one study reported the isolation of *A. cryaerophilus* from a natural infection of a fish namely rainbow trout (*Oncorhynchus mykiss*). The *in vivo* pathogenicity experiment of the isolated strain revealed infection of intestine, liver and kidney and also caused death of the infected fish (Yildiz & Aydin 2006Yildiz H, Aydin S. 2006. Pathological effects *of Arcobacter cryaerophilus*infection in rainbow trout (*Oncorhynchus mykissWalbaum*). Acta Vet Hung. 54:191–199. Collado & Figueras 2011Collado L, Figueras MJ. 2011. Taxonomy, epidemiology and clinical relevance of the genus *Arcobacter*. Clin Microbiol Rev. 24:174–192 ,). So far, laboratory animal models for studying *A. butzleri* pathogenesis are lacking because its pathogenesis depends on species and breed of the host. Similarly, *A. butzleri* did not induce disease in conventional chicken, but can cause disease in turkey strains (Wesley & Baetz 1999Wesley IV, Baetz AL. 1999. Natural and experimental infections of*Arcobacter* in poultry. Poult Sci. 78:536–545 ,). Similarly, experimental infection of adult rats with *Arcobacter* led to watery diarrhea and imbalanced serum electrolytes which depends on the bacterial load, while neonatal albino rats suffered from self-limiting diarrhea, small intestinal and hepatic necrosis (Adesiji 2010Adesiji YO. 2010. Faecal shedding of *Arcobacter* species following experimental infection in rats: public health implications. Cent Eur J Med. 5:470–474., ; Adesiji et al. 2012Adesiji YO, Seibu E, Emikpe BO, Moriyonu BT, Oloke JK, Coker AO. 2012. Serum biochemistry and heamatological changes associated with graded doses of experimental Arcobacter infection in rats. West Afr J Med. 31:186–191.,). Açik et al. (2016Açik MN, Yüksel H, Ulucan A, Çetinkaya B. 2016. The first experimental research on the pathogenicity of *Arcobacter butzleri*in zebrafish. Vet Microbiol. 189:32–38 ,) for the first time experimentally studied *A. butzleri* infection in wild zebrafish. The zebrafish were inoculated with mean infective dose (ID_{50}) of *A. butzleri* as 1.3×10^8 CFU/mL by immersion and 1×10^5 CFU/mL by intraperitoneal injection. Histopathological examination revealed acute inflammation characterized by neutrophils and plasma cells, local necrosis and congestion in various visceral organs. These findings suggested that zebrafish could be used as a laboratory animal model to study the pathogenicity of *Arcobacters*.

Immunological parameters after *Arcobacter* infection in a murine model have been investigated. After oral administration of *A. butzleri*, the bacteria readily colonize the murine intestinal tract. However, infected mice did not show any overt clinical signs, the bacteria not only induced intestinal immune responses but also systemic immune responses in extra-intestinal places depending on the strain and time course of infection (Gölz et al. 2015aGölz G, Karadas G, Alutis ME, Fischer A, Kuhl AA, Breithaupt A, Gobel UB, Alter T, Bereswill S, Heimesaat MM. 2015a. *Arcobacter butzleri* induce colonic, extra-intestinal and systemic inflammatory responses in gnotobiotic IL-10 deficient mice in a strain-dependent manner. PloS One. 10:e0139402. Heimesaat et al. 2015aHeimesaat MM, Karadas G, Alutis M, Fischer A, Kuhl AA, Breithaupt A, Gobel UB, Alter T, Bereswill S, Gölz G. 2015a. Survey of small intestinal and systemic immune responses following murine *Arcobacter butzleri* infection. Gut Pathog. 7:28 ,). These immune responses were TLR-4 dependent recognized by LOS and LPS derived from the cell walls of Gram-negative bacteria in gnotobiotic IL-$10^{-/-}$ mice lacking TLR-4 with less pronounced intestinal and systemic immune responses (Gölz et al. 2015bGölz G, Karadas G, Fischer A, Gobel UB, Alter T, Bereswill S, Heimesaat MM. 2015b. Toll-like receptor-4 is essential for *Arcobacter butzleri*-induced colonic and systemic immune responses in gnotobiotic IL-10–/–mice. Eur J Microbiol Immunol. 5:321–332., ; Heimesaat et al. 2015bHeimesaat MM, Karadas G, Fischer A, Gobel UB, Alter T, Bereswill

S, Gölz G. 2015b. Toll-like receptor-4 dependent small intestinal immune responses following murine *Arcobacter butzleri* infection. Eur J Microbiol Immunol. 5:333–342., ; Heimesaat et al. 2016Heimesaat MM, Alter T, Bereswill S, Gölz G. 2016. Intestinal expression of genes encoding inflammatory mediators and gelatinases during *Arcobacter butzleri* infection of gnotobiotic Il-10 deficient mice. Eur J Microbiol Immunol. 6:56–66.,). The mucin-2 is a pivotal constituent of the intestinal mucosa providing first line defense against intestinal pathogens (Strugala et al. 2003Strugala V, Allen A, Dettmar PW, Pearson JP. 2003. Colonic mucin: methods of measuring mucus thickness. P Nutr Soc. 62:237–243 ,) and was downregulated in the colon, but not in the ileum of *A. butzleri* infected mice (Gölz et al. 2016bGölz G, Alter T, Bereswill S, Heimesaat MM. 2016b. Toll-like receptor-4 dependent intestinal gene expression during *Arcobacter butzleri* infection of gnotobiotic Il-10 deficient mice. Eur J Microbiol Immunol. 6:67–80., ; Heimesaat et al. 2016Heimesaat MM, Alter T, Bereswill S, Gölz G. 2016. Intestinal expression of genes encoding inflammatory mediators and gelatinases during *Arcobacter butzleri* infection of gnotobiotic Il-10 deficient mice. Eur J Microbiol Immunol. 6:56–66.,). *A. butzleri* induces antimicrobial host immunity and cytokines expression such as IL-17A, IL-22, IFN-γ and their master regulator IL-23 in intestine that are secreted by T helper (Th) 17 cells. During *A. butzleri* infection in murine, IL-22 and IL-18 in intestine were upregulated in a strain-dependent manner indicating that the IL-22/IL-18 axis might be responsible for regulating the immune responses during arcobacteriosis (Gölz et al. 2016aGölz G, Alter T, Bereswill S, Heimesaat MM. 2016a. The Immunopathogenic Potential of *Arcobacter butzleri* – lessons from a meta-analysis of murine infection studies. PLoS One. 11:e0159685 , ,2016bGölz G, Alter T, Bereswill S, Heimesaat MM. 2016b. Toll-like receptor-4 dependent intestinal gene expression during *Arcobacter butzleri* infection of gnotobiotic Il-10 deficient mice. Eur J Microbiol Immunol. 6:67–80., ; Heimesaat et al. 2016Heimesaat MM, Alter T, Bereswill S, Gölz G. 2016. Intestinal expression of genes encoding inflammatory mediators and gelatinases during *Arcobacter butzleri* infection of gnotobiotic Il-10 deficient mice. Eur J Microbiol Immunol. 6:56–66.,). *A. butzleri*infection in gnotobiotic IL-10$^{-/-}$ mice showed up-regulation of mRNA levels of pro-inflammatory cytokines such as TNF and IL-1β, and MMP-2 and MMP-9 in the intestine (Heimesaat et al. 2016Heimesaat MM, Alter T, Bereswill S, Gölz G. 2016. Intestinal expression of genes encoding inflammatory mediators and gelatinases during *Arcobacter butzleri* infection of gnotobiotic Il-10 deficient mice. Eur J Microbiol Immunol. 6:56–66.,).

HUMAN INFECTIONS WITH *ARCOBACTERS*

A. cryaerophilus was the first *Arcobacter* species to be identified from humans in the year 1988 (Tee et al. 1988Tee W, Baird R, Dyall-Smith M, Dwyer B. 1988. *Campylobacter cryaerophila* isolated from a human. J Clin Microbiol. 26:2469–2473 ; Collado et al. 2011Collado L, Levican A, Perez J, Figueras MJ. 2011. *Arcobacter defluvii* sp. nov., isolated from sewage samples. Int J Syst Evol Microbiol. 61:2155–2161 ,). *Arcobacter* has been isolated from feces and blood samples of humans and symptoms may range from diarrhea to septicemia (Fisher et al. 2014Fisher JC, Levican A, Figueras MJ, McLellan SL. 2014. Population dynamics and ecology of *Arcobacter* in sewage. Front Microbiol. 5:1–9 ,). Gastrointestinal manifestations are the common signs in humans and may be exhibited as watery diarrhea in case of *A. butzleri* infection, while bloody diarrhea is usually noticed in *Campylobacter jejuni* (Collado et al. 2011Collado L, Levican A, Perez J, Figueras MJ. 2011. *Arcobacter defluvii* sp. nov., isolated from sewage samples. Int J Syst Evol Microbiol. 61:2155–2161 ,). Several studies indicate the prevalence of *Arcobacter* in humans from different countries of the world. *Arcobacter* spp. has also been attributed to one of the bacterial agents of traveler's diarrhea along with *E. coli, Shigella, Salmonella* and *Campylobacter* (McGrego & Wright 2015McGregor AC, Wright SG. 2015. Gastrointestinal symptoms in travellers. Clin Med. 15:93–95 ,). Recently, an invasive form of *Arcobacter* infection in an 85-year-old immunocompromised man was documented (Arguello et al. 2015Arguello E, Otto CC, Mead P, Babady NE. 2015. Bacteremia caused by *Arcobacter butzleri* in an immunocompromised host. J Clin Microbiol. 53:1448–1451 ,).

Ways of Transmission in Humans

Arcobacter in humans mainly causes enteric problem and there are several modes of transmission by which the pathogen can colonize the host cells (Collado & Figueras 2011Collado L, Figueras MJ. 2011. Taxonomy, epidemiology and clinical relevance of the genus *Arcobacter*. Clin Microbiol Rev. 24:174–192 ,). Transmission route can be feco-oral or direct from animals (Jacob et al. 1998Jacob J, Woodward D, Feuerpfeil I, Johnson WM. 1998. Isolation of *Arcobacter butzleri* in raw water and drinking water treatment plants in Germany. Zbl Hyg. 201:189–198., ; Shah et al. 2011Shah AH, Saleha AA, Zunita Z, Murugaiyah M. 2011. *Arcobacter*– an emerging threat to animals and animal origin food products? Tr Food Sci

Technol. 22:225–236., ; Ferreira et al. 2016Ferreira S, Queiroz JA, Oleastro M, Domingues FC. 2016. Insights in the pathogenesis and resistance of *Arcobacter*: a review. Crit Rev Microbiol. 42:364–383). *Arcobacters* are mainly transmitted through contaminated food (vegetables, chicken and pork meat) and water sources which may be contaminated through sewage. Several reports regarding the presence of *Arcobacter* in water have been documented and hence consumption of contaminated water acts as an efficient source of infection (Jacob et al. 1993Jacob J, Loir H, Feuerpfeil I. 1993. Isolation of *Arcobacter butzleri*from drinking water reservoir in Eastern Germany. Int J Hyg Env Med. 193:557–562. ; González et al. 2010González A, Suski J, Ferrus MA. 2010. Rapid and accurate detection of *Arcobacter* contamination in commercial chicken products and waste water samples by real-time polymerase chain reaction. Foodborne Pathog Dis. 7:327–338. Collado et al. 2010Collado L, Kasimir G, Perez U, Bosch A, Pinto R, Saucedo G, Huguet JM, Figueras MJ. 2010. Occurrence and diversity of *Arcobacter* spp. along the Llobregat river catchment, at sewage effluents and in a drinking water treatment plant. Water Res. 44:3696–3702. Lee & Choi 2013Lee MH, Choi, C. 2013. Survival of *Arcobacter butzleri* in apple and pear purees. J Food Safety. 33:333–339.,). Sewage contamination is another possible way of infection (Collado et al. 2010Collado L, Kasimir G, Perez U, Bosch A, Pinto R, Saucedo G, Huguet JM, Figueras MJ. 2010. Occurrence and diversity of *Arcobacter* spp. along the Llobregat river catchment, at sewage effluents and in a drinking water treatment plant. Water Res. 44:3696–3702. Merga et al. 2014Merga JY, Royden A, Pandey AK, Williams NJ. 2014. Arcobacter spp. isolated from untreated domestic effluent. Lett Appl Microbiol. 59:122–126 ,). Among the bacterial population in sewage, *Arcobacter* shares in the region of 5%–11% (Fisher et al. 2014Fisher JC, Levican A, Figueras MJ, McLellan SL. 2014. Population dynamics and ecology of *Arcobacter* in sewage. Front Microbiol. 5:1–9 ,). Chicken meat, pork and sea foods which are consumed uncooked or partially cooked can lead to infection as these are the major source where *Arcobacter* was isolated by several workers (Collado et al. 2009aCollado L, Guarro J, Figueras M. 2009a. Prevalence of *Arcobacter* in meat and shellfish. J Food Protect. 72:1102–1110 ,).

Food-borne Transmission and Prevalences

Arcobacters have been isolated from a variety of meat samples of chicken, pork, beef and seafood (Colado & Figueras 2011Collado L, Figueras MJ. 2011. Taxonomy, epidemiology and clinical relevance of the genus

Arcobacter. Clin Microbiol Rev. 24:174–192 ,). Chicken meat has been reported with highest prevalence for *Arcobacter* followed by pork and beef (Shah et al. 2011Shah AH, Saleha AA, Zunita Z, Murugaiyah M. 2011. *Arcobacter*– an emerging threat to animals and animal origin food products? Tr Food Sci Technol. 22:225–236.,). Different prevalence rates have been recorded in chicken meat samples in India ranging from 12% to 33% (Patyal et al. 2011Patyal A, Rathore RS, Mohan HV, Dhama K, Kumar A. 2011. Prevalence of *Arcobacter* spp. in humans, animals and foods of animal origin including sea food from India. Transbound Emerg Dis. 58:402–410. Mohan et al. 2014Mohan HV, Rathore RS, Dhama K, Ramees TP, Patyal A, Bagalkot PS, Wani MY, Bhilegaonkar KN, Kumar A. 2014. Prevalence of *Arcobacter*spp. in humans, animals and foods of animal origin in India based on cultural isolation, antibiogram, PCR and multiplex PCR detection. Asian J Anim Vet Adv. 9:452–466., ; Ramees et al. 2014bRamees TP, Rathore RS, Bagalkot PS, Mohan HV, Kumar A, Dhama K. 2014b. Detection of *Arcobacter butzleri* and *Arcobacter cryaerophilus* in clinical samples of humans and foods of animal origin by cultural and multiplex PCR based methods. Asian J Anim Vet Adv. 9:243–252.,). Recently, different *Arcobacter* spp. were isolated (22 samples (11%) of the 200 samples examined) from poultry in Costa Rica (Bogantes et al. 2015Bogantes EV, Fallas-Padilla KL, Rodríguez-Rodríguez CE, Jaramillo HF, Echandi ML. 2015. Zoonotic species of the genus arcobacter in poultry from different regions of Costa Rica. J Food Prot. 78:808–811 ,). Highest incidence of *A. butzleri* was reported with 83% from chicken meat followed by beef (20%) and pork (15%) in Poland (Zacharow et al. 2015Zacharow I, Bystroń J, Wałecka-Zacharska E, Podkowik M, Bania J. 2015. Prevalence and antimicrobial resistance of *Arcobacter butzleri*and *Arcobacter cryaerophilus*isolates from retail meat in Lower Silesia region, Poland. Pol J Vet Sci. 18:63–69 ,). Villalobos et al. (2013Villalobos EG, Jaramillo HF, Ulate CC, Echandi ML. 2013. Isolation and identification of zoonotic species of genus arcobacter from chicken viscera obtained from retail distributors of the metropolitan area of San José, Costa Rica. J Food Prot. 76:879–882 ,) reported isolation of *Arcobacter* as 17% from poultry visceral samples sold in Costa Rica and most of these isolates were resistant to chloramphenicol and ampicillin. Badilla-Ramírez et al. (2016Badilla-Ramírez Y, Fallas-Padilla KL, Fernández-Jaramillo H, Arias-Echandi ML. 2016. Survival capacity of *Arcobacter butzleri* inoculated in poultry meat at two different refrigeration temperatures. Rev Inst Med Trop Sao Paulo. 58:22 ,) reported that *A. butzleri* had the capacity to grow at 4 °C and 10 °C. They suggested that *A. butzleri* is a powerful bacterium for surviving in various storage temperatures, and is a potential health risk for poultry meat consumers. A study conducted in Germany by Lehmann et al. (2015Lehmann D, Alter T, Lehmann L, Uherkova S, Seidler T, Gölz G. 2015. Prevalence, virulence gene distribution and genetic diversity of *Arcobacter* in food samples in Germany. Berl Munch Tierarztl Wochenschr. 128:163–168) reported the prevalence of *Arcobacter* spp. as 34% from fish meat, 27% from poultry meat and 2% from minced meat. Similarly, De Smet et al. (2010De Smet S, De Zutter L, Van Hende J, Houf K. 2010. Arcobacter contamination on pre- and post-chilled bovine carcasses and in minced beef at retail. J Appl Microbiol. 108:299–305 ,) reported the presence of *Arcobacter* from pre- and post-chilled bovine carcasses indicating the need for hygienic practices to break the transmission cycle. At various time period reports of *Arcobacter* identification from different meat sources around the world has been documented (Vytrasová et al. 2003Vytrasová J, Pejchalová M, Harsová K, Bínová S. 2003. Isolation of *Arcobacter butzleri* and *A. cryaerophilus* in samples of meats and from meat-processing plants by a culture technique and detection by PCR. Folia Microbiol. 48:227–232. Rivas et al. 2004Rivas L, Fegan N, Vanderlinde P. 2004. Isolation and characterisation of *Arcobacter butzleri* from meat. Int J Food Microbiol. 91:31–41. Van Driessche & Houf 2007aVan Driessche E, Houf K. 2007a. Characterization of the *Arcobacter*contamination on Belgian pork carcasses and raw retail pork. Int J Food Microbiol. 118:20–26 ,). Several contraindications remain in the isolation of *Arcobacter* from intestine of poultry carcasses during slaughter house processing though it is believed that time and way of sample collection have a positive role in result interpretation (Houf et al. 2002aHouf K, De Zutter L, Van Hoof J, Vandamme P. 2002a. Occurrence and distribution of *Arcobacter*species in poultry processing. J Food Prot. 65:1233–1239. Van Driessche & Houf 2007bVan Driessche E, Houf K. 2007b. Discrepancy between the occurrence of *Arcobacter* in chickens and broiler carcass contamination. Poult Sci. 86:744–751 ,). Water treatment plant and the environment of the slaughter house were implicated for the contamination of poultry carcasses with *Arcobacter* in Iran (Khoshbakht et al. 2014Khoshbakht R, Tabatabaei M, Shirzad Aski H, Seifi S. 2014. Occurrence of *Arcobacter* in Iranian poultry and slaughterhouse samples implicates contamination by processing equipment and procedures. Br Poult Sci. 55:732–736). *Arcobacter* was isolated from rabbit meat in Egypt (Suelam 2012Suelam II A. 2012. Isolation and identification of *Arcobacter* species recovered from rabbits in Zagazig, Egypt. Int J of Micr Res. 3:87–

92.) and also from Spain (Collado et al. 2009aCollado L, Guarro J, Figueras M. 2009a. Prevalence of *Arcobacter* in meat and shellfish. J Food Protect. 72:1102–1110 ,). Four *Arcobacter*spp. (*A. butzleri, A. cryaerophilus, A. skirrowii* and *A. cibarius*) have been isolated in high numbers from meat, especially from chicken carcasses. Isolation of *Arcobacter* from sea foods like oysters (Romero et al. 2002Romero J, Varela GM, Laclette JP, Espejo RT. 2002. Bacterial 16S rRNA gene analysis revealed that bacteria related to *Arcobacter* spp. constitute an abundant and common component of the oyster microbiota (*Tiostrea chilensis*). Microb Ecol. 44:365–371 ,), fishes (Patyal et al. 2011Patyal A, Rathore RS, Mohan HV, Dhama K, Kumar A. 2011. Prevalence of *Arcobacter* spp. in humans, animals and foods of animal origin including sea food from India. Transbound Emerg Dis. 58:402–410 ,), shellfish and clams (Collado et al. 2009aCollado L, Guarro J, Figueras M. 2009a. Prevalence of *Arcobacter* in meat and shellfish. J Food Protect. 72:1102–1110 ,) are also available.

Reports of *Arcobacter* isolation from milk samples are also available from different parts of the world, indicating the possible role of milk in its transmission. In a study from Selangor, Malaysia, 5.8% prevalence of *Arcobacter* spp. from cow milk samples, wherein *A. butzleri* was the foremost species (60%), followed by *A. cryaerophilus* (40%), but goat samples were found negative for *Arcobacter* was reported (Shah et al. 2012aShah AH, Saleha AA, Murugaiyah M, Zunita Z, Memon AA. 2012a. Prevalence and distribution of *Arcobacter* spp. in raw milk and retail raw beef. J Food Prot. 75:1474–1478 ,).Similarly, Ertas et al. (2010Ertas N, Dogruer Y, Gonulalan Z, Guner A, Ulger I. 2010. Prevalence of *Arcobacter* species in drinking water, spring water, and raw milk as determined by multiplex PCR. J Food Prot. 73:2099–2102 ,) reported 6% prevalence from raw milk samples in Turkey and Revez et al. (2013Revez J, Huuskonen M, Ruusunen M, Lindström M, Hänninen ML. 2013. *Arcobacter* species and their pulsed-field gel electrophoresis genotypes in Finnish raw milk during summer 2011. J Food Prot. 76:1630–1632 ,) reported 15% prevalence in Finland. Study conducted by Giacometti et al. (2014Giacometti F, Serraino A, Pasquali F, De Cesare A, Bonerba E, Rosmini R. 2014. Behavior of *Arcobacter butzleri* and *Arcobacter cryaerophilus* in ultrahigh-temperature, pasteurized, and raw cow's milk under different temperature conditions. Foodborne Pathog Dis. 11:15–20 ,) showed that *A. butzleri* and *A. cryaerophilus* survived for a period of six days in milk stored at 4 °C and 10 °C. Hence, it was concluded that milk can serve as an efficient source of *Arcobacter* infection to humans. The organism also survived during processing and storage of water buffalo mozzarella cheese, fresh village cheese and sheep ricotta cheese (Serraino et al. 2013aSerraino A, Giacometti F, Daminelli P, Losio MN, Finazzi G, Marchetti G, Zambrini AV, Rosmini R. 2013a. Survival of *Arcobacter butzleri*during production and storage of artisan water buffalo mozzarella cheese. Foodborne Pathog Dis. 10:820–824. Yesilmen et al. 2014Yesilmen S, Vural A, Erkan ME, Yildirim IH. 2014. Prevalence and antimicrobial susceptibility of *Arcobacter* species in cow milk, water buffalo milk and fresh village cheese. Int J Food Microbiol. 188:11–14. Scarano et al. 2014Scarano C, Giacometti F, Manfreda G, Lucchi A, Pes E, Spanu C, De Santis EP, Serraino A. 2014. *Arcobacter butzleri* in sheep ricotta cheese at retail and related sources of contamination in an industrial dairy plant. Appl Environ Microbiol. 80:7036–7041 ,). Studies showed that *Arcobacter* was also isolated from inline milk filters used in farms (Serraino et al. 2013bSerraino A, Florio D, Giacometti F, Piva S, Mion D, Zanoni RG. 2013b. Presence of *Campylobacter* and *Arcobacter* species in in-line milk filters of farms authorized to produce and sell raw milk and of a water buffalo dairy farm in Italy. J Dairy Sci. 96:2801–2807 ,). Fecal shedding of the organism by cattle can lead to contamination of milk and beef carcasses leading to infection of humans (Shah et al. 2012bShah AH, Saleha AA, Zunita Z, Cheah YK, Murugaiyah M, Korejo NA. 2012b. Genetic characterization of Arcobacter isolates from various sources. Vet Microbiol. 160:355–361 ,). Several outbreaks of human gastroenteritis associated with *Arcobacter* have been linked to the consumption of contaminated fresh vegetables and thus could act as an important source of infection (González & Ferrus 2011González AM, Ferrus MA. 2011. Study of *Arcobacter* spp. contamination in fresh lettuces detected by different cultural and molecular methods. Int J Food Microbiol. 145:311–314 , : Lee & Choi 2013Lee MH, Choi, C. 2013. Survival of *Arcobacter butzleri* in apple and pear purees. J Food Safety. 33:333–339.,). González et al. (2010González A, Suski J, Ferrus MA. 2010. Rapid and accurate detection of *Arcobacter* contamination in commercial chicken products and waste water samples by real-time polymerase chain reaction. Foodborne Pathog Dis. 7:327–338 ,) reported the presence of *A. butzleri* in 10 out of 50 fresh lettuce samples screened (20%). Recently, some *Arcobacter* species were reported on lettuce (González & Ferrús 2011González AM, Ferrus MA. 2011. Study of *Arcobacter* spp. contamination in fresh lettuces detected by different cultural and molecular methods. Int J Food Microbiol. 145:311–314 ,) and from carrot-processing plant (Hausdorf et al. 2011Hausdorf L, Fröhling A, Schlüter O, Klocke M. 2011. Analyses of the bacterial community within carrot wash water. Can J

Microbiol. 57:447–452 ,).Mottola et al. (2016Mottola A, Bonerba E, Bozzo G, Marchetti P, Celano GV, Colao V, Terio V, Tantillo G, Figueras MJ, Di Pinto A. 2016. Occurrence of emerging food-borne pathogenic Arcobacter spp. isolated from pre-cut (ready-to-eat) vegetables. Int J Food Microbiol. 236:33–37 ,) reported that the occurrence of *Arcobacter* spp. on pre-cut ready-to-eat (RTE) vegetables was 28% (44/160), of which 91% (40/44) isolates were *A. butzleri*and 9.1% (4/44) isolates were *A. cryaerophilus*. The study also revealed widespread distribution of VAGs among the *Arcobacter* isolates tested. These results indicated that the possibility of health risks associated with the direct consumption of raw vegetables. Food-borne infection due to *Arcobacter* spp. was reported from Wisconsin, USA, where attendees of a wedding reception who consumed roasted chicken suffered from the illness. PCR and sequence analysis confirmed two species of *Arcobacter* namely *A. butzleri* and *A. cryaerophilus* from the patients (Lappi et al. 2013Lappi V, Archer JR, Cebelinski E, Leano F, Besser JM, Klos RF, Medus C, Smith KE, Fitzgerald C, Davis JP. 2013. An outbreak of foodborne illness among attendees of a wedding reception in Wisconsin likely caused by *Arcobacter butzleri*. Foodborne Pathog Dis. 10:250–255 ,). Person-to-person transmission of *A. butzleri* was reported in an Italian school with symptoms of recurrent abdominal cramps (Vandamme et al. 1992aVandamme P, Pugina P, Benzi G, Van Etterijck R, Vlaes L, Kersters K, Butzler JP, Lior H, Lauwers S. 1992a. Outbreak of recurrent abdominal cramps associated with *Arcobacter butzleri* in an Italian school. J Clin Microbiol. 30:2335–2337 ; González & Ferrús 2011González AM, Ferrus MA. 2011. Study of *Arcobacter* spp. contamination in fresh lettuces detected by different cultural and molecular methods. Int J Food Microbiol. 145:311–314 ,).

Water-borne Transmission and Prevalences

Contaminated water is considered as an important source of *Arcobacter* infection to human beings (Jacob et al. 1998Jacob J, Woodward D, Feuerpfeil I, Johnson WM. 1998. Isolation of *Arcobacter butzleri* in raw water and drinking water treatment plants in Germany. Zbl Hyg. 201:189–198., ; Ho et al. 2006aHo HT, Lipman LJ, Gaastra W. 2006a. *Arcobacter*, what is known and unknown about a potential food-borne zoonotic agent.Vet Microbiol. 115:1–13. Collado et al. 2010Collado L, Kasimir G, Perez U, Bosch A, Pinto R, Saucedo G, Huguet JM, Figueras MJ. 2010. Occurrence and diversity of *Arcobacter* spp. along the Llobregat river catchment, at sewage effluents and in a drinking water treatment plant. Water Res. 44:3696–3702

,). It has been estimated that 63% of *A. butzleri* infections are acquired in humans by the consumption of or contact with contaminated water (De Smet et al. 2011De Smet S, De Zutter L, Houf K. 2011. Small ruminants as carriers of the emerging food-borne pathogen *Arcobacter* on small and medium farms. Small Rumin Res. 97:124–129.,). *Arcobacters* have been recovered from rivers, lakes, groundwater and seawater, as well as from plankton (Rice et al. 1999Rice EW, Rodgers MR, Wesley IV, Johnson CH, Tanner SA. 1999. Isolation of *Arcobacter butzleri*from ground water. Lett Appl Microbiol. 28:31–35. Moreno et al. 2004Moreno Y, Alonso JL, Botella S, Ferrus MA, Hernandez J. 2004. Survival and injury of *Arcobacter*after artificial inoculation into drinking water. Res Microbiol. 155:726–730. Collado et al. 2008Collado L, Inza I, Guarro J, Figueras M J. 2008. Presence of *Arcobacter*spp. in environmental waters correlates with high levels of fecal pollution. Environ Microbiol. 10:1635–1640. Fera et al. 2010Fera MT, Gugliandolo C, Lentini V, Favaloro A, Bonanno D, La Camera E, Maugeri TL. 2010. Specific detection of *Arcobacter* spp. in estuarine waters of Southern Italy by PCR and fluorescent *in situ*hybridization. Lett Appl Microbiol. 50:65–70 ,). A 4% prevalence of *Arcobacter* was recorded among 175 samples of drinking water collected at Kayseri, Turkey (Ertas et al. 2010Ertas N, Dogruer Y, Gonulalan Z, Guner A, Ulger I. 2010. Prevalence of *Arcobacter* species in drinking water, spring water, and raw milk as determined by multiplex PCR. J Food Prot. 73:2099–2102 ,). Several water-borne *Arcobacter* outbreaks associated with drinking water have been documented including those reported from Idaho, USA (Prouet-Mauleon et al. 2006Prouzet-Mauleon V, Labadi L, Bouges N, Menard A, Megraud F. 2006. *Arcobacter butzleri*: underestimated enteropathogen. Emerg Infect Dis. 12:307–309 ,), Slovenia (Kopilovic et al. 2008Kopilovic B, Ucakar V, Koren N, Krek M, Kraigher A. 2008. Waterborne outbreak of acute gastroentritis in a coastal area in Slovenia in June and July 2008. Eurosurveillance. 13:1–3.,) and Ohio, USA (Fong et al. 2007Fong TT, Mansfield LS, Wilson DL, Schwab DJ, Molly SL, Rose JB. 2007. Massive microbiological ground water contamination associated with a waterborne outbreak in lake Eri South Bass Island, Ohio. Environ Health Perspect. 115:856–864 ,). Same species of *Arcobacters* were detected from drinking water treatment plants and infected humans in Germany showing that water acts as an important source of infection (Jacob et al. 1993Jacob J, Loir H, Feuerpfeil I. 1993. Isolation of *Arcobacter butzleri*from drinking water reservoir in Eastern Germany. Int J Hyg Env Med. 193:557–562. ; Jacob et al. 1998Jacob J, Woodward D, Feuerpfeil I, Johnson WM. 1998. Isolation of *Arcobacter butzleri* in raw water and drinking water treatment plants in Germany. Zbl Hyg. 201:189–198., ; Collado & Figueras 2011Collado L, Figueras MJ. 2011. Taxonomy, epidemiology and clinical relevance of the genus *Arcobacter*. Clin Microbiol Rev. 24:174–192 ,). Seasonal variations in the prevalence rate of *Arcobacter* in water treatment plant in Catalonia, Spain, has also been reported with 92% prevalence in spring, 83% in summer and 75% in winter (Collado et al. 2010Collado L, Kasimir G, Perez U, Bosch A, Pinto R, Saucedo G, Huguet JM, Figueras MJ. 2010. Occurrence and diversity of *Arcobacter* spp. along the Llobregat river catchment, at sewage effluents and in a drinking water treatment plant. Water Res. 44:3696–3702 ,). Reports regarding isolation of *Arcobacters* from sea water have also been presented (Ansari et al. 2015Ansari MI, Harb M, Jones B, Hong PY. 2015. Molecular-based approaches to characterize coastal microbial community and their potential relation to the trophic state of Red Sea. Sci Rep. 5:1–10., ; Levican et al. 2015Levican A, Rubio-Arcos S, Martinez-Murcia A, Collado L, José Figueras M. 2015. *Arcobacter ebronensis* sp. nov. and *Arcobacter aquimarinus*sp. nov., two new species isolated from marine environment. Syst Appl Microbiol. 38:30–35 ,). In this regard, contradictory findings were also stated by Ottaviani et al. (2013Ottaviani D, Chierichetti S, Rocchegiani E, Bartolini C, Masini L, Santarelli S, Leoni F. 2013. Bioaccumulation experiments in mussels contaminated with the food-borne pathogen *Arcobacter butzleri*: preliminary data for risk assessment. Biomed Res Int. 2013:153419) where they found that *A. butzleri* can neither grow experimentally in sea water nor bioaccumualte in mussels. Levican et al. (2014Levican A, Collado L, Yustes C, Aguilar C, Figueras MJ. 2014. Higher water temperature and incubation under aerobic and microaerobic conditions increase the recovery and diversity of *Arcobacter* spp. from shellfish. Appl Environ Microbiol. 80:385–391 ,) found that prevalence of the organism in water sources increased in summer period as the water temperature increased. An interesting study with dishcloth showed the presence of various ciliates along with other food borne pathogen including *Arcobacter* spp. which urge for adopting good hygienic practices while cooking (Chavatte et al. 2014Chavatte N, Baré J, Lambrecht E, Van Damme I, Vaerewijck M, Sabbe K, Houf K. 2014. Co-occurrence of free-living protozoa and foodborne pathogens on dishcloths: implications for food safety. Int J Food Microbiol. 191:89–96 ,). Talay et al. (2016Talay F, Molva C, Atabay HI. 2016. Isolation and identification of *Arcobacter* species from environmental and drinking water samples. Folia Microbiol. 61:479–484. doi:10.1007/s12223-016-0460-0) reported that 36% (41/115) of water samples from various sources (66 sewages, 25 rivers, 16 spring waters and 8 drinking waters) were showed positive for *Arcobacter* spp. in Izmir, Turkey. Among them, *A. butzleri* was reported in 34% of samples (39/115; 24 sewages, 13 rivers and 2 spring waters). The occurrence of Arcobacter spp. in finfish was 19% (8/42), shellfish 15% (5/34) and 21% (5/24) from water samples (Laishram et al. 2016Laishram M, Rathlavath S, Lekshmi M, Kumar S, Nayak BB. 2016. Isolation and characterization of Arcobacter spp. from fresh seafood and the aquatic environment. Int J Food Microbiol. 232:87–89 ,). These results indicated that environmental water samples are common sources for *Arcobacter* spp.

Disease in Humans and Public Health Concerns

A study in South Africa showed *A. butzleri* to be the third most prevalent species in feces of man while in Belgium and France it was reported to be the fourth most prevalent species (Van Driessche et al. 2003Van Driessche E, Houf K, Van Hoof J, De Zutter L, Vandamme P. 2003. Isolation of *Arcobacter* species from animal feces. FEMS Microbiol. 229:243–248., ; Prouzet-Mauleon et al. 2006Prouzet-Mauleon V, Labadi L, Bouges N, Menard A, Megraud F. 2006. *Arcobacter butzleri*: underestimated enteropathogen. Emerg Infect Dis. 12:307–309. Samie et al. 2007Samie A, Obi CL, Barrett LJ, Powell SM, Guerrant RL. 2007. Prevalence of *Campylobacter species*, *Helicobacter pylori* and *Arcobacter*species in stool samples from the Venda region, Limpopo, South Africa: studies using molecular diagnostic methods. J Infect. 54:558–566. Van den Abeele et al. 2014Van den Abeele AM, Vogelaers D, Van Hende J, Houf K. 2014. Prevalence of *Arcobacter* species among humans, Belgium, 2008–2013. Emerg Infect Dis. 20:1731–1734 ,). *Acrobacter* spp. has been associated with enteritis and bacteremia in humans, but still it has not been recognized as a potential human pathogen (Hsueh et al. 1997Hsueh PR, Teng LJ, Yang, PC, Wang SK, Chang SC, Ho SW, Hsieh WC, Luh KT. 1997. Bacteremia caused by *Arcobacter cryaerophilus* 1B. J Clin Microbiol. 35:489–491 ; Engberg et al. 2000Engberg J, On SLW, Harrington CS, Gerner-Smidt P. 2000. Prevalence of *Campylobacter*, *Arcobacter*, *Helicobacter* and *Sutterella* spp. in human fecal samples as estimated by a reevaluation of isolation methods for campylobacters. J Clin Microbiol. 38:286–291 ; Wybo et al. 2004Wybo I, Breynaert J, Lauwers S, Lindenburg F, Houf K. 2004. Isolation of *Arcobacter skirrowii*from a patient with chronic diar-rhea. J Clin Microbiol. 42:1851–1852. Figueras et al. 2014Figueras MJ, Levican A, Pujol I, Ballester F, Rabada, Quilez MJ, Gomez-Bertomeu F. 2014. A severe case of persistent diarrhoea associated with *Arcobacter cryaerophilus* but attributed to *Campylobacter* sp. and a review of the clinical incidence of *Arcobacter*spp. New Microbes New Infect. 2:31–37). *Arcobacter* are also detected in blood samples of patients with clinical conditions like liver cirrhosis and appendiculitis (Yan et al. 2000Yan JJ, Ko WC, Huang AH, Chen HM, Jin YT, Wu JJ. 2000. *Arcobacter butzleri* bacteremia in a patient with liver cirrhosis. J Formos Med Assoc. 99:166–169 ; Lau et al. 2002Lau SKP, Woo PCY, Teng J, Leung KW, Yuen KY. 2002. Identification by 16S ribosomal RNA gene sequencing of *Arcobacter*bacteraemia in a patient with acute gangrenous appendicitis. Mol Pathol. 55:182–185 ; Aydin et al. 2007Aydin F, Gumussoy KS, Atabay HI, Ica T, Abay S. 2007. Prevalence and distribution of *Arcobacter* species in various sources in Turkey and molecular analysis of isolated strains by ERIC-PCR. J Appl Microbiol. 103:27–35 ,). The enteritis caused by *Arcobacter* is an acute diarrhea lasting for 3–15 days, sometimes becoming persistent or recurrent for more than two weeks or even as long as two months (Vandenberg et al. 2004Vandenberg O, Dediste A, Houf K, Ibekwem S, Souayah H, Cadranel S, Douat N, Zissis G, Butzler JP, Vandamme P. 2004. *Arcobacter*species in humans. Emerg Infect Dis. 10:1863–1867 ,). The condition is often accompanied by abdominal pain and nausea and some patients also experience fever, chills, vomiting and weakness (Vandamme et al. 1992aVandamme P, Pugina P, Benzi G, Van Etterijck R, Vlaes L, Kersters K, Butzler JP, Lior H, Lauwers S. 1992a. Outbreak of recurrent abdominal cramps associated with *Arcobacter butzleri* in an Italian school. J Clin Microbiol. 30:2335–2337).

A. butzleri was frequently isolated more from diarrheic cases as compared to non-diarrheic patients. With a prevalence of 8%, *A. butzleri* was found to be the etiological agent of traveller's diarrhea acquired by USA and European travellers to Guatemala, Mexico and India (Jaing et al. 2010Jiang ZD, DuPont HL, Brown EL, Nandy RK, Ramamurthy T, Sinha A, Ghosh, S, Guin S, Gurleen K, Rodrigues S, et al. 2010. Microbial etiology of travelers' diarrhea in Mexico, Guatemala, and India: importance of enterotoxigenic *Bacteroides fragilis* and *Arcobacter*species. J Clin Microbiol. 48:1417–1419. Shah et al. 2011Shah AH, Saleha AA, Zunita Z, Murugaiyah M. 2011. *Arcobacter*– an emerging threat to animals and animal origin food products? Tr Food Sci Technol. 22:225–236., ; McGregor & Wright 2015McGregor AC, Wright SG. 2015. Gastrointestinal symptoms in travellers. Clin Med. 15:93–95 ,). *A. butzleri* shows similar features of *C. jejuni* with watery diarrhea. But the diarrhea associated with *A. butzleri* seems more persistent and is less acute than that caused by *C. jejuni* (Vandenberg et al.

2004Vandenberg O, Dediste A, Houf K, Ibekwem S, Souayah H, Cadranel S, Douat N, Zissis G, Butzler JP, Vandamme P. 2004. *Arcobacter*species in humans. Emerg Infect Dis. 10:1863–1867 ,). Lerner et al. (1994Lerner J, Brumberger V, Preac-Mursic V. 1994. Severe diarrhea associated with *Arcobacter butzleri*. Eur J Clin Microbiol Infect Dis. 13:660–662 ,) isolated *A. butzleri* from the stool samples of two patients with persistent diarrhea and severe abdominal cramps. *A. butzleri* was the only potential enteric pathogen isolated during an outbreak among nursery and primary school children in Italy, in this case, 10 infected children suffered from recurrent abdominal cramps without diarrhea (Vandamme et al. 1992aVandamme P, Pugina P, Benzi G, Van Etterijck R, Vlaes L, Kersters K, Butzler JP, Lior H, Lauwers S. 1992a. Outbreak of recurrent abdominal cramps associated with *Arcobacter butzleri* in an Italian school. J Clin Microbiol. 30:2335–2337). A study conducted on the prevalence of *Helicobacter, Campylobacter*, and *Arcobacter* with 322 stool specimens from patients with and without HIV revealed that *A. butzleri* was the third most prevalent species after *H. pylori* and *C. jejuni* in South Africa (Samie et al. 2007Samie A, Obi CL, Barrett LJ, Powell SM, Guerrant RL. 2007. Prevalence of *Campylobacter species, Helicobacter pylori* and *Arcobacter*species in stool samples from the Venda region, Limpopo, South Africa: studies using molecular diagnostic methods. J Infect. 54:558–566 ,). *A. butzleri* is considered as the fourth most common *Campylobacter*-like organism next to *Campylobacter jejuni, C. coli* and *C. upsaliensis* isolated from the stool of human patients in France and Belgium (Vandenberg et al. 2004Vandenberg O, Dediste A, Houf K, Ibekwem S, Souayah H, Cadranel S, Douat N, Zissis G, Butzler JP, Vandamme P. 2004. *Arcobacter*species in humans. Emerg Infect Dis. 10:1863–1867. Prouzet-Mauleon et al. 2006Prouzet-Mauleon V, Labadi L, Bouges N, Menard A, Megraud F. 2006. *Arcobacter butzleri*: underestimated enteropathogen. Emerg Infect Dis. 12:307–309. Van den Abeele et al. 2014Van den Abeele AM, Vogelaers D, Van Hende J, Houf K. 2014. Prevalence of *Arcobacter* species among humans, Belgium, 2008–2013. Emerg Infect Dis. 20:1731–1734 ,). Houf and Stephan (2007Houf K, Stephan R. 2007. Isolation and characterization of the emerging food-borne pathogen *Arcobacter* from human stool. J Microbiol Methods. 68:408–413 ,) isolated *Arcobacter*from seven of 500 (1.4%) stool samples of healthy people with *A. cryaerophilus* being the only species detected in an Australian adult with intermittent diarrhea and abdominal pain (Tee et al. 1988Tee W, Baird R, Dyall-Smith M, Dwyer B. 1988. *Campylobacter cryaerophila* isolated from a human. J Clin Microbiol. 26:2469–2473). Although the role of *A. skirrowii* in human disease is not yet clear in 2004, the first report of *A. skirrowii* isolation from an elderly person with chronic diarrhea has been published (Wybo et al. 2004Wybo I, Breynaert J, Lauwers S, Lindenburg F, Houf K. 2004. Isolation of *Arcobacter skirrowii*from a patient with chronic diar-rhea. J Clin Microbiol. 42:1851–1852 ,).

Cases of bacteremia common for both *A. butzleri* and *A. cryaerophilus* have infrequently been documented (On et al. 1995On SLW, Stacey A, Smyth J. 1995. Isolation of *Arcobacter butzleri*from a neonate with bacteremia. J Infect. 31:225–227. Hsueh et al. 1997Hsueh PR, Teng LJ, Yang, PC, Wang SK, Chang SC, Ho SW, Hsieh WC, Luh KT. 1997. Bacteremia caused by *Arcobacter cryaerophilus* 1B. J Clin Microbiol. 35:489–491 ; Yan et al. 2000Yan JJ, Ko WC, Huang AH, Chen HM, Jin YT, Wu JJ. 2000. *Arcobacter butzleri* bacteremia in a patient with liver cirrhosis. J Formos Med Assoc. 99:166–169 ; Lau et al. 2002Lau SKP, Woo PCY, Teng J, Leung KW, Yuen KY. 2002. Identification by 16S ribosomal RNA gene sequencing of *Arcobacter*bacteraemia in a patient with acute gangrenous appendicitis. Mol Pathol. 55:182–185). Recently, a case of *A. butzleri* bacteremia was reported from a patient with chronic lymphocytic leukemia (Arguello et al. 2015Arguello E, Otto CC, Mead P, Babady NE. 2015. Bacteremia caused by *Arcobacter butzleri* in an immunocompromised host. J Clin Microbiol. 53:1448–1451 ,). *A. butzleri* was isolated from blood of a 69-year-old woman (Lau et al. 2002Lau SKP, Woo PCY, Teng J, Leung KW, Yuen KY. 2002. Identification by 16S ribosomal RNA gene sequencing of *Arcobacter*bacteraemia in a patient with acute gangrenous appendicitis. Mol Pathol. 55:182–185). Woo et al. (2001Woo PCY, Chong KTK, Leung KW, Que TL, Yuen KY. 2001. Identification of *Arcobacter cryaerophilus* isolated from a traffic accident victim with bacteremia by 16S ribosomal RNA gene sequencing. Diagn Microbiol Infect Dis. 40:125–127 ,) reported isolation of *A. cryaerophilus* from a blood sample of a boy who developed acute respiratory distress and succumbed. Also in a case of a newborn bacteremia wherein vertical/transplacental transmission was assumed as route of transmission (On et al. 1995On SLW, Stacey A, Smyth J. 1995. Isolation of *Arcobacter butzleri*from a neonate with bacteremia. J Infect. 31:225–227 ,).

Reports of *Arcobacter* occurrence were few from North America earlier, but the current status shows that the pathogen has been reported from Argentina, Brazil, Chile, Costa Rica and also from Mexico (Arias et al. 2011Arias ML, Cid A, Fernandéz H. 2011. *Arcobacter butzleri*: first isolation report from chicken carcasses in Costa Rica. Braz J Microbiol. 42:703–

706 ,). The use of various molecular tools in Portugal has revealed that *Arcobacter* was prevalent in pediatric people (Ferreira et al. 2014cFerreira S, Júlio C, Queiroz JA, Domingues FC, Oleastro M. 2014c. Molecular diagnosis of *Arcobacter*and *Campylobacter* in diarrhoeal samples among Portuguese patients. Diagn Microbiol Infect Dis. 78:220–225 ,). *A. butzleri* induced more secretion of pro-inflammatory cytokines, namely TNF, IFN-γ, IL-6 and MCP-1 in small and large intestines. *A. butzleri* also induces more pronounced local and systemic immune responses than commensal *E. coli* in a strain-dependent manner. These data suggest that *A. butzleri* is more than a commensal in vertebrate hosts (Gölz et al. 2016aGölz G, Alter T, Bereswill S, Heimesaat MM. 2016a. The Immunopathogenic Potential of *Arcobacter butzleri* – lessons from a meta-analysis of murine infection studies. PLoS One. 11:e0159685 ,). Zur Bruegge et al. (2016Zur Bruegge J, Backes C, Gölz G, Hemmrich-Stanisak G, Scharek-Tedin L, Franke A, Alter T, Einspanier R, Keller A, Sharbati S. 2016. MicroRNA response of primary human macrophages to *Arcobacter butzleri* infection. Eur J Microbiol Immunol. 6:99–108.,) studied the role of miRNAs in the immune signaling of macrophages infected with *A. butzleri*. Reports showed several miRNA expressions during bacterial infection but authors also found the up-regulated expression of novel miRNAs like miR-155, miR-212 and miR-125 which has a key role in Toll-like receptor signaling (involved in production of immune response). This study reported the interaction of human innate immune cells with *Arcobacter* during the course of infection..

DIAGNOSIS

Identification of *Arcobacter* by different biochemical tests is difficult as these organisms are metabolically inert (Collado & Figueras 2011Collado L, Figueras MJ. 2011. Taxonomy, epidemiology and clinical relevance of the genus *Arcobacter*. Clin Microbiol Rev. 24:174–192 ,), thus isolation of these organisms remains gold standard method for reaching to diagnostic conclusion. Due to lack of accurate detection methods there is no clear data regarding the exact prevalence of *Arcobacter* from different parts of the world (Fallas-Padilla et al. 2014Fallas-Padilla KL, Rodríguez-Rodríguez CE, Fernández Jaramillo H, Arias Echandi ML. 2014. *Arcobacter*: comparison of isolation methods, diversity, and potential pathogenic factors in commercially retailed chicken breast meat from Costa Rica. J Food Prot. 77:880–884 ,). Development of efficient diagnostic methods could lead to more accurate reporting of human infections and the exact prevalence rate of *Arcobacters* (Prouzet-Mauleon et al. 2006Prouzet-Mauleon V, Labadi L, Bouges N, Menard A, Megraud F. 2006. *Arcobacter butzleri*: underestimated enteropathogen. Emerg Infect Dis. 12:307–309. Merga et al. 2011Merga JY, Leatherbarrow AJH, Winstanley C, Bennett M, Hart CA, Miller WG, Williams NJ. 2011. Comparison of *Arcobacter* isolation methods, and diversity of *Arcobacter* spp. in Cheshire, United Kingdom. Appl Environ Microbiol. 77:1646–1650 ,). Current advances in the area of developing rapid and confirmatory diagnostics have provided various molecular methods for rapid and specific detection and differentiation of the *Arcobacters* at genus and species level with higher sensitivity and specificity.

Media for Isolation of Arcobacters

Arcobacters were first recovered and isolated from bovine and porcine fetal tissues cultured in semi-solid Ellinghausen—McCullough–Johnson–Harris (EMJH) medium supplemented with 5-fluorouracil (Ellis et al. 1977Ellis WA, Neill SD, O' Brien JJ, Ferguson, HW, Hanna J. 1977. Isolation of *Spirillum*/*Vibrio*-like organisms from bovine fetuses. Vet Record. 100:451–452. Merga et al. 2011Merga JY, Leatherbarrow AJH, Winstanley C, Bennett M, Hart CA, Miller WG, Williams NJ. 2011. Comparison of *Arcobacter* isolation methods, and diversity of *Arcobacter* spp. in Cheshire, United Kingdom. Appl Environ Microbiol. 77:1646–1650 ,). Johnson and Murano (1999aJohnson LG, Murano EA. 1999a. Development of a new medium for the isolation of *Arcobacter* spp. J Food Prot. 62:456–462 ,) developed Johnson–Murano (JM) broth and plates, and Houf et al. (2001Houf K, Devriese LA, De zutter L, Van Hoof J, Vandamme P. 2001. Susceptibility of *Arcobacter butzleri, Arcobacter cryaerophilus*, and *Arcobacter skirrowii* to antimicrobial agents used in selective media. J Clin Microbiol. 39:1654–1656 ,) developed an *Arcobacter*-specific isolation method involving the use of *Arcobacter* medium (Oxoid, United Kingdom) with addition of five antimicrobials (cefoperazone, trimethoprim, amphotericin, novobiocin and 5-fluorouracil). Other isolation methods included EMJH p-80 and *Brucella* broth (Ongor et al. 2004Ongor H, Cetinkaya B, Açik MN, Atabay HI. 2004. Investigation of Arcobacters in meat and fecal samples of clinically healthy cattle in Turkey. Lett Appl Microbiol. 38:339–344 ,) and direct filtration onto agar without antibiotics via a membrane (Merga et al. 2011Merga JY, Leatherbarrow AJH, Winstanley C, Bennett M, Hart CA, Miller WG, Williams NJ. 2011. Comparison of *Arcobacter* isolation methods, and diversity of *Arcobacter* spp. in Cheshire, United Kingdom. Appl Environ Microbiol. 77:1646–1650 ,). Houf et al. (2001aHouf K,

Devriese LA, De zutter L, Van Hoof J, Vandamme P. 2001. Susceptibility of *Arcobacter butzleri, Arcobacter cryaerophilus*, and *Arcobacter skirrowii* to antimicrobial agents used in selective media. J Clin Microbiol. 39:1654–1656 ,) proposed media with cefoperazone, amphotericin B and teicoplanin (CAT) supplement and EMJH P80 with 5-fluorouracil were sufficient to isolate three *Arcobacter*species, namely *A. butzleri, A. cryaerophilus* and *A. skirrowii. Arcobacter* broth supplemented with CAT was also evaluated for the recovery of these three species (Atabay & Corry 1998Atabay HI, Corry JEL. 1998. Evaluation of a new *Arcobacter*enrichment medium and comparison with new media developed for enrichment for *Campylobacter spp*. Int J Food Microbiol. 41:53–58. Rahimi 2014Rahimi E. 2014. Prevalence and antimicrobial resistance of *Arcobacter* species isolated from poultry meat in Iran. Br Poult Sci. 55:174–180). With 5%–7% defibrinised blood, the samples were incubated on to the *Arcobacter* media for 48 h at 30 °C under microaerophilic conditions (Aydin et al. 2007Aydin F, Gumussoy KS, Atabay HI, Ica T, Abay S. 2007. Prevalence and distribution of *Arcobacter* species in various sources in Turkey and molecular analysis of isolated strains by ERIC-PCR. J Appl Microbiol. 103:27–35. Ramees et al. 2014bRamees TP, Rathore RS, Bagalkot PS, Mohan HV, Kumar A, Dhama K. 2014b. Detection of *Arcobacter butzleri* and *Arcobacter cryaerophilus* in clinical samples of humans and foods of animal origin by cultural and multiplex PCR based methods. Asian J Anim Vet Adv. 9:243–252.,).

Modified charcoal cefoperazone deoxycholate (mCCDA) agar supplemented with CAT antibiotics was used for the isolation of *Arcobacter* from poultry carcasses (Lammerding et al. 1994Lammerding AM, Harris JE, Loir H, Muckel CA, Irwin RJ. 1994. An isolation method for *Arcobacter butzleri* from poultry. Dairy Food Environ Sanitation. 14:600. , 1996Lammerding AM, Harris JE, Lior H, Woodward D, Cole L, Muckle CA. 1996. Isolation method for recovery of *Arcobacter butzleri* from fresh poultry and poultry products. In Newell DG, Ketley J, Feldman RA, editors. 8th International Workshop on '*Campylobacters, Helicobacters*and Related Organisms'. New York (NY): Plenum Press; p. 329–334.,). de Boer et al. (1996de Boer E, Tilburg JJHC, Woodward DL, Loir H, Johnson WM. 1996. A selective medium for the isolation of *Arcobacter* from meats. Lett Appl Microbiol. 23:64–66 ,) developed an *Arcobacter* selective enrichment broth (ASB) and *Arcobacter* selective semisolid medium (ASM) for the recovery of *Arcobacter* from food. Collins et al. (1996Collins CI, Wesley IV, Murano EA. 1996. Detection of *Arcobacter* spp. in ground pork by modified plating methods. J Food Protect. 59:448–452.,) developed modified cefsulodinirgasan-novobiocin (mCIN) agar with CAT supplement for the recovery of *Arcobacter* from poultry. Atabay et al. (1996Atabay HI, Corry JEL, Prost DE. 1996. Comparison of the productivity of a variety of selective media for *Campylobacter* and *Arcobacter* species In: Newell DJ, Ketley JM, Felman RA, editors. *Campylobacters, Helicobacters* and related organisms New York (NY): Plenum Press; p. 19–23.,) tested the suitability of Karmali medium, mCCDA agar, semi-solid blood-free selectivity-motility (SSM) medium and CAT medium to grow *Arcobacters* and *Campylobacters*. Two strains of *A. butzleri* grew well only on Karmali medium and SSM. *A. skirrowii* grew poorly on all the selective media used, while *A. cryaerophilus* did not grow on any of the selective media used. Atabay and Corry (1997Atabay HI, Corry JEL. 1997. The prevalence of *Campylobacters* and *Arcobacters* in broiler chickens. J Appl Microbiol. 83:619–626 ,) compared CAT agar and mCCDA for the growth of various *Arcobacter* strains and found that CAT agar performed better. Merga et al. (2011Merga JY, Leatherbarrow AJH, Winstanley C, Bennett M, Hart CA, Miller WG, Williams NJ. 2011. Comparison of *Arcobacter* isolation methods, and diversity of *Arcobacter* spp. in Cheshire, United Kingdom. Appl Environ Microbiol. 77:1646–1650 ,) compared different *Arcobacter* spp. isolation methods from feces like Houf broth and Houf plates (HH), Houf broth with mCCDA-CAT plates (HCC), *Arcobacter* broth-CAT broth with Houf plates (ACH), *Arcobacter* broth-CAT broth with mCCDA-CAT plates (ACCC) and *Campylobacter*-specific broth and *Campylobacter*-specific plates (CC). This study revealed that HCC method had higher sensitivity (71%) and specificity (64%).

Johnson and Murano (1999aJohnson LG, Murano EA. 1999a. Development of a new medium for the isolation of *Arcobacter* spp. J Food Prot. 62:456–462 , , 1999bJohnson LG, Murano EA. 1999b. Comparison of three protocols for the isolation of *Arcobacter* from poultry. J Food Prot. 62:610–614 ,) developed an enrichment broth and agar medium to eliminate the growth of *Campylobacter* species. *A. butzleri, A. cryaerophilus*and *A. nitrofigilis* produced red colonies on the medium. Engberg et al. (2000Engberg J, On SLW, Harrington CS, Gerner-Smidt P. 2000. Prevalence of *Campylobacter*, *Arcobacter*, *Helicobacter* and *Sutterella* spp. in human fecal samples as estimated by a reevaluation of isolation methods for campylobacters. J Clin Microbiol. 38:286–291) and Atabay et al. (2002Atabay HI, Aydin F, Houf K, Sahin M, Vandamme P. 2002. The prevalance of *Arcobacter* spp. on chicken carcasses sold in retail markets in Turkey, and identification of the isolates using SDS-PAGE. Int J Food Microbiol. 81:21–28.,) found that the membrane filtration method is superior to the isolation method without filtration as the percentage of positive samples increased after filtration. Van Driessche et al. (2003Van Driessche E, Houf K, Van Hoof J, De Zutter L, Vandamme P. 2003. Isolation of *Arcobacter* species from animal feces. FEMS Microbiol. 229:243–248.,) proposed *Arcobacter* selective isolation broth (ASIB) and *Arcobacter* selective isolation agar (ASIA) for selective isolation of *Arcobacter*from feces.

Hamill et al. (2008Hamill S, Neill SD, Madden RH. 2008. A comparison of media for the isolation of *Arcobacter* spp. from retail packs of beef. J Food Protect. 71:850–854 ,) compared different media for the isolation of *Arcobacter* spp. from retail packs of beef and reported that the modified enrichment and selective media of Johnson and Murano gave the highest recovery (15%) of *Arcobacters*. However, direct detection from the CAT broth by m-PCR and identification of isolates recovered by culture after passive filtration of the broth on blood agar (without any antibiotic supplement) produced more or less the same results in other studies (Collado et al. 2008Collado L, Inza I, Guarro J, Figueras M J. 2008. Presence of *Arcobacter*spp. in environmental waters correlates with high levels of fecal pollution. Environ Microbiol. 10:1635–1640 , , 2009aCollado L, Guarro J, Figueras M. 2009a. Prevalence of *Arcobacter* in meat and shellfish. J Food Protect. 72:1102–1110 , , 2010Collado L, Kasimir G, Perez U, Bosch A, Pinto R, Saucedo G, Huguet JM, Figueras MJ. 2010. Occurrence and diversity of *Arcobacter* spp. along the Llobregat river catchment, at sewage effluents and in a drinking water treatment plant. Water Res. 44:3696–3702 ,). A new culture approach was recently adopted for enhanced isolation of *Arcobacter* from marine samples. Addition of NaCl to the growth media enhances the number of positive samples (Salas-Massó et al. 2016Salas-Massó N, Andree KB, Furones MD, Figueras MJ. 2016. Enhanced recovery of Arcobacter spp. using NaCl in culture media and re-assessment of the traits of *Arcobacter marinus* and *Arcobacter halophilus* isolated from marine water and shellfish. Sci Total Environ. 566–567:1355–1361 ,). Possibility of isolation of *A. butzleri* increased when the samples are pre-enriched before plating (Levican et al. 2016Levican A, Collado L, Figueras MJ. 2016. The use of two culturing methods in parallel reveals a high prevalence and diversity of *Arcobacter* spp. in a wastewater treatment plant. Biomed Res Int. 2016:8132058).

Detection of Arcobacter spp. by molecular methods

Cultural differentiation is very difficult between *Arcobacter* and

Campylobacter because of their phenotypic similarity and so molecular methods are more useful (Douidah et al. 2010Douidah L, Zutter LD, Vandamme P, Houf K. 2010. Identification of five human and mammal associated *Arcobacter* species by a novel multiplex-PCR assay. J Microbiol Methods. 80:281–286. Collado & Figueras 2011 Collado L, Figueras MJ. 2011. Taxonomy, epidemiology and clinical relevance of the genus *Arcobacter*. Clin Microbiol Rev. 24:174–192 ,). Similarly, PCR combined with cultural methods has shown good results in drafting the picture of *Arcobacter* from sea water in Italy (Maugeri et al. 2005Maugeri TL, Irrera GP, Lentini V, Carbone M, Fera MT, Gugliandolo C. 2005. Detection and enumeration of *Arcobacter* spp. in the coastal environment of the straits of Messina (Italy). New Microbiol. 28:177–182). This has led to the development of many different molecular detection methods like PCR, multiplex PCR (m-PCR) and real-time PCR, restriction fragment length polymorphism (RFLP), matrix-assisted laser desorption ionization mass spectrometry (MALDITOF MS), denaturing gradient gel electrophoresis PCR (DGGE-PCR) and fluorescence *in situ* hybridization (FISH) (Petersen et al. 2007Petersen RF, Harrington CS, Kortegaard HE, On SL. 2007 . A PCR-DGGE method for detection and identification of *Campylobacter*, *Helicobacter*, *Arcobacter* and related Epsilobacteria and its application to saliva samples from humans and domestic pets. J Appl Microbiol. 103:2601–2615. Alispahic et al. 2010Alispahic M, Hummel K, Jandreski-Cvetkovic D, Nöbauer K, Razzazi-Fazeli E, Hess M, Hess C. 2010. Species-specific identification and differentiation of *Arcobacter*, *Helicobacter* and *Campylobacter* by full-spectral matrix-associated laser desorption/ionization time of flight mass spectrometry analysis. J Med Microbiol. 59:295–301. Collado & Figueras 2011Collado L, Figueras MJ. 2011. Taxonomy, epidemiology and clinical relevance of the genus *Arcobacter*. Clin Microbiol Rev. 24:174–192. Patyal et al. 2011Patyal A, Rathore RS, Mohan HV, Dhama K, Kumar A. 2011. Prevalence of *Arcobacter* spp. in humans, animals and foods of animal origin including sea food from India. Transbound Emerg Dis. 58:402–410. Mohan et al. 2014Mohan HV, Rathore RS, Dhama K, Ramees TP, Patyal A, Bagalkot PS, Wani MY, Bhilegaonkar KN, Kumar A. 2014. Prevalence of *Arcobacter*spp. in humans, animals and foods of animal origin in India based on cultural isolation, antibiogram, PCR and multiplex PCR detection. Asian J Anim Vet Adv. 9:452–466.,).

Detection of *Arcobacter* in chicken meat at genus level by PCR was developed by González et al. (2014González I, Fernández-Tomé S, García T, Martín R. 2014. Genus-specific PCR assay for screening Arcobacter spp. in chicken meat. J Sci Food Agric. 94:1218–1224 ,) which has been suggested to be a sensitive tool for epidemiological survey. A nested PCR targeting the 16S rRNA gene to detect *Arcobacter* spp. and *Helicobacter*spp. was developed by Suarez et al. (1997Suarez DL, Wesley IV, Larson DJ. 1997. Detection of *Arcobacter*species in gastric samples from swine. Vet Microbiol. 57:325–336 ,). Multiplex PCR to detect both *C. jejuni* and *A. butzleri* in the food products was proposed by Winters and Slavik (2000Winters DK, Slavik MF. 2000. Multiplex PCR detection of *Campylobacter jejuni* and *Arcobacter butzleri* in food products. Mol Cell Probes. 14:95–99 ,). González et al. (2000González I, Garcia T, Antonlin A, Hernandez PE, Martin R. 2000. Development of a combined PCR-culture technique for the rapid detection of *Arcobacter* spp. in chicken meat. Lett Appl Microbiol. 30:207–212 ,) developed combined PCR-culture technique for the rapid detection of *Arcobacter* spp. in chicken meat following a short selective enrichment of samples. Multiplex real-time PCR was developed for detection of *A. butzleri* and several *Campylobacter* spp. from stool samples (de Boer et al. 2013de Boer RF, Ott A, Güren P, van Zanten E, van Belkum A, Kooistra-Smid AM. 2013. Detection of *Campylobacter* species and *Arcobacter butzleri* in stool samples by use of real-time multiplex PCR. J Clin Microbiol. 51:253–259 ,). Neubauer and Hess (2006Neubaure C, Hess M. 2006. Detection and identification of food borne pathogens of the genera *Campylobacter, Arcobacter* and *Helicobacter* by multiplex PCR in poultry and poultry products. J Vet Med. 53:376–381.,) developed a multiplex PCR protocol which can detect the three closely related genera, i.e. *Campylobacter, Helicobacter* and *Arcobacter* in a single reaction cycle from poultry products using primers targeted to amplify the 16S rRNA genes. Wesley et al.(1995Wesley IV, Schroder-Tucker L, Baetz AL, Dewhurst FE, Paster BJ. 1995. *Arcobacter butzleri*-specific 16S rRNA based probes. J Clin Microbiol. 33:1691–1698) have designed *Arcobacter* genus and species specific 16S and 23S rRNA probes. Bastyns et al. (1995Bastyns K, Cartuyvels D, Chapelle S, Vandamme P, Goossens H, De Wachter R. 1995. A variable 23S rDNA region is a useful discriminating target for genus-specific and species-specific PCR amplification in *Arcobacter* species. System Appl Microbiol. 18:353–356.,) developed a genus-specific PCR assay for *Arcobacter* strains using the variable regions of 23S rRNA. A multiplex PCR assay was developed by Harmon and Wesley (1997Harmon KM, Wesley IV. 1997. Multiplex PCR for the identification of Arcobacter and differentiation of *Arcobacter butzleri* from other Arcobacters. Vet Microbiol. 58:215–227 ,) to identify *Arcobacter* isolates, in particular

A. butzleri. Harmon and Wesley (1996Harmon KM, Wesley IV. 1996. Identification of *Arcobacter* isolates by PCR. Lett Appl Microbiol. 23:241–244 ,) developed a PCR assay targeting 16S rRNA for the identification of *A. butzleri, A. cryaerophilus* and *A. skirrowii*. Using a variable 16S rRNA and 23S rRNA region, Houf et al. (2000Houf K, Tutenel A, De Zutter L, Van Hoof J, Vandamme P. 2000. Development of a multiplex PCR assay for the simultaneous detection and identification of *Arcobacter butzleri, Arcobacter cryaerophilus* and *Arcobacter skirrowii*. FEMS Immunol Med Microbiol. 49:337–345.) developed a species-specific multiplex PCR assay for the simultaneous detection and identification of *A. butzleri, A. cryaerophilus* and *A. skirrowii*. Kabeya et al. (2003bKabeya H, Koboyashi Y, Maruyama S, Mikami T. 2003b. One step polymerase chain reaction-based typing of *Arcobacter* species. Int J Food Microbiol. 81:163–168 ,) developed an *Arcobacter* species-specific PCR assay for the detection of *A. butzleri, A. cryaerophilus 1A* and *1B* and *A. skirrowii*. Levican and Figueras (2013Levican A, Figueras MJ. 2013. Performance of five molecular methods for monitoring *Arcobacter*spp. BMC Microbiol. 13:220 ,) compared and also reviewed commonly used molecular methods for detection of *Arcobacter* spp. Authors reported that mPCR of Houf et al. (2000Houf K, Tutenel A, De Zutter L, Van Hoof J, Vandamme P. 2000. Development of a multiplex PCR assay for the simultaneous detection and identification of *Arcobacter butzleri, Arcobacter cryaerophilus* and *Arcobacter skirrowii*. FEMS Immunol Med Microbiol. 49:337–345.) was most commonly referred followed by Figueras et al. (2008Figueras MJ, Collado L, Guarro J. 2008. A new 16S rDNA-RFLP method for the discrimination of the accepted species of *Arcobacter*. Diag Microbiol Infect Dis. 62:11–15 ,) which employed 16S rDNA-RFLP and then m-PCR of Douidah et al. (2010Douidah L, Zutter LD, Vandamme P, Houf K. 2010. Identification of five human and mammal associated *Arcobacter* species by a novel multiplex-PCR assay. J Microbiol Methods. 80:281–286 ,). However, Levican and Figueras (2013Levican A, Figueras MJ. 2013. Performance of five molecular methods for monitoring *Arcobacter*spp. BMC Microbiol. 13:220 ,) found that none of the five PCR developed by different authors (Houf et al. 2000Houf K, Tutenel A, De Zutter L, Van Hoof J, Vandamme P. 2000. Development of a multiplex PCR assay for the simultaneous detection and identification of *Arcobacter butzleri, Arcobacter cryaerophilus* and *Arcobacter skirrowii*. FEMS Immunol Med Microbiol. 49:337–345. ; Kabeyaet al. 2003aKabeya H, Maruyama S, Morita Y, Kubo M, Yamamoto K, Arai S, Izumi T, Kobayashi Y, Katsube Y, Mikami T. 2003a. Distribution of *Arcobacter*species among livestock in Japan. Vet Microbiol. 93:153–158. Figueras et al. 2008Figueras MJ, Collado L, Guarro J. 2008. A new 16S rDNA-RFLP method for the discrimination of the accepted species of *Arcobacter*. Diag Microbiol Infect Dis. 62:11–15. Pentimalli et al. 2009Pentimalli D, Pegels N, Garcia T, Martin R, González I. 2009. Specific PCR detection of *Arcobacter butzleri, Arcobacter cryaerophilus, Arcobacter skirrowii* and *Arcobacter cibarius* in chicken meat. J Food Protect. 72:1491–1495. Douidah et al. 2010Douidah L, Zutter LD, Vandamme P, Houf K. 2010. Identification of five human and mammal associated *Arcobacter* species by a novel multiplex-PCR assay. J Microbiol Methods. 80:281–286 ,) were reliable. Authors concluded that result was bad for identification of *A. cryaerophilus* and *A. butzleri* using 23S rRNA gene as the target (Levican & Figueras 2013Levican A, Figueras MJ. 2013. Performance of five molecular methods for monitoring *Arcobacter*spp. BMC Microbiol. 13:220 ,). Samarajeewa et al. (2015Samarajeewa AD, Hammad A, Masson L, Khan IU, Scroggins R, Beaudette LA. 2015. Comparative assessment of next-generation sequencing, denaturing gradient gel electrophoresis, clonal restriction fragment length polymorphism and cloning-sequencing as methods for characterizing commercial microbial consortia. J Microbiol Methods. 108:103–111 ,) compared four diagnostic assays namely next-generation sequencing (Ion torrent technology), DGGE, clonal restriction fragment length polymorphism (C/RFLP) and cloning-sequencing methods for detection of microbes and concluded that all the methods detected *Arcobacter* spp. Sequencing of PCR products and use of chaperonin sequence database (cpnDB) can help in differentiation of *Campylobacter, Arcobacter* and *Helicobacter* isolates (Hill et al. 2006Hill JE, Paccagnella A, Law K, Melito PL, Woodward DL, Price L, Leung AH, Ng LK, Hemmingsen SM, Goh SH. 2006. Identification of Campylobacter spp. and discrimination from Helicobacter and Arcobacter spp. by direct sequencing of PCR-amplified cpn60 sequences and comparison to cpnDB, a chaperonin reference sequence database. J Med Microbiol. 55:393–399 ,).

PCR plus ELISA (Antolin et al. 2001Antolin A, González I, Garcia T, Hernandez PE, Martin R. 2001. *Arcobacter* spp. enumeration in poultry meat using a combined PCR-ELISA assay. Meat Sci. 59:169–174 ,) and real-time PCR assays (qRT-PCR) (Abdelbaqi et al. 2007aAbdelbaqi K, Buissonniere A, Prouzet-Mauleon V, Gresser J, Wesley I, Megraud F, Menard A. 2007a. Development of a realtime fluorescence resonance energy transfer PCR to detect *Arcobacter*species. J Clin Microbiol. 45:3015–3021. Brightwell et al. 2007Brightwell G, Mowat E, Clemens R, Boerema J, Pulford DJ, On SL. 2007. Development of a multiplex and real time PCR assay for the specific detection of *Arcobacter butzleri*and *Arcobacter cryaerophilus*. J Microbiol Methods. 68:318–325 ,) have also been developed for detection of *Arcobacter*. A SYBR green real-time PCR assay was developed for rapid detection of *Arcobacter* in food and wastewater samples at the genus level (González et al. 2010González A, Suski J, Ferrus MA. 2010. Rapid and accurate detection of *Arcobacter* contamination in commercial chicken products and waste water samples by real-time polymerase chain reaction. Foodborne Pathog Dis. 7:327–338 ,). PCR-RFLP was developed for identification of three different pathogens namely *Campylobacter*, *Arcobacter* and *Helicobacter* (Marshall et al. 1999Marshall SM, Melito PL, Woodward DL, Johnson WM, Rodgers FG, Mulvey MR. 1999. Rapid identification of *Campylobacter*, *Arcobacter*, and *Helicobacter*isolates by PCR-restriction fragment length polymorphism analysis of the 16S rRNA gene. J Clin Microbiol. 37:4158–4160). Vergis et al. (2013Vergis J, Negi M, Poharkar K, Das DP, Malik SV, Kumar A, Doijad SP, Barbuddhe SB, Rawool DB. 2013. 16S rRNA PCR followed by restriction endonuclease digestion: a rapid approach for genus level identification of important enteric bacterial pathogens. J Microbiol Methods. 95:353–356 ,) developed a 16s rRNA PCR combined with RE digestion using *Eco*RI, *Hind*III and *Sal*I enzymes which differentiated eight food-borne pathogens (*Listeria, Escherichia, Salmonella, Shigella, Vibrio, Campylobacter, Yesinia* and *Arcobacter*). A 16S rDNA-RFLP method was developed using *Mse*I enzyme for six important species of *Arcobacter*including two hybridization groups of *A. cryaerophilus* (Figueras et al. 2008Figueras MJ, Collado L, Guarro J. 2008. A new 16S rDNA-RFLP method for the discrimination of the accepted species of *Arcobacter*. Diag Microbiol Infect Dis. 62:11–15 ,). Later as the number of species in the genus increased to 17 as per the record of 2012, RFLP studies using the same enzyme *Mse*I yielded distinct pattern for 10 species (*A. nitrofigilis, A. cryaerophilus, A. skirrowii, A. cibarius, A. halophilus, A. mytili, A. molluscorum, A. ellisii, A. bivalviorum* and *A. cloacae*) while other species namely *A. marinus, A. venerupis, A. defluvii, A. suis, A. butzleri, A. thereius and A. trophiarum* showed similar pattern. Hence, different enzymes namely *Mnl*I and *Bfa*I were used to differentiate these species. This study also documented the microheterogenicity among *Arcobacter* spp. (Figueras et al. 2012Figueras MJ, Levican A, Collado L. 2012. Updated 16S rRNA-RFLP method for the identification of all currently characterised *Arcobacter*spp. BMC Microbiol. 12:292 ,).

Pentimalli et al. (2009Pentimalli D, Pegels N, Garcia T, Martin R, González I. 2009. Specific PCR detection of *Arcobacter butzleri, Arcobacter cryaerophilus, Arcobacter skirrowii* and *Arcobacter cibarius* in chicken meat. J Food Protect. 72:1491–1495 ,) developed a PCR assay using species-specific primers for the detection of *A. butzleri, A. cryaerophilus, A. skirrowii* and *A. cibarius* in chicken meat. A multiplex-PCR assay with seven primers was developed by Douidah et al. (2010Douidah L, Zutter LD, Vandamme P, Houf K. 2010. Identification of five human and mammal associated *Arcobacter* species by a novel multiplex-PCR assay. J Microbiol Methods. 80:281–286 ,) for the identification of the five human and mammal related species of *Arcobacter* (*A. butzleri, A. cryaerophilus, A. skirrowii, A. thereius* and *A. cibarius*). Recently, Webb et al. (2016Webb AL, Boras VF, Kruczkiewicz P, Selinger LB, Taboada EN, Inglis GD. 2016. Comparative detection and quantification of *Arcobacter butzleri* in stools from diarrheic and nondiarrheic people in Southwestern Alberta, Canada. J Clin Microbiol. 54:1082–1088 ,) developed the PCR primers for direct detection and quantification of *A. butzleri* DNA in microbiologically complex matrices. Densities of *A. butzleri* DNA in diarrheic stools were higher when compared to non-diarrheic stools in southwestern Alberta, Canada. The study also reported that among 892 *A. butzleri* positive diarrheic samples, 74% were not positive for other bacterial or viral pathogens. Miller et al. (2014Miller WG, Yee E, Jolley KA, Chapman MH. 2014. Use of improved *atpA* amplification and sequencing method to identify members of the Campylobacteraceae and Helicobacteraceae. Lett Appl Microbiol. 58:582–590 ,) formulated a new method for identification of *Arcobacter* and *Campylobacter* by using single primer pair to amplify the region of *atpA* gene followed by sequencing. MALDI-TOF MS has been developed for fast and accurate identification of *Arcobacter*, *Helicobacter* and *Campylobacter* species approving to be a reliable assay (Alispahic et al. 2010Alispahic M, Hummel K, Jandreski-Cvetkovic D, Nöbauer K, Razzazi-Fazeli E, Hess M, Hess C. 2010. Species-specific identification and differentiation of *Arcobacter*, *Helicobacter* and *Campylobacter* by full-spectral matrix-associated laser desorption/ionization time of flight mass spectrometry analysis. J Med Microbiol. 59:295–301 ,). Whiteduck-Léveillée et al. (2016aWhiteduck-Léveillée J, Cloutier M, Topp E, Lapen DR, Talbot G, Villemur R, Khan IU. 2016a. Development and evaluation of multiplex PCR assays for rapid detection of virulence-associated genes in *Arcobacter* species. J Microbiol Methods. 121:59–65 ,) developed and evaluated the multiplex PCR technique for quick detection of VAGs in *Arcobacter* spp. This technique could be useful for the detection and prevalence of VAGs in the *Arcobacter* spp.

Recently developed diagnostic techniques at field level like loop-mediated isothermal amplification (LAMP) test is very useful in pathogen detection with speed and accuracy (Karthik et al. 2014aKarthik K, Rathore R, Thomas P, Arun TR, Viswas KN, Agarwal RK, Manjunathachar HV, Dhama K. 2014a. Loop-mediated isothermal amplification (LAMP) test for specific and rapid detection of *Brucella abortus* in cattle. Vet Quart. 34:174–179). LAMP has been designed for various pathogens and has also been developed for detection of *Arcobacter* spp. (Dhama et al. 2014aDhama K, Karthik K, Chakraborty S, Tiwari R, Kapoor S, Kumar A, Thomas P. 2014a. Loop-mediated isothermal amplification of DNA (LAMP) – a new diagnostic tool lights the world of diagnosis of animal and human pathogens: a review. Pak J Biol Sci. 17:151–166 ; Karthik et al. 2014bKarthik K, Rathore R, Thomas P, Arun TR, Viswas KN, Dhama K, Agarwal RK. 2014b. New closed tube loop mediated isothermal amplification assay for prevention of product cross contamination. MethodsX. 1:e137–e143 ; Wang et al. 2014Wang X, Seo DJ, Lee MH, Choi C. 2014. Comparison of conventional PCR, multiplex PCR and loop-mediated isothermal amplification assays for rapid detection of *Arcobacter* species. J Clin Microbiol. 52:557–563 ,). Results showed that the developed LAMP assay was more sensitive than the other molecular assays like multiplex PCR and conventional PCR that were compared in the study (Wang et al. 2014Wang X, Seo DJ, Lee MH, Choi C. 2014. Comparison of conventional PCR, multiplex PCR and loop-mediated isothermal amplification assays for rapid detection of *Arcobacter* species. J Clin Microbiol. 52:557–563 ,). Chicken meat samples were used in the study and LAMP assay could detect *Arcobacter* spp. specifically and sensitively. The major issue with molecular methods for detection of pathogen is that these assays cannot guarantee the viability of the organisms hence status of infection cannot be assessed clearly (Samarajeewa et al. 2015Samarajeewa AD, Hammad A, Masson L, Khan IU, Scroggins R, Beaudette LA. 2015. Comparative assessment of next-generation sequencing, denaturing gradient gel electrophoresis, clonal restriction fragment length polymorphism and cloning-sequencing as methods for characterizing commercial microbial consortia. J Microbiol Methods. 108:103–111 ,).

An overview of different molecular diagnostic assays for detection of *Arcobacters* along their advantages and disadvantages/limitations is presented in Table 3.

Table 3. An overview of molecular diagnostic assays for detection of *Arcobacters*.

S. no.	Diagnostic test	Gene targeted	Advantage	Limitation	References
(1)	Multiplex PCR	23S rRNA gene	• Single step differentiation of *Arcobacter butzleri, A. cryaerophilus 1A and 1B, and A. skirrowii*	• Higher chances of misidentification of non-targeted species • Subgroup level identification of *A. cryaerophilus*was not proper	Kabeya et al. (2003bKabeya H, Koboyashi Y, Maruyama S, Mikami T. 2003b. One step polymerase chain reaction-based typing of *Arcobacter*species. Int J Food Microbiol. 81:163–168. Levican and Figueras (2013Levican A, Figueras MJ. 2013. Performance of five molecular methods for monitoring *Arcobacter* spp. BMC Microbiol. 13:220.
(2)		16S rRNA and 23S rRNA	• *Arcobacter butzleri, A. cryaerophilus and A. skirrowii* • 100% reliable identification of *A. butzleri*	• False positive reaction occurs for *A. cryaerophilus*and *A. skirrowii* • Could not differentiate other *Arcobacter* spp.	Houf et al. (2000Houf K, Tutenel A, De Zutter L, Van Hoof J, Vandamme P. 2000. Development of a multiplex PCR assay for the simultaneous detection and identification of *Arcobacter butzleri, Arcobacter cryaerophilus* and *Arcobacter skirrowii*. FEMS Immunol Med Microbiol. 49:337–345. Levican and Figueras (2013Levican A, Figueras MJ. 2013. Performance of five molecular methods for monitoring *Arcobacter* spp. BMC Microbiol. 13:220.
(3)		*gyrA* gene	• Differentiates A. butzleri, A. cryaerophilus, A. skirrowii, and A. cibarius	• Misidentification of A. cibarius as A. butzleri and A. skirrowii	Pentimalli et al. (2009Pentimalli D, Pegels N, Garcia T, Martin R, González I. 2009. Specific PCR detection of *Arcobacter butzleri, Arcobacter cryaerophilus, Arcobacter skirrowii* and *Arcobacter cibarius* in chicken meat. J Food Protect. 72:1491–1495.; Levican and Figueras (2013Levican A, Figueras MJ. 2013. Performance of five molecular methods for monitoring *Arcobacter* spp. BMC Microbiol. 13:220.
(4)		23S RNA	• Differentiation of *A. butzleri, A. cryaerophilus, A. skirrowii, A. cibarius and A. thereuis* • 100% reliable for *A. skirrowii, A. cibarius*, and *A. thereius*identification	• Could not detect A. trophiarum • Misidentifies *A. defluvii, A. ellisii, A. venerupis, and A. suis* as *A. butzleri*	Douidah et al. (2010Douidah L, Zutter LD, Vandamme P, Houf K. 2010. Identification of five human and mammal associated *Arcobacter* species by a novel multiplex-PCR assay. J Microbiol Methods. 80:281–286.
(5)		*A. butzleri* (rpoB/C), *A. cryaerophilus* (23s rRNA)	• Differentiates *A. butzleri, A. cryaerophilus*	• Could not differentiate other *Arcobacter* spp.	Brightwell et al. (2007Brightwell G, Mowat E, Clemens R, Boerema J, Pulford DJ, On SL. 2007. Development of a multiplex and real time PCR assay for the specific detection of *Arcobacter butzleri* and *Arcobacter cryaerophilus*. J Microbiol Methods. 68:318–325.
(6)	Real time PCR	23S rRNA	• Sensitive and specific compared to conventional PCR assay • Detection level up to 5.6 X 10 CFU/mL	• Needs sophisticated instruments • Needs skilled persons	González et al. (2010González A, Suski J, Ferrus MA. 2010. Rapid and accurate detection of *Arcobacter*contamination in commercial chicken products and waste water samples by real-time polymerase chain reaction. Foodborne Pathog Dis. 7:327–338.
(7)	Multiplex real time PCR	Campylobacter spp. (16S rRNA gene), *Arcobacter butzleri* (hsp60)	• Differentiates *Arcobacter butzleri* and several *Campylobacter* spp.	• Needs skilled persons	de Boer et al. (2013de Boer RF, Ott A, Güren P, van Zanten E, van Belkum A, Kooistra-Smid AM. 2013. Detection of *Campylobacter* species and *Arcobacter butzleri* in stool samples by use of real-time multiplex PCR. J Clin Microbiol. 51:253–259.

GENOTYPING AND DETERMINING GENETIC DIVERSITY OF ARCOBACTER SPP.

In epidemiological surveillance, typing plays a crucial role to study risk factors and routes of transmission. Several techniques have been tried for strain differentiation of *Arcobacters* and to trace the source of infection or routes of transmission. These include the use of different PCR-based methods like ERIC-PCR which is most commonly used to trace the different outbreaks (Vandamme et al. 1993Vandamme P, Giesendorf AJ, Van Belkum P, Lauwers S, Kersters K, Butzler P, Goossens H, Quint WG. 1993. Discrimination of epidemic and sporadic isolates of *Arcobacter butzleri* by polymerase chain reaction-mediated DNA fingerprinting. J Clin Microbiol. 31:3317–3319), to characterize different isolates (Aydin et al. 2007Aydin F, Gumussoy KS, Atabay HI, Ica T, Abay S. 2007. Prevalence and distribution of *Arcobacter* species in various sources in Turkey and molecular analysis of isolated strains by ERIC-PCR. J Appl Microbiol. 103:27–35 ,) and to study genetic diversity among them (Houf et al. 2002bHouf K, De Zutter L, Van Hoof J, Vandamme P. 2002b. Assessment of the genetic diversity among arcobacters isolated from poultry products by using two PCR-based typing methods. Appl Environ Microbiol. 68:2172–2178. Collado et al. 2010Collado L, Kasimir G, Perez U, Bosch A, Pinto R, Saucedo G, Huguet JM, Figueras MJ. 2010. Occurrence and diversity of *Arcobacter* spp. along the Llobregat river catchment, at sewage effluents and in a drinking water treatment plant. Water Res. 44:3696–3702. Suelam 2012Suelam II A. 2012. Isolation and identification of *Arcobacter* species recovered from rabbits in Zagazig, Egypt. Int J of Micr Res. 3:87–92.). Other methods include RAPD-PCR, AFLP and PFGE (Houf et al. 2002bHouf K, De Zutter L, Van Hoof J, Vandamme P. 2002b. Assessment of the genetic diversity among arcobacters isolated from poultry products by using two PCR-based typing methods. Appl Environ Microbiol. 68:2172–2178. On et al. 2002On SLW, Jensen TK, Bille-Hansen V, Jorsal SE, Vandamme P. 2002. Prevalence and diversity of *Arcobacter* spp. isolated from the internal organs of spontaneous porcine abortions in Denmark. Vet Microbiol. 85:159–167. Ongor et al. 2004Ongor H, Cetinkaya B, Açik MN, Atabay HI. 2004. Investigation of Arcobacters in meat and fecal samples of clinically healthy cattle in Turkey. Lett Appl Microbiol. 38:339–344 ,). ERIC-PCR and RAPD-PCR are both valuable techniques for characterizing *A. butzleri*, *A. cryaerophilus* and *A. skirrowii*isolates. Both methods have satisfactory type ability and discriminatory power. The fingerprints generated with ERIC-PCR were more reproducible and complex than the fingerprints generated with RAPD-PCR (Houf et al. 2002bHouf K, De Zutter L, Van Hoof J, Vandamme P. 2002b. Assessment of the genetic diversity among arcobacters isolated from poultry products by using two PCR-based typing methods. Appl Environ Microbiol. 68:2172–2178 ,). Other recently developed nucleic acid approaches increasingly being considered for detection, identification and monitoring of *Arcobacter* in foods include PCR plus RFLP (Neubauer et al. 2003Neubauer J, Jauk V, Szolgyenyi W. 2003. Phenotypic and genotypic differentiation of *Campylobacter*spp. isolated from Austrian broiler farms: a comparison. Avian Pathol. 32:33–37), PCR plus RAPD (Atabay et al. 2002Atabay HI, Aydin F, Houf K, Sahin M, Vandamme P. 2002. The prevalance of *Arcobacter* spp. on chicken carcasses sold in retail markets in Turkey, and identification of the isolates using SDS-PAGE. Int J Food Microbiol. 81:21–28., ; Houf et al. 2002bHouf K, De Zutter L, Van Hoof J, Vandamme P. 2002b. Assessment of the genetic diversity among arcobacters isolated from poultry products by using two PCR-based typing methods. Appl Environ Microbiol. 68:2172–2178 ,) and PCR plus DNA sequencing (Karenlampi et al. 2004Kärenlampi RI, Tolvanen TP, Hänninen ML. (2004). Phylogenetic analysis and PCR-restriction fragment length polymorphism identification of campylobacter species based on partial groEL gene sequences. J Clin Microbiol. 42:5731–5738 ,).

Quinones et al. (2007Quinones B, Parker CT, Janda JM, Miller WG, Mandrell RE. 2007. Detection and genotyping of *Arcobacter* and *Campylobacter*isolates from retail chicken samples by use of DNA oligonucleotide arrays. Appl Environ Microbiol. 73:3645–3655 ,) used DNA oligonucleotide arrays to simultaneously determine the presence of *Arcobacter* and *Campylobacter* in retail chicken samples by targeting VAGs. *Arcobacter* having wide genetic diversity among the species and between the species have been isolated from different sources (Houf et al. 2002bHouf K, De Zutter L, Van Hoof J, Vandamme P. 2002b. Assessment of the genetic diversity among arcobacters isolated from poultry products by using two PCR-based typing methods. Appl Environ Microbiol. 68:2172–2178. Merga et al. 2013Merga JY, Williams NJ, William G. Miller WG, Leatherbarrow AJH, Bennett M, Hall N, Ashelford KE, Winstanley C. 2013. Exploring the diversity of *Arcobacter butzleri*from cattle in the UK using MLST and whole genome sequencing. PLoS One. 8:e55240. Gilbert et al. 2014Gilbert MJ, Kik M, Timmerman AJ, Severs TT, Kusters JG, Duim B, Wagenaar JA. 2014. Occurrence, diversity, and host association of intestinal *Campylobacter*, *Arcobacter* and *Helicobacter* in Reptiles. PloS One. 9:e101599 ,). Several researchers stated that there is a high level of diversity among *A. butzleri* isolates obtained from both human and animal sources as detected by micro restriction PFGE (Hume et al. 2001; Rivas et al 2004Rivas L, Fegan N, Vanderlinde P. 2004. Isolation and characterisation of *Arcobacter butzleri* from meat. Int J Food Microbiol. 91:31–41. González et al. 2007aGonzález A, Botella S, Montes RM, Moreno Y, Ferrus MA. 2007a. Direct detection and identification *of Arcobacter* species by multiplex PCR in chicken and waste water samples from Spain. J Food Prot. 70:341–347. Merga et al. 2013Merga JY, Williams NJ, William G. Miller WG, Leatherbarrow AJH, Bennett M, Hall N, Ashelford KE, Winstanley C. 2013. Exploring the diversity of *Arcobacter butzleri*from cattle in the UK using MLST and whole genome sequencing. PLoS One. 8:e55240 ,) and ERIC–PCR (Houf et al. 2002bHouf K, De Zutter L, Van Hoof J, Vandamme P. 2002b. Assessment of the genetic diversity among arcobacters isolated from poultry products by using two PCR-based typing methods. Appl Environ Microbiol. 68:2172–2178. Van Driessche et al. 2005Van Driessche E, Houf K, Vangroenweghe F, De Zutter L, Van Hoof J. 2005. Prevalence, enumeration and strain variation of *Arcobacter* species in the faeces of healthy cattle in Belgium. Vet Microbiol. 105:149–154 , , Van Driessche & Houf 2007aVan Driessche E, Houf K. 2007a. Characterization of the *Arcobacter*contamination on Belgian pork carcasses and raw retail pork. Int J Food Microbiol. 118:20–26. De Smet et al. 2010De Smet S, De Zutter L, Van Hende J, Houf K. 2010. Arcobacter contamination on pre- and post-chilled bovine carcasses and in minced beef at retail. J Appl Microbiol. 108:299–305 ,). PFGE using restriction enzyme digestion with *Kpn*I has been suggested as the primary method for typing the isolate but huge isolates cannot be typed using this technique. Douidah et al. (2014bDouidah L, De Zutter L, Baré J, Houf K. 2014b. Towards a typing strategy for *Arcobacter* species isolated from humans and animals and assessment of the in vitro genomic stability. Foodborne Pathog Dis. 11:272–280 ,) proposed for the use of ERIC-PCR for typing large number of isolates instead of PFGE analysis with *Kpn*I. Son et al. (2006Son I, Englen MD, Berrang ME, Fedorka-Cray PJ, Harrison MA. 2006. Genetic diversity of *Arcobacter* and *Campylobacter* on broiler carcasses during processing. J Food Prot. 69:1028–1033 ,) studied the diversity of *Arcobacter* and *Campylobacter* spp. during the processing of broiler carcasses and found that *Arcobacter*was more diverse genetically than *Campylobacter* spp. Previous reports suggest that different genotypes have been found in the feces of a single animal (Van Driessche et al. 2004Van Driessche E, Houf K, Vangroenweghe F, Nollet N, De Zutter L, Vandamme P, Van Hoof J.

2004. Occurrence and strain diversity of *Arcobacter* species isolated from healthy Belgian pigs. Res Microbiol. 155:662–666 , , 2005Van Driessche E, Houf K, Vangroenweghe F, De Zutter L, Van Hoof J. 2005. Prevalence, enumeration and strain variation of *Arcobacter* species in the faeces of healthy cattle in Belgium. Vet Microbiol. 105:149–154 ,). In a case from a pig, 7 *A. butzleri,* 10 *A. cryaerophilus* and 6 *A. skirrowii*genotypes were documented, whereas 6 *A. cryaerophilus* and 2 *A. skirrowii* genotypes were reported from cows (Van Driessche et al. 2004Van Driessche E, Houf K, Vangroenweghe F, Nollet N, De Zutter L, Vandamme P, Van Hoof J. 2004. Occurrence and strain diversity of *Arcobacter* species isolated from healthy Belgian pigs. Res Microbiol. 155:662–666 , , 2005Van Driessche E, Houf K, Vangroenweghe F, De Zutter L, Van Hoof J. 2005. Prevalence, enumeration and strain variation of *Arcobacter* species in the faeces of healthy cattle in Belgium. Vet Microbiol. 105:149–154 ,). First MLST scheme and the first website database of MLST (http://pubmlst.org/arcobacter/) for *Arcobacter* spp. was done by Miller et al. (2009Miller WG, Wesley IV, On SLW, Houf K, Megraud F, Wang G, Yee E, Srijan A, Mason CJ. 2009. First multi-locus sequence typing scheme for *Arcobacter* spp. BMC Microbiol. 14:196.,), set of 374 isolates of *Arcobacter* (275 *A. butzleri*, 72 A. *cryaerophilus*, 15 *A. skirrowii* and 8 *A. cibarius*) were screened and concluded that there is no correlation of MLST alleles with host or geographical source. *Arcobacter* MultiLocus Sequence Typing website (http://pubmlst.org/arcobacter/) provides valuable information for primers and sequencing conditions for the seven genes (*atpA, aspA*, *gltA, glnA*, *pgm*, *tkt* and *glyA*) and for submitting new sequences (Miller et al. 2009Miller WG, Wesley IV, On SLW, Houf K, Megraud F, Wang G, Yee E, Srijan A, Mason CJ. 2009. First multi-locus sequence typing scheme for *Arcobacter* spp. BMC Microbiol. 14:196.,). MLST study using 45 isolates from food products like 13 from raw milk, 12 isolates from poultry, 10 from clams, 5 from pork meats, 4 from mussels and 1 from minced beef in Northern Spain revealed great diversity among *A. butzleri* isolates and also reported the lateral gene transfer at *glyA* locus from *A. skirrowi* to *A. butzleri* (Alonso et al. 2014Alonso R, Girbau C, Martinez-Malaxetxebarria I, Fernández-Astorga A. 2014. Multilocus sequence typing reveals genetic diversity of foodborne *Arcobacter butzleri* isolates in the North of Spain. Int J Food Microbiol. 191:125–128 ,). Single-enzyme amplified fragment length polymorphism (s-AFLP) technique and PFGE used in tandem could differentiate *A. butzleri* from *A. cryaerophilus* and *A. skirrowii* isolated from water and chicken (González et al. 2007bGonzález A, Ferrús MA, González R, Hernández J. 2007b. Molecular fingerprinting of *Campylobacter*and *Arcobacter* isolated from chicken and water. Int Microbiol. 10:85–90).

Isolates of *Arcobacter* spp. from water samples along the Llobregat River catchment, at sewage effluents and in a drinking water treatment plant showed high genotypic diversity as revealed by the ERIC-PCR (Collado et al. 2010Collado L, Kasimir G, Perez U, Bosch A, Pinto R, Saucedo G, Huguet JM, Figueras MJ. 2010. Occurrence and diversity of *Arcobacter* spp. along the Llobregat river catchment, at sewage effluents and in a drinking water treatment plant. Water Res. 44:3696–3702 ,). The workers here reported that out of 339 investigated isolates, 308 had different genotypes among which 248 belonged to the 275 isolates of *A. butzleri* and 60 to the 63 isolates of *A. cryaerophilus* were found genotypically diverse. MLST of 32 isolates from swabs of boiler processing line resulted in 10 new sequence types (STs) (Rasmussen et al. 2013Rasmussen LH, Jette K, Jens PC, Hanne I. 2013. Multilocus sequence typing and biocide tolerance of *Arcobacter butzleri* from Danish broiler carcasses. BMC Res Notes. 6:322). A total of 49 *Arcobacter* isolates (27 *A. butzleri* and 22 *A. cryaerophilus*) recovered from a 506 samples (chicken meat, poultry skin, dairy cow milk and human stool) when tested by ERIC-PCR revealed *A. butzleri* isolates to be grouped into 18 subtypes and *A. cryaerophilus* isolates grouped into 14 subtypes (Ramees et al. 2014dRamees TP, Rathore RS, Bagalkot PS, Sailo Blessa, Mohan HV, Kumar A, Dhama K, Singh RK. 2014d. Genotyping and genetic diversity of *Arcobacter butzleri* and *Arcobacter cryaerophilus* isolated from different sources by using ERIC-PCR from India. Vet Quart. 34:211–217). Genotyping of *A. butzleri* from cattle in the United Kingdom using MLST and whole genome sequencing revealed an unexpected amount of diversity (Merga et al. 2013Merga JY, Williams NJ, William G. Miller WG, Leatherbarrow AJH, Bennett M, Hall N, Ashelford KE, Winstanley C. 2013. Exploring the diversity of *Arcobacter butzleri*from cattle in the UK using MLST and whole genome sequencing. PLoS One. 8:e55240 ,).

In another study, isolates of *Campylobacter, Arcobacter* and *Helicobacter* from reptiles were compared using AFLP *atpA* and 16S rRNA sequencing (Gilbert et al. 2014Gilbert MJ, Kik M, Timmerman AJ, Severs TT, Kusters JG, Duim B, Wagenaar JA. 2014. Occurrence, diversity, and host association of intestinal *Campylobacter*, *Arcobacter* and *Helicobacter* in Reptiles. PloS One. 9:e101599 ,). Recently, a rapid, high-throughput, high-resolution assay, comparative genomic fingerprinting analysis, has been developed for genotyping of *A. butzleri* isolates by the use of which reproducible subtyping of these organisms is possible (Webb et al. 2015Webb AL, Kruczkiewicz P,

Selinger LB, Inglis GD, Taboada EN. 2015. Development of a comparative genomic fingerprinting assay for rapid and high resolution genotyping of *Arcobacter butzleri*. BMC Microbiol. 15:94 ,). Development of rapid diagnostic assays will aid in early diagnosis of this food-borne pathogen so that further transmission of this pathogen into the food chain can be prevented.

Advancements in developing rapid, reliable and confirmatory diagnostics along with adapting effective surveillance and monitoring strategies need to be explored to their full potential for countering *Arcobacters* and their food-borne zoonotic impacts. For this purpose, further strengthening of the immunodiagnostics, molecular detection and characterization, genotyping and differentiation of *Arcobacters* is suggested by employing advances in PCR and its allied versions, LAMP, recombinant diagnostics, lateral flow assay, biochips, biosensors, microarrays, gene sequencing and phylogenetic analysis, genomic fingerprinting and nanotechnology-based diagnostics (Schmitt & Henderson 2005Schmitt B, Henderson L. 2005. Diagnostic tools for animal diseases. Rev Sci Tech. 24:243–250. Deb & Chakraborty 2012Deb R, Chakraborty S. 2012. Trends in veterinary diagnostics. J Vet Sci Technol. 3:1., ; Dhama et al. 2012Dhama K, Wani MY, Tiwari R, Kumar D. 2012. Molecular diagnosis of animal diseases: the current trends and perspectives. Livestock Sphere. 1:6–10. , 2014aDhama K, Karthik K, Chakraborty S, Tiwari R, Kapoor S, Kumar A, Thomas P. 2014a. Loop-mediated isothermal amplification of DNA (LAMP) – a new diagnostic tool lights the world of diagnosis of animal and human pathogens: a review. Pak J Biol Sci. 17:151–166 ; Num & Useh 2013Num SM, Useh NM. 2013. Nanotechnology applications in veterinary diagnostics and therapeutics. Sokoto J Vet Sci. 11:10–14. ; Gilbert et al. 2014Gilbert MJ, Kik M, Timmerman AJ, Severs TT, Kusters JG, Duim B, Wagenaar JA. 2014. Occurrence, diversity, and host association of intestinal *Campylobacter*, *Arcobacter* and *Helicobacter* in Reptiles. PloS One. 9:e101599. Ramees et al. 2014dRamees TP, Rathore RS, Bagalkot PS, Sailo Blessa, Mohan HV, Kumar A, Dhama K, Singh RK. 2014d. Genotyping and genetic diversity of *Arcobacter butzleri* and *Arcobacter cryaerophilus* isolated from different sources by using ERIC-PCR from India. Vet Quart. 34:211–217 ; Webb et al. 2015Webb AL, Kruczkiewicz P, Selinger LB, Inglis GD, Taboada EN. 2015. Development of a comparative genomic fingerprinting assay for rapid and high resolution genotyping of *Arcobacter butzleri*. BMC Microbiol. 15:94 , , 2016Webb AL, Boras VF, Kruczkiewicz P, Selinger LB, Taboada EN, Inglis GD. 2016. Comparative detection and quantification of *Arcobacter butzleri* in stools from diarrheic and nondiarrheic people in Southwestern Alberta, Canada. J Clin Microbiol. 54:1082–1088. Whiteduck-Léveillée et al. 2016aWhiteduck-Léveillée J, Cloutier M, Topp E, Lapen DR, Talbot G, Villemur R, Khan IU. 2016a. Development and evaluation of multiplex PCR assays for rapid detection of virulence-associated genes in *Arcobacter* species. J Microbiol Methods. 121:59–65 ,).

ANTIBIOTICSENSITIVITYANDEMERGINGDRUG RESISTANCE

Regarding antimicrobial susceptibility of *Arcobacter,* very limited data are available and *A. butzleri* is the only species which is mostly discussed (Kiehlbauch et al. 1992Kiehlbauch JA, Baker C, Wachsmuth IK. 1992. *In vitro* susceptibilities of aerotolerant *Campylobacter*isolates to 22 antimicrobial agents. Antimicrob Agents Chemother. 36:717–722. Atabay et al. 2003Atabay HI, Aydin F, Houf K, Sahin M, Vandamme P. 2003. The prevalence of *Arcobacter* spp. on chicken carcass sold in retail markets in Turkey and identification of the isolates using SDS-PAGE. Int J Food Microbiol. 25:21–28., ; Houf et al. 2001Houf K, Devriese LA, De zutter L, Van Hoof J, Vandamme P. 2001. Susceptibility of *Arcobacter butzleri, Arcobacter cryaerophilus*, and *Arcobacter skirrowii* to antimicrobial agents used in selective media. J Clin Microbiol. 39:1654–1656. Fera et al. 2003Fera MT, Maugeri TL, Giannone M, Gugliandolo C, La Camera E, Blandino G, Carbone M. 2003. *In vitro* susceptibility of *Arcobacter butzleri* and *Arcobacter cryaerophilus* to different antimicrobial agents. Int J Antimicrob Agents. 21:488–491. Kabeya et al. 2004Kabeya H, Maruyama S, Morita Y, Ohsuga T, Ozawa S, Kobayashi Y, Abe M, Katsube Y Mikami T. 2004. Prevalence of *Arcobacter* species in retail meats and antimicrobial susceptibility of the isolates in Japan. Int J Food Microbiol. 90:303–308 ,). Available reports suggest that there is an increase in antimicrobial resistance against this emerging food-borne pathogen leading to treatment failures with commonly used antimicrobials. *A. butzleri* is comparatively more resistant than *A. cryophilus* and *A. skirrowi* (Houf et al. 2001Houf K, Devriese LA, De zutter L, Van Hoof J, Vandamme P. 2001. Susceptibility of *Arcobacter butzleri, Arcobacter cryaerophilus*, and *Arcobacter skirrowii* to antimicrobial agents used in selective media. J Clin Microbiol. 39:1654–1656. Kabeya et al. 2004Kabeya H, Maruyama S, Morita Y, Ohsuga T, Ozawa S, Kobayashi Y, Abe M, Katsube Y Mikami T. 2004. Prevalence of *Arcobacter* species in retail meats and antimicrobial susceptibility of the isolates in Japan. Int J Food Microbiol. 90:303–308. Abay et al. 2012Abay S, Kayman T, Hizlisoy H, Aydin F. 2012. In vitro antibacterial susceptibility of *Arcobacter butzleri*isolated from different sources. J Vet Med Sci. 74:613–616 ,). In a study conducted by Kabeya et al. (2004Kabeya H, Maruyama S, Morita Y, Ohsuga T, Ozawa S, Kobayashi Y, Abe M, Katsube Y Mikami T. 2004. Prevalence of *Arcobacter* species in retail meats and antimicrobial susceptibility of the isolates in Japan. Int J Food Microbiol. 90:303–308 ,), all the *Arcobacter* strains tested were found susceptible to ampicillin. Several other studies indicated that *Arcobacter* are susceptible to aminoglycosides and tetracycline (Fera et al. 2003Fera MT, Maugeri TL, Giannone M, Gugliandolo C, La Camera E, Blandino G, Carbone M. 2003. *In vitro* susceptibility of *Arcobacter butzleri* and *Arcobacter cryaerophilus* to different antimicrobial agents. Int J Antimicrob Agents. 21:488–491. Son et al. 2007Son I, Englen MD, Berrang ME, Fedorka-Carry PJ, Harrison MA. 2007. Prevalence of *Arcobacter* and *Campylobacter* on broiler carcases during processingInt. J Food Microbiol. 113:16–22. Collado & Figueras 2011Collado L, Figueras MJ. 2011. Taxonomy, epidemiology and clinical relevance of the genus *Arcobacter*. Clin Microbiol Rev. 24:174–192. Abay et al. 2012Abay S, Kayman T, Hizlisoy H, Aydin F. 2012. In vitro antibacterial susceptibility of *Arcobacter butzleri*isolated from different sources. J Vet Med Sci. 74:613–616 ,). Shah et al. (2013Shah AH, Saleha AA, Zunita Z, Murugaiyah M, Aliyu AB, Jafri N. 2013. Prevalence, distribution and antibiotic resistance of emergent Arcobacter spp. from clinically healthy cattle and goats. Transbound Emerg Dis. 60:9–16 ,) showed that *A. butzleri* were resistant to ampicillin (56%), followed by cefotaxime (33%) and ciprofloxacin (33%) and susceptible to enrofloxacin and gentamicin, the drugs of choice for the treatment for *Arcobacter* infection. But in a study conducted by Unver et al. (2013Ünver A, Atabay Hİ, Şahin M, Çelebi Ö. 2013. Antimicrobial susceptibilities of various *Arcobacter* species. Turk J Med Sci. 43:548–552.,) all of the *A. skirrowii* and most of the *A. cryaerophilus*isolates were found susceptible to amoxicillin/clavulanic acid and resistant towards optochin, vancomycin, fusidic acid, cloxacillin and cefazolin with moderate susceptibility to amikacin, enrofloxacin, ofloxacin, oxytetracycline, chloramphenicol, nitrofurantoin, erythromycin, ampicillin sulbactam and amoxicillin. Resistance pattern against antibiotics like cephalothin, novobiocin and vancomycin has been reported by several workers whereas in most cases it was found sensitive against azithromycin, nalidixic acid and gentamicin (Mohan et al. 2014Mohan HV, Rathore RS, Dhama K, Ramees TP, Patyal A, Bagalkot PS, Wani MY, Bhilegaonkar

KN, Kumar A. 2014. Prevalence of *Arcobacter*spp. in humans, animals and foods of animal origin in India based on cultural isolation, antibiogram, PCR and multiplex PCR detection. Asian J Anim Vet Adv. 9:452–466.,). Antimicrobial susceptibilities of 71 *Arcobacter* isolates were tested against 14 drugs using the disk diffusion method, all of the *Arcobacter*isolates tested were found to be resistant to one or more antimicrobial agents. Resistance to cephalothin and vancomycin (96%) was the most common finding, followed by resistance to methicillin, azithromycin and ampicillin. All the *Arcobacter* isolates were susceptible to gentamicin, streptomycin, tetracyclin and kanamycin (Rahimi et al. 2014).

Van den Abeele et al. (2016Van den Abeele AM, Vogelaers D, Vanlaere E, Houf K. 2016. Antimicrobial susceptibility testing of *Arcobacter butzleri* and *Arcobacter cryaerophilus* strains isolated from Belgian patients. J Antimicrob Chemother. 71:1241–1244 ,) reported the antimicrobial susceptibility of *A. butzleri* and *A. cryaerophilus* strains isolated from Belgian patients. Majority of the *Arcobacter* strains were susceptible to gentamicin (99%) and tetracycline (89%). However, erythromycin (78%), ciprofloxacin (72%) and doxycycline (76%) showed moderate activity against *Arcobacter* spp. Only 9% of the strains were susceptible to ampicillin. Majority of the *A. butzleri* strains were susceptible to ciprofloxacin (87%), whereas half of the *A. cryaerophilus* isolates (51%) showed high-level resistance (MIC >32 mg/L). Ciprofloxacin-resistant strains harbored an identical mutation in *gyrA*. These results suggest that *A. cryaerophilus* showed an acquired resistance to fluoroquinolones. Macrolides are not first-choice empirical antibiotics for *Arcobacter* infections. Tetracyclines can be suggested for treatment of *Arcobacter* induced gastrointestinal infections. Shirzad Aski et al. (2016Shirzad Aski H, Tabatabaei M, Khoshbakht R, Raeisi M. 2016. Occurrence and antimicrobial resistance of emergent Arcobacter spp. isolated from cattle and sheep in Iran. Comp Immunol Microbiol Infect Dis. 44:37–40 ,) conducted an antimicrobial susceptibility study of *Arcobacter* spp. isolated from clinically healthy food animals. All the *Arcobacter* isolates were resistant to vancomycin, rifampicin, trimethoprim, ceftriaxone and cephalothin, whereas isolates were highly susceptible to oxytetracycline, tetracycline, ciprofloxacin, erythromycin, kanamycin, amikacin, enrofloxacin and gentamicin. Tetracycline and aminoglycosides can be used for treatment of human *Arcobacter* infections. The resistance mechanisms of *Arcobacter* for antibiotics were of mainly chromosomal in nature (Abdelbaqi et al. 2007bAbdelbaqi K, Me´nard A, Prouzet-Mauleon V, Bringaud F, Lehours P, Mégraud F. 2007b. Nucleotide sequence of the *gyrA* gene of *Arcobacter* species and characterization of human ciprofloxacin-resistant clinical isolates. FEMS Immunol Med Microbiol. 49:337–345 ; Miller et al. 2007Miller WG, Parker CT, Rubenfield M, Mendz GL, Wösten MM, Ussery DW, Stolz JF, Binnewies TT, Hallin PF, Wang G, et al. 2007. The complete genome sequence and analysis of the epsilonproteobacterium *Arcobacter butzleri*. PLoS One. 2:1358. Douidah et al. 2014aDouidah L, De Zutter L, Van Nieuwerburgh F, Deforce D, Ingmer H, Vandenberg O, Van den Abeele AM, Houf K. 2014a. Presence and analysis of plasmids in human and animal associated *Arcobacter*species. PloS One. 9:e85487. Ferreira et al. 2016Ferreira S, Queiroz JA, Oleastro M, Domingues FC. 2016. Insights in the pathogenesis and resistance of *Arcobacter*: a review. Crit Rev Microbiol. 42:364–383).

PREVENTION AND CONTROL

For food-borne pathogens the control measures are very important to reduce the disease burden (Dhama et al. 2013aDhama K, Rajagunalan S, Chakraborty S, Verma AK, Kumar A, Tiwari R, Kapoor S. 2013a. Food-borne pathogens of animal origin-diagnosis, prevention, control and their zoonotic significance: a review. Pak J Biol Sci. 16:1076–1085). Improperly cooked and contaminated food of animal origin like meat, milk, sea food could act as an important source of *Arcobacter* infection to humans (Collado et al. 2009aCollado L, Guarro J, Figueras M. 2009a. Prevalence of *Arcobacter* in meat and shellfish. J Food Protect. 72:1102–1110. Ertaset al. 2010Ertas N, Dogruer Y, Gonulalan Z, Guner A, Ulger I. 2010. Prevalence of *Arcobacter* species in drinking water, spring water, and raw milk as determined by multiplex PCR. J Food Prot. 73:2099–2102. Iwamoto et al. 2010Iwamoto M, Ayers T, Mahon BE, Swerdlow DL. 2010. Epidemiology of seafood-associated infections in the United States. Clin Microbiol Rev. 23:399–411. Shah et al. 2011Shah AH, Saleha AA, Zunita Z, Murugaiyah M. 2011. *Arcobacter*– an emerging threat to animals and animal origin food products? Tr Food Sci Technol. 22:225–236.,). Therefore, thorough and good cooking practices help in prevention of food-borne infections. *A. butzleri* is more resistant in food samples than *C. jejuni* to irradiation or cold pasteurization with average D10 value of 0.27 kGy compared with 0.19 kGy for *C. jejuni* (Phillips 2001Phillips CA. 2001. *Arcobacter* spp in food: isolation, identification, and control. Trends Food Sci Technol. 12:263–275.,). Therefore, radiation doses recommended for pork in United States is 0.3–1.0 kGy. Citric acid and lactic acid can eliminate the growth of *Arcobacter*.

At a concentration of 1.0% and 2.0%, citric and lactic acid, respectively can inhibit the growth of *A. butzleri,* and 2% sodium lactate also effectively checks *Arcobacter* (Phillips 1999Phillips CA. 1999. The effect of citric acid, lactic acid, sodium citrate and sodium lactate, alone and in combination with nisin, on the growth of *Arcobacter butzleri*. Lett Appl Microbiol. 29:424–428. Lehner et al. 2005Lehner A, Tasara T, Stephan R. 2005. Relevant aspects of *Arcobacter* spp. as potential food-borne pathogens. Int J Food Microbiol. 102:127–135 ,). In a study conducted with nisin alone or in combination with agents like citric acid, lactic acid, sodium citrate or sodium lactate showed that approximately 50% survivability of *Arcobacter* can be reduced by the use of nisin alone at 500 IU/ml (Phillips 1999Phillips CA. 1999. The effect of citric acid, lactic acid, sodium citrate and sodium lactate, alone and in combination with nisin, on the growth of *Arcobacter butzleri*. Lett Appl Microbiol. 29:424–428 ,). Trisodium phosphate and EDTA have been found to be effective in reducing survival of *A. butzleri* in pure cultures (Phillips & Duggan 2001Phillips CA, Duggan J. 2001. The effect of EDTA and trisodium phosphate, alone and in combination with nisin, on the growth of *Arcobacter butzleri* in culture. Food Microbiol. 18:547–554.,). Low temperature storage can also reduce *Arcobacter* growth (Phillips 2001Phillips CA. 2001. *Arcobacter* spp in food: isolation, identification, and control. Trends Food Sci Technol. 12:263–275., ; Phillips & Long 2001Phillips CA, Long C. 2001. The survival of *Arcobacter butzleri* on chicken. Int J Med Microbiol. 291:93.). A study on the effect of 17 organic acids (acetic, propionic, butyric, caproic, caprylic, capric, lauric, myristic, palmitic, phenylace-tic, sorbic, benzoic, fumaric, succinic, lactic, malic and citric acid) on *Arcobacter* spp. showed, that the maximum inhibitory activities were observed with benzoic, citric, malic and sorbic acids (Skřivanová et al. 2011Skrivanová E, Molatová Zuzana, Matenová Michaela, Houf Kurt, Marounek Milan. 2011. Inhibitory effect of organic acids on arcobacters in culture and their use for control of *Arcobacter butzleri*on chicken skin. Int J Food Microbiol. 144:367–371 ,). *Arcobacters* can effectively colonize the surface of pipes such as copper, stainless steel, and plastics (Assanta et al. 2002Assanta MA, Roy D, Lemay MJ, Montpetit D. 2002. Attachment of *Arcobacter butzleri*, a new waterborne pathogen, to water distribution pipe surfaces. J Food Prot. 65:1240–1247 ,). But regardless of this bacterium's susceptibility to chlorine (Rice et al. 1999Rice EW, Rodgers MR, Wesley IV, Johnson CH, Tanner SA. 1999. Isolation of *Arcobacter butzleri*from ground water. Lett Appl Microbiol. 28:31–35. Moreno et al. 2004Moreno Y, Alonso JL, Botella S, Ferrus MA, Hernandez J. 2004. Survival and injury of *Arcobacter*after artificial inoculation into drinking water. Res Microbiol. 155:726–730 ,) it is still doubtful that whether conventional treatments at a drinking water treatment plant can effectively remove *Arcobacter,* as failure of treatment could lead to infection (Ho et al. 2006aHo HT, Lipman LJ, Gaastra W. 2006a. *Arcobacter*, what is known and unknown about a potential food-borne zoonotic agent.Vet Microbiol. 115:1–13. Collado et al. 2010Collado L, Kasimir G, Perez U, Bosch A, Pinto R, Saucedo G, Huguet JM, Figueras MJ. 2010. Occurrence and diversity of *Arcobacter* spp. along the Llobregat river catchment, at sewage effluents and in a drinking water treatment plant. Water Res. 44:3696–3702 ,). Since contaminated water acts as an important source of infection, hence the effective treatment of water resources is imperative (Snelling et al. 2006Snelling WJ, Matsuda M, Moore JE, Dooley JS. 2006. Under the microscope: *Arcobacter*. Lett Appl Microbiol. 42:7–14. Collado et al. 2010Collado L, Kasimir G, Perez U, Bosch A, Pinto R, Saucedo G, Huguet JM, Figueras MJ. 2010. Occurrence and diversity of *Arcobacter* spp. along the Llobregat river catchment, at sewage effluents and in a drinking water treatment plant. Water Res. 44:3696–3702 ,). Meat also acts as an important source of *Arcobacter* transmission, therefore maintenance of slaughter hygiene is crucial for prevention and control of such meat-borne infection. Appropriate hygiene in the slaughter house and dairy plants combined with use of proper disinfectant is needed to eliminate *Arcobacter* as they could survive even after use of common sanitization protocol (Giacometti et al. 2013Giacometti F, Lucchi A, Manfreda G, Florio D, Zanoni RG, Serraino A. 2013. Occurrence and genetic diversity of *Arcobacter butzleri* in an artisanal dairy plant in Italy. Appl Environ Microbiol. 79:6665–6669. Serraino & Giacometti 2014Serraino A, Giacometti F. 2014. Short communication: occurrence of *Arcobacter* species in industrial dairy plants. J Dairy Sci. 97:2061–2065 ,). Adherence to practices like strict microbiological monitoring of carcasses, use of verification systems, HACCP and good manufacturing practices are needed to be executed (Lehner et al. 2005Lehner A, Tasara T, Stephan R. 2005. Relevant aspects of *Arcobacter* spp. as potential food-borne pathogens. Int J Food Microbiol. 102:127–135 ,).

In the era of emerging and rising antibiotic resistance against several commonly used antibiotics, there is a demanding need for exploring alternative and novel therapeutic options for countering infection with *Arcobacters* in a better way, as like being explored to tackle other food-borne pathogens (Dhama et al. 2013aDhama K, Rajagunalan S, Chakraborty S, Verma AK, Kumar A, Tiwari R, Kapoor S. 2013a. Food-borne pathogens of animal origin-diagnosis, prevention, control and their zoonotic significance: a review. Pak J Biol Sci. 16:1076–1085 , 2013bDhama K, Chakraborty S, Mahima, Wani MY, Verma AK, Deb R, Tiwari R, Kapoor S. 2013b. Novel and emerging therapies safeguarding health of humans and their companion animals: a review. Pak J Biol Sci. 16:101–111 ; Tiwari et al. 2013Tiwari R, Chakraborty S, Dhama K, Rajagunalan S, Singh SV. 2013. Antibiotic resistance – an emerging health problem: causes, worries, challenges and solutions – a review. Int J Curr Res. 5:1880–1892. ; Yadav et al. 2016Yadav AS, Kolluri G, Gopi M, Karthik K, Malik YS, Dhama K. 2016. Exploring alternatives to antibiotics as health promoting agents in poultry- a review. J Exp Biol Agri Sci. 4:368–383.). In this direction, there is an essential need to shift the attention towards various therapeutic regimens like bacteriophages, probiotics, avian egg yolk antibodies, cytokines, RNAi technology, quorum sensing inhibitors, herbs, essential oils, plant derivatives and others, which have been recently documented to possess potent applications against various pathogenic bacteria having high food-borne concerns (Dhama et al. 2008Dhama K, Mahendran M, Tomar S, Chauhan RS. 2008. Beneficial effects of probiotics and prebiotics in livestock and poultry: the current perspectives. Intas Polivet. 9:1–12. , 2013bDhama K, Chakraborty S, Mahima, Wani MY, Verma AK, Deb R, Tiwari R, Kapoor S. 2013b. Novel and emerging therapies safeguarding health of humans and their companion animals: a review. Pak J Biol Sci. 16:101–111 , 2014bDhama K, Tiwari R, Chakraborty S, Saminathan M, Kumar A, Karthik K, Wani MY, Amarpal, Singh SV, Rahal A. 2014b. Evidence based antibacterial potentials of medicinal plants and herbs countering bacterial pathogens especially in the era of emerging drug resistance: an integrated update. Int J Pharmacol. 10:1–43., ; Sudhakar et al. 2013Sudhakar NR, Manjunathachar HV, Karthik K, Sahu S, Gopi M, Shanthaveer SB, Nagaraja KH, Shinde S, Tamilmahan P, Madhu DN, Maurya PS. 2013. RNA interference in parasites: prospects and pitfalls. Adv Anim Vet Sci. 1:1–6. ; Tiwari et al. 2014Tiwari R, Dhama K, Chakraborty S, Kumar A, Rahal A, Kapoor S. 2014. Bacteriophage therapy for safeguarding animal and human health: a review. Pak J Biol Sci. 17:301–315 , 2016Tiwari R, Karthik K, Rana R, Malik YS, Dhama K, Joshi SK. 2016. Quorum sensing inhibitors/antagonists countering food spoilage bacteria-need molecular and pharmaceutical intervention for protecting current issues of food safety. Int J Pharmacol. 12:262–271., ; Gopi et al. 2014Gopi M, Karthik K, Manjunathachar HV, Tamilmahan P, Kesavan M, Dashprakash M, Balaraju BL, Purushothaman MR. 2014. Essential oils as a feed additive in poultry nutrition. Adv Anim Vet Sci. 2:1–7., ; Khan et al.

2016Khan RU, Naz S, Dhama K, Karthik K, Tiwari R, Abdelrahman MM, Alhidary IA, Zahoor A. 2016. Direct-fed microbial: beneficial applications, modes of action and prospecas safe tool for enhancing ruminant production and safeguarding health. Int J Pharmacol. 12:220–231.,). Bacteriophage therapy has been used successfully for prevention of various food-borne pathogen like *Salmonella*, *Listeria,* etc. Hence, this therapy can also be helpful for prevention and control of *Arcobacter* infection with good perspectives at slaughter houses thereby checking the entry of this organism into the human food chain (Karthik et al. 2014cKarthik K, Muneeswaran NS, Manjunathachar HV, Gopi M, Elamurugan A, Kalaiyarasu S. 2014c. Bacteriophages: effective alternative to antibiotics. Adv Anim Vet Sci. 2:1–7.,). Experimental trials with medicinal herb extracts and spices with respect to their antimicrobial potential have demonstrated that chamomile, sage cinnamon, rosemary, bearberry extracts, which are conventionally practiced for gastric trouble, hinder the growth and development of *Arcobacter* (Cervenka et al. 2006Cervenka L, Peskova I, Foltynova E, Pejchalova M, Brozkova I, Vytrasova J. 2006. Inhibitory effects of some spice and herb extracts against *Arcobacter butzleri*, *A. cryaerophilus*, and *A. skirrowii*. Curr Microbiol. 53:435–439 ,). Seventeen essential oils (rosemary, garlic, sage, bearberry, allspice, black pepper, cumin, cinnamon, coriander, thyme, clove, fennel, licorice, St. Johne's wart, ginger, chamomile and mint) when tested against *Arcobacters* namely *A. butzleri*, *A. cryaerophilus* and *A. skirrowii,* showed inhibition of these organisms. Similarly, various plant derivatives (lemon, sweet orange and bergamot oils) were documented to prevent *Arcobacterosis* (Fisher et al. 2007Fisher K, Rowe C, Phillips CA. 2007. The survival of three strains of *Arcobacter butzleri* in the presence of lemon, orange and bergamot essential oils and their components in vitro and on food. Lett Appl Microbiol. 44:495–499 ,). Usage of rosemary essential oil while cooking of minced beef has been reported to cause substantial reduction of *A. butzleri* (Irkin et al. 2011Irkin R, Abay S, Aydin F. 2011. Inhibitory effects of some plant essential oils against *Arcobacter butzleri* and potential for rosemary oil as a natural food preservative. J Med Food. 14:291–296 ,). Resveratrol plant extract has been found to show good bactericidal effect on *A. butzleri* and *A. cryaerophilus* via inhibiting the metabolic activity, development and cellular functions of these bacteria, particularly affects DNA in the *Arcobacter* thereby blocking its cell division, hence can be used for designing potential strategies to control this food-borne pathogen (Ferreira et al. 2014aFerreira S, Silva F, Queiroz JA, Oleastro, M, Domingues FC. 2014a. Resveratrol against *Arcobacter butzleri* and *Arcobacter cryaerophilus*: activity and effect on cellular functions. Int J Food Microbiol. 180:62–68. Alagawany et al. 2015Alagawany M, Farag MR, Dhama K, El-Hack MEA, Tiwari R, Alam GM. 2015. Mechanisms and beneficial applications of resveratrol as feed additive in animal and poultry nutrition: a review. Int J Pharmacol. 11:213–221.,).

CONCLUSION AND FUTURE PERSPECTIVES

Arcobacter is an emerging food-borne pathogen causing serious infections in humans and animals. Nonetheless, various reports on infection due to *Arcobacters* are becoming continuously available from different countries, still there is no clear documentation on virulence genes/factors that are directly involved in the causation of infection and pathogenesis of this pathogen. Studies are warranted on this aspect so that targeting of the virulence genes could help in preventing spread of infection. Complete genome of different species in the genus needs to be analyzed so as to draw a conclusion regarding the virulence factors and also the pathogenesis of *Arcobacter*. Hence, focus of research should be aimed in the near future for unraveling the unknown aspects of pathology and pathogenesis of these organisms, which would pave the way for developing suitable control strategies. Exact status of the incidence of *Arcobacter* in wild animals is presently lacking which can show the role played by these animals in transmission. This will also help in exploring the detailed epidemiology and knowing the real magnitude of infection, understanding various modes of transmission and spread of *Arcobacters* in a better way, tracing the source of origin, carriers, and revealing the evolutionary mechanisms associated with their diversity and continuous emergence in humans and animals. Veterinarians and medical professionals should be aware of the route of transmission of *Arcobacters* so as to educate the human community regarding hygienic practices while cooking and also in abattoirs which would further eliminate the chances of spread of these organisms.

There is also an increase in the development of antimicrobial resistance among *Arcobacters* against commonly used antimicrobials which is worrisome from the public health perspective. Increasing drug resistant scenario demands attention of researchers towards finding novel and alternative therapeutic options to prevent and control *Arcobacters* in an effective way, which would be also beneficial with regards to public health and food safety viewpoints. Use of herbal active principles, essential oils, bacteriophages, probiotics, avian egg antibodies, RNAi technologies and other novel therapeutic avenues can help to prevent the development of antibiotic resistance and also the spread of *Arcobacters*. Strengthening and exploiting the full potential of advanced molecular diagnostic tools and techniques is needed for early and confirmatory diagnosis, differentiation and genotyping of *Arcobacters*. Research should also be directed towards the development of specific selective media as isolation is always the gold standard for confirmation of this bacterial infection. Diagnosis of *Arcobacter* has presently improved with the development of various molecular methods giving prompt and confirmatory diagnosis of the pathogen with high sensitivity and specificity. Nowadays, PCR based methods and their allied versions are being increasingly used due the rapidity and accuracy of the result obtained. Techniques like microarray expression profile and proteomic studies would further elaborate on in-depth understanding of *Arcobacter* infection.

ACKNOWLEDGMENTS

The authors of this manuscript thank and acknowledge the support of their respective institutes/universities and the Indian Council of Agriculture Research, New Delhi, India.

REFERENCES

1. Abay S, Kayman T, Hizlisoy H, Aydin F. 2012. In vitro antibacterial susceptibility of *Arcobacter butzleri* isolated from different sources. J Vet Med Sci. 74:613–616.

2. Abdelbaqi K, Buissonniere A, Prouzet-Mauleon V, Gresser J, Wesley I, Megraud F, Menard A.2007a. Development of a realtime fluorescence resonance energy transfer PCR to detect*Arcobacter* species. J Clin Microbiol. 45:3015–3021.

3. Abdelbaqi K, Me´nard A, Prouzet-Mauleon V, Bringaud F, Lehours P, Mégraud F. 2007b.Nucleotide sequence of the *gyrA* gene of *Arcobacter* species and characterization of human ciprofloxacin-resistant clinical isolates. FEMS Immunol Med Microbiol. 49:337–345.

4. Açik MN, Yüksel H, Ulucan A, Çetinkaya B. 2016. The first experimental research on the pathogenicity of *Arcobacter butzleri* in zebrafish. Vet Microbiol. 189:32–38.

5. Adam Z, WhiteduckLeveillee K, Cloutier M, Chen W, Lewis CT, Lévesque CA, Topp E, Lapen DR,Tambong JT, Talbot G, Khan IU. 2014a. Draft genome sequence of *Arcobacter cibarius* Strain LMG21996T, isolated from Broiler Carcasses. Genome Announc. 2:e00034–14.

6. Adam Z, Whiteduck-Leveillee K, Cloutier M, Tambong JT, Chen W, Lewis CT, Lévesque CA, Topp E,Lapen DR, Talbot G, Khan IU. 2014b. Draft genome sequences of three *Arcobacter* strains of pig and dairy cattle manure origin. Genome Announc. 2:e00377–14.

7. Adam Z, Whiteduck-Léveillée K, Cloutier M, Chen W, Lewis CT, Lévesque CA, Topp E, Lapen DR,Tambong JT, Talbot G, Khan IU. 2014c. Draft genome sequences of two *Arcobacter* strains isolated from human feces. Genome Announc. 2:e00113–14.

8. Adesiji YO, Seibu E, Emikpe BO, Moriyonu BT, Oloke JK, Coker AO. 2012. Serum biochemistry and heamatological changes associated with graded doses of experimental Arcobacter infection in rats. West Afr J Med. 31:186–191.

9. Adesiji YO. 2010. Faecal shedding of *Arcobacter* species following experimental infection in rats: public health implications. Cent Eur J Med. 5:470–474.

10. Alagawany M, Farag MR, Dhama K, El-Hack MEA, Tiwari R, Alam GM. 2015. Mechanisms and beneficial applications of resveratrol as

feed additive in animal and poultry nutrition: a review. Int J Pharmacol. 11:213–221.

11. Alispahic M, Hummel K, Jandreski-Cvetkovic D, Nöbauer K, Razzazi-Fazeli E, Hess M, Hess C.2010. Species-specific identification and differentiation of *Arcobacter*, *Helicobacter* and*Campylobacter* by full-spectral matrix-associated laser desorption/ionization time of flight mass spectrometry analysis. J Med Microbiol. 59:295–301.

12. Alonso R, Girbau C, Martinez-Malaxetxebarria I, Fernández-Astorga A. 2014. Multilocus sequence typing reveals genetic diversity of foodborne *Arcobacter butzleri* isolates in the North of Spain. Int J Food Microbiol. 191:125–128.

13. Amare LB, Sahela AA, Zunita Z, Jalila A, Hassan L. 2011. Prevalence of *Arcobacter* spp. on chicken meat at retail markets and in farm chickens in Selengor, Malaysia. Food Control. 22:732–736.

14. Anderson KF, Kiehlbauch JA, Anderson DC. 1993. *Arcobacter* (*Campylobacter*) *butzleri*-associated diarrheal illness in a nonhuman primate population. Infect Immun. 61:2220–2223.

15. Ansari MI, Harb M, Jones B, Hong PY. 2015. Molecular-based approaches to characterize coastal microbial community and their potential relation to the trophic state of Red Sea. Sci Rep. 5:1–10.

16. Antolin A, González I, Garcia T, Hernandez PE, Martin R. 2001. *Arcobacter* spp. enumeration in poultry meat using a combined PCR-ELISA assay. Meat Sci. 59:169–174.

17. Arguello E, Otto CC, Mead P, Babady NE. 2015. Bacteremia caused by *Arcobacter butzleri* in an immunocompromised host. J Clin Microbiol. 53:1448–1451.

18. Arias ML, Cid A, Fernandéz H. 2011. *Arcobacter butzleri*: first isolation report from chicken carcasses in Costa Rica. Braz J Microbiol. 42:703–706.

19. santa MA, Roy D, Lemay MJ, Montpetit D. 2002. Attachment of *Arcobacter butzleri*, a new waterborne pathogen, to water distribution pipe surfaces. J Food Prot. 65:1240–1247.

20. Atabay HI, Aydin F, Houf K, Sahin M, Vandamme P. 2002. The prevalance of *Arcobacter* spp. on chicken carcasses sold in retail markets in Turkey, and identification of the isolates using SDS-PAGE. Int J Food Microbiol. 81:21–28.

21. Atabay HI, Aydin F, Houf K, Sahin M, Vandamme P. 2003. The

prevalence of *Arcobacter* spp. on chicken carcass sold in retail markets in Turkey and identification of the isolates using SDS-PAGE. Int J Food Microbiol. 25:21–28.

22. Atabay HI, Aydin F. 2001. Susceptibility of *Arcobacter butzleri* isolates to 23 antimicrobial agents. Lett Appl Microbiol. 33:430–433.
23. Atabay HI, Corry JEL, On SLW. 1998. Diversity and prevalence of Arcobacter spp. in broiler chickens. J Appl Microbiol. 84:1007–1016.
24. Atabay HI, Corry JEL, Prost DE. 1996. Comparison of the productivity of a variety of selective media for *Campylobacter* and *Arcobacter* species In: Newell DJ, Ketley JM, Felman RA, editors.*Campylobacters, Helicobacters* and related organisms New York (NY): Plenum Press; p. 19–23.
25. Atabay HI, Corry JEL. 1997. The prevalence of *Campylobacters* and *Arcobacters* in broiler chickens. J Appl Microbiol. 83:619–626.
26. Atabay HI, Corry JEL. 1998. Evaluation of a new *Arcobacter* enrichment medium and comparison with new media developed for enrichment for *Campylobacter spp*. Int J Food Microbiol. 41:53–58.
27. Atabay HI, Unver A, Sahin M, Otlu S, Elmali M, Yaman H. 2008. Isolation of various *Arcobacter*species from domestic geese (Anser anser). Vet Microbiol. 128:400–405.
28. Atabay HI, Wain M, Madsen M. 2006. Detection and diversity of various *Arcobacter* species in Danish poultry. Int J Food Microbiol. 9:139–145.
29. Aydin F, Gumussoy KS, Atabay HI, Ica T, Abay S. 2007. Prevalence and distribution of *Arcobacter*species in various sources in Turkey and molecular analysis of isolated strains by ERIC-PCR. J Appl Microbiol. 103:27–35.
30. Badilla-Ramírez Y, Fallas-Padilla KL, Fernández-Jaramillo H, Arias-Echandi ML. 2016. Survival capacity of *Arcobacter butzleri* inoculated in poultry meat at two different refrigeration temperatures. Rev Inst Med Trop Sao Paulo. 58:22.
31. Bagalakote PS, Rathore RS, Ramees TP, Mohan HV, Sumankumar M, Agarwal RK, Kumar A,Dhama K. 2014. Molecular characterization of *Arcobacter* isolates using randomly amplified polymorphic DNA-polymerase chain reaction (RAPD-PCR). Asian J Anim Vet Adv. 9:543–555.
32. Bastyns K, Cartuyvels D, Chapelle S, Vandamme P, Goossens H, De

Wachter R. 1995. A variable 23S rDNA region is a useful discriminating target for genus-specific and species-specific PCR amplification in *Arcobacter* species. System Appl Microbiol. 18:353–356.

33. Bath GF, Leask R, Pettey KP, Coetzee DJ. 2013. Abortions in sheep associated with *Arcobacter skirrowii* infection. J South Afr Vet Assoc. 84:4. Available from:https://doi.org/10.4102/jsava.v84i1.952
34. Bogantes EV, Fallas-Padilla KL, Rodríguez-Rodríguez CE, Jaramillo HF, Echandi ML. 2015.Zoonotic species of the genus arcobacter in poultry from different regions of Costa Rica. J Food Prot. 78:808–811.
35. Brann OS. 2001. Infectious complications of cirrhosis. Curr Gastroenterol Rep. 3:285–292.
36. Brightwell G, Mowat E, Clemens R, Boerema J, Pulford DJ, On SL. 2007. Development of a multiplex and real time PCR assay for the specific detection of *Arcobacter butzleri* and*Arcobacter cryaerophilus*. J Microbiol Methods. 68:318–325.
37. Bucker R, Troeger H, Kleer J, Fromm M, Schulzke JD. 2009. *Arcobacter butzleri* induces barrier dysfunction in intestinal HT-29/B6 cells. J Infect Dis. 200:756–764.
38. Calvo G, Arias ML, Fernández H. 2013. *Arcobacter*: a foodborne emerging pathogen. Arch Latinoam Nutr. 63:164–172.
39. Cervenka L, Peskova I, Foltynova E, Pejchalova M, Brozkova I, Vytrasova J. 2006. Inhibitory effects of some spice and herb extracts against *Arcobacter butzleri*, *A. cryaerophilus*, and *A. skirrowii*. Curr Microbiol. 53:435–439.
40. Cervenka L 2007. Survival and inactivation of *Arcobacter* spp., a current status and future prospect. Crit Rev Microbiol. 33:101–108.
41. Chavatte N, Baré J, Lambrecht E, Van Damme I, Vaerewijck M, Sabbe K, Houf K. 2014. Co-occurrence of free-living protozoa and foodborne pathogens on dishcloths: implications for food safety. Int J Food Microbiol. 191:89–96.
42. Chinivasagam HN, Corney BG, Wright LL, Diallo IS, Blackall PJ. 2007. Detection of *Arcobacter* spp. in piggery effluent and effluent-irrigated soils in southeast Queensland. J Appl Microbiol. 103:418–426.
43. Collado L, Cleenwerck I, Van Trappen S, De Vos P, Figueras MJ. 2009b. *Arcobacter mytili* sp. nov., an indoxyl acetate-hydrolysis-negative bacterium isolated from mussels. Int J Syst Evol Microbiol.

59:1391–1396.

44. Collado L, Figueras MJ. 2011. Taxonomy, epidemiology and clinical relevance of the genus*Arcobacter*. Clin Microbiol Rev. 24:174–192.
45. Collado L, Guarro J, Figueras M. 2009a. Prevalence of *Arcobacter* in meat and shellfish. J Food Protect. 72:1102–1110.
46. Collado L, Inza I, Guarro J, Figueras M J. 2008. Presence of *Arcobacter* spp. in environmental waters correlates with high levels of fecal pollution. Environ Microbiol. 10:1635–1640.
47. Collado L, Jara R, Vásquez N, Telsaint C. 2014. Antimicrobial resistance and virulence genes of*Arcobacter* isolates recovered from edible bivalve mollusks. Food Control. 46:508–512.
48. Collado L, Kasimir G, Perez U, Bosch A, Pinto R, Saucedo G, Huguet JM, Figueras MJ. 2010.Occurrence and diversity of *Arcobacter* spp. along the Llobregat river catchment, at sewage effluents and in a drinking water treatment plant. Water Res. 44:3696–3702.
49. Collado L, Levican A, Perez J, Figueras MJ. 2011. *Arcobacter defluvii* sp. nov., isolated from sewage samples. Int J Syst Evol Microbiol. 61:2155–2161.
50. Collins CI, Wesley IV, Murano EA. 1996. Detection of *Arcobacter* spp. in ground pork by modified plating methods. J Food Protect. 59:448–452.
51. Corry JE, Atabay HI. 2001. Poultry as a source of *Campylobacter* and related organisms. J Appl Microbiol. 90:96S–114S.
52. D'Sa EM, Harrison MA. 2005. Effect of pH, NaCl content, and temperature on growth and survival of *Arcobacter* spp. J Food Protect. 68:18–25.
53. de Boer E, Tilburg JJHC, Woodward DL, Loir H, Johnson WM. 1996. A selective medium for the isolation of *Arcobacter* from meats. Lett Appl Microbiol. 23:64–66.
54. de Boer RF, Ott A, Güren P, van Zanten E, van Belkum A, Kooistra-Smid AM. 2013. Detection of*Campylobacter* species and *Arcobacter butzleri* in stool samples by use of real-time multiplex PCR. J Clin Microbiol. 51:253–259.
55. de Oliveira SJ, Wesley I, Harmon KM. 1997. Classification of *Arcobacter* species isolated from aborted pig fetuses and sows with reproductive problems in Brazil. Vet Microbiol. 57:347–354.
56. De Smet S, De Zutter L, Houf K. 2011. Small ruminants as carriers of

the emerging food-borne pathogen *Arcobacter* on small and medium farms. Small Rumin Res. 97:124–129.

57. De Smet S, De Zutter L, Van Hende J, Houf K. 2010. Arcobacter contamination on pre- and post-chilled bovine carcasses and in minced beef at retail. J Appl Microbiol. 108:299–305.

58. Deb R, Chakraborty S. 2012. Trends in veterinary diagnostics. J Vet Sci Technol. 3:1.

59. Debruyne L, Gevers D, Vandamme P. 2008. Taxonomy of the family *Campylobacteraceae*. In:Nachamkin I, Szymanski C, Blaser M, editors. *Campylobacter*. Vol. 3. Washington (DC): ASM press; p. 3–25.

60. Dhama K, Chakraborty S, Mahima, Wani MY, Verma AK, Deb R, Tiwari R, Kapoor S. 2013b. Novel and emerging therapies safeguarding health of humans and their companion animals: a review. Pak J Biol Sci. 16:101–111.

61. Dhama K, Karthik K, Chakraborty S, Tiwari R, Kapoor S, Kumar A, Thomas P. 2014a. Loop-mediated isothermal amplification of DNA (LAMP) – a new diagnostic tool lights the world of diagnosis of animal and human pathogens: a review. Pak J Biol Sci. 17:151–166.

62. Dhama K, Mahendran M, Tomar S, Chauhan RS. 2008. Beneficial effects of probiotics and prebiotics in livestock and poultry: the current perspectives. Intas Polivet. 9:1–12.

63. Dhama K, Rajagunalan S, Chakraborty S, Verma AK, Kumar A, Tiwari R, Kapoor S. 2013a. Food-borne pathogens of animal origin-diagnosis, prevention, control and their zoonotic significance: a review. Pak J Biol Sci. 16:1076–1085.

64. Dhama K, Tiwari R, Chakraborty S, Saminathan M, Kumar A, Karthik K, Wani MY, Amarpal, SinghSV, Rahal A. 2014b. Evidence based antibacterial potentials of medicinal plants and herbs countering bacterial pathogens especially in the era of emerging drug resistance: an integrated update. Int J Pharmacol. 10:1–43.

65. Dhama K, Wani MY, Tiwari R, Kumar D. 2012. Molecular diagnosis of animal diseases: the current trends and perspectives. Livestock Sphere. 1:6–10.

66. Di Francesco A, Delogu M, Giacometti F, Stancampiano L, Grilli E, Guarniero I, Serraino A. 2014.First detection of Arcobacter sp. in Eurasian collared doves (Streptopelia decaocto). Vet Ital. 50:313–315.

67. Diéguez AL, Balboa S, Magnesen T, Romalde JL. 2017. Arcobacter

lekithochrous sp. nov., a new species isolated from a molluscan hatchery in Norway. Int J Syst Evol Microbiol. doi:10.1099/ijsem.0.001809.

68. Donachie SP, Bowman JP, On SLW, Alam M. 2005. *Arcobacter halophilus* sp. nov., the first obligate halophile in the genus Arcobacter. Int J Syst Evol Microbiol. 55:1271–1277.

69. Douidah L, De Zutter L, Baré J, Houf K. 2014b. Towards a typing strategy for *Arcobacter* species isolated from humans and animals and assessment of the in vitro genomic stability. Foodborne Pathog Dis. 11:272–280.

70. Douidah L, De Zutter L, Van Nieuwerburgh F, Deforce D, Ingmer H, Vandenberg O, Van den Abeele AM, Houf K. 2014a. Presence and analysis of plasmids in human and animal associated*Arcobacter* species. PloS One. 9:e85487.

71. Douidah L, Zutter LD, Vandamme P, Houf K. 2010. Identification of five human and mammal associated *Arcobacter* species by a novel multiplex-PCR assay. J Microbiol Methods. 80:281–286.

72. Douidah L, de Zutter L, Baré J, De Vos P, Vandamme P, Vandenberg O, Van den Abeele AM, HoufK. 2012. Occurrence of putative virulence genes in *Arcobacter* species isolated from humans and animals. J Clin Microbiol. 50:735–741.

73. Ellis WA, Neill SD, O' Brien JJ, Ferguson, HW, Hanna J. 1977. Isolation of *Spirillum*/*Vibrio*-like organisms from bovine fetuses. Vet Record. 100:451–452.

74. Ellis WA, Neill SD, O' Brien JJ, Hannna J. 1978. Isolation of *Spirullum*-like organisms from pig fetuses. Vet Record. 102:106.

75. Engberg J, On SLW, Harrington CS, Gerner-Smidt P. 2000. Prevalence of *Campylobacter*,*Arcobacter*, *Helicobacter* and *Sutterella* spp. in human fecal samples as estimated by a reevaluation of isolation methods for campylobacters. J Clin Microbiol. 38:286–291.

76. Ertas N, Dogruer Y, Gonulalan Z, Guner A, Ulger I. 2010. Prevalence of *Arcobacter* species in drinking water, spring water, and raw milk as determined by multiplex PCR. J Food Prot. 73:2099–2102.

77. Fallas-Padilla KL, Rodríguez-Rodríguez CE, Fernández Jaramillo H, Arias Echandi ML. 2014.*Arcobacter*: comparison of isolation methods, diversity, and potential pathogenic factors in commercially retailed chicken breast meat from Costa Rica. J Food Prot. 77:880–884.

78. Fera MT, Gugliandolo C, Lentini V, Favaloro A, Bonanno D, La

Camera E, Maugeri TL. 2010.Specific detection of *Arcobacter* spp. in estuarine waters of Southern Italy by PCR and fluorescent *in situ* hybridization. Lett Appl Microbiol. 50:65–70.

79. Fera MT, La Camera E, Carbone H, Malara D, Pennisi MG. 2009. Pet cats as carriers of *Arcobacter*spp. in Southern Italy. J Appl Microbiol. 106:1661–1666.
80. Fera MT, Maugeri TL, Giannone M, Gugliandolo C, La Camera E, Blandino G, Carbone M. 2003. *In vitro* susceptibility of *Arcobacter butzleri* and *Arcobacter cryaerophilus* to different antimicrobial agents. Int J Antimicrob Agents. 21:488–491.
81. Ferna´ndez H, Eller G, Paillacar J, Gajardo T, Riquelme A. 1995. Toxigenic and invasive capacities: possible pathogenic mechanisms in *Arcobacter cryaerophilus*. Mem Inst Oswaldo. 90:633–634.
82. Ferreira S, Júlio C, Queiroz JA, Domingues FC, Oleastro M. 2014c. Molecular diagnosis of*Arcobacter* and *Campylobacter* in diarrhoeal samples among Portuguese patients. Diagn Microbiol Infect Dis. 78:220–225.
83. Ferreira S, Queiroz JA, Oleastro M, Domingues FC. 2014b. Genotypic and phenotypic features of*Arcobacter butzleri* pathogenicity. Microb Pathog. 76:19–25.
84. Ferreira S, Queiroz JA, Oleastro M, Domingues FC. 2016. Insights in the pathogenesis and resistance of *Arcobacter*: a review. Crit Rev Microbiol. 42:364–383.
85. Ferreira S, Silva F, Queiroz JA, Oleastro, M, Domingues FC. 2014a. Resveratrol against *Arcobacter butzleri* and *Arcobacter cryaerophilus*: activity and effect on cellular functions. Int J Food Microbiol. 180:62–68.
86. Figueras MJ, Collado L, Guarro J. 2008. A new 16S rDNA-RFLP method for the discrimination of the accepted species of *Arcobacter*. Diag Microbiol Infect Dis. 62:11–15.
87. Figueras MJ, Collado L, Levican A, Perez J, Solsona MJ, Yustes C. 2011a. *Arcobacter molluscorum*sp. nov., a new species isolated from shellfish. Syst Appl Microbiol. 34:105–109.
88. Figueras MJ, Levican A, Collado L, Inza MI, Yustes C. 2011b. *Arcobacter ellisii* sp. nov., isolated from mussels. Syst Appl Microbiol. 34:414–418.
89. Figueras MJ, Levican A, Collado L. 2012. Updated 16S rRNA-RFLP

method for the identification of all currently characterised *Arcobacter* spp. BMC Microbiol. 12:292.

90. Figueras MJ, Levican A, Pujol I, Ballester F, Rabada, Quilez MJ, Gomez-Bertomeu F. 2014. A severe case of persistent diarrhoea associated with *Arcobacter cryaerophilus* but attributed to*Campylobacter* sp. and a review of the clinical incidence of *Arcobacter* spp. New Microbes New Infect. 2:31–37.
91. Fisher JC, Levican A, Figueras MJ, McLellan SL. 2014. Population dynamics and ecology of*Arcobacter* in sewage. Front Microbiol. 5:1–9.
92. Fisher K, Rowe C, Phillips CA. 2007. The survival of three strains of *Arcobacter butzleri* in the presence of lemon, orange and bergamot essential oils and their components in vitro and on food. Lett Appl Microbiol. 44:495–499.
93. Flanagan RC, Neal-McKinney JM, Dhillon AS, Miller WG, Konkel ME. 2009. Examination of*Campylobacter jejuni* putative adhesins leads to the identification of a new protein, designated FlpA, required for chicken colonization. Infect Immun. 77:2399–2407.
94. Fong TT, Mansfield LS, Wilson DL, Schwab DJ, Molly SL, Rose JB. 2007. Massive microbiological ground water contamination associated with a waterborne outbreak in lake Eri South Bass Island, Ohio. Environ Health Perspect. 115:856–864.
95. Giacometti F, Lucchi A, Di Francesco A, Delogu M, Grilli E, Guarniero I, Stancampiano L,Manfreda G, Merialdi G, Serraino A. 2015. *Arcobacter butzleri, Arcobacter cryaerophilus* and*Arcobacter skirrowii* circulation in a dairy farm and sources of milk contamination. Appl Environ Microbiol. 81:5055–5063.
96. Giacometti F, Lucchi A, Manfreda G, Florio D, Zanoni RG, Serraino A. 2013. Occurrence and genetic diversity of *Arcobacter butzleri* in an artisanal dairy plant in Italy. Appl Environ Microbiol. 79:6665–6669.
97. Giacometti F, Serraino A, Pasquali F, De Cesare A, Bonerba E, Rosmini R. 2014. Behavior of*Arcobacter butzleri* and *Arcobacter cryaerophilus* in ultrahigh-temperature, pasteurized, and raw cow's milk under different temperature conditions. Foodborne Pathog Dis. 11:15–20.
98. Gilbert MJ, Kik M, Timmerman AJ, Severs TT, Kusters JG, Duim B, Wagenaar JA. 2014. Occurrence, diversity, and host association of intestinal *Campylobacter*, *Arcobacter* and *Helicobacter* in Reptiles.

PloS One. 9:e101599.

99. Gill KP.1983. Aerotolerant *campylobacter* strain isolated from a bovine preputial sheath washing. Vet Rec. 112:459.

100. Golla SC, Murano EA, Johnson LG, Tipton NC, Cureington EA, Savell JW. 2002. Determination of the occurrence of *Arcobacter butzleri* in beef and dairy cattle from Texas by various isolation methods. J Food Prot. 65:1849–1853.

101. González A. Suski J, Ferrus MA. 2010. Rapid and accurate detection of *Arcobacter* contamination in commercial chicken products and waste water samples by real-time polymerase chain reaction. Foodborne Pathog Dis. 7:327–338.

102. GonzálezAM, FerrusMA. 2011. Study of *Arcobacter* spp. contamination in fresh lettuces detected by different cultural and molecular methods. Int J Food Microbiol. 145:311–314.

103. González I, Garcia T, Antonlin A, Hernandez PE, Martin R. 2000. Development of a combined PCR-culture technique for the rapid detection of *Arcobacter* spp. in chicken meat. Lett Appl Microbiol. 30:207–212.

104. González A, Botella S, Montes RM, Moreno Y, Ferrus MA. 2007a. Direct detection and identification *of Arcobacter* species by multiplex PCR in chicken and waste water samples from Spain. J Food Prot. 70:341–347.

105. González A, Moreno Y, González R, Herna´ndez J, Ferrus MA. 2006. Development of a simple and rapid method based on polymerase chain reaction-based restriction fragment length polymorphism analysis to differentiate *Helicobacter*, *Campylobacter* and *Arcobacter* species. Curr Microbiol. 53:416–421.

106. González A, Ferrús MA, González R, Hernández J. 2007b. Molecular fingerprinting of*Campylobacter* and *Arcobacter* isolated from chicken and water. Int Microbiol. 10:85–90.

107. González I, Fernández-Tomé S, García T, Martín R. 2014. Genus-specific PCR assay for screening Arcobacter spp. in chicken meat. J Sci Food Agric. 94:1218–1224.

108. Gopi M, Karthik K, Manjunathachar HV, Tamilmahan P, Kesavan M, Dashprakash M, Balaraju BL,Purushothaman MR. 2014. Essential oils as a feed additive in poultry nutrition. Adv Anim Vet Sci. 2:1–7.

109. Gölz G, Alter T, Bereswill S, Heimesaat MM. 2016a. The

Immunopathogenic Potential of*Arcobacter butzleri* – lessons from a meta-analysis of murine infection studies. PLoS One. 11:e0159685.

110. Gölz G, Alter T, Bereswill S, Heimesaat MM. 2016b. Toll-like receptor-4 dependent intestinal gene expression during *Arcobacter butzleri* infection of gnotobiotic Il-10 deficient mice. Eur J Microbiol Immunol. 6:67–80.

111. Gölz G, Karadas G, Alutis ME, Fischer A, Kuhl AA, Breithaupt A, Gobel UB, Alter T, Bereswill S,Heimesaat MM. 2015a. *Arcobacter butzleri* induce colonic, extra-intestinal and systemic inflammatory responses in gnotobiotic IL-10 deficient mice in a strain-dependent manner. PloS One. 10:e0139402.

112. Gölz G, Karadas G, Fischer A, Gobel UB, Alter T, Bereswill S, Heimesaat MM. 2015b. Toll-like receptor-4 is essential for *Arcobacter butzleri*-induced colonic and systemic immune responses in gnotobiotic IL-10–/–mice. Eur J Microbiol Immunol. 5:321–332.

113. Gugliandolo C, Irrera GP, Lentini V, Maugeri TL. 2008. Pathogenic *Vibrio*, *Aeromonas* and*Arcobacter* spp. associated with copepods in the straits of Messina (Italy). Mar Pollut Bull. 56:600–606.

114. Hamill S, Neill SD, Madden RH. 2008. A comparison of media for the isolation of *Arcobacter* spp. from retail packs of beef. J Food Protect. 71:850–854.

115. Hamir AN, Sonn RJ, Franklin S, Wesley IV. 2004. *Campylobacter* jejuni and *Arcobacter* species associated with intussusceptions in a raccoon (Procyon lotor). Vet Record. 155:338–340.

116. Hänel I, Tomaso H, Neubauer H. 2016. Arcobacter - an underestimated zoonotic pathogen?Bundesgesundheitsblatt Gesundheitsforschung Gesundheitsschutz. 59:789–794.

117. Harmon KM, Wesley IV. 1996. Identification of *Arcobacter* isolates by PCR. Lett Appl Microbiol. 23:241–244.

118. Harmon KM, Wesley IV. 1997. Multiplex PCR for the identification of Arcobacter and differentiation of *Arcobacter butzleri* from other Arcobacters. Vet Microbiol. 58:215–227.

119. Harrab B, Schwarz S, Wenzel S. 1998. Identi¢cation and characterization of *Arcobacter* isolates from broilers by biochemical tests, antimicrobial resistance patterns and plasmid analysis. J Vet Med. 45:87–94.

120. Hausdorf L, Fröhling A, Schlüter O, Klocke M. 2011. Analyses of the bacterial community within carrot wash water. Can J Microbiol.

57:447–452.

121. Hausdorf L, Neumann M, Bergmann I, Sobiella K, Mundt K, Fröhling A, Schlüter O, Klocke M.2013. Occurrence and genetic diversity of *Arcobacter* spp. in a spinach-processing plant and evaluation of two *Arcobacter*-specific quantitative PCR assay. Syst Appl Microbiol. 36:235–243.

122. Heimesaat MM, Alter T, Bereswill S, Gölz G. 2016. Intestinal expression of genes encoding inflammatory mediators and gelatinases during *Arcobacter butzleri* infection of gnotobiotic Il-10 deficient mice. Eur J Microbiol Immunol. 6:56–66.

123. Heimesaat MM, Karadas G, Alutis M, Fischer A, Kuhl AA, Breithaupt A, Gobel UB, Alter T,Bereswill S, Gölz G. 2015a. Survey of small intestinal and systemic immune responses following murine *Arcobacter butzleri* infection. Gut Pathog. 7:28.

124. Heimesaat MM, Karadas G, Fischer A, Gobel UB, Alter T, Bereswill S, Gölz G. 2015b. Toll-like receptor-4 dependent small intestinal immune responses following murine *Arcobacter butzleri*infection. Eur J Microbiol Immunol. 5:333–342.

125. Higgins R, Messier S, Daignault D, Lorange M. 1999. *Arcobacter butzleri* isolated from a diarrhoeic non-human primate. Lab Anim. 33:87–90.

126. Hill JE, Paccagnella A, Law K, Melito PL, Woodward DL, Price L, Leung AH, Ng LK, HemmingsenSM, Goh SH. 2006. Identification of Campylobacter spp. and discrimination from Helicobacter and Arcobacter spp. by direct sequencing of PCR-amplified cpn60 sequences and comparison to cpnDB, a chaperonin reference sequence database. J Med Microbiol. 55:393–399.

127. Hilton CL, Mackey BM, Hargreaves AJ, Forsythe SJ. 2001. The recovery of *Arcobacteri* NCTC 12481 from various temperature treatments. J Appl Microbiol. 91:929–932.

128. Ho HT, Lipman LJ, Gaastra W. 2006a. *Arcobacter*, what is known and unknown about a potential food-borne zoonotic agent.Vet Microbiol. 115:1–13.

129. Ho HT, Lipman LJ, Gaastra W. 2008a. The introduction of *Arcobacter* spp. in poultry slaughterhouses. Int J Food Microbiol. 125:223–229.

130. Ho HT, Lipman LJ, Hendriks HG, Tooten PC, Ultee T, Gaastra W. 2007. Interaction of *Arcobacter*spp. with human and porcine intestinal

epithelial cells. FEMS Immunol Med Microbiol. 50:51–58.

131. Ho HT, Lipman LJ, Wosten MM, van Asten AJ, Gaastra W. 2008b. *Arcobacter* spp. possess two very short flagellins of which FlaA is essential for motility. FEMS Immunol Med Microbiol. 53:85–95.

132. Ho HT, Lipman LJA, Van der Graaf-van Bloois L, Van Bergen M, Gaastra W. 2006b. Potential routes of acquisition of *Arcobacter* species by piglets. Vet Microbiol. 114:122–133.

133. Houf K, De Zutter L, Van Hoof J, Vandamme P. 2002a. Occurrence and distribution of *Arcobacter*species in poultry processing. J Food Prot. 65:1233–1239.

134. Houf K, De Zutter L, Van Hoof J, Vandamme P. 2002b. Assessment of the genetic diversity among arcobacters isolated from poultry products by using two PCR-based typing methods. Appl Environ Microbiol. 68:2172–2178.

135. Houf K, Devriese LA, De Zutter L, Verbeke B, Van Hoof J, Vandamme P. 2003. Molecular characterization of *Arcobacter* isolates collected in a poultry slaughterhouse. J Food Prot. 66:364–369.

136. Houf K, Devriese LA, De zutter L, Van Hoof J, Vandamme P. 2001. Susceptibility of *Arcobacter butzleri, Arcobacter cryaerophilus*, and *Arcobacter skirrowii* to antimicrobial agents used in selective media. J Clin Microbiol. 39:1654–1656.

137. Houf K, On SL, Coenye T, Debruyne L, De Smet S, Vandamme P. 2009. *Arcobacter thereius* sp. nov., isolated from pigs and ducks. Int J Syst Evol Microbiol. 59:2599–2604.

138. Houf K, Sarah DS, Julie B, Sylvie D. 2008. Dogs as carriers of the emerging pathogen; *Arcobacter*. J Vet Microbiol. 130:208–213.

139. Houf K, Stephan R. 2007. Isolation and characterization of the emerging food-borne pathogen*Arcobacter* from human stool. J Microbiol Methods. 68:408–413.

140. Houf K, Tutenel A, De Zutter L, Van Hoof J, Vandamme P. 2000. Development of a multiplex PCR assay for the simultaneous detection and identification of *Arcobacter butzleri, Arcobacter cryaerophilus* and *Arcobacter skirrowii*. FEMS Immunol Med Microbiol. 49:337–345.

141. Hsu TT, Lee J. 2015. Global distribution and prevalence of arcobacter in food and water. Zoonoses Public Health. 62:579–589.

142. Hsueh PR, Teng LJ, Yang, PC, Wang SK, Chang SC, Ho SW, Hsieh WC, Luh KT. 1997. Bacteremia caused by *Arcobacter cryaerophilus*

1B. J Clin Microbiol. 35:489–491.

143. Hu Q, Lyu D, Shi X, Jiang Y, Lin Y, Li Y, Qiu Y, He L, Zhang R, Li Q. 2014. A modified molecular beacons-based multiplex real-time pcr assay for simultaneous detection of eight foodborne pathogens in a single reaction and its application. Foodborne Pathog Dis. 11:207–214.
144. Irkin R, Abay S, Aydin F. 2011. Inhibitory effects of some plant essential oils against *Arcobacter butzleri* and potential for rosemary oil as a natural food preservative. J Med Food. 14:291–296.
145. Iwamoto M, Ayers T, Mahon BE, Swerdlow DL. 2010. Epidemiology of seafood-associated infections in the United States. Clin Microbiol Rev. 23:399–411.
146. Jacob J, Loir H, Feuerpfeil I. 1993. Isolation of *Arcobacter butzleri* from drinking water reservoir in Eastern Germany. Int J Hyg Env Med. 193:557–562.
147. Jacob J, Woodward D, Feuerpfeil I, Johnson WM. 1998. Isolation of *Arcobacter butzleri* in raw water and drinking water treatment plants in Germany. Zbl Hyg. 201:189–198.
148. Jiang ZD, DuPont HL, Brown EL, Nandy RK, Ramamurthy T, Sinha A, Ghosh, S, Guin S, Gurleen K,Rodrigues S, et al. 2010. Microbial etiology of travelers' diarrhea in Mexico, Guatemala, and India: importance of enterotoxigenic *Bacteroides fragilis* and *Arcobacter* species. J Clin Microbiol. 48:1417–1419.
149. Johnson LG, Murano EA. 1999a. Development of a new medium for the isolation of *Arcobacter*spp. J Food Prot. 62:456–462.
150. Johnson LG, Murano EA. 1999b. Comparison of three protocols for the isolation of *Arcobacter*from poultry. J Food Prot. 62:610–614.
151. Kabeya H, Koboyashi Y, Maruyama S, Mikami T. 2003b. One step polymerase chain reaction-based typing of *Arcobacter* species. Int J Food Microbiol. 81:163–168.
152. Kabeya H, Maruyama S, Morita Y, Kubo M, Yamamoto K, Arai S, Izumi T, Kobayashi Y, Katsube Y,Mikami T. 2003a. Distribution of *Arcobacter* species among livestock in Japan. Vet Microbiol. 93:153–158.
153. Kabeya H, Maruyama S, Morita Y, Ohsuga T, Ozawa S, Kobayashi Y, Abe M, Katsube Y Mikami T.2004. Prevalence of *Arcobacter* species in retail meats and antimicrobial susceptibility of the isolates in Japan. Int J Food Microbiol. 90:303–308.

154. Karadas G, Sharbati S, Hänel I, Messelhäußer U, Glocker E, Alter T, Gölz G. 2013. Presence of virulence genes, adhesion and invasion of *Arcobacter butzleri*. J Appl Microbiol. 115:583–590.

155. Kärenlampi RI, Tolvanen TP, Hänninen ML. (2004). Phylogenetic analysis and PCR-restriction fragment length polymorphism identification of campylobacter species based on partial groEL gene sequences. J Clin Microbiol. 42:5731–5738.

156. Karthik K, Muneeswaran NS, Manjunathachar HV, Gopi M, Elamurugan A, Kalaiyarasu S. 2014c.Bacteriophages: effective alternative to antibiotics. Adv Anim Vet Sci. 2:1–7.

157. Karthik K, Rathore R, Thomas P, Arun TR, Viswas KN, Agarwal RK, Manjunathachar HV, Dhama K.2014a. Loop-mediated isothermal amplification (LAMP) test for specific and rapid detection of*Brucella abortus* in cattle. Vet Quart. 34:174–179.

158. Karthik K, Rathore R, Thomas P, Arun TR, Viswas KN, Dhama K, Agarwal RK. 2014b. New closed tube loop mediated isothermal amplification assay for prevention of product cross contamination. MethodsX. 1:e137–e143.

159. Khan RU, Naz S, Dhama K, Karthik K, Tiwari R, Abdelrahman MM, Alhidary IA, Zahoor A. 2016.Direct-fed microbial: beneficial applications, modes of action and prospecas safe tool for enhancing ruminant production and safeguarding health. Int J Pharmacol. 12:220–231.

160. Khoshbakht R, Tabatabaei M, Shirzad Aski H, Seifi S. 2014. Occurrence of *Arcobacter* in Iranian poultry and slaughterhouse samples implicates contamination by processing equipment and procedures. Br Poult Sci. 55:732–736.

161. Kiehlbauch JA, Baker C, Wachsmuth IK. 1992. *In vitro* susceptibilities of aerotolerant*Campylobacter* isolates to 22 antimicrobial agents. Antimicrob Agents Chemother. 36:717–722.

162. Kiehlbauch JA, Brenner DJ, Nicholson MA, Baker CN, Patton CM, Steigerwalt AG, Wachsmuth IK.1991. *Campylobacter butzleri* sp. nov. isolated from humans and animals with diarrheal illness. J Clin Microbiol. 29:376–385.

163. Kim HM, Hwang CY, Cho BC. 2010. *Arcobacter marinus* sp. nov. Int J Syst Evol Microbiol. 60:531–536.

164. Kjeldgaard J, Jorgensen K, Ingmer H. 2009. Growth and survival

at chiller temperature of*Arcobacter butzleri*. Int J Food Microbiol. 131:256–259.

165. Kolling G, Wu M, Guerrant RL. 2012. Enteric pathogens through life stages. Front Cell Infect Microbiol. 2:114.
166. Kopilovic B, Ucakar V, Koren N, Krek M, Kraigher A. 2008. Waterborne outbreak of acute gastroentritis in a coastal area in Slovenia in June and July 2008. Eurosurveillance. 13:1–3.
167. Laishram M, Rathlavath S, Lekshmi M, Kumar S, Nayak BB. 2016. Isolation and characterization of Arcobacter spp. from fresh seafood and the aquatic environment. Int J Food Microbiol. 232:87–89.
168. Lammerding AM, Harris JE, Lior H, Woodward D, Cole L, Muckle CA. 1996. Isolation method for recovery of *Arcobacter butzleri* from fresh poultry and poultry products. In Newell DG, Ketley J, Feldman RA, editors. 8th International Workshop on '*Campylobacters, Helicobacters* and Related Organisms'. New York (NY): Plenum Press; p. 329–334.
169. Lammerding AM, Harris JE, Loir H, Muckel CA, Irwin RJ. 1994. An isolation method for *Arcobacter butzleri* from poultry. Dairy Food Environ Sanitation. 14:600.
170. Lappi V, Archer JR, Cebelinski E, Leano F, Besser JM, Klos RF, Medus C, Smith KE, Fitzgerald C,Davis JP. 2013. An outbreak of foodborne illness among attendees of a wedding reception in Wisconsin likely caused by *Arcobacter butzleri*. Foodborne Pathog Dis. 10:250–255.
171. Lau SKP, Woo PCY, Teng J, Leung KW, Yuen KY. 2002. Identification by 16S ribosomal RNA gene sequencing of *Arcobacter* bacteraemia in a patient with acute gangrenous appendicitis. Mol Pathol. 55:182–185.
172. Lee MH, Cheon DS, Choi S, Lee BH, Jung JY, Choi C. 2010. Prevalence of *Arcobacter* species isolated from retail meats in Korea. J Food Protect. 73:1313–1316.
173. Lee MH, Choi, C. 2013. Survival of *Arcobacter butzleri* in apple and pear purees. J Food Safety. 33:333–339.
174. Lehmann D, Alter T, Lehmann L, Uherkova S, Seidler T, Gölz G. 2015. Prevalence, virulence gene distribution and genetic diversity of *Arcobacter* in food samples in Germany. Berl Munch Tierarztl Wochenschr 128:163–168.
175. Lehner A, Tasara T, Stephan R. 2005. Relevant aspects of *Arcobacter* spp. as potential food-borne pathogens. Int J Food Microbiol. 102:127–135.

176. Lerner J, Brumberger V, Preac-Mursic V. 1994. Severe diarrhea associated with *Arcobacter butzleri*. Eur J Clin Microbiol Infect Dis. 13:660–662.

177. Levican A, Alkeskas A, Günter C, Forsythe SJ, Figueras MJ. 2013b. Adherence to and invasion of human intestinal cells by *Arcobacter* species and their virulence genotypes. Appl Environ Microbiol. 79:4951–4957.

178. Levican A, Collado L, Aguilar C, Yustes C, Dieguez AL, Romalde JL, Figueras MJ. 2012. *Arcobacter bivalviorum* sp. nov. and *Arcobacter venerupis* sp. nov., new species isolated from shellfish. Syst Appl Microbiol. 35:133–138.

179. Levican A, Collado L, Figueras MJ. 2013a. *Arcobacter cloacae* sp. nov. and *Arcobacter suis* sp. nov., two new species isolated from food and sewage. Syst Appl Microbiol. 36:22–27.

180. Levican A, Collado L, Figueras MJ. 2016. The use of two culturing methods in parallel reveals a high prevalence and diversity of *Arcobacter* spp. in a wastewater treatment plant. Biomed Res Int. 2016:8132058.

181. Levican A, Collado L, Yustes C, Aguilar C, Figueras MJ. 2014. Higher water temperature and incubation under aerobic and microaerobic conditions increase the recovery and diversity of*Arcobacter* spp. from shellfish. Appl Environ Microbiol. 80:385–391.

182. Levican A, Figueras MJ. 2013. Performance of five molecular methods for monitoring *Arcobacter*spp. BMC Microbiol. 13:220.

183. Levican A, Rubio-Arcos S, Martinez-Murcia A, Collado L, José Figueras M. 2015. *Arcobacter ebronensis* sp. nov. and *Arcobacter aquimarinus* sp. nov., two new species isolated from marine environment. Syst Appl Microbiol. 38:30–35.

184. Logan EF, Neill SD, Mackie DP. 1982. Mastitis in dairy cows associated with an aerotolerant*Campylobacter*. Vet Res. 110:229–230.

185. Manke TR, Wesley IV, Dickson JS, Harmon KM. 1998. Prevalence and genetic variability of*Arcobacter* species in mechanically separated turkey. J Food Prot. 61:1623–1628.

186. Marshall SM, Melito PL, Woodward DL, Johnson WM, Rodgers FG, Mulvey MR. 1999. Rapid identification of *Campylobacter*, *Arcobacter*, and *Helicobacter* isolates by PCR-restriction fragment length polymorphism analysis of the 16S rRNA gene. J Clin Microbiol. 37:4158–4160.

187. Maugeri TL, Irrera GP, Lentini V, Carbone M, Fera MT, Gugliandolo C. 2005. Detection and enumeration of *Arcobacter* spp. in the coastal environment of the straits of Messina (Italy). New Microbiol. 28:177–182.

188. McClung CR, Patriquin DG, Davis RE. 1983. Campylobacter nitrofigilis sp. nov., a nitrogen-fixing bacterium associated with roots of Spartina alterniflora. Int J Syst Bacteriol. 33:605–612.

189. McGregor AC, Wright SG. 2015. Gastrointestinal symptoms in travellers. Clin Med. 15:93–95.

190. Medina G, Flores-Martin S, Fonseca B, Otth C, Fernandez H. 2014. Mechanisms associated with phagocytosis of *Arcobacter butzleri* by *Acanthamoeba castellanii*. Parasitol Res. 113:1933–1942.

191. Merga JY, Leatherbarrow AJH, Winstanley C, Bennett M, Hart CA, Miller WG, Williams NJ. 2011.Comparison of *Arcobacter* isolation methods, and diversity of *Arcobacter* spp. in Cheshire, United Kingdom. Appl Environ Microbiol. 77:1646–1650.

192. Merga JY, Royden A, Pandey AK, Williams NJ. 2014. Arcobacter spp. isolated from untreated domestic effluent. Lett Appl Microbiol. 59:122–126.

193. Merga JY, Williams NJ, William G. Miller WG, Leatherbarrow AJH, Bennett M, Hall N, Ashelford KE,Winstanley C. 2013. Exploring the diversity of *Arcobacter butzleri* from cattle in the UK using MLST and whole genome sequencing. PLoS One. 8:e55240.

194. Mesbah M, Premachandran U, Whitman WB. 1989. Precise measurement of the G+C content of deoxyribonucleicacid by high-performance liquid chromatography. Int J Syst Bacteriol. 39:159–167.

195. Miller WG, Parker CT, Rubenfield M, Mendz GL, Wösten MM, Ussery DW, Stolz JF, Binnewies TT,Hallin PF, Wang G, et al. 2007. The complete genome sequence and analysis of the epsilonproteobacterium *Arcobacter butzleri*. PLoS One. 2:1358.

196. Miller WG, Wesley IV, On SLW, Houf K, Megraud F, Wang G, Yee E, Srijan A, Mason CJ. 2009. First multi-locus sequence typing scheme for *Arcobacter* spp. BMC Microbiol. 14:196.

197. Miller WG, Yee E, Jolley KA, Chapman MH. 2014. Use of improved *atpA* amplification and sequencing method to identify members of the Campylobacteraceae and Helicobacteraceae. Lett Appl Microbiol. 58:582–590.

198. Mohan HV, Rathore RS, Dhama K, Ramees TP, Patyal A, Bagalkot PS, Wani MY, Bhilegaonkar KN,Kumar A. 2014. Prevalence of *Arcobacter* spp. in humans, animals and foods of animal origin in India based on cultural isolation, antibiogram, PCR and multiplex PCR detection. Asian J Anim Vet Adv. 9:452–466.

199. Moreno Y, Alonso JL, Botella S, Ferrus MA, Hernandez J. 2004. Survival and injury of *Arcobacter*after artificial inoculation into drinking water. Res Microbiol. 155:726–730.

200. Moreno Y, Botella S, Alonso JL, Ferrus MA, Hernandez M, Hernandez J. 2003. Specific detection of *Arcobacter* and *Campylobacter* strains in water and sewage by PCR and fluorescent *in situ*hybridization. Appl Environ Microbiol. 69:1181–1186.

201. Mottola A, Bonerba E, Bozzo G, Marchetti P, Celano GV, Colao V, Terio V, Tantillo G, Figueras MJ,Di Pinto A. 2016. Occurrence of emerging food-borne pathogenic Arcobacter spp. isolated from pre-cut (ready-to-eat) vegetables. Int J Food Microbiol. 236:33–37.

202. Neill SD, Cambell JN, O'brien JJ, Weatherup STC, Ellis WA. 1985. Taxonomic position of*Campylobacter cryaerophila* sp. Nov. Int J Systemat Bacteriol. 35:342–356.

203. Neill SD, Ellis WA, O'brien JJ. 1979. Designation of aerotolerant *Campylobacter* like organisms from porcine and bovine abortions to the genus *Campylobacter*. Res Vet Sci. 27:180–186.

204. Neubauer J, Jauk V, Szolgyenyi W. 2003. Phenotypic and genotypic differentiation of*Campylobacter* spp. isolated from Austrian broiler farms: a comparison. Avian Pathol. 32:33–37.

205. Neubaure C, Hess M. 2006. Detection and identification of food borne pathogens of the genera*Campylobacter, Arcobacter* and *Helicobacter* by multiplex PCR in poultry and poultry products. J Vet Med. 53:376–381.

206. Num SM, Useh NM. 2013. Nanotechnology applications in veterinary diagnostics and therapeutics. Sokoto J Vet Sci. 11:10–14.

207. Ohlendorf DS, Murano EA. 2002. Prevalence of *Arcobacter* spp. in raw ground pork from several geographical regions according to various isolation methods. J Food Protect. 65:1700–1705.

208. On SLW, Jensen TK, Bille-Hansen V, Jorsal SE, Vandamme P. 2002. Prevalence and diversity of*Arcobacter* spp. isolated from the internal organs of spontaneous porcine abortions in Denmark. Vet Microbiol.

85:159–167.

209. On SLW, Stacey A, Smyth J. 1995. Isolation of *Arcobacter butzleri* from a neonate with bacteremia. J Infect. 31:225–227.

210. Ongor H, Cetinkaya B, Açik MN, Atabay HI. 2004. Investigation of Arcobacters in meat and fecal samples of clinically healthy cattle in Turkey. Lett Appl Microbiol. 38:339–344.

211. Ottaviani D, Chierichetti S, Rocchegiani E, Bartolini C, Masini L, Santarelli S, Leoni F. 2013.Bioaccumulation experiments in mussels contaminated with the food-borne pathogen*Arcobacter butzleri*: preliminary data for risk assessment. Biomed Res Int. 2013:153419.

212. Park S, Jung YT, Kim S, Yoon JH. 2016. Arcobacter acticola sp. nov., isolated from seawater on the East Sea in South Korea. J Microbiol. 54:655–659.

213. Patyal A, Rathore RS, Mohan HV, Dhama K, Kumar A. 2011. Prevalence of *Arcobacter* spp. in humans, animals and foods of animal origin including sea food from India. Transbound Emerg Dis. 58:402–410.

214. Pejchalova M, Dostali´kova, E. Slamova M, Brozkova I, Vytrasova J. 2008. Prevalence and diversity of *Arcobacter* spp. in the Czech Republic. J Food Protect. 71:719–727.

215. Pentimalli D, Pegels N, Garcia T, Martin R, González I. 2009. Specific PCR detection of *Arcobacter butzleri, Arcobacter cryaerophilus, Arcobacter skirrowii* and *Arcobacter cibarius* in chicken meat. J Food Protect. 72:1491–1495.

216. Petersen RF, Harrington CS, Kortegaard HE, On SL. 2007 . A PCR-DGGE method for detection and identification of *Campylobacter, Helicobacter*, *Arcobacter* and related Epsilobacteria and its application to saliva samples from humans and domestic pets. J Appl Microbiol. 103:2601–2615.

217. Phillips CA, Duggan J. 2001. The effect of EDTA and trisodium phosphate, alone and in combination with nisin, on the growth of *Arcobacter butzleri* in culture. Food Microbiol. 18:547–554.

218. Phillips CA, Long C. 2001. The survival of *Arcobacter butzleri* on chicken. Int J Med Microbiol. 291:93.

219. Phillips CA. 1999. The effect of citric acid, lactic acid, sodium citrate and sodium lactate, alone and in combination with nisin, on the growth of *Arcobacter butzleri*. Lett Appl Microbiol. 29:424–428.

220. Phillips CA. 2001. *Arcobacter* spp in food: isolation, identification, and control. Trends Food Sci Technol. 12:263–275.

221. Prouzet-Mauleon V, Labadi L, Bouges N, Menard A, Megraud F. 2006. *Arcobacter butzleri*: underestimated enteropathogen. Emerg Infect Dis. 12:307–309.

222. Quinones B, Parker CT, Janda JM, Miller WG, Mandrell RE. 2007. Detection and genotyping of*Arcobacter* and *Campylobacter* isolates from retail chicken samples by use of DNA oligonucleotide arrays. Appl Environ Microbiol. 73:3645–3655.

223. Rahimi E. 2014. Prevalence and antimicrobial resistance of *Arcobacter* species isolated from poultry meat in Iran. Br Poult Sci. 55:174–180.

224. Ramees TP, Rathore RS, Bagalakote PS, Mohan HV, Kumar A, Dhama K. 2014a. Multiplex PCR detection of *Arcobacter butzleri* and *Arcobacter cryaerophilus* in skin of poultry. J Pure Appl Microbiol. 8:1755–1758.

225. Ramees TP, Rathore RS, Bagalkot PS, Mohan HV, Kumar A, Dhama K. 2014b. Detection of*Arcobacter butzleri* and *Arcobacter cryaerophilus* in clinical samples of humans and foods of animal origin by cultural and multiplex PCR based methods. Asian J Anim Vet Adv. 9:243–252.

226. Ramees TP, Rathore RS, Bagalkot PS, Ravi Kumar GVPPS, Mohan HV, Anoopraj R, Kumar A,Dhama K. 2014c. Real-time PCR detection of *Arcobacter butzleri* and *Arcobacter cryaerophilus* in chicken meat samples. J Pure Appl Microbiol. 8:3165–3169.

227. Ramees TP, Rathore RS, Bagalkot PS, Sailo Blessa, Mohan HV, Kumar A, Dhama K, Singh RK.2014d. Genotyping and genetic diversity of *Arcobacter butzleri* and *Arcobacter cryaerophilus*isolated from different sources by using ERIC-PCR from India. Vet Quart. 34:211–217.

228. Rasmussen LH, Jette K, Jens PC, Hanne I. 2013. Multilocus sequence typing and biocide tolerance of *Arcobacter butzleri* from Danish broiler carcasses. BMC Res Notes. 6:322.

229. Revez J, Huuskonen M, Ruusunen M, Lindström M, Hänninen ML. 2013. *Arcobacter* species and their pulsed-field gel electrophoresis genotypes in Finnish raw milk during summer 2011. J Food Prot. 76:1630–1632.

230. Ricci JCD, Hernandez ME. 2000. Plasmid effects on *Escherichia coli* metabolism. Crit Rev Biotechnol. 20:79–108.

231. Rice EW, Rodgers MR, Wesley IV, Johnson CH, Tanner SA. 1999. Isolation of *Arcobacter butzleri*from ground water. Lett Appl Microbiol. 28:31–35.

232. Rivas L, Fegan N, Vanderlinde P. 2004. Isolation and characterisation of *Arcobacter butzleri* from meat. Int J Food Microbiol. 91:31–41.

233. Romero J, Varela GM, Laclette JP, Espejo RT. 2002. Bacterial 16S rRNA gene analysis revealed that bacteria related to *Arcobacter* spp. constitute an abundant and common component of the oyster microbiota (*Tiostrea chilensis*). Microb Ecol. 44:365–371.

234. Ruiz N. 2008. Bioinformatics identification of MurJ (MviN) as the peptidoglycan lipid II flippase in Escherichia coli. Proc Natl Acad Sci USA. 105:15553–15557.

235. Salas-Massó N, Andree KB, Furones MD, Figueras MJ. 2016. Enhanced recovery of Arcobacter spp. using NaCl in culture media and re-assessment of the traits of *Arcobacter marinus* and*Arcobacter halophilus* isolated from marine water and shellfish. Sci Total Environ. 566–567:1355–1361.

236. Samarajeewa AD, Hammad A, Masson L, Khan IU, Scroggins R, Beaudette LA. 2015. Comparative assessment of next-generation sequencing, denaturing gradient gel electrophoresis, clonal restriction fragment length polymorphism and cloning-sequencing as methods for characterizing commercial microbial consortia. J Microbiol Methods. 108:103–111.

237. Samie A, Obi CL, Barrett LJ, Powell SM, Guerrant RL. 2007. Prevalence of *Campylobacter species,Helicobacter pylori* and *Arcobacter* species in stool samples from the Venda region, Limpopo, South Africa: studies using molecular diagnostic methods. J Infect. 54:558–566.

238. Sasi Jyothsna TS, Rahul K, Ramaprasad EV, Sasikala Ch, Ramana ChV. 2013. Arcobacter anaerophilus sp. nov., isolated from an estuarine sediment and emended description of the genus *Arcobacter*. Int J Syst Evol Microbiol. 63:4619–4625.

239. Scallan E, Griffin PM, Angulo FJ, Tauxe RV, Hoekstra RM. 2011. Foodborne illness acquired in the United States – unspecified agents. Emerg Infect Dis. 17:16.

240. Scarano C, Giacometti F, Manfreda G, Lucchi A, Pes E, Spanu C, De Santis EP, Serraino A. 2014.*Arcobacter butzleri* in sheep ricotta cheese at retail and related sources of contamination in an industrial dairy

plant. Appl Environ Microbiol. 80:7036–7041.

241. Schmitt B, Henderson L. 2005. Diagnostic tools for animal diseases. Rev Sci Tech. 24:243–250.
242. Scullion R, Harrington CS, Madden RH. 2004. A comparison of three methods for the isolation of*Arcobacter* spp. from retail raw poultry in Northern Ireland. J Food Prot. 67:799–804.
243. Serraino A, Florio D, Giacometti F, Piva S, Mion D, Zanoni RG. 2013b. Presence of *Campylobacter*and *Arcobacter* species in in-line milk filters of farms authorized to produce and sell raw milk and of a water buffalo dairy farm in Italy. J Dairy Sci. 96:2801–2807.
244. Serraino A, Giacometti F, Daminelli P, Losio MN, Finazzi G, Marchetti G, Zambrini AV, Rosmini R.2013a. Survival of *Arcobacter butzleri* during production and storage of artisan water buffalo mozzarella cheese. Foodborne Pathog Dis. 10:820–824.
245. Serraino A, Giacometti F. 2014. Short communication: occurrence of *Arcobacter* species in industrial dairy plants. J Dairy Sci. 97:2061–2065.
246. Shah AH, Saleha AA, Murugaiyah M, Zunita Z, Memon AA. 2012a. Prevalence and distribution of*Arcobacter* spp. in raw milk and retail raw beef. J Food Prot. 75:1474–1478.
247. Shah AH, Saleha AA, Zunita Z, Cheah YK, Murugaiyah M, Korejo NA. 2012b. Genetic characterization of Arcobacter isolates from various sources. Vet Microbiol. 160:355–361.
248. Shah AH, Saleha AA, Zunita Z, Murugaiyah M, Aliyu AB, Jafri N. 2013. Prevalence, distribution and antibiotic resistance of emergent Arcobacter spp. from clinically healthy cattle and goats. Transbound Emerg Dis. 60:9–16.
249. Shah AH, Saleha AA, Zunita Z, Murugaiyah M. 2011. *Arcobacter*– an emerging threat to animals and animal origin food products? Tr Food Sci Technol. 22:225–236.
250. Shirzad Aski H, Tabatabaei M, Khoshbakht R, Raeisi M. 2016. Occurrence and antimicrobial resistance of emergent Arcobacter spp. isolated from cattle and sheep in Iran. Comp Immunol Microbiol Infect Dis. 44:37–40.
251. Silipo A, Sturiale L, Perino V, Garozzo D, Lanzetta R, Parrilli M, Molinaro A. 2010. The structure of the carbohydrate backbone of

the lipooligosaccharide from the halophilic bacterium *Arcobacter halophilus*. Carbohydr Res. 345:850–853.

252. Skrivanová E, Molatová Zuzana, Matenová Michaela, Houf Kurt, Marounek Milan. 2011.Inhibitory effect of organic acids on arcobacters in culture and their use for control of*Arcobacter butzleri* on chicken skin. Int J Food Microbiol. 144:367–371.

253. Snelling WJ, Matsuda M, Moore JE, Dooley JS. 2006. Under the microscope: *Arcobacter*. Lett Appl Microbiol. 42:7–14.

254. Son I, Englen MD, Berrang ME, Fedorka-Carry PJ, Harrison MA. 2007. Prevalence of *Arcobacter*and *Campylobacter* on broiler carcases during processingInt. J Food Microbiol. 113:16–22.

255. Son I, Englen MD, Berrang ME, Fedorka-Cray PJ, Harrison MA. 2006. Genetic diversity of*Arcobacter* and *Campylobacter* on broiler carcasses during processing. J Food Prot. 69:1028–1033.

256. Stirling J, Griffith M, Blair I, Cormican M, Dooley JS, Goldsmith CE, Glover SG, Loughrey A, LoweryCJ, Matsuda M, et al. 2008. Prevalence of gastrointestinal bacterial pathogens in a population of zoo animals. Zoonoses Public Health. 55:166–172.

257. Strugala V, Allen A, Dettmar PW, Pearson JP. 2003. Colonic mucin: methods of measuring mucus thickness. P Nutr Soc. 62:237–243.

258. Suarez DL, Wesley IV, Larson DJ. 1997. Detection of *Arcobacter* species in gastric samples from swine. Vet Microbiol. 57:325–336.

259. Sudhakar NR, Manjunathachar HV, Karthik K, Sahu S, Gopi M, Shanthaveer SB, Nagaraja KH,Shinde S, Tamilmahan P, Madhu DN, Maurya PS. 2013. RNA interference in parasites: prospects and pitfalls. Adv Anim Vet Sci. 1:1–6.

260. Suelam II A. 2012. Isolation and identification of A*rcobacter* species recovered from rabbits in Zagazig, Egypt. Int J of Micr Res. 3:87–92.

261. Tabatabaei M, Aski HS, Shayegh H, Khoshbakht R. 2014. Occurrence of six virulence-associated genes in *Arcobacter* species isolated from various sources in Shiraz, Southern Iran. Microb Pathog. 66:1–4.

262. Talay F, Molva C, Atabay HI. 2016. Isolation and identification of *Arcobacter* species from environmental and drinking water samples. Folia Microbiol. 61:479–484. doi:10.1007/s12223-016-0460-0

263. Tee W, Baird R, Dyall-Smith M, Dwyer B. 1988. *Campylobacter cryaerophila* isolated from a human. J Clin Microbiol. 26:2469–2473.

264. Tiwari R, Chakraborty S, Dhama K, Rajagunalan S, Singh SV. 2013. Antibiotic resistance – an emerging health problem: causes, worries, challenges and solutions – a review. Int J Curr Res. 5:1880–1892.

265. Tiwari R, Dhama K, Chakraborty S, Kumar A, Rahal A, Kapoor S. 2014. Bacteriophage therapy for safeguarding animal and human health: a review. Pak J Biol Sci. 17:301–315.

266. Tiwari R, Karthik K, Rana R, Malik YS, Dhama K, Joshi SK. 2016. Quorum sensing inhibitors/antagonists countering food spoilage bacteria-need molecular and pharmaceutical intervention for protecting current issues of food safety. Int J Pharmacol. 12:262–271.

267. Tsang RS, Luk JM, Woodward DL, Johnson WM. 1996. Immunochemical characterization of a haemagglutinating antigen of *Arcobacter* spp. FEMS Microbiol Lett. 136:209–213.

268. Ünver A, Atabay Hİ, Şahin M, Çelebi Ö. 2013. Antimicrobial susceptibilities of various *Arcobacter*species. Turk J Med Sci. 43:548–552.

269. Van Driessche E, Houf K, Van Hoof J, De Zutter L, Vandamme P. 2003. Isolation of *Arcobacter*species from animal feces. FEMS Microbiol. 229:243–248.

270. Van Driessche E, Houf K, Vangroenweghe F, De Zutter L, Van Hoof J. 2005. Prevalence, enumeration and strain variation of *Arcobacter* species in the faeces of healthy cattle in Belgium. Vet Microbiol. 105:149–154.

271. Van Driessche E, Houf K, Vangroenweghe F, Nollet N, De Zutter L, Vandamme P, Van Hoof J.2004. Occurrence and strain diversity of *Arcobacter* species isolated from healthy Belgian pigs. Res Microbiol. 155:662–666.

272. Van Driessche E, Houf K. 2007a. Characterization of the *Arcobacter* contamination on Belgian pork carcasses and raw retail pork. Int J Food Microbiol. 118:20–26.

273. Van Driessche E, Houf K. 2007b. Discrepancy between the occurrence of *Arcobacter* in chickens and broiler carcass contamination. Poult Sci. 86:744–751.

274. Vandamme P, De Ley J. 1991. Proposal for a new family *Campylobacteraceae*. Int J Syst Bacteriol. 41:451–455.

275. Vandamme P, Falsen E, Rossau R, Hoste B, Segers P, Tytgat R, De Ley J. 1991. Revision of*Campylobacter*, *H elicobacter* and *Wolinella*

taxonomy: emendation of generic descriptions and proposal of *Arcobacter* genera. Int J Syst Bacteriol. 41:88–103.

276. Vandamme P, Giesendorf AJ, Van Belkum P, Lauwers S, Kersters K, Butzler P, Goossens H, QuintWG. 1993. Discrimination of epidemic and sporadic isolates of *Arcobacter butzleri* by polymerase chain reaction-mediated DNA fingerprinting. J Clin Microbiol. 31:3317–3319.

277. Vandamme P, Pugina P, Benzi G, Van Etterijck R, Vlaes L, Kersters K, Butzler JP, Lior H, Lauwers S.1992a. Outbreak of recurrent abdominal cramps associated with *Arcobacter butzleri* in an Italian school. J Clin Microbiol. 30:2335–2337.

278. Vandamme P, Vanvanneyt M, Pot B, Mels L, Hoste B, Dewettinck D, Vlaes L, Van Den Borre C,Higgins R, Hommez J, et al. 1992b. Polyphasic taxonomic study of the emended genus*Arcobacter* with *Arcobacter butzleri* comb. Nov. and *Arcobacter skirrowii* sp. Nov., and aerotolerant bacterium isolated from veterinary specimens. Int J Syst Bacteriol. 42:344–356.

279. Van den Abeele AM, Vogelaers D, Van Hende J, Houf K. 2014. Prevalence of *Arcobacter* species among humans, Belgium, 2008–2013. Emerg Infect Dis. 20:1731–1734.

280. Van den Abeele AM, Vogelaers D, Vanlaere E, Houf K. 2016. Antimicrobial susceptibility testing of*Arcobacter butzleri* and *Arcobacter cryaerophilus* strains isolated from Belgian patients. J Antimicrob Chemother. 71:1241–1244.

281. Vandenberg O, Dediste A, Houf K, Ibekwem S, Souayah H, Cadranel S, Douat N, Zissis G, ButzlerJP, Vandamme P. 2004. *Arcobacter* species in humans. Emerg Infect Dis. 10:1863–1867.

282. Vergis J, Negi M, Poharkar K, Das DP, Malik SV, Kumar A, Doijad SP, Barbuddhe SB, Rawool DB.2013. 16S rRNA PCR followed by restriction endonuclease digestion: a rapid approach for genus level identification of important enteric bacterial pathogens. J Microbiol Methods. 95:353–356.

283. Vilar MJ, Pena FJ, Pe´rez I, Die´guez FJ, Sanjuan ML, Rodri´guez-Otero JL, Yus E. 2010. Presence of *Listeria*, *Arcobacter*, and *Campylobacter* spp. in dairy farms in Spain. Berl Munch Tierarztl Wochenschr. 23:58–62.

284. Villalobos EG, Jaramillo HF, Ulate CC, Echandi ML. 2013. Isolation

and identification of zoonotic species of genus arcobacter from chicken viscera obtained from retail distributors of the metropolitan area of San José, Costa Rica. J Food Prot. 76:879–882.

285. Villanueva MP, Medina G, Fernández H. 2016. *Arcobacter butzleri* survives within trophozoite of*Acanthamoeba castellanii*. Rev Argent Microbiol. 48:105–109.

286. Villarruel-López A, Márquez-González M, Garay-Martínez LE, Zepeda H, Castillo A, Mota de la Garza L, Murano EA, Torres-Vitela R. 2003. Isolation of *Arcobacter* spp. from retail meats and cytotoxic effects of isolates against vero cells. J Food Prot. 66:1374–1378.

287. Vytrasová J, Pejchalová M, Harsová K, Bínová S. 2003. Isolation of *Arcobacter butzleri* and *A. cryaerophilus* in samples of meats and from meat-processing plants by a culture technique and detection by PCR. Folia Microbiol. 48:227–232.

288. Wang X, Seo DJ, Lee MH, Choi C. 2014. Comparison of conventional PCR, multiplex PCR and loop-mediated isothermal amplification assays for rapid detection of *Arcobacter* species. J Clin Microbiol. 52:557–563.

289. Webb AL, Boras VF, Kruczkiewicz P, Selinger LB, Taboada EN, Inglis GD. 2016. Comparative detection and quantification of *Arcobacter butzleri* in stools from diarrheic and nondiarrheic people in Southwestern Alberta, Canada. J Clin Microbiol. 54:1082–1088.

290. Webb AL, Kruczkiewicz P, Selinger LB, Inglis GD, Taboada EN. 2015. Development of a comparative genomic fingerprinting assay for rapid and high resolution genotyping of*Arcobacter butzleri*. BMC Microbiol. 15:94.

291. Wesley IV, Baetz AL. 1999. Natural and experimental infections of *Arcobacter* in poultry. Poult Sci. 78:536–545.

292. Wesley IV, Schroder-Tucker L, Baetz AL, Dewhurst FE, Paster BJ. 1995. *Arcobacter butzleri*-specific 16S rRNA based probes. J Clin Microbiol. 33:1691–1698.

293. Wesley IV, Schroeder-Tucker L, Franklin SL. 2003. Recovery of *Arcobacter* species from exotic animal species. Int J Med Microbiol. 293:57.

294. Wesley IV, Wells SJ, Harmon KM, Green A, Schroeder-Tucker L, Glover M, Siddique I. 2000. Fecal shedding of *Campylobacter* and *Arcobacter* spp. in dairy cattle. Appl Environ Microbiol. 66:1994–

2000.

295. Wesley IV. 1994. *Arcobacter* infections. In: Beran GW, Steele JH, editors. In handbook of zoonoses. CRC Press; p. 181–190.

296. Wesley IV. 1997. *Helicobacter and Arcobacter*: potential human foodborne pathogens? Trends Food Sci Technol. 8:293–299.

297. Whiteduck-Léveillée J, Cloutier M, Topp E, Lapen DR, Talbot G, Villemur R, Khan IU. 2016a.Development and evaluation of multiplex PCR assays for rapid detection of virulence-associated genes in *Arcobacter* species. J Microbiol Methods. 121:59–65.

298. Whiteduck-Léveillée K, Whiteduck-Léveillée J, Cloutier M, Tambong JT, Xu R, Topp E, Arts MT,Chao J, Adam Z, Lévesque CA, et al. 2015. *Arcobacter lanthieri* sp. nov., isolated from pig and dairy cattle manure. Int J Syst Evol Microbiol. 65:2709–2716.

299. Whiteduck-Léveillée K, Whiteduck-Léveillée J, Cloutier M, Tambong JT, Xu R, Topp E, Arts MT,Chao J, Adam Z, Lévesque CA, et al. 2016b. Identification, characterization and description of Arcobacter faecis sp. nov., isolated from a human waste septic tank. Syst Appl Microbiol. 39:93–99.

300. Winters DK, Slavik MF. 2000. Multiplex PCR detection of *Campylobacter jejuni* and *Arcobacter butzleri* in food products. Mol Cell Probes. 14:95–99.

301. Wirsen CO, Sievert SM, Cavanaugh CM, Molyneaux SJ, Ahmad A, Taylor LT, DeLong EF, Taylor CD.2002. Characterization of an autotrophic sulfide-oxidizing marine Arcobacter sp. that produces filamentous sulfur. Appl Environ Microbiol. 68:316–325.

302. Woo PCY, Chong KTK, Leung KW, Que TL, Yuen KY. 2001. Identification of *Arcobacter cryaerophilus* isolated from a traffic accident victim with bacteremia by 16S ribosomal RNA gene sequencing. Diagn Microbiol Infect Dis. 40:125–127.

303. Wybo I, Breynaert J, Lauwers S, Lindenburg F, Houf K. 2004. Isolation of *Arcobacter skirrowii*from a patient with chronic diar-rhea. J Clin Microbiol. 42:1851–1852.

304. Yadav AS, Kolluri G, Gopi M, Karthik K, Malik YS, Dhama K. 2016. Exploring alternatives to antibiotics as health promoting agents in poultry- a review. J Exp Biol Agri Sci. 4:368–383.

305. Yan JJ, Ko WC, Huang AH, Chen HM, Jin YT, Wu JJ. 2000. *Arcobacter butzleri* bacteremia in a patient with liver cirrhosis. J Formos Med

Assoc. 99:166–169.

306. Yesilmen S, Vural A, Erkan ME, Yildirim IH. 2014. Prevalence and antimicrobial susceptibility of*Arcobacter* species in cow milk, water buffalo milk and fresh village cheese. Int J Food Microbiol. 188:11–14.

307. Yildiz H, Aydin S. 2006. Pathological effects *of Arcobacter cryaerophilus* infection in rainbow trout (*Oncorhynchus mykiss Walbaum*). Acta Vet Hung. 54:191–199.

308. Zacharow I, Bystroń J, Wałecka-Zacharska E, Podkowik M, Bania J. 2015. Prevalence and antimicrobial resistance of *Arcobacter butzleri* and *Arcobacter cryaerophilus* isolates from retail meat in Lower Silesia region, Poland. Pol J Vet Sci. 18:63–69.

309. Zhang Z, Yu C, Wang X, Yu S, Zhang XH. 2016. *Arcobacter pacificus* sp. nov., isolated from seawater of the South Pacific Gyre. Int J Syst Evol Microbiol. 66:542–547.

310. Zur Bruegge J, Backes C, Gölz G, Hemmrich-Stanisak G, Scharek-Tedin L, Franke A, Alter T,Einspanier R, Keller A, Sharbati S. 2016. MicroRNA response of primary human macrophages to*Arcobacter butzleri* infection. Eur J Microbiol Immunol. 6:99–108.

311. Zur Bruegge J, Hanisch C, Einspanier R, Alter T, Gölz G, Sharbati S. 2014. *Arcobacter butzleri*induces a pro-inflammatory response in THP-1 derived macrophages and has limited ability for intracellular survival. Int J Med Microbiol. 304:1209–1217.

CHAPTER 6

APPLICATION OF NEXT GENERATION SEQUENCING IN CLINICAL MICROBIOLOGY AND INFECTION PREVENTION

Ruud H.Deurenberg[a], ErikBathoorn[a], Monika A.Chlebowicz[a], NatachaCouto[a], MithilaFerdous[a], SilviaGarcía-Cobos[a], Anna M.D.Kooistra-Smid[ab], Erwin C.Raangs[a], SigridRosema[a], Alida C.M.Veloo[a], KaiZhou[c], Alexander W.Friedrich[a], John W.A.Rossen[a]

[a]Department of Medical Microbiology, University of Groningen, University Medical Center Groningen, The Netherlands
[b]Certe, Department of Medical Microbiology, Groningen, The Netherlands
[c]State Key Laboratory for Diagnosis and Treatment of Infectious Diseases, Collaborative Innovation Center for Diagnosis and Treatment of Infectious Diseases, The First Affiliated Hospital of Medicine School, Zhejiang University, Hangzhou, China

CITATION: Deurenberg, R.H., et al., et al., "Application of next generation sequencing in clinical microbiology and infection prevention", Journal of Biotechnology (2017), http://dx.doi.org/10.1016/j.jbiotec.2016.12.022.

https://www.sciencedirect.com/science/article/pii/S0168165616316686

ABSTRACT

Current molecular diagnostics of human pathogens provide limited information that is often not sufficient for outbreak and transmission investigation. Next generation sequencing (NGS) determines the DNA sequence of a complete bacterial genome in a single sequence run, and from these data, information on resistance and virulence, as well as information for typing is obtained, useful for outbreak investigation. The obtained genome data can be further used for the development of an outbreak-specific screening test. In this review, a general introduction to NGS is presented, including the library preparation and the major characteristics of the most common NGS platforms, such as the MiSeq (Illumina) and the Ion PGM™ (ThermoFisher). An overview of the software used for NGS data analyses used at the medical microbiology diagnostic laboratory in the University Medical Center Groningen in The Netherlands is given. Furthermore, applications of NGS in the clinical setting are described, such as outbreak management, molecular case finding, characterization and surveillance of pathogens, rapid identification of bacteria using the 16S-23S rRNA region, taxonomy, metagenomics approaches on clinical samples, and the determination of the transmission of zoonotic micro-organisms from animals to humans. Finally, we share our vision on the use of NGS in personalised microbiology in the near future, pointing out specific requirements.

Keywords: Infection, prevention, Ion PGM, Clinical Microbiology, MiSeq, Next Generation Sequencing, Whole genome sequencin

INTRODUCTION

Identification and characterization of micro-organisms that cause infections are crucial for successful treatment, recovery and safety of patients. However, not every bacterial species can be successfully cultured in the diagnostic laboratory, and the available molecular tests are unable to detect emerging genetic features in successfully evolving pathogens that spread in humans, animals and the environment. Unrecognized pathogens can easily cause hospital outbreaks, putting patients at risk during their hospital admissions.

During the last two decades, molecular diagnostic methods have experienced a rapid development and played an increasingly important role in medical microbiology laboratories (Buchan and Ledeboer, 2014). These methods have reduced the turnaround time from receiving the sample to the final result, and made it possible to detect non-cultivable pathogens. However, molecular methods need a priori knowledge of the likely pathogenic species that could be present in the sample. One of the molecular methods used in medical microbiology laboratories is the sequence analyses of genes or the whole genome of pathogens.

Sequence analyses can be used to answer different diagnostic questions, such as the genetic relationship of either bacteria or viruses, the detection of mutations in viral or bacterial genomes leading to resistance against antivirals or antibiotics, identification of fungi through sequence analyses of the 18S ribosomal deoxyribonucleic acid (rDNA) of the internal transcribed spacer (ITS) region and identification of bacteria through sequence analyses of the 16S rDNA (Bush, 2013, Deurenberg and Stobberingh, 2008, Liu et al., 2012, Reiss et al., 2000). In general, Sanger sequencing is used for this, preceded by amplification of each gene or genomic region using specific primers. The same method can be applied for the identification of pathogens in clinical material. However, this approach becomes problematic when clinical material is more complex and contains multiple species, such as faecal samples. In such cases, results obtained by Sanger sequencing are not reliable and make it hard or even impossible to identify specific pathogens. Furthermore, the cost of Sanger sequencing for these tasks is high, and the turnaround time is long.

The University Medical Center Groningen (UMCG) is one of the largest university hospitals in The Netherlands with 1339 beds and more than 12,000 employees. The clinical microbiology diagnostic laboratory at the UMCG receives around 5750 samples per year for detailed molecular analysis, of which approximately 1500 are subjected to next generation sequencing (NGS) using two Illumina MiSeq and one Life Technologies Ion PGM™ sequencers. NGS was introduced for routine diagnostics in 2014, and the majority of indications are outbreak investigation and genotyping of highly resistant micro-organisms. NGS is requested by clinical microbiologists or infectious disease specialists in collaboration with molecular microbiologists and infection control practioners.

NEXT GENERATION SEQUENCING

NGS allows sequencing of the whole genome of numerous pathogens in one sequence run, either from bacterial isolates of (different) patients, or from multiple species present in patient material from one individual (metagenomics). Both the investment- and the running costs of NGS have decreased dramatically during the last decade (Dark, 2013, Sboner et al., 2011). A great advantage of NGS is that, in contrast to Sanger sequencing, a single protocol can be used for all pathogens for both identification and typing applications. Therefore, this technology has been proven to be useful in medical microbiology laboratories and for infection prevention measures (Zhou et al., 2016). A schematic overview of the general workflow used for NGS analyses at the UMCG is shown in Fig. 1.

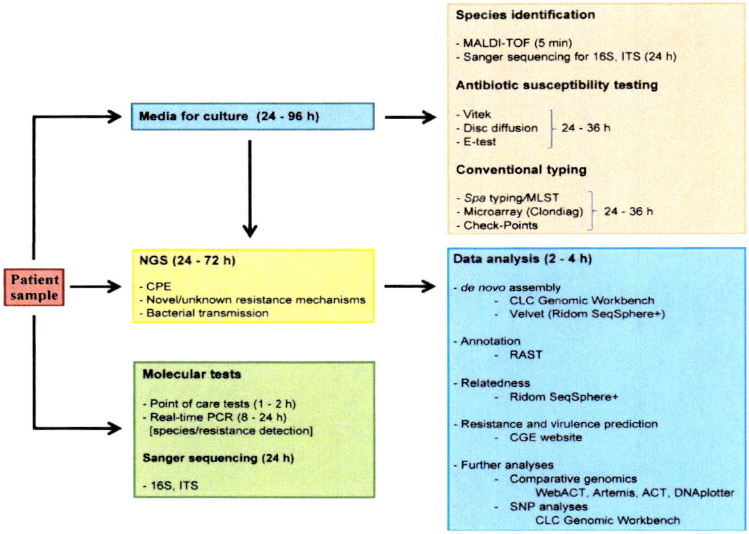

Figure 1. A schematic overview of the general workflow of diagnostic procedures including NGS in our laboratory.

For NGS, there is no need for target specific primers, which are needed for Sanger sequencing. In a single run, the whole genome of a pathogen is sequenced at random. Before sequencing, fragmentation of the genome is performed, since the maximum length a benchtop sequencer can sequence varies between 100 and 1000 bases and thus the genome cannot be sequenced in one part (Junemann et al., 2013, Loman et al., 2012). An exception to this are the third generation of sequencers, such as the MinION (Oxford Nanopore) and the Sequel (Pacific Biosciences), which can generate larger fragments (more than 200 kb). However, these sequencers are not yet used in the clinical microbiology laboratory, due to their lack of affordability, the lower quality of the sequences, and the low throughput. Therefore, NGS still requires the preparation of libraries, in which fragments of DNA or RNA are fused to adapters and barcodes to distinguish the DNA of the sequenced isolates after sequencing, followed by a clonal amplification, normalization and sequencing. For this, a robust preparation of the libraries, which contains a representative source of the DNA or RNA of the genome under investigation, is needed (Head et al., 2014).

Fragmentation can be performed in several ways, either mechanical, using, e.g., the Adaptive Focused Acoustics (AFA) technology from Covaris, followed by adaptor ligation, or enzymatic, such as with transposons as used in the Nextera XT Library Preparation kit from Illumina. This method has the advantage that fragmentation and fusion of the adaptors to the DNA or RNA fragments are performed in one step, which makes it easier to automate it. Besides that, less input DNA is needed. Mechanical fragmentation has the advantage that the generation of the appropriate fragment length is less influenced by factors present in the sample that inhibit the enzymes used during the library preparation, and is therefore very suitable for library preparations of direct sample material, such as biopsies and faeces samples (Head et al., 2014).

NGS libraries may contain errors that decrease the data quality, and thus can disrupt the data interpretation. Detailed knowledge of the kind of errors is important to find ways to avoid the introduction of such errors and for the correct interpretation of the NGS data. Almost all separate steps in the sequence procedure can introduce errors. This is especially true with RNA sequencing that is technically more challenging compared to DNA sequencing (Junemann et al., 2013, Loman et al., 2012).

At the moment, a number of NGS platforms are available. The most important properties of several NGS platforms, such as output and fragment length, are presented in Table 1 (Bertelli and Greub, 2013, Dark, 2013, Junemann et al., 2013, Loman et al., 2012). The different NGS platforms use different sequencing technologies. Illumina sequencers use sequencing by synthesis of fluorescent, reversible terminators, and ThermoFisher sequencers use semiconductor sequencing that measure a change in pH during the incorporation of nucleotides. Pacific Biosciences use fluorescent nucleotides in their single molecule real-time (SMRT) technology, and Oxford Nanopore platforms use ionic current sensing, in which DNA is guided through nano-pores, thereby changing the current in a way that is specific for the type of nucleotide. Extensive information on the different NGS platforms and their method of sequencing are available on the companies' websites. Due to the technological developments, the cost of NGS decreased between 2001 and 2015, while the speed of sequencing has increased (Dark, 2013, Sboner et al., 2011).

Table 1. Properties of current NGS platforms.

Company	Equipment	Output/run (Gb)	Maximum read length (bp)	Reads ($x10^6$)	Running time
Illumina	MiniSeq	0.6–7.5	2 × 150	25	4–24 h
Illumina	Miseq	0.3–15	2 × 300	25	5–55 h
Illumina	NextSeq	20–120	2 × 150	130/400	12–30 h
Illumina	HiSeq 3000	125–700	2 × 150	2500	< 1 – 3.5 days
ThermoFisher	Ion PGM™	0.03–2	200–400	0.4–5.5	2–7 h
ThermoFisher	Ion 5S™	0.6–15	200–400	3–80	2.5–4 h
ThermoFisher	Ion 5S™ XL	0.6–15	200–400	3–80	<24 h
Oxford Nanopore	MinION	21–42	230,000–300,000	2.2–4.4	1 min–48 h
Pacific Biosciencesa	Sequel	0.75–1.25	>20,000	370,000	30 min–6 h
Pacific Biosciences[a]	RSII	0.5–1	>20,000	55,000	30 min–4 h

aThe Pacific Biosciences data are per smart cell; both the Sequel and the RSII can run 1–16 smart cells in one run.

SOFTWARE FOR DATA ANALYSES

The biggest challenge concerning the introduction of NGS in the clinical microbiology laboratory is the data analyses. Nonetheless, even with little knowledge of bioinformatics, it is possible to perform NGS data analyses for diagnostic purposes, using the numerous user-friendly software packages available (Edwards and Holt, 2013). However, for more in-depth analysis, scientific knowledge is required on the genomic features and the biological background of the micro-organism under investigation.

After sequencing, the sequenced fragments (reads) can be *de novo* assembled (genome assembly). Hereby, the reads are aligned against each other without the use of a reference organism. In general, the larger the fragments, the easier and more accurate the genome assembly will be. Software packages (Table 2), such as CLC Genomic Workbench (Qiagen), SPAdes and Velvet, are used in our laboratory to assemble the genomes. The genetic relationship between isolates can be investigated by using a gene-by-gene comparison using a multi-locus sequencing typing (MLST) approach, either by studying the conserved core genome (cgMLST), or the whole genome (wgMLST), which includes a set of variable accessory genes. Several software packages, such as SeqSphere (Ridom) and BioNumerics (Applied Maths, Biomérieux), or online tools, such as EnteroBase and BIGSdb (Bacterial Isolate Genome Sequence Database) (Jolley and Maiden, 2010), can be used for this approach. Furthermore, the use of an established cgMLST scheme allows the introduction of a common nomenclature for genetically related strains (de Been et al., 2015, Kohl et al., 2014, Ruppitsch et al., 2015). At the moment, it is not clear how many alleles two genomes may differ to call them (close to being) identical. The same problem applies when comparing two genomes using single-nucleotide polymorphisms (SNP) typing (Maiden et al., 2013). However, the term genetic distance (the proportion of different alleles, calculated by dividing the number of allele differences by the total number of genes shared by two sequences) has been recently introduced, and enables unbiased comparisons for different cgMLST or wgMLST schemes as well as the definition of thresholds by studying collections of epidemiologically and non-epidemiologically related strains (Kluytmans-van den Bergh et al., 2016b). An advantage of cgMLST and wgMLST is that there is compatibility between cgMLST, wgMLST and older typing methods, since both BioNumerics and SeqSphere give the sequence type (ST) as determined by conventional MLST and the *spa* type (in case of *Staphylococcus aureus*).

Table 2. Software packages frequently used for NGS data analyses in our laboratory.

Application	Software	Link	Note
Annotation	Prokka	www.vicbioinformatics.com	
	RAST	http://rast.nmpdr.org	
Assembly	BioNumerics	www.applied-maths.com	Commercial software
	CLC Genomic Workbench	www.clcbio.com	Commercial software
	SeqSphere	www.ridom.de	Commercial software
	SPAdes	http://bioinf.spbau.ru/spades	Unix-based
	Velvet	www.ebi.ac.uk/~zerbino/velvet	Unix-based

Data quality check	BaseSpace	https://basespace.illumina.com	Commercial software
	BioNumerics	www.applied-maths.com	Commercial software
	CLC Genomic Workbench	www.clcbio.com	Commercial software
	FastQC	www.bioinformatics.babraham.ac.uk	
Identification	K-merFinder	www.genomicepidemiology.org	
	NCBI BLAST	www.ncbi.nlm.nih.gov/blast	
Metagenomics	MEGAN	http://ab.inf.uni-tuebingen.de/software/malt	
Phylogeny	FastTree	www.microbesonline.org/fasttree	
	RAxML	http://sco.h-its.org/exelixis/software.html	
	SeqSphere	www.ridom.de	Commercial software
	SNPTree	www.genomicepidemiology.org	
Resistance	ARDB	https://ardb.cbcb.umd.edu	
	CARD	https://card.mcmaster.ca	
	ResFinder	www.genomicepidemiology.org	
SNP calling	BioNumerics	www.applied-maths.com	Commercial software
	CLC Genomic Workbench	www.clcbio.com	Commercial software
	Samtools	www.htslib.org	
	SeqSphere	www.ridom.de	Commercial software

Typing (wgMLST)	BIGSdb	http://bigsdb.readthedocs.io	
	BioNumerics	www.applied-maths.com	Commercial software
	CLC Genomic Workbench	www.clcbio.com	Commercial software
	EnteroBase	https://enterobase.warwick.ac.uk	
	SeqSpere	www.ridom.de	Commercial software
Virulence	VFDB	www.mgc.ac.cn/VFs	
	VirulenceFinder	www.genomicepidemiology.org	
Visualisation &	ACT	www.sanger.ac.uk/science/tools	
comparative study	Artemis	www.sanger.ac.uk/science/tools	
	BRIG	https://sourceforge.net/projects/brig/	
	ClustalW	www.genome.jp/tools/clustalw	
	DNA plotter	www.sanger.ac.uk/science/tools	
	WebACT	www.webact.org	

Several possibilities to analyse NGS data can be found on the website of the Center for Genomic Epidemiology. For the detection of virulence- and resistance genes, VirulenceFinder and ResFinder, can be used. Alternatively, the Comprehensive Antibiotic Resistance Database (CARD) and the Virulence Factor Database (VFDB) can be used to obtain data on resistance and virulence genes. With these online tools, both the non-assembled sequence data and the assembled genome can be uploaded. However, the results obtained through these websites needs confirmation using other methods. In case of *S. aureus*, it has been reported that there is a good correlation between the presence of resistance genes and its phenotypic resistance pattern (Aanensen et al., 2016).

Further comparative genome studies are possible using Artemis Comparison Tool (ACT) (Carver et al., 2005), Artemis (Carver et al., 2012), and DNA plotter (Carver et al., 2009) from the Sanger Institute.

Specific research questions require knowledge of Unix-systems and to handle the large diversity of bioinformatics software packages available for it. Furthermore, software for the analyses of metagenomics data is available, such as the MEGAN Alignment Tool (Huson et al., 2007). In contrast to the decreased sequencing costs, the costs for data storage and data analyses have increased due the generation of large amounts of data, and the complexity of it.

NGS IN CLINICAL MICROBIOLOGY

NGS is already applied in several medical microbiology laboratories, including our laboratory at the UMCG, where it is used for outbreak management, molecular case finding, characterization and surveillance of pathogens, rapid identification of bacteria using the 16S-23S rRNA region, taxonomy, metagenomics approaches on clinical samples, and the determination of the transmission of zoonotic micro-organisms from animals to humans.

Outbreak Management

The advantages of using whole genome sequencing (WGS)-based typing is promoting the implementation of NGS for epidemiological studies and public health investigations. It is especially important and helpful in outbreak detection and monitoring the evolution and dynamics of multi-drug resistant pathogens (ECDC, 2016). Several studies illustrated the usefulness of WGS-based typing for disclosing and tracing the dissemination of emerging pathogens. Indeed, it was used in our hospital to characterize a newly emerging CTX-M-15 producing K. pneumoniae clone with sequence type (ST) 1427 (Zhou et al., 2015b). In addition, the transmission of a CTX-M-15-producing ST15 Klebsiella pneumoniae between patients treated in a single centre and the subsequent inter-institutional spread by patient referral has been traced by genomic phylogenetic analysis (Fig. 2). The investigation allowed the early detection of a K. pneumoniae high-risk-clone (HiRiC) with prolonged circulation in the regional patient population (Zhou et al., 2016). Furthermore, this study showed the usefulness of a unique marker approach, in which a clone-specific PCR was developed to investigate the dissemination of the HiRiC between healthcare centres.

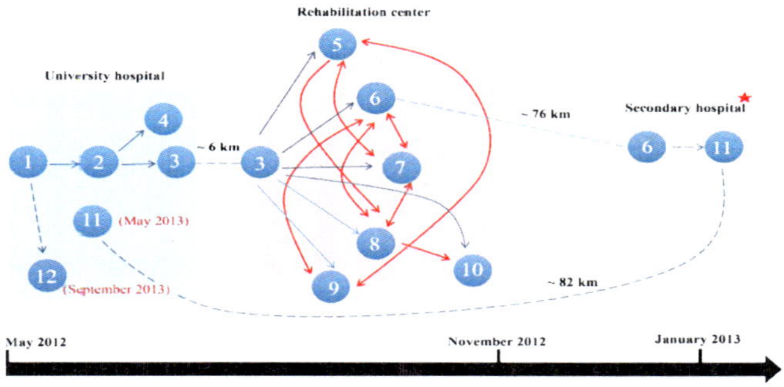

Figure 2. The transmission route was reconstructed by epidemiological and genomic data. Each node represents a patient, and an arrow indicates a possible transmission event from one patient to another. The blue arrow with solid line represents a direct transmission event supported by both epidemiological data and genetic data, the blue arrow with dash line represents an indirect transmission (e.g. via environment) supported by epidemiological data, and the red arrow indicates the equally parsimonious transmission link which cannot be resolved by neither epidemiological data nor genetic data. The inter-institutional transfer of the patient is shown by dash lines, on which the distance between institutions is indicated. The red star represents an outbreak at a secondary hospital, but the isolates were unavailable for further research (Zhou et al., 2016). (For interpretation of the references to color in this figure legend, the reader is referred to the web version of this article.)

In addition to outbreak tracing and characterization, the use of WGS also allows the implementation of control measures to avoid the spread of resistant bacterial clones. An outbreak of a colistin-resistant carbapenemase-producing K. pneumoniae (KPC) with inter-institutional spread in The Netherlands, was controlled by transferring all positive residents to a separate location outside the institution, where a dedicated team cared for the patients (Weterings et al., 2015).

Apart from multidrug-resistant bacteria, WGS is also useful and applicable to characterize highly-virulent bacteria, such as shiga toxin-producing Escherichia coli(STEC) O104:H4. This bacterium has emerged as an important pathogen and has been responsible for large outbreaks. However, there has been little information about the evolutionary history and genomic diversity of it. Phylogenetic analysis of outbreak and non-outbreak related isolates using WGS provided an evolutionary context and revealed lineage-specific markers, indicative for selective pressure and niche adaptation (Zhou et al., 2015a). In addition, core-genome phylogenetic analysis of shiga toxin-producing Enteroaggregative E. coli (EAEC Stx2a+) O104:H4 showed different clustering and different resistance and virulence patterns depending on the time of isolation (Ferdous et al., 2015). These studies reflect the importance of NGS as a high discriminatory power tool to differentiate between clones with specific properties and to use the obtained knowledge for patient management, infection prevention and evolutionary studies.

Molecular Case Finding

Molecular case definitions of outbreak isolates are commonly used in outbreak investigations. NGS databases can retrospectively be searched for cases in complex and comprehensive outbreaks. This may result in detection of cases that would not have been found by traditional epidemiological investigation. In a study, a New Delhi Metallo-ß-lactamase-5 (NDM-5)-producing K. pneumoniae ST16 strain was isolated from a Dutch patient from a long-term care facility without a recent history of travel abroad. Molecular case finding showed that the Dutch strain was clonally related to strains isolated from four patients in Denmark in 2014 (Bathoorn et al., 2015), but there were no obvious epidemiological links between the cases in the Danish and Dutch hospitals. European national surveillance centres were contacted for molecular case finding in their NGS databases, but no additional cases were detected.

Other examples of molecular case finding in NGS databases have been reported after the discovery of the plasmid resistance gene mcr-1, responsible for colistin resistance, in life-stock and hospitalised patients. Its introduction in European countries was investigated using a retrospectively search for the mcr-1 gene in NGS databases. This resulted in the detection of cases in Denmark, Germany, and The Netherlands (Falgenhauer et al., 2016, Hasman et al., 2015, Kluytmans-van den Bergh et al., 2016a). In The Netherlands, more than 2000 Dutch Enterobacteriaceae isolates were screened within a few hours to reveal the presence of the mcr-1 gene (Kluytmans-van den Bergh et al., 2016a). So, NGS data already present can be used to screen for the presence of new (antibiotic resistance) genes (in silico screening).

Characterization and Surveillance of Pathogens

The current routine procedure for pathogen characterization is based on a large variety of bacteriological, biochemical and molecular methods, making this procedure laborious, time-consuming, and expensive. NGS may serve as a perfect one-step tool to study a broad range of pathogen characteristics and is applicable on a wide range of pathogens (Aanensen et al., 2016, Fournier et al., 2014, Hasman et al., 2014). Knowledge of the virulence profile of a pathogen is crucial to predict the disease severity, outcome of the infection and to allow risk assessment during the early onset of the disease. WGS has the potential to make a substantial contribution to determine the presence of virulence factors using several online tools, since it is not restricted to a specific gene (Franz et al., 2014, Laabei et al., 2014).

In a large cohort study, WGS was used for molecular characterization of STEC, resulting in a clear understanding about the population structure and genomic plasticity of STEC in the regions around the cities Groningen and Rotterdam in The Netherlands (Ferdous et al., 2016). All relevant information could be extracted *in silico*from the sequence data, including genotype, serotype, MLST profile, virulence and antibiotic resistance profiles, and the phylogenetic background to obtain the overall molecular features with a high discrimination among closely related strains. NGS allowed to characterize and compare many strains in detail within a relatively short time span. Thus, for a rapid and improved molecular epidemiological surveillance of pathogens at regional and national scale, the role of WGS is undeniable.

NGS is also helpful in detection of novel resistance genes in bacteria, both in current as well as in historical strain collections. Novel variants of antibiotic resistance genes (ARG) can be identified using NGS, and further experiments can be performed to determine if these genes are indeed responsible for the observed antibiotic resistance pattern (Nijhuis et al., 2015).

Targeted NGS of the 16S-23S rRNA Cluster Region for rapid Bacterial Identification in Clinical Specimen

NGS allows culture-free detection of a theoretically unlimited number of pathogens and thus provides insight in the full microbiome. Metagenomics will be the ultimate approach in detecting all micro-organisms (e.g. bacteria, viruses, fungi) in a clinical sample (Hasman et al., 2014). However, analysis of large datasets requires a combination of bioinformatics skills and computational resources that is nowadays mostly absent in diagnostic medical microbiological laboratories. Furthermore, metagenomics approaches are time consuming as the turnaround time is approximately four to five days.

To fill the gap between the conventional methods (culture and PCR) and metagenomics, a culture-free approach using targeted NGS appears to be an excellent approach to detect and identify bacterial species. Compared to metagenomics, it is faster, less complicated and cheaper, and therefore more likely to get implemented in diagnostic laboratories within a short timeframe. The 16S rRNA gene sequence has been proven to be a reliable genetic marker as it is present in all bacteria, and the function has not changed over time (Patel, 2001). It can be applied directly on clinical materials and has proven to be a valuable supplementary test in daily clinical practice (Schuurman et al., 2004, Srinivasan et al., 2012). However, high sequence similarities in this gene between certain bacterial species do not always lead to an unequivocal identification (Kalia et al., 2016).

Recently, we developed an innovative culture-free 16S-23S rRNA NGS approach for the detection and identification of bacterial species in clinical samples. The method proved to be superior to other commonly used identification methods and correctly identified pathogens in urine samples that were also identified as the cause for urinary tract infections with conventional culture (Sabat et al., 2016). Furthermore, the method allows simultaneous identification of several pathogens in clinical materials that previously would have remained uncultured and PCR negative. Clearly, this will have an enormous clinical impact and will have consequences for patient treatment, including improved antibiotic treatment. Finally, this method will allow clinical microbiology laboratories to implement NGS in their routine diagnostic laboratory and to keep up with technological and bioinformatics developments required to be able to implement metagenomics in diagnostics in the future.

WGS and Taxonomy

In the eighteenth century, Linnaeus (Linnaeus, 1735) provided guidelines for classification of living creatures based on their phenotypic features. A century later, Darwin added the phylogenetic component to the taxonomy (Darwin, 1859). The taxonomy of bacterial species changed dramatically by the introduction of 16S rRNA gene sequencing. Nowadays, WGS is also used to identify and describe new species. By comparing the whole genome sequences of different species with each other, the limited taxonomic resolution of only the 16S rRNA gene can be overcome (Tindall et al., 2010). Another change in taxonomy may be expected when WGS is used for revealing the taxonomy of bacteria.

Indeed, using WGS for taxonomy purposes allows to include more genes to delineate between species than the classical DNA–DNA hybridization or 16S rRNA sequencing methods thereby improving the resolution. Furthermore, as WGS can be used to calculate taxonomic trees based on the whole genome-sequence alignment of all the genes present in the core genome, a more robust tree will be obtained (Daubin et al., 2001). It has already been proposed that descriptions of new taxa should also include a draft genome sequence, with at least 20 times coverage (Thompson et al., 2013).

Metagenomics in Clinical Microbiology

As already mentioned, NGS can be applied directly to clinical specimens. Not only by using a targeted NGS approach, but also by sequencing the DNA or RNA from patient samples by shotgun metagenomics sequencing (Fig. 3). Using this method, it is possible to investigate the presence of pathogens and the presence of virulence and/or resistance genes in one sequence run.

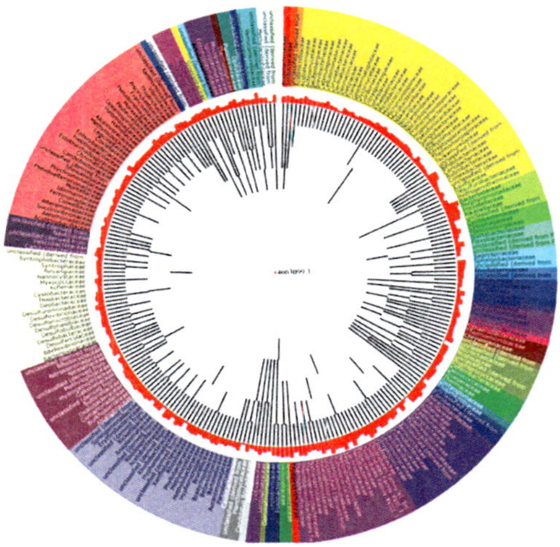

Figure 3. An example of an output of a metagenomics approach of a faecal sample. The different colours represent different bacterial families.

A recent study compared the detection of viruses in known respiratory virus-positive samples and not previously analysed nasopharyngeal swabs by an RNA sequencing-based metagenomics approach with a more conventional molecular method. The data-analyses was performed using Taxonomer, a rapid and interactive, web-based metagenomics data-analyses tool (Flygare et al., 2016). Overall, the metagenomics approach had a high agreement with the molecular method, detected viruses not targeted by the molecular method, and yielded epidemiologically and clinically relevant sequence information (Graf et al., 2016).

Apart from identifying pathogens, a metagenomics approach can also be used to study the resistome. The gut is a known reservoir for antibiotic resistance genes (ARG), and treatment with antibiotics has an impact on the intestinal resistome, which can lead to horizontal gene transfer and the selection of resistant bacteria. A study at the University of Tübingen investigated the presence of ARGs in the gut over a six-day period of ciprofloxacin treatment in two individuals using metagenomics. Furthermore, this study presented a novel method for analysing the determination of antibiotic selection pressure, which can be used in hospitals to compare therapeutic regimens and their effect on the intestinal resistome. This information is important for clinicians to choose antibiotic therapy with a low selective antibiotic pressure on the patient's bacteria in the gut, possibly resulting in a decreased dissemination of antibiotic resistant bacteria (Willmann et al., 2015).

Determining the Transmission of Zoonotic Micro-organisms from Animals to Humans

NGS will also reveal more knowledge on zoonotic transmission of micro-organisms. The first studies on this topic were based on low discriminatory methods, such as serotyping (Tenover et al., 1997). More recently, studies using higher discriminatory techniques, such as pulsed-field gel electrophoresis or multi-locus variable number tandem repeat analysis, were used to detect specific bacterial clones in animals and humans (Sabat et al., 2013). However, much remains to be understood, especially when it comes to the frequency of transmission (e.g. single contact or repeated contact with animals or animal products), risk factors associated with the acquisition of a zoonotic microorganism (e.g. risk conducts, such as animal kissing in companion animals, or stool handling in farm animals) and how the use of antibiotics in animals affects the transfer of pathogenic bacteria to humans.

NGS brings a new perspective to these topics. A higher discriminatory power will reveal differences in previously indistinguishable animal and human bacterial strains. This together with epidemiological information allows source tracing of potential zoonotic infections (Harrison et al., 2013). In addition, NGS allows a comprehensive analysis of how antibiotic use manipulates specific microbiota and the consequences for interspecies transmission and will increase the knowledge on microbial evolution through the analysis of bacterial genomes, namely the variable regions, which usually determine host-adaptation and the potential of spread to different hosts (Harrison et al., 2014, Price et al., 2012).

As the patients' safety is depending on its environment, including their contact with food and animals, research projects are currently performed in the UMCG to understand the dynamics of transmission of bacteria between humans, animals and the environment. These studies are performed in collaboration with veterinary research groups and focus on anti-microbial resistant bacteria. In one such study, the mcr-1 gene was detected by WGS to be present in three E. coli strains isolated from retail chicken meat. Although none of the human strains carried this gene, two of the three strains belonged to ST117, a common clone in both poultry and humans, representing a potential public health concern (Kluytmans-van den Bergh et al., 2016a).

CONCLUSION AND OUTLOOK

For generating NGS data from samples originating from humans, animals, food and the environment, the same laboratory protocol for library preparation can be used, and, after data analyses, information on the presence of specific antibiotic resistance and virulence genes is obtained. Furthermore, NGS makes it possible to standardise typing methods for pathogens ("one test fits all"). The role of NGS in medical microbiology laboratories will increase during the next years, not only for research, but also, and more importantly, for molecular diagnostics, infection prevention, the investigation of outbreaks by the use of a unique outbreak marker approach, the characterization and surveillance of pathogens, the detection of novel resistance genes and for the application of a metagenomics approach on clinical samples.

However, further studies are required to improve the workflow for NGS, in particular shorten the turnaround time for the library preparation and the runs on the NGS platforms, and, at the same time, further reducing costs. Next, automatic pipelines for data-analyses and easy-to-use software for metagenomics have to be developed. Additionally, more established typing schemes for pathogens and cut-off values for these typing schemes have to be established, leading to reference databases with genetic and metadata, and (inter)regional and international collaborations. Importantly, external quality controls for proficiency testing have to be developed. Only then will patient guidance and infection control management at local, (inter)regional and international level, as well as targeted antibiotic therapy using NGS data become a possibility, leading to personalised microbiology.

REFERENCES

1. Aanensen, D.M., Feil, E.J., Holden, M.T., Dordel, J., Yeats, C.A., Fedosejev, A., Goater, R., Castillo-Ramirez, S., Corander, J., Colijn, C., Chlebowicz, M.A., Schouls, L., Heck, M., Pluister, G., Ruimy, R., Kahlmeter, G., Ahman, J., Matuschek, E., Friedrich, A.W., Parkhill, J., Bentley, S.D., Spratt, B.G., Grundmann, H., European,
2. S.R.L.W.G., 2016. Whole-genome sequencing for routine pathogen surveillance in public health: a population snapshot of invasive Staphylococcus aureus in Europe. MBio 7, e00444-16.
3. Bathoorn, E., Rossen, J.W., Lokate, M., Friedrich, A.W., Hammerum, A.M., 2015. Isolation of an NDM-5-producing ST16 Klebsiella pneumoniae from a Dutch patient without travel history abroad, August 2015. Euro Surveil. 20, 30040.
4. Bertelli, C., Greub, G., 2013. Rapid bacterial genome sequencing: methods and applications in clinical microbiology. Clin. Microbiol. Infect. 19, 803–813.
5. Buchan, B.W., Ledeboer, N.A., 2014. Emerging technologies for the clinical microbiology laboratory. Clin. Microbiol. Rev. 27, 783–822.
6. Bush, K., 2013. Proliferation and significance of clinically relevant beta-lactamases. Ann. N. Y. Acad. Sci. 1277, 84–90.
7. Carver, T.J., Rutherford, K.M., Berriman, M., Rajandream, M.A., Barrell, B.G., Parkhill, J., 2005. ACT: the artemis comparison tool. Bioinformatics 21, 3422–3423.
8. Carver, T., Thomson, N., Bleasby, A., Berriman, M., Parkhill, J., 2009. DNAPlotter: circular and linear interactive genome visualization. Bioinformatics 25, 119–120.
9. Carver, T., Harris, S.R., Berriman, M., Parkhill, J., McQuillan, J.A., 2012. Artemis: an integrated platform for visualization and analysis of high-throughput sequence-based experimental data. Bioinformatics 28, 464–469.
10. de Been, M., Pinholt, M., Top, J., Bletz, S., Mellmann, A., van Schaik, W., Brouwer, E., Rogers, M., Kraat, Y., Bonten, M., Corander, J., Westh, H., Harmsen, D., Willems, R.J., 2015. Core genome multilocus sequence typing scheme for high-resolution typing of Enterococcus faecium. J. Clin. Microbiol. 53, 3788–3797.
11. Dark, M.J., 2013. Whole-genome sequencing in bacteriology: state of

the art. Infect. Drug Resist. 6, 115–123.

12. Darwin, C., 1859. On the Origin of Species by Means of Natural Selection, or the Preservation of Races in the Struggle for Life. John Murray, London. Daubin, V., Gouy, M., Perriere, G., 2001. Bacterial molecular phylogeny using supertree approach. Genome Inform. 12, 155–164.
13. Deurenberg, R.H., Stobberingh, E.E., 2008. The evolution of Staphylococcus aureus. Infect. Genet. Evol. 8, 747–763.
14. ECDC, 2016. Expert Opinion on Whole Genome Sequencing for Public Health Surveillance. Edwards, D.J., Holt, K.E., 2013. Beginner's guide to comparative bacterial genome analysis using next-generation sequence data. Microb. Inform. Exp. 3, 2.
15. Falgenhauer, L., Waezsada, S.E., Yao, Y., Imirzalioglu, C., Kasbohrer, A., Roesler, U., Michael, G.B., Schwarz, S., Werner, G., Kreienbrock, L., Chakraborty, T., 2016.
16. Colistin resistance gene mcr-1 in extended-spectrum beta-lactamase-producing and carbapenemase-producing Gram-negative bacteria in Germany. Lancet Infect. Dis. 16, 282–283.
17. Ferdous, M., Zhou, K., de Boer, R.F., Friedrich, A.W., Kooistra-Smid, A.M., Rossen, J.W., 2015. Comprehensive characterization of Escherichia coli O104:H4 isolated from patients in the Netherlands. Front. Microbiol. 6, 1348.
18. Ferdous, M., Friedrich, A.W., Grundmann, H., de Boer, R.F., Croughs, P.D., Islam, M.A., Kluytmans-van den Bergh, M.F., Kooistra-Smid, A.M., Rossen, J.W., 2016.
19. Molecular characterization and phylogeny of Shiga toxin-producing Escherichia coli isolates obtained from two Dutch regions using whole genome sequencing. Clin. Microbiol. Infect. 22, 642.e1–642.e9.
20. Flygare, S., Simmon, K., Miller, C., Qiao, Y., Kennedy, B., Di Sera, T., Graf, E.H., Tardif, K.D., Kapusta, A., Rynearson, S., Stockmann, C., Queen, K., Tong, S., Voelkerding, K.V., Blaschke, A., Byington, C.L., Jain, S., Pavia, A., Ampofo, K., Eilbeck, K., Marth, G., Yandell, M., Schlaberg, R., 2016. Taxonomer: an interactive metagenomics analysis portal for universal pathogen detection and host mRNA expression profiling. Genome Biol. 17, 111.
21. Fournier, P.E., Dubourg, G., Raoult, D., 2014. Clinical detection and characterization of bacterial pathogens in the genomics era. Genome

Med. 6, 114.

22. Franz, E., Delaquis, P., Morabito, S., Beutin, L., Gobius, K., Rasko, D.A., Bono, J., French, N., Osek, J., Lindstedt, B.A., Muniesa, M., Manning, S., LeJeune, J., Callaway, T., Beatson, S., Eppinger, M., Dallman, T., Forbes, K.J., Aarts, H., Pearl, D.L., Gannon, V.P., Laing, C.R., Strachan, N.J., 2014. Exploiting the explosion of information associated with whole genome sequencing to tackle Shiga toxin-producing Escherichia coli (STEC) in global food production systems. Int. J. Food Microbiol. 187, 57–72.
23. Graf, E.H., Simmon, K.E., Tardif, K.D., Hymas, W., Flygare, S., Eilbeck, K., Yandell, M., Schlaberg, R., 2016. Unbiased detection of respiratory viruses by use of RNA sequencing-based metagenomics: a systematic comparison to a commercial PCR panel. J. Clin. Microbiol. 54, 1000–1007.
24. Harrison, E.M., Paterson, G.K., Holden, M.T., Larsen, J., Stegger, M., Larsen, A.R., Petersen, A., Skov, R.L., Christensen, J.M., Bak Zeuthen, A., Heltberg, O., Harris, S.R., Zadoks, R.N., Parkhill, J., Peacock, S.J., Holmes, M.A., 2013. Whole genome sequencing identifies zoonotic transmission of MRSA isolates with the novel mecA homologue mecC. EMBO Mol. Med. 5, 509–515.
25. Harrison, E.M., Weinert, L.A., Holden, M.T., Welch, J.J., Wilson, K., Morgan, F.J., Harris, S.R., Loeffler, A., Boag, A.K., Peacock, S.J., Paterson, G.K., Waller, A.S., Parkhill, J., Holmes, M.A., 2014. A shared population of epidemic methicillin-resistant Staphylococcus aureus 15 circulates in humans and companion animals. MBio 5, e00985–00913.
26. Hasman, H., Saputra, D., Sicheritz-Ponten, T., Lund, O., Svendsen, C.A., Frimodt-Moller, N., Aarestrup, F.M., 2014. Rapid whole-genome sequencing for detection and characterization of microorganisms directly from clinical samples. J. Clin. Microbiol. 52, 139–146.
27. Hasman, H., Hammerum, A.M., Hansen, F., Hendriksen, R.S., Olesen, B., Agerso, Y., Zankari, E., Leekitcharoenphon, P., Stegger, M., Kaas, R.S., Cavaco, L.M., Hansen, D.S., Aarestrup, F.M., Skov, R.L., 2015. Detection of mcr-1 encoding plasmid-mediated colistin-resistant Escherichia coli isolates from human bloodstream infection and imported chicken meat, Denmark 2015. Euro Surveil. 20, 30085.
28. Head, S.R., Komori, H.K., LaMere, S.A., Whisenant, T., Van Nieuwerburgh, F., Salomon, D.R., Ordoukhanian, P., 2014. Library construction for next-generation sequencing: overviews and challenges.

Biotechniques 56, 61–64.

29. Huson, D.H., Auch, A.F., Qi, J., Schuster, S.C., 2007. MEGAN analysis of metagenomics data. Genome Res. 17, 377–386.

30. Jolley, K.A., Maiden, M.C., 2010. BIGSdb: scalable analysis of bacterial genome variation at the population level. BMC Bioinform. 11, 595.

31. Junemann, S., Sedlazeck, F.J., Prior, K., Albersmeier, A., John, U., Kalinowski, J., Mellmann, A., Goesmann, A., von Haeseler, A., Stoye, J., Harmsen, D., 2013.

32. Updating benchtop sequencing performance comparison. Nat. Biotechnol. 31, 294–296.

33. Kalia, V.C., Kumar, R., Kumar, P., Koul, S., 2016. A genome-wide profiling strategy as an aid for searching unique identification biomarkers for Streptococcus. Indian J. Microbiol. 56, 46–58.

34. Kluytmans-van den Bergh, M.F., Huizinga, P., Bonten, M.J., Bos, M., De Bruyne, K., Friedrich, A.W., Rossen, J.W., Savelkoul, P.H., Kluytmans, J.A., 2016a. Presence of mcr-1-positive Enterobacteriaceae in retail chicken meat but not in humans in the Netherlands since 2009. Euro Surveill. 21, 30149.

35. Kluytmans-van den Bergh, M.F., Rossen, J.W., Bruijning-Verhagen, P.C., Bonten, M.J., Friedrich, A.W., Vandenbroucke-Grauls, C.M., Willems, R.J., Kluytmans, J.A., 2016b. Whole genome multilocus sequence typing of extended-spectrum beta-lactamase-producing Enterobacteriaceae. J. Clin. Microbiol. 54, 2919–2927.

36. Kohl, T.A., Diel, R., Harmsen, D., Rothganger, J., Walter, K.M., Merker, M., Weniger, T., Niemann, S., 2014. Whole-genome-based Mycobacterium tuberculosis surveillance: a standardized, portable, and expandable approach. J. Clin. Microbiol. 52, 2479–2486.

37. Laabei, M., Recker, M., Rudkin, J.K., Aldeljawi, M., Gulay, Z., Sloan, T.J., Williams, P., Endres, J.L., Bayles, K.W., Fey, P.D., Yajjala, V.K., Widhelm, T., Hawkins, E., Lewis, K., Parfett, S., Scowen, L., Peacock, S.J., Holden, M., Wilson, D., Read, T.D., van den Elsen, J., Priest, N.K., Feil, E.J., Hurst, L.D., Josefsson, E., Massey, R.C., 2014. Predicting the virulence of MRSA from its genome sequence. Genome Res. 24, 839–849.

38. Linnaeus, C., 1735. Systema Naturæ, Sive Regna Tria Naturæ Systematice Proposita Per Classes, Ordines, Genera, & Species. Apud

Theodorum Haak, Leiden. Liu, W., Li, L., Khan, M.A., Zhu, F., 2012. Popular molecular markers in bacteria. Mol. Gen. Mikrobiol. Virusol., 14–17.

39. Loman, N.J., Misra, R.V., Dallman, T.J., Constantinidou, C., Gharbia, S.E., Wain, J., Pallen, M.J., 2012. Performance comparison of benchtop high-throughput sequencing platforms. Nat. Biotechnol. 30, 434–439.
40. Maiden, M.C., Jansen van Rensburg, M.J., Bray, J.E., Earle, S.G., Ford, S.A., Jolley, K.A., McCarthy, N.D., 2013. MLST revisited: the gene-by-gene approach to bacterial genomics. Nat. Rev. Microbiol. 11, 728–736.
41. Nijhuis, R.H., Oueslati, S., Zhou, K., Bosboom, R.W., Rossen, J.W., Naas, T., 2015. OXY-2-15, a novel variant showing increased ceftazidime hydrolytic activity. J. Antimicrob. Chemother. 70, 1429–1433.
42. Patel, J.B., 2001. 16S rRNA gene sequencing for bacterial pathogen identification in the clinical laboratory. Mol. Diagn. 6, 313–321.
43. Price, L.B., Stegger, M., Hasman, H., Aziz, M., Larsen, J., Andersen, P.S., Pearson, T., Waters, A.E., Foster, J.T., Schupp, J., Gillece, J., Driebe, E., Liu, C.M., Springer, B., Zdovc, I., Battisti, A., Franco, A., Zmudzki, J., Schwarz, S., Butaye, P., Jouy, E., Pomba, C., Porrero, M.C., Ruimy, R., Smith, T.C., Robinson, D.A., Weese, J.S., Arriola, C.S., Yu, F., Laurent, F., Keim, P., Skov, R., Aarestrup, F.M., 2012. Staphylococcus aureus CC398: host adaptation and emergence of methicillin resistance in livestock. MBio 3, e00305–00311.
44. Reiss, E., Obayashi, T., Orle, K., Yoshida, M., Zancope-Oliveira, R.M., 2000. Non-culture based diagnostic tests for mycotic infections. Med. Mycol. 38 (Suppl. 1), 147–159.
45. Ruppitsch, W., Pietzka, A., Prior, K., Bletz, S., Fernandez, H.L., Allerberger, F., Harmsen, D., Mellmann, A., 2015. Defining and evaluating a core genome multilocus sequence typing scheme for whole-genome sequence-based typing of Listeria monocytogenes. J. Clin. Microbiol. 53, 2869–2876.
46. Sabat, A.J., Budimir, A., Nashev, D., Sa-Leao, R., van Dijl, J., Laurent, F., Grundmann, H., Friedrich, A.W., 2013. Overview of molecular typing methods for outbreak detection and epidemiological surveillance. Euro Surveil. 18, 20380.
47. Sabat, A.J., Zanten van, E., Akkerboom, V., Wisselink, G., Slochteren

van, K., Boer de, R.F., Friedrich, A.W., Rossen, J.W.A., Kooistra-Smid, A.M D., 2016. Targeted amplification for bacterial identification at the species-level using next-generation sequencing—increased discrimination of closely related species. ECCMID, E-poster EP0219.

48. Sboner, A., Mu, X.J., Greenbaum, D., Auerbach, R.K., Gerstein, M.B., 2011. The real cost of sequencing: higher than you think! Genome Biol. 12, 125.

49. Schuurman, T., de Boer, R.F., Kooistra-Smid, A.M., van Zwet, A.A., 2004. Prospective study of use of PCR amplification and sequencing of 16S ribosomal DNA from cerebrospinal fluid for diagnosis of bacterial meningitis in a clinical setting. J. Clin. Microbiol. 42, 734–740.

50. Srinivasan, L., Pisapia, J.M., Shah, S.S., Halpern, C.H., Harris, M.C., 2012. Can broad-range 16S ribosomal ribonucleic acid gene polymerase chain reactions improve the diagnosis of bacterial meningitis? A systematic review and meta-analysis. Ann. Emerg. Med. 60, 609–620.

51. Tenover, F.C., Arbeit, R.D., Goering, R.V., 1997. How to select and interpret molecular strain typing methods for epidemiological studies of bacterial infections: a review for healthcare epidemiologists. Molecular Typing Working Group of the Society for Healthcare Epidemiology of America. Infect. Control Hosp. Epidemiol. 18, 426–439.

52. Thompson, C.C., Chimetto, L., Edwards, R.A., Swings, J., Stackebrandt, E., Thompson, F.L., 2013. Microbial genomic taxonomy. BMC Genom. 14, 913.

53. Tindall, B.J., Rossello-Mora, R., Busse, H.J., Ludwig, W., Kampfer, P., 2010. Notes on the characterization of prokaryote strains for taxonomic purposes. Int. J. Syst. Evol. Microbiol. 60, 249–266.

54. Weterings, V., Zhou, K., Rossen, J.W., van Stenis, D., Thewessen, E., Kluytmans, J., Veenemans, J., 2015. An outbreak of colistin-resistant Klebsiella pneumonia carbapenemase-producing Klebsiella pneumoniae in the Netherlands (July–December 2013), with inter-institutional spread. Eur. J. Clin. Microbiol. Infect. Dis. 34, 1647–1655.

55. Willmann, M., El-Hadidi, M., Huson, D.H., Schutz, M., Weidenmaier, C., Autenrieth, I.B., Peter, S., 2015. Antibiotic selection pressure determination through sequence-based metagenomics. Antimicrob. Agents Chemother. 59, 7335–7345.

56. Zhou, K., Ferdous, M., de Boer, R.F., Kooistra-Smid, A.M., Grundmann, H., Friedrich, A.W., Rossen, J.W., 2015a. The mosaic genome structure

and phylogeny of Shiga toxin-producing Escherichia coli O104:H4 is driven by short-term adaptation. Clin. Microbiol. Infect. 21 (468), e467–418.

57. Zhou, K., Lokate, M., Deurenberg, R.H., Arends, J., Lo-Ten Foe, J., Grundmann, H., Rossen, J.W., Friedrich, A.W., 2015b. Characterization of a CTX-M-15 producing Klebsiella pneumoniae outbreak strain assigned to a novel Sequence Type (1427). Front. Microbiol. 6, 1250.
58. Zhou, K., Lokate, M., Deurenberg, R.H., Tepper, M., Arends, J.P., Raangs, E.G., Lo-Ten-Foe, J., Grundmann, H., Rossen, J.W., Friedrich, A.W., 2016. Use of whole-genome sequencing to trace, control and characterize the regional expansion of extended-spectrum beta-lactamase producing ST15 Klebsiella pneumoniae. Sci. Rep. 6, 20840.

CHAPTER 7

NEXT GENERATION MICRO-BIOLOGICAL RISK ASSESS-MENT: OPPORTUNITIES OF WHOLE GENOME SEQUENC-ING (WGS) FOR FOODBORNE PATHOGEN SURVEILLANCE, SOURCE TRACKING AND RISK ASSESSMENT

Kalliopi Rantsiou[a], Sophia Kathariou[b], Annet Winkler[c], Panos Skandamis[d], Manuel Jimmy Saint-Cyr[g], Katia Rouzeau-Szynalski[e], Alejandro Amézquita[f]

CITATION: Rantsiou, K., et al., "Next generation microbiological risk assessment: opportunities of whole genome sequencing (WGS) for foodborne pathogen surveillance, source tracking and risk assessment", International Journal of Food Microbiology (2017), https://doi.org/10.1016/j.ijfoodmicro.2017.11.007.

[a]Department of Agriculture, Forest and Food Sciences, University of Turin, Largo Paolo Braccini 2, 10095 Grugliasco, Italy
[b]North Carolina State University, Department of Food, Bioprocessing and Nutrition Sciences, Raleigh, NC, USA
[c]Mondelez International, Kraft Foods R&D, Unterbiberger St. 15, D-81737 Munich, Germany
[d]Laboratory of Food Quality Control & Hygiene, Department of Food Science & Technology, Agricultural University of Athens, Iera Odos 75, 118 55 Athens, Greece
[e]Food Safety Microbiology, Nestec Ltd, Nestlé Research Center, Lausanne, Switzerland
[f]Safety and Environmental Assurance Centre, Unilever, Colworth Science Park, Sharnbrook, Bedfordshire MK44 1LQ, UK
[g]SECALIM Unit UMR1014, Oniris, INRA, Université Bretagne Loire, Nantes, France

ABSTRACT

Whole genome sequencing (WGS) of important foodborne pathogens is a technology under development, but is already employed in routine surveillance by public health agencies and is being increasingly exploited in tracing transmission routes and identifying contamination events (source tracking) that take place in the farm-to-fork continuum. Furthermore, data generated from WGS, complemented by other –omics data, have the potential to be integrated into and strengthen microbiological risk assessment. In this paper, we discuss the contribution of WGS in diverse areas important to food safety and public health. Additionally, an outlook of future WGS applications, which should contribute to our understanding of the ecology and physiology of foodborne microorganisms, is presented.

Keywords: Whole genome sequencing, Surveillance. Epidemiology, Foodborne pathogens, Source tracking

INTRODUCTION

In the past few years, the intensive use of high-throughput Next Generation Sequencing (NGS) technologies has led to an unprecedented increase in speed and cost-effectiveness of the acquisition of very high volumes of nucleic acid sequence data (Deng et al., 2016). The technology has markedly increased the feasibility and routine implementation of a number of previously challenging applications. One such currently widely used application of NGS is whole genome sequencing (WGS), an analytical approach to determine the sequence of the whole genomic content of an organism (Mardis, 2008, Ståhl and Lundeberg, 2012).

In several industrialized countries, public health agencies and regulatory bodies [e.g. Food and Drug Administration (FDA), Centers for Disease Control and Prevention (CDC), Public Health England, European Center for Disease Prevention and Control (ECDC)] are now using WGS routinely to characterize clinical isolates of selected foodborne pathogens and to support epidemiological investigations. For instance, in the United States the CDC is performing WGS on all isolates from human listeriosiscases and retrospectively on stored isolates from earlier cases of disease (Jackson et al., 2016). This would ensure that isolates can be compared after different time periods. The CDC has announced the intention to phase out current typing techniques and fully move to WGS by 2018. WGS is also included as a routine tool in the National Antimicrobial Resistance Surveillance System (NARMS). The FDA is routinely sequencing all *L. monocytogenes* isolates and has established the GenomeTrakr Network to facilitate sharing of genomic and other isolate-related information between laboratories. WGS is now used in parallel with other typing techniques in many countries of the European Union, with Denmark having already transitioned completely to WGS for certain foodborne pathogens (ECDC, 2016a). To further promote application of WGS, the ECDC will oversee and coordinate efforts to overcome current challenges and is aiming to establish WGS as the method of choice for foodborne pathogen typing and surveillance within the next five years (ECDC, 2016a).

WGS typically includes several steps: obtaining a pure culture of the organism of interest, DNA extraction, library preparation and amplification, and DNA sequencing, followed by possible alignment, data analysis, and finally biological interpretation. Today, WGS is not necessarily followed by annotation, data analysis or functional (biological) interpretation. It is generally limited to obtaining and ideally depositing a file with the entire genome sequence in a database such as the National Center for Biotechnology Information (NCBI). Furthermore, in most instances WGS data are in the form of contigs, and in the absence of a closed genome sequence the available data are expected to contain sequence gaps. As WGS evolves from a research tool to a routine surveillance instrument, different pipelines, software and platforms are used to perform sequencing data analysis. Many of the pipelines are operational as research tools and under continuous development; for instance, FDA CFSAN provides periodic updates on the single nucleotide polymorphism Pipeline (https://github.com/CFSAN-Biostatistics/snp-pipeline; Davis et al., 2015). An overview of different bioinformatics approaches to WGS data analysis is provided by

Franz et al. (2016). Standardization, i.e. validation and articulation of key quality requirements and criteria, is needed to be able to compare results produced by different pipelines, software and platforms.

In addition to epidemiological surveillance and clinical applications, WGS is slowly being explored by the food industry to improve food safety. In the context of food safety management systems (FSMS) and the implementation of measures to prevent foodborne pathogen contamination, food business operators (FBO) routinely monitor the production environment. Such monitoring programs rely on classical microbiological methods but may be complemented by molecular-based methods to obtain a more precise view of the hygienic state of the environment and a much more extended view of the ecological richness. The information collected from such approach may provide insight into contamination routes. Currently, the development of WGS is not at the same level in food industry compared to public health agencies, which are using it routinely. Food industry started to explore WGS for tracking the source of microbial contamination and for other purposes, such as determination of virulence and antibiotic resistance genes for pathogenic strains. Delays in adoption of WGS are mainly due to high costs, time and expertise/resources/capacities requirements. Extensive application by the food industry will reveal the potential of WGS to differentiate between strains persisting in a facility and those repeatedly introduced from an outside source.

The aim of this paper is to present the current status and the opportunities of WGS applications in foodborne pathogen surveillance, in understanding reservoirs, delineating transmission routes and integrating genomic data into risk assessment.

FOODBORNE PATHOGEN SURVEILLANCE

For the purposes of this paper, foodborne pathogen surveillance is defined as the systematic collection, analysis and interpretation of data essential to the planning, implementation and evaluation of public health practice, and the timely dissemination of this information for public health action (WHO, 2008). The main goals of surveillance programs are to identify, control and prevent foodborne outbreaks, determine the causes of foodborne disease, and monitor trends in occurrence of foodborne disease (Potter et al., 2000). Information gathered during surveillance programs generally contributes to the development of prevention strategies and to the assessment of their effectiveness (Braden and Tauxe, 2006). Moreover, information on the epidemiology of foodborne pathogens is highly valuable for risk assessment and HACCP plan development and implementation (Braden and Tauxe, 2006, Potter et al., 2000).

Although WGS can be used for all foodborne biological agents, it is currently most intensely used for bacteria. WGS information has been used to characterize, subtype or compare isolates of several bacterial species (Frey and Bishop-Lilly, 2015, Stasiewicz et al., 2015). To date, agents most extensively investigated by WGS are the bacteria *Listeria monocytogenes*, *Salmonella* enterica and *Escherichia* coli, partly because of their importance as human foodborne pathogens and the relatively small size of their genomes (*L. monocytogenes*, approx. 3 Mb; *Salmonella* and *E. coli*, approx. 4.6 Mb each). Numerous WGS investigations have addressed other pathogens from food animals or food production environment. For instance, WGS of *Campylobacter* jejuni and *C. coli* from food animals and farm ecosystem have already yielded insights on potential host-adapted attributes and novel antimicrobial resistancedeterminants (Chen et al., 2013, Dutta et al., 2016, Miller et al., 2016, Qin et al., 2014, Zhao et al., 2015).

Routine use or complete replacement of other typing techniques by WGS will need adequate attention to overcome existing challenges and limitations such as costs, manpower, laboratory and bioinformatics infrastructure, safeguarding of findings, standardization and validation of the methodology, data interpretation requiring expert knowledge (both from the bioinformatics and the microbiology perspective). The lack of standards and harmonization is one of the challenges to overcome to enable full exploitation of WGS for routine epidemiological uses. On a more global level, WGS will face a different set of constrains including legal aspects of data sharing, international trade, integration of data into food safety decision making, and capabilities to perform WGS (European Food Safety Authority Scientific Colloquium Summar, Food and Agriculture Organization of the United Nations (FA).

Foodborne pathogen surveillance programs have long been implemented through networks bringing together several laboratories, such as FoodNet, PulseNet, the European Surveillance System (TESSy) (FDA, CDC, van Walle, 2013, ECDC). Until recently, the standard molecular subtyping and fingerprinting methods were pulsed-field gel electrophoresis (PFGE) and multiple-locus variable-number tandem repeat analysis (MLVA). PFGE lacks the resolution to effectively pinpoint the source of an outbreak while MLVA generally demonstrates a higher discriminatory power than PFGE

(Best et al., 2007). However, MLVA has a notable limitation in relation to source attribution studies because many MLVA schemes are specific for certain clones only (Nadon et al., 2013). For instance, the MLVA scheme effective for one *Salmonella*serovar is frequently not equally discriminative for others (Ross et al., 2011). WGS has the potential to overcome challenges related to discriminatory power.

The integration of WGS in surveillance programs has already started to significantly contribute to the investigation of foodborne outbreaks involving bacterial agents. In the United States, the FDA and the CDC are intensively using this technology in outbreak investigation by comparing isolates from patients to those from food or food production environments. WGS data are uploaded to the open-access database GenomeTrakr. The GenomeTrakr network is currently comprising of numerous laboratories in the US and elsewhere, leading to a dramatic increase in the number of sequences in the database (Allard et al., 2016). In the European Union, the ECDC is also putting in place a network of public health laboratories employing WGS, with at least 19 countries enrolled in surveillance programs and outbreak investigations. A common European database for WGS data and metadata is under development (ECDC, 2016b).

In the context of outbreak investigations, GenomeTrakr and equivalent databases will make it increasingly possible to accurately establish links between sequences from outbreaks and from food or environmental sources. Furthermore, the high discriminatory power of WGS allows for detection of diffuse outbreaks by linking infrequent cases, which would otherwise have been considered sporadic cases without a common source. Such capacities are expected to help to contain outbreaks due to earlier outbreak detection and intervention at the source(s). Recent examples are given in "Applications of WGS in Food Safety management" (FAO, 2016) and are increasingly encountered in the literature (Butcher et al., 2016, Dallman et al., 2016, Holmes et al., 2015).

The development of such databases combined with epidemiological investigations can enhance the possibility to share sequences among countries and thus identify and monitor global outbreaks (Allard et al., 2016). The expansion of the databases will allow more links between present and past outbreaks or contamination events, thus potentially identifying strains persistent through time in certain production facilities or other ecosystems. Thus, WGS data may be employed to retrospectively investigate unsolved outbreaks or contamination events. However, this type of analysis can only be supported by complete metadata providing the relevant context. The metadata should include information, such as where strains were isolated, from which kind of sample, country of origin and date of isolation.

It is assumed that all identical or nearly identical strains belonging to the same species have a common ancestry. Finding nearly identical strains at different processing facilities raises questions about the expected genetic variability within pathogens. Studies exploring overall genetic diversity among strains of a certain species are needed, especially for highly clonal microorganisms such as lineages within *L. monocytogenes*, *Salmonella* and others (Leekitcharoenphon et al., 2014). This evidence poses a challenge for the food industry; (nearly)-identical strains can be found in multiple food environments i.e. an isolate of one factory does not necessarily mean a causal link with a clinical isolate and thus an isolate of a human patient. Stasiewicz et al. (2015) showed that *L. monocytogenes* isolated in a retail deli-specific setting differed by very limited Single Nucleotide Polymorphisms (median 2 to 11 SNPs) and have concluded that identical or nearly identical strains of *L. monocytogenes* can occur at different deli environments at different locations, with no evident clear links in transmission. WGS data provided evidence for repeated introduction of *L. monocytogenes* from an external source to multiple retail establishments. Similarly, Fagerlund et al. (2016) reported nearly identical *L. monocytogenes* MLST Sequence Type 8 isolates in Danish and Norwegian salmon processing plants spanning a period of more than 10 years. In another case, nearly identical *L. monocytogenes* ST 8 isolates were found in two poultry processing plants in Norway (Fagerlund et al., 2016). These examples emphasized that linking disease back to isolates from processing environments can be challenging as almost identical isolates can be detected in more than one food processing plant. To resolve situations of uncertainty it is important that metadata are carefully collected and evaluated in combination with the WGS data.

Sharing of WGS data and metadata across countries will allow development of novel surveillance programs and facilitate global perspectives of pathogen trends and emergence of new pathogens or food safety issues. The need to globally monitor important foodborne pathogens was recently highlighted by Moura et al. (2016) who subjected 1696 strains of *L. monocytogenes* to WGS and demonstrated that circulation of isolates between continents/countries takes place. These approaches, can only be fully effective with harmonization of methods and data-sharing platforms (Deng et al., 2016). Specific pilot initiatives are already in place with other pathogens, e.g. molecular surveillance of multidrug-resistant *Mycobacterium tuberculosis* within the European Union (van Walle, 2013). International initiatives such as the Global Microbial Identifier (GMI) as well as tripartite expert groups that involve industry, academia and government (e.g. ILSI Europe Microbiological Food Safety Task Force), are making efforts in this direction, in particular for method standardization and data sharing among public health organizations, academia and industry, areas that currently present substantial challenges.

UNDERSTANDING RESERVOIRS AND DELINEATING TRANSMISSION ROUTES

The WGS data collected from surveillance programs can be used to perform phylogenetic analyses of the population for various foodborne pathogens. Compared to previously employed methods to infer phylogenetic relationships among pathogenic isolates (such as PFGE or MLST), WGS presents the unprecedented characteristics of providing sequence data that span the whole genome and a discriminatory power that can be adapted, based on the downstream data analysis performed, to the genome characteristics of the pathogen in question. Hence, WGS is a data generation tool that can be considered "universal" i.e. can be similarly applied to all pathogens, independently of the level of genome diversity that their population exhibits. Based on the structure of the population (monophyletic vs. highly diverse), the discrimination level of the downstream data analysis can be attuned (for example SNP based vs. genome wide-MLST). WGS has therefore the potential to gradually replace all pre-existing typing methods, as long as robust standards of harmonization are put in place in all steps of the process, i.e. before, during and after the acquisition of the sequencing data.

The genetic relations inferred from phylogenetic studies can be used in source attribution. In source attribution the goal is to "quantify the relative importance of specific food sources and animal reservoirs for human cases of foodborne illness" (EFSA, 2008). The genetic relations may indicate associations with specific hosts (therefore potentially identifying important reservoirs of pathogens) or habitats (resident populations at specific locations) as well as point to the geographical distribution of a pathogen's subtypes at the local, regional or global level (therefore identifying putative routes of transmission). Several subtyping methods can provide information that may be used to calculate frequencies of isolation of certain subtypes from certain foods (EFSA, 2013). This information can be correlated to frequency of occurrence of the subtypes among clinical isolates to infer the most important food source leading to disease by a certain subtype of the pathogen. For example, multilocus sequence typing (MLST) has been used with success for source attribution and discrimination of *Campylobacter* (Sheppard et al., 2009, Guyard-Nicodème et al., 2015) whereas it appears to be less appropriate for *Salmonella* (Barco et al., 2013).

WGS is a powerful tool to identify transmission pathways (i.e. epidemiological links between reservoirs or sources and infections) and complement epidemiological data (i.e. spatial, temporal, and host characteristics) rather than replace them. This technology is highly suitable for *Salmonella* and several other foodborne pathogens. For *Salmonella* strains, WGS is replacing conventional tools such as PFGE, which often is inadequate in accurately tracking the source of contamination (Allard et al., 2012). This was already shown by Hoffmann et al. (2016) in the context of a *Salmonella* Bareilly outbreak with scraped tuna. WGS determined links between the outbreak isolates in the US and isolates from a fish processing facility in India, thus yielding a geo (spatio-)-temporal transmission map for this outbreak. It was also demonstrated that WGS from "historical" archived isolates was useful in establishing links between isolates from 2003 and 2012 (Hoffmann et al., 2016). Furthermore, results were compared with PFGE, and WGS demonstrated a higher resolution power. WGS analysis of S. Enteritidis isolates from an egg outbreak in the UK revealed a clear link between human, egg and environmental S. Enteritidis isolates specific to the outbreak (Inns et al., 2015). Concerning *Campylobacter*, molecular epidemiology of *C. jejuni* and *C. coli* has remained challenging due to the extensive genomic and phenotypic diversity within these species (Bronnec et al., 2016, Haddad et al., 2010, Rivoal et al., 2005, Rodrigues et al., 2015). The rapid evolution of bacterial genomes has important consequences for interpretation of molecular typing information (Lomonaco et al., 2015). Outbreak isolates may be missed in cases where small genomic changes result in large changes of molecular profiles by conventional tools (Barton et al., 2007). Conversely, some strains appear to be clonal and may be linked by conventional typing methods despite significant differences in gene content among isolates and absence of epidemiological linkage (Taboada et al., 2008).

Apart from clearly defining genetic relationships among subtypes within pathogenic species and permitting associations to be made with specific reservoirs or habitats, WGS data can also be used to obtain basic biological insights that could explain such associations. Moreover, considering both genetic and functional (e.g. virulence, survival and antimicrobial resistance) relationships among isolates will provide added value to the source attribution models (EFSA, 2013). In this context it should be underlined that WGS data can be used to predict functional properties of isolates and it is foreseen that it will be possible to discover biomarker genes that indicate association with a specific reservoir or habitat (addressed also in Cocolin et al., den Besten et al., this issue).

CONTRIBUTION TO RISK ASSESSMENT: "UP-GRADING" HAZARD IDENTIFICATION

Microbiological Risk Assessment (MRA) is a structured, systematic, science-based approach that evaluates the likelihood of adverse human health effects occurring following exposure to a pathogenic microorganism or its products, e.g. a toxin. MRA is intended to support the understanding and management of microbiological risks, so that it informs the need for possible risk management options and enables evaluation of management practices in the farm-to-fork continuum. MRA consists of four steps (CAC (Codex Alimentarius Commission), 2014, US EPA and USDA/FSIS (Environmental Protection Agency and D): (Aanensen et al., 2016) *hazard identification* – knowledge about the microorganism and its association with adverse health effects; (Allard et al., 2012) *hazard characterization* – likelihood of infection given the level of exposure (i.e. dose-response relationship) and the consequences of infection; (Allard et al., 2016) *exposure assessment* – which considers the quantities of microorganisms in raw materials and the impact of preservation and processing systems to arrive at likely quantities in finished products, the product use and hence consumer exposure; and (Barco et al., 2013) *risk characterization* – which provides an estimate of the level of risk to the exposed individuals, allowing for risk management decisions to be made on the basis of risk instead of solely on the potential for the organism to be present.

Whilst the use of MRA in food safety has been established for almost two decades, the application of WGS in MRA is largely unexplored. Many challenges remain, particularly in reconciling the different scales of data necessary to inform a decision based on an estimate of risk to human health – i.e. from tens of thousands of SNPs, to thousands of genes, to tens of biologically relevant phenotypic traits, to a single measure of risk (Franz et al., 2016, Pielaat et al., 2015). Two 'sister' papers in this issue discuss the potential application of WGS and other omics technologies in '*hazard characterization*' and '*exposure assessment*'. Therefore, this section focuses on its potential application in *hazard identification*.

Phenotypic diversity of foodborne pathogens concerns numerous traits, especially those related to (i) virulence potential based on epidemiological evidence; certain strains of a species may be strongly associated with human disease while others are encountered rarely, or not at all, in clinical cases; (ii) behavior within foods or in the food environment, for example different abilities to resist low pH, low a_w, heat, biocides or to form biofilm and persist in the processing environment; (iii) association with specific habitats or animal hosts (for agents with environmental or animal reservoirs). Such phenotypic traits may allow sorting of strains based on phenotypic differences, but may not readily correspond to population structure of the pathogens. Furthermore, phenotypic observations do not provide insights on the underlying molecular mechanisms responsible for a certain (phenotypic) behavior. On the other hand, understanding the differences among members of the same species, but also within clades or phylogenetic lineages of the same species, and discerning their unique genetic features has food safety relevance, for example in terms of defining pathogenic potential, in addition to providing ecological and evolutionary insights.

The advent of molecular biology has provided tools such as PFGE and multilocus sequence typing (MLST), which allowed phylogenetic studies, albeit of limited resolution (Klemm and Dougan, 2016). WGS, on the other hand, is proving to be a flexible instrument to map genetic relationships among isolates of both monophyletic, highly clonal species/lineages and polyphyletic, highly diverse species, based on post-sequencing analysis (Deng et al., 2010, Gordienko et al., 2013). Understanding of the population structure of a bacterial species or other taxonomic unit provides insights into mechanisms of evolution. Gene or gene function loss and horizontal gene transfer(HGT) are key mechanisms that drive bacterial evolution and can frequently underlie the emergence of either pathogenic or attenuated variants, adaptation to specific environments or specific animal hosts, tolerance to specific stresses and tissue or organ tropism. WGS and comparative genomics have highlighted the role of mobile genetic elements and HGT in the evolution of major foodborne pathogens such as *S. enterica* and *Shigella* spp. (Foley et al., 2013, The et al., 2016). Phylogenetic analysis based on WGS data will also help in the understanding of the evolution and spread of multidrug resistance phenotypes or antibiotic-resistant clades within pathogenic species (Klemm and Dougan, 2016).

By comparing a large set of genomic data it is possible to associate genomic sequences with specific phenotypic traits and identify the molecular basis of a phenotype (Falush and Bowden, 2006). Such genome–wide association studies (GWAS), validated by other –omics (transcriptomics, proteomics) approaches, can eventually lead to the identification of genomic sequences that may serve as markers or indicators of a specific phenotype (e.g. virulence, tolerance to specific stress, host association, environmental distribution) and can also have a predictive value when a bacterial strain or species is unknown or "emerging". In the future, a *hazard identification* approach can be assisted by the use of such indicator sequences. Whilst the idea of linking genes to phenotypic traits in bacteriology is not new, the technical feasibility of bacterial GWAS remains a challenge due to the clonal nature of bacterial reproduction and population structure; association tests that do not take this into account may result in non-causal relationships (Earle et al., 2016).

A clear understanding of diversity at the genome level and of the population structure of a given pathogen is critical when "representative" or "reference" strains are subjected to further analysis. This is of primary importance in *hazard identification* and an informed choice regarding the strains to be included or considered will impact the fulfillment of the criteria regarding MRA and its success. The study of Pielaat et al. (2015) presents a first attempt to use GWAS in the context of *hazard identification*, showing promising results. The authors coupled genomic (SNP) data with *in vitro*measurements of adherence of *E. coli* O157:H7 to Caco-2 cells (as a proxy for virulence), and highlighted the importance of considering population structure to establish causality between virulence phenotypes and thus, capacity of causing a disease, and genetic disposition, suggesting the need to use mixed-effects models. This is consistent with the findings of a more recent study (Earle et al., 2016), which reports good performance of GWAS in 26 association studies concerning the resistance of 4 bacterial species to 17 antimicrobials. Specifically, in 25 of the 26 associations the leading candidate was a validated antibiotic resistance gene. Similar efforts are underway in order to associate genotypes of foodborne pathogens with tolerance to common food stresses, such as cold, salt or acid (Hingston et al., 2017).

As the cost of sequencing continues to decrease and public databases for WGS data for foodborne pathogens continue to develop (e.g. GenomeTrakr), the amount of WGS data is dramatically increasing, and is expected to continue to do so in the coming years. The potential use of these 'big data' in *hazard identification* will enable risk assessors and risk managers to move from a 'taxonomic' understanding of the hazard to one based on genotype-phenotype associations and on a much deeper understanding of virulence at the strain level. An example of effective integration of bacterial population genomics with clinical and epidemiological data was recently presented with *L. monocytogenes*, with certain hypervirulent clones found to be over-represented among human disease cases, while hypovirulent ones were over-represented in foods (Maury et al., 2016). Interestingly, this analysis also revealed that commonly employed reference strains of *L. monocytogenes* actually belonged to hypovirulent clones (Maury et al., 2016). In addition to public WGS databases, further information about relevant factors for *hazard identification* needs to be developed. Especially needed are data to inform gene family profiling, as well as the establishment and agreement on relevant reference genomes for specific pathogens that reflect strains actually involved in human illness.

As a systematic approach to support decision-making, MRA requires that all stakeholders in the food safety continuum have confidence in the robustness of the data, approaches and scientific principles underpinning the output of the risk assessment. Current paradigms for pathogens, such as *L. monocytogenes* in existing regulatory frameworks, has resulted in policies which are based at differentiation at the species level (i.e. ignoring differences in virulence among strains). Therefore, challenges to fully realize the potential for WGS data in *hazard identification* can be witnessed not only at the technical and scientific level but also at the decision-making level, if decisions are to be based on interpretation of high-resolution genomic data such as virulence variation at the strain level.

OTHER APPLICATIONS OF WGS INFOOD SAFETY

WGS has high discriminatory power and will in the future have short turn-around time, and overall lower costs, compared to conventional methods. For example, the determination of multidrug resistance and virulence of a pathogen can be time consuming and costly, as it is generally linked with culture-based analysis of the isolates. WGS will allow detection of genes in a one-step analysis, and cosequentely determination of resistance and virulence (Allard et al., 2016, Köser et al., 2016, Kwong et al., 2015, Reuter et al., 2014). Chattaway et al. (2016) evaluated the use of WGS for the subtyping of Stx-encoding *E. coli* isolates. The results showed that WGS was able, under a routine surveillance set-up, to provide information of the pathogenic potential of each isolate, enabling the prediction of clinical outcomes and the monitoring of emergence of hypervirulent isolates. In another study (Aanensen et al., 2016), WGS data of invasive *Staphylococcus aureus* was combined with epidemiological and resistance data (based on specific genes content). It was then possible to identify high-risk isolates and map their spread through Europe. Data obtained will help to elucidate and follow evolutionary events leading to the emergence and dissemination of hypervirulent strains or antimicrobial resistancegenes.

As previously discussed, genomic information gained from WGS should be interpreted in the context of epidemiological analyses. For other applications, e.g. antimicrobial resistance, virulence and other functional assessments, such as tolerance to environmental stress, WGS data should be validated by functional analyses and implementation of additional technologies, including other omics-based tools such as transcriptomics, proteomics and metabolomics.

In addition to its use in surveillance and outbreak investigation, WGS-based information can also provide other subtyping data of relevance to the target microorganism, such as *in silico* determination of serotype and sequence type (ST) (via *in silico* multi-locus sequence typing [MLST]). In fact, such determinations are not only more rapid and accurate but ultimately more cost-effective via WGS-based analysis (Kwong et al., 2016).

Besides the application of WGS on pure cultures of specific isolates, it is starting also to be used to study microbial populations. Whole genomes of microorganisms can be fully or partially reconstructed from metagenomics data (Cocolin et al. this issue). Metagenomics is a culture-independent application of NGS whereby the community DNA from a sample is sequenced, including DNA from all microbes in the sample. If a reference genome is available and the target microorganism being investigated is present at a certain detection level in the sample, its genome can be assembled, allowing culture-independent detection. Culture-independent detection of foodborne pathogens in food, clinical, veterinary or environmental samples circumvents issues related with the "culturability" of microorganisms, e.g. microorganisms that are slow, difficult or impossible to culture, or that cannot be recovered in conventional selective enrichments due to injury or the presence of competing microbiota. In addition it reduces the time necessary for detection since it bypasses the culture steps.

A further contribution of NGS to the process of understanding the association of pathogenic microorganisms with various environments is envisioned. Through a metagenomic approach, the microbial communities of a sample can be described in a detail never before achieved. Metagenomic analysis of a sample that is associated (in space or time) with a specific WGS subtype of a pathogen, can shed light to events related with co-presence or co-exclusion, co-evolution, horizontal gene transfer and other microbial inter-relations that are, as of yet, poorly studied in nature. This has, among others, the potential to yield relevant indicators for pathogens in samples from specific habitats. Ultimately, it will be possible to understand (predict) the biotic factors that underlie the microbial propensity for a specific habitat.

CONCLUSIONS AND FUTURE PERSPECTIVES

The potential of WGS as a typing tool to be used in surveillance and outbreak investigation and to contribute in risk assessment is indisputable. Scientists in academia, the regulatory sector and the food industry increasingly recognize WGS as the method of choice for basic research applications and epidemiological investigations. Nevertheless, complete replacement of other typing techniques by WGS will need adequate attention to overcome existing challenges and limitations. These include expenses related to initial investment and running costs, human resources, laboratory and bioinformatics infrastructure but also the integration of WGS data into risk assessment and decision-making processes. A gradual phasing out of existing methodologies is foreseen as technical capacity is building up in, and becoming harmonized among, different countries. In this process, key issues that need to be addressed concern standardization, robustness and validation of the analytical methodology, from sample preparation to data interpretation. Specifically in data interpretation, a multidisciplinary approach is/will be required to match and integrate expertise in bioinformatics and biological/microbiological/epidemiological sciences. Initiatives to bring food industry, authorities from food safety and public health together to achieve common agreed understanding and approach of applications, in particular source tracking analysis, is urgently needed. Besides discussions on the technical aspects of the WGS workflow, the impact of WGS on the daily operation in the food industry needs to be addressed. Considering the worldwide food trade, source tracking needs to be performed at the global level, and authorities from different zones need to be involved in related initiatives.

Finally, the future may also open new doors to high-resolution whole-genome profiling of specific microbes from metagenomics datasets, markedly lowering the need for culture-based pathogen detection and characterization.

GLOSSARY AND ABBREVIATIONS

The glossary and abbreviations list are presented in Supplementary Tables 1 and 2, respectively, and are reproduced in the four joint papers on "Next generation Microbiological Risk Assessment".

ACKNOWLEDGEMENTS

The paper is the result of a working group within the ILSI workshop on "Next generation MRA (Microbiological Risk Assessment); integration of Omics data into assessment" that took place in Athens, Greece, May 13-14th 2016. This workshop was organised with the help of ILSI-Europe, IAFP and the ICFMH. This paper forms part of a series of four papers that address the contribution of NGS and omics approaches in MRA

REFERENCES

1. Aanensen, D.M., Feil, E.J., Holden, M.T., Dordel, J., Yeats, C.A., Fedosejev, A., Goater, R., Castillo-Ramírez, S., Corander, J., Colijn, C., Chlebowicz, M.A., Schouls, L., Heck, M., Pluister, G., Ruimy, R., Kahlmeter, G., Åhman, J., Matuschek, E., Friedrich, A.W., Parkhill, J., Bentley, S.D., Spratt, B.G., Grundmann, H., European SRL Working Group, 2016. Whole-genome sequencing for routine pathogen surveillance in public health: a population snapshot of invasive Staphylococcus aureus in Europe. MBio 7, e00444–16.
2. Allard, M.W., Luo, Y., Strain, E., Li, C., Keys, C.E., Son, I., Brown, E.W., 2012. High resolution clustering of Salmonella enterica serovar Montevideo strains using a nextgeneration sequencing approach. BMC Genomics 13, 1.
3. Allard, M.W., Strain, E., Melka, D., Bunning, K., Musser, S.M., Brown, E.W., Timme, R., 2016. The practical value of food pathogen traceability through building a wholegenome sequencing network and database. J. Clin. Microbiol. 54, 1975–1983.
4. Barco, L., Barrucci, F., Olsen, L.E., Ricci, A., 2013. Salmonella source attribution based on microbial subtyping. Int. J. Food Microbiol. 163, 193–203.
5. Barton, C., Ng, L.K., Tyler, S.D., Clark, C.G., 2007. Temperate bacteriophages affect pulsed-field gel electrophoresis patterns of Campylobacter jejuni. J. Clin. Microbiol. 45, 386–391.
6. Best, E.L., Lindstedt, B.A., Cook, A., Clifton Hadley, F.A., Threlfall, E.J., Liebana, E., 2007. Multiple-locus variable-number tandem repeat analysis of Salmonella enterica subsp. enterica serovar Typhimurium: comparison of isolates from pigs, poultry and cases of human gastroenteritis. J. Appl. Microbiol. 103, 565–572.
7. Braden, C.R., Tauxe, R.V., 2006. Surveillance for emerging pathogens in the United States. In: Motarjemi, Y., Adams, M. (Eds.), Emerging Foodborne Pathogens. CRC Press, Washington, DC. Bronnec, V., Haddad, N., Cruveiller, S., Hernould, M., Tresse, O., Zagorec, M., 2016. Draft genome sequence of Campylobacter jejuni Bf, an atypical strain able to grow under aerobiosis. Genome Announc. 4, e00120–16.
8. Butcher, H., Elson, R., Chattaway, M.A., Featherstone, C.A., Willis, C., Jorgensen, F., Dallman, T.J., Jenkins, C., McLauchlin, J., Beck, C.R., Harrison, S., 2016. Whole genome sequencing improved case

ascertainment in an outbreak of Shiga toxinproducing Escherichia coli O157 associated with raw drinking milk. Epidemiol. Infect. 144, 2812–2823.

9. CAC (Codex Alimentarius Commission), 2014. Principles and Guidelines for the Conduct of Microbiological Risk Assessment. CAC/GL 30-1999. Food and Agriculture Organization of the United Nations/World Health Organization, Rome/Geneva Available at: http://www.codexalimentarius.org/download/standards/357/CXG_ 030e_2014.pdf, Accessed date: 11 January 2016.

10. Chattaway, M.A., Dallman, T.J., Gentle, A., Wright, M.J., Long, S.E., Ashton, P.M., Perry, N.T., Jenkins, C., 2016. Whole genome sequencing for public health surveillance of shiga toxin-producing Escherichia coli other than serogroup 0157. Front. Microbiol. 7, 258.

11. Chen, Y., Mukherjee, S., Hoffmann, M., Kotewicz, M.L., Young, S., Abbott, J., Luo, Y., Davidson, M.K., Allard, M., McDermott, P., Zhao, S., 2013. Whole-genome sequencing of gentamicin-resistant Campylobacter coli isolated from U.S. retail meats reveals novel plasmid-mediated aminoglycoside resistance genes. Antimicrob. Agents Chemother. 57, 5398–5405.

12. Dallman, T., Inns, T., Jombart, T., Ashton, P., Loman, N., Chatt, C., Messelhaeusser, U., Rabsch, W., Simon, T., Nikisins, S., Bernard, H., Hello, S.I., Jourdan da-Silva, N., Kornschober, C., Mossong, J., Hawkey, P., de Pinna, E., Grant, K., Cleary, P., 2016.

13. Phylogenetic structure of European Salmonella Enteritidis outbreak correlates with national and international egg distribution network. Microb. Genom. 2, e000070.

14. Davis, S., Pettengill, J.B., Luo, Y., Payne, J., Shpuntoff, A., Rand, H., Strain, E., 2015. CFSAN pipeline: an automated method for constructing SNP matrices from nextgeneration sequence data. Peer J. Comput. Sci. 1, e20. https://doi.org/10.7717/ peerj-cs.20.

15. Deng, Z., Phillippy, A.M., Li, Z., Salzberg, S.L., Zhang, W., 2010. Probing the pan-genome of Listeria monocytogenes: new insights into intraspecific niche expansion and genomic diversification. BMC Genomics 11, 500.

16. Deng, X., den Bakker, H.C., Hendriksen, R.S., 2016. Genomic epidemiology: wholegenome-sequencing-powered surveillance and outbreak investigation of foodborne bacterial pathogens. Ann review food. Sci. Technol. 7, 353–374.

17. Dutta, V., Altermann, E., Olson, J., Wray, G.A., Siletzky, R.M., Kathariou, S., 2016. Wholegenome sequences of agricultural, host-associated Campylobacter coli and Campylobacter jejuni strains. Genome Announc. 4 (4) (pii: e00833-16).

18. Earle, S.G., Wo, C.-H., Charlesworth, J., Stoesser, N., Gordon, N.C., Walker, T.M., Spencer, C.C.A., Iqbal, Z., Clifton, D.A., Hopkins, K.L., Woodford, N., Smith, E.G., Ismail, N., Llewelyn, M.J., Peto, T.E., Crook, D.W., McVean, G., Walker, A.S., Wilson, D.J., 2016.

19. Identifying lineage effects when controlling for population structure improves power in bacterial association studies. Nat Microbiol. http://dx.doi.org/10.1038/nmicrobiol.2016.41.

20. European Centre for Disease Prevention and Control, 2016a. ECDC roadmap for integration of molecular and genomic typing into European-level surveillance and epidemic preparedness – Version 2.1, 2016–19. ECDC, Stockholmhttp://ecdc.europa. eu/en/publications/Publications/molecular-typing-EU-surveillance-epidemicpreparedness-2016-19-roadmap.pdf.

21. European Centre for Disease Prevention and Control, 2016b. Expert Opinion on Whole Genome Sequencing for Public Health Surveillance. ECDC, Stockholm. http://ecdc. europa.eu/en/publications/Publications/whole-genome-sequencing-for-publichealth-surveillance.pdf. European Food Safety Authority BIOHAZ Panel, 2008. Scientific opinion of the panel on biological hazards on a request from EFSA on overview of methods for source attribution for human illness from food-borne microbiological hazards. EFSA J. 764, 1–43.

22. European Food Safety Authority BIOHAZ Panel, 2013. Scientific opinion on the evaluation of molecular typing methods for major food-borne microbiological hazards and their use for attribution modeling, outbreak investigation and scanning surveillance: part 1 (evaluation of methods and applications). EFSA J. 11, 3502. http://dx.doi.org/10.2903/j.efsa.2013.3502.

23. EuropeanFoodSafetyAuthorityScientificColloquiumSummaryReport, 2014European Food Safety Authority Scientific Colloquium Summar. Use of Whole Genome Sequencing (WGS) of Foodborne Pathogens for Public Health Protection. http://dx. doi.org/10.2805/66246.

24. Fagerlund, A., Langsrud, S., Schirmer, B.C.T., Møretrø, T., Heir, E., 2016. Genome analysis of Listeria monocytogenes sequence type 8 strains persisting in salmon and poultry processing environments and

comparison with related strains. PLoS One. http://dx. doi.org/10.1371/journal.pone.0151117.

25. Falush, D., Bowden, R., 2006. Genome-wide association mapping in bacteria? Trends Microbiol. 14, 353–355.
26. Foley, S.L., Johnson, T.J., Ricke, S.C., Nayak, R., Danzeisen, J., 2013. Salmonella pathogenicity and host adaptation in chicken-associated serovars. Microbiol. Mol. Biol. Rev. 77, 582–607.
27. Food and Agriculture Organization of the United Nations (FAO), 2016Food and Agriculture Organization of the United Nations (FA. Applications of WGS in food safety management, Rome. http://www.fao.org/3/a-i5619e.pdf.
28. Franz, E., Mughini Gras, L., Dallman, T., 2016. Significance of whole genome sequencing for surveillance, source attribution and microbial risk assessment of foodborne pathogens. Curr. Opin. Food Sci. 8, 74–79.
29. Frey, K.G., Bishop-Lilly, K.A., 2015. Next-generation sequencing for pathogen detection and identification. Methods Microbiol. 42, 525–554.
30. Gordienko, E.N., Kazanov, M.D., Gelfand, M.S., 2013. Evolution of pan-genomes of Escherichia coli, Shigella spp. and Salmonella enterica. J. Bacteriol. 195, 2786–2792.
31. Guyard-Nicodème, M., Rivoal, K., Houard, E., Rose, V., Quesne, S., Mourand, G., Rouxel, S., Kempf, I., Guillier, L., Gauschard, F., Chemaly, M., 2015. Prevalence and characterization of Campylobacter jejuni from chicken meat sold in French retail outlets. Int. J. Food Microbiol. 203, 8–14.
32. Haddad, N., Maillart, G., Garénaux, A., Jugiau, F., Federighi, M., Cappelier, J.M., 2010. Adhesion ability of Campylobacter jejuni to HT-29 cells increases with the augmentation of oxidant agent concentration. Curr. Microbiol. 61, 500–505.
33. Hingston, P., Chen, J., Dhillon, B.K., Laing, C., Bertelli, C., Gannon, V., Tasara, T., Allen, K., Brinkman, F.S.L., Hansen, T.L., Wang, S., 2017. Genotypes associated with Listeria monocytogenes isolates displaying impaired or enhanced tolerances to cold, salt, acid, or desiccation stress. Front. Microbiol. 8, 369. http://dx.doi.org/10.3389/fmicb. 2017.00369.
34. Hoffmann, M., Luo, Y., Monday, S.R., Gonzalez-Escalona, N.,

Ottesen, A.R., Muruvanda, T., Senturk, I.F., 2016. Tracing origins of the Salmonella Bareilly strain causing a foodborne outbreak in the United States. J. Infect. Dis. 213, 502–508.

35. Holmes, A., Allison, L., Ward, M., Dallman, T.J., Clark, R., Fawkes, A., Murphy, L., Hanson, M., 2015. Utility of whole-genome sequencing of Escherichia coli O157 for outbreak detection and epidemiological surveillance. J. Clin. Microbiol. 53, 3565–3573.

36. Inns, T., Lane, C., Peters, T., Dallman, T., Chatt, C., McFarland, N., Crook, P., Bishop, T., Edge, J., Hawker, J., Elson, R., Neal, K., Adak, G.K., Cleary, P., Outbreak control team, 2015. A multi-country Salmonella Enteritidis phage type 14b outbreak associated with eggs from a German producer: 'near real-time' application of whole genome sequencing and food chain investigations, United Kingdom, May to September 2014. Euro Surveill. 20, 21098.

37. Jackson, B.R., Tarr, C., Strain, E., Jackson, K.A., Conrad, A., Carleton, H., Katz, L.S., Stroika, S., Gould, L.H., Mody, R.K., Silk, B.J., Beal, J., Chen, Y., Timme, R., Doyle, M., Fields, A., Wise, M., Tillman, G., Defibaugh-Chavez, S., Kucerova, Z., Sabol, A., Roache, K., Trees, E., Simmons, M., Wasilenko, J., Kubota, K., Pouseele, H., Klimke, W., Besser, J., Brown, E., Allard, M., Gerner-Smidt, P., 2016. Implementation of nationwide real-time whole genome sequencing to enhance listeriosis outbreak detection and investigation. Clin. Infect. Dis. 63, 380–386.

38. Klemm, E., Dougan, G., 2016. Advances in understanding bacterial pathogenesis gained from whole-genome sequencing and phylogenetics. Cell Host Microbe 19, 599–610.

39. Köser, C.U., Ellington, M.J., Peacock, S.J., 2016. Whole-genome sequencing to control antimicrobial resistance. Trends Genet. 30, 401–407.

40. Kwong, J.C., McCallum, N., Sintchenko, V., Howden, B.P., 2015. Whole genome sequencing in clinical and public health microbiology. Pathology 47, 199–210.

41. Kwong, J.C., Mercoulia, K., Tomita, T., Easton, M., Li, H.Y., Bulach, D.M., Stinear, T.P., Seemann, T., Howden, B.P., 2016. Prospective whole-genome sequencing enhances national surveillance of Listeria monocytogenes. J. Clin. Microbiol. 54, 333–342. http://dx.doi.org/10.1128/JCM.02344-15. (Epub 2015 Nov 25).

42. Leekitcharoenphon, P., Nielsen, E.M., Kaas, R.S., Lund, O., Aarestrup,

F.M., 2014. Evaluation of whole genome sequencing for outbreak detection of Salmonella enterica. PLoS One 9, e87991. http://dx.doi.org/10.1371/journal.pone.0087991.

43. Lomonaco, S., Nucera, D., Filipello, V., 2015. The evolution and epidemiology of Listeria monocytogenes in Europe and the United States. Infect. Genet. Evol. 35, 172–183. http://dx.doi.org/10.1016/j.meegid.2015.08.008.
44. Mardis, E.R., 2008. Next-generation DNA sequencing methods. Ann. Rev. Genom. Human Genet. 9, 387–402.
45. Maury, M.M., Tsai, Y.-H., Charlier, C., Touchon, M., Chenal-Francisque, V., Leclercq, A., Criscuolo, A., Gaultier, C., Roussel, S., Brisabois, A., Disson, O., Rocha, E.P.C., Brisse, S., Lecuit, M., 2016. Uncovering Listeria mocytogenes hypervirulence by harnessing biodiversity. Nat. Genet. 48, 308–313.
46. Miller, W.G., Huynh, S., Parker, C.T., Niedermeyer, J.A., Kathariou, S., 2016. Complete genome sequences of multidrug-resistant Campylobacter jejuni strain 14980A (turkey feces) and Campylobacter coli strain 14983A (housefly from a turkey farm), harboring a novel gentamicin resistance mobile element. Genome Announc. 4 (pii: e01175-16).
47. Moura, A., Criscuolo, A., Pouseele, H., Maury, M.M., Leclercq, A., Tarr, C., Björkman, J.T., Dallman, T., Reimer, A., Enouf, V., Larsonneur, E., Carleton, H., Bracq-Dieye, H., Katz, L.S., Jones, L., Touchon, M., Tourdjman, M., Walker, M., Stroika, S., Cantinelli, T., Chenal-Francisque, V., Kucerova, Z., Rocha, E.P.C., Nadon, C., Grant, K., Nielsen, E.M., Pot, B., Gerner-Smidt, P., Lecuit, M., Sylvain Brisse, S., 2016. Whole genomebased population biology and epidemiological surveillance of Listeria monocytogenes. Nat Microbiol 2, 16185. http://dx.doi.org/10.1038/nmicrobiol.2016.185.
48. Nadon, C.A., Trees, E., Ng, L.K., Møller Nielsen, E., Reimer, A., Maxwell, N., Kubota, K.A., Gerner-Smidt, P., the MLVA Harmonization Working Group, 2013. Development and application of MLVA methods as a tool for inter-laboratory surveillance. Euro Surveill. 18 (pii=20565).
49. NARMS available at: http://www.fda.gov/AnimalVeterinary/NewsEvents/CVMUpdates/ ucm498038.htm, Accessed date: 28 November 2016.

50. Pielaat, A., Boer, M.P., Wijnands, L.M., van Hoek, A.H.A.M., Bouw, E., Barker, G.C., Teunis, P.F.M., Aarts, H.J.M., Franz, E., 2015. First step in using molecular data for microbial food safety risk assessment; hazard identification for of Escherichia coli O157:H7 by coupling genomic data with in vitro adherence to human epithelial cells. Int. J. Food Microbiol. 213, 130–138.
51. Potter, M., Archer, D., Benson, A., Busta, F., Dickson, J., Doyle, M., Farber, J., Finlay, B.B., Goldblatt, M., Hedberg, C., Hoober, D., Jahncke, M., Jaykus, L.-A., Kaspar, C., Liang, A.P., Lindsay, J., Pestka, J., Pierson, M., Slade, P., Tompkin, R.B., Tortorello, M.L., 2000. Emerging Microbiological Food Safety Issues: Implications for Control in the 21st Century. Institute of Food Technology, Chicago.
52. Qin, S., Wang, Y., Zhang, Q., Zhang, M., Deng, F., Shen, Z., Wu, C., Wang, S., Zhang, J., Shen, J., 2014. Report of ribosomal RNA methylase gene erm (B) in multidrug-resistant Campylobacter coli. J. Antimicrob. Chemother. 69, 964–968.
53. Reuter, S., Ellington, M.J., Cartwright, E.J.P., Köser, C.U., Török, M.E., Gouliouris, T., Harris, S.R., Brown, N.M., Holden, M.T.G., Quail, M., Parkhill, J., Smith, G.P., Bentley, S.D., Peacock, S.J., 2014. Rapid bacterial whole-genome sequencing to enhance diagnostic and public health microbiology. JAMA Intern. Med. 173, 1397–1404.
54. Rivoal, K., Ragimbeau, C., Salvat, G., Colin, P., Ermel, G., 2005. Genomic diversity of Campylobacter coli and Campylobacter jejuni isolates recovered from free-range broiler farms and comparison with isolates of various origins. Appl. Environ. Microbiol. 71, 6216–6227.
55. Rodrigues, R.C., Pocheron, A.L., Hernould, M., Haddad, N., Tresse, O., Cappelier, J.M., 2015. Description of Campylobacter jejuni Bf, an atypical aero-tolerant strain. Gut. Pathog. 7, 1.
56. Ross, I.L., Davos, D.E., Mwanri, L., Raupach, J., Heuzenroeder, M.W., 2011. MLVA and phage typing as complementary tools in the epidemiological investigation of Salmonella enterica serovar Typhimurium clusters. Curr. Microbiol. 62, 1034–1038.
57. Sheppard, S.K., Dallas, J.F., Strachan, N.J., MacRae, M., McCarthy, N.D., Wilson, D.J., Forbes, K.J., 2009. Campylobacter genotyping to determine the source of human infection. Clin. Infect. Dis. 48, 1072–1078.
58. Ståhl, P.L., Lundeberg, J., 2012. Towards the single-hour high-quality genome. Annu. Rev. Biochem. 81, 359–378.

59. Stasiewicz, M.J., Oliver, H.F., Wiedmann, M., den Bakker, H.C., 2015. Whole-genome sequencing allows for improved identification of persistent Listeria monocytogenes in food-associated environments. Appl. Environ. Microbiol. 81, 6024–6037.

60. Taboada, E.N., MacKinnon, J.M., Luebbert, C.C., Gannon, V.P., Nash, J.H., Rahn, K., 2008. Comparative genomic assessment of multi-locus sequence typing: rapid accumulation of genomic heterogeneity among clonal isolates of Campylobacter jejuni. BMC Evol. Biol. 8, 229.

61. The, H.C., Thanh, D.P., Holt, K.E., Thomson, N.R., Baker, S., 2016. The genomic signatures of Shigella evolution, adaptation and geographical spread. Nat. Rev. 14, 235–250.

62. US EPA and USDA/FSIS (Environmental Protection Agency and Department of Agriculture - Food Safety and Inspection Service), 2012US EPA and USDA/FSIS (Environmental Protection Agency and D. Microbial risk assessment guideline: pathogenic microorganisms with focus on food and water. Available at: http://www2.

63. epa.gov/osa/microbial-risk-assessment-guideline-pathogenic-microorganisms-focusfood-and-water, Accessed date: 1 November 2016.

64. van Walle, I., 2013. ECDC starts pilot phase for collection of molecular typing data. Euro Surveill. 18 (pii 20357).

65. World Health Organisation (WHO), 2008. Foodborne disease outbreaks: guidelines for investigation and control. http://www.who.int/foodsafety/publications/foodborne_ disease/outbreak_guidelines.pdf.

66. S. Zhao, G.H. Tyson, Y. Chen, C. Li, S. Murkerjee, S. Young, C. Lam, J.P. Folster, J.M.Whichard, P.F. McDermottWhole-genome sequencing analysis accurately predicts antimicrobial resistance phenotypes in Campylobacter spp. Appl. Environ. Microbiol., 82 (2015), pp. 459-466

CHAPTER 8

DETECTION OF FOODBORNE PATHOGENS BY QPCR: A PRACTICAL APPROACH FOR FOOD INDUSTRY APPLICATIONS

María-José Chapela[1], Alejandro Garrido-Maestu[2,3] and Ana G. Cabado[3]

[1]New Preservation Technologies Division, ANFACO-CECOPESCA, Campus Univ. 16, 36310 Vigo, Pontevedra,
Spain.
[2]Microbiology and Bioassays Laboratory, ANFACO-CECOPESCA, Campus Univ. 16, 36310 Vigo, Pontevedra, Spain.
[3]Food Safety Division, ANFACO-CECOPESCA, Campus Univ. 16, 36310 Vigo, Pontevedra, Spain.

CITATION: Chapela, M.-J., A. Garrido-Maestu and A. G. Cabado, "Detection of foodborne pathogens by qPCR: A practical approach for food industry applications", Cogent Food & Agriculture (2015), http://dx.doi.org/10.1080/23311932.2015.1013771.

ABSTRACT

Microbiological analysis of food is an integrated part of microbial safety management in the food chain. Monitoring and controlling foodborne pathogens are traditionally carried out by conventional microbiological methods based on culture-dependent approaches in control laboratories and private companies. However, polymerase chain reaction (PCR) has revolutionized microbiological analysis allowing detection of pathogenic microorganisms in food, without the necessity of classical isolation and identification. However, at present, PCR and quantitative polymerase chain reaction (qPCR) are essential analytical tools for researchers working in the field of foodborne pathogens. This manuscript reviews recently described qPCR methods applied for foodborne bacteria detection, serving as economical, safe, and reliable alternatives for application in the food industry and control laboratories. Multiplex qPCR, which allows the simultaneous detection of more than one pathogen in one single reaction, saving considerable effort, time, and money, is emphasized in the article.

Keywords: PCR, qPCR, pathogen, detection, industry

Public Interest Statement

This manuscript is intended to provide readers, especially for those who develop their professional activity related in the food industry, a general idea about how molecular methods, particularly qPCR, may be applied for controlling foodborne pathogens. Two main fields are covered, on the one hand, a more scientifically oriented overview of PCR/qPCR is included, summarizing from the technique fundamentals to primer/probe design and advanced multiplex applications. On the other hand, a detailed description of available methodologies is provided for the different bacterial pathogens covered, intended to be helpful for people looking for direct industry applications.

INTRODUCTION

In order to ensure food safety, it is necessary to consider all aspects of the food production chain. The free movement of safe and wholesome food is an essential aspect of the internal market and contributes significantly to the health and well-being of citizens, and to their social and economic interests. A high level of protection of human life and health should be assured in the pursuit of community policies (EC, 2002(EC), C. R. (2002). Regulation (EC) No. 178/2002 of the European Parliament and of the Council of 28 January2002. Laying down the general principles and requirements of food law, establishing the European food safety authorityand laying down procedures in matters of food safety. *Official Journal of the European Communities, Official Journal of the European Union, 178/2002.*).

During the year 2012, the number of food alerts and notifications received through the European alert information system or Rapid Alert system for Food and Feed (http://www.eftasurv.int) has risen to a total of 2,801 for food products. This number includes different contaminants, of which 462 (16%) have been alerts, 869 (31.5%) have been information notifications, and 1476 (52.5%) were rejections in border. Microbiological contamination corresponds to 21% of notifications, that *Salmonella* spp. has the highest risk and causes 49% of the microbiological contamination (Agroalimentaria, 2012Agroalimentaria, E. F. V. p. l. S. (2012). *Informe RASFF.*). All these notifications and risks have led to extensive analyses of foodstuffs before their release and distribution to the market.

Networking of laboratories of excellence, at regional and/or interregional level, with the aim of ensuring continuous monitoring of food safety, could play an important role in the prevention of potential health risks for consumers as well as of unnecessary economic costs. In this sense, the availability of quick and quality assays allows giving precise results in a few hours and results as a great economic and safe advantage. Among microbiological risks that can lead to foodborne diseases, *Salmonella*, *Listeria monocytogenes,* and *Escherichia coli* are pathogenic micro-organisms that keep rising in numbers through recent years (Commission, 2011Commission, E. (2011). *RASFF-Annual Report.*). According to several ISOs: ISO 6579, ISO 11290-1 and -2, and ISO 16654 for *Salmonella*, *L. monocytogenes,* and *E. coli,* respectively (ISO, 1996ISO. (1996). *Microbiology of food and animal feeding stuffs—Horizontal method for the detection and enumeration of Listeria monocytogenes-Part 1: Detection method* (Vol. 11290-1:1996). , 1998ISO. (1998). *Microbiology of food and animal feeding stuffs—Horizontal method for the detection and enumeration of Listeria monocytogenes-Part 2: Enumeration method*(Vol. ISO 11290-2:1998). , 2001ISO. (2001). *Microbiology of food and animal feeding stuffs—Horizontal method for the detection of Escherichia coli O157* (Vol. 16654:2001). , 2003ISO. (2003). *Microbiology of food and animal feeding stuffs—Horizontal method for the detection of Salmonella spp.* (Vol. 6579:2002).) complete analyses for these pathogens can take from two to seven days since the sample arrives to the laboratory. This period can be longer if the pathogenic serotype have to be determined, what means that foodstuffs have to be immobilized and waiting for the release until obtaining the final results. It is worth to mention that if the food is fresh, it can be completely spoiled and even if is a ready-to-eat product, it can loss the quality and not be allowed for consumption. In the case of products with a short shelf life, if the immobilization is very long, the foodstuffs can reach the expiry date or be close to that with all the consequent commercial problems that can prevent to place these products in the market and all derived economic losses included those that have to be paid due to the destruction of food.

All these facts highlight the necessity of the development of fast, feasible, and accurate analytical methods in order to allow a product to be released immediately or, at least, in a very short period of time, to prevent or decrease no necessary costs and food spoilage. Figure 1 shows a schematic comparison of traditional culture methods and quantitative polymerase chain reaction (qPCR) detection methods.

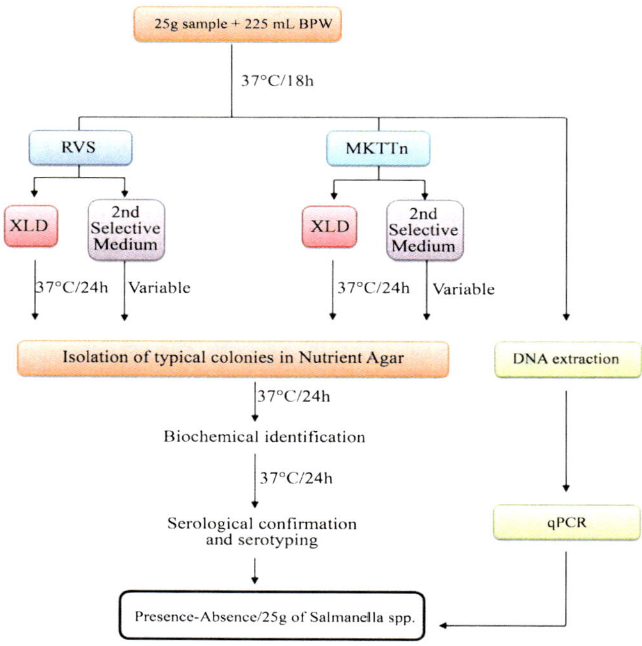

Figure 1. Schematic comparison of traditional culture methods and qPCR detection methods.

Notes: BPW: Buffered Peptone Water; RVS: Rappaport-Vassiliadis Soya; XLD: Xylose Lysine Deoxycholate agar; MKTTn: Muller–Kauffmann Tetrathionate-Novobiocin Broth.

IMPACT OF MULTIPLEX REAL-TIME POLYMERASE CHAIN REACTION IN ROUTINE FOOD CONTROL

The food microbiological testing is experiencing an exciting momentum due to the development in the analytical field of new and accurate approaches. Polymerase chain reaction (PCR) has revolutionized traditional microbiological analysis by allowing detection of pathogenic microorganisms directly in the food, without the necessity of classical isolation and identification. Since its discovery it has became an essential analytical tool for researchers working in the foodborne pathogens field (Wernars, Delfgou, Soentoro, & Notermans, 1991Wernars, K., Delfgou, E., Soentoro, P. S., & Notermans, S. (1991). Successful approach for detection of low numbers of enterotoxigenic *Escherichia coli* in minced meat by using the polymerase chain-reaction. *Applied and Environmental Microbiology, 57*, 1914–1919., , ; Wernars, Heuvelman, Chakraborty, & Notermans, 1991Wernars, K., Heuvelman, C. J., Chakraborty, T., & Notermans, S. H. W. (1991). Use of the polymerase chain reaction for direct detection of *Listeria monocytogenes* in soft cheese. *Journal of Applied Bacteriology, 70*, 121–126.10.1111/jam.1991.70.issue-2, ,). This technique was used in more than 300,000 scientific publications, and was applied in different areas of research, including food safety, due to its versatility, specificity, and sensitivity. In recent years, PCR has been successfully applied in food science for microorganisms identification (Rodriguez-Lazaro et al., 2007Rodriguez-Lazaro, D., Lombard, B., Smith, H., Rzezutka, A., D'Agostino, M., Helmuth, R., ... Cook, N. (2007). Trends in analytical methodology in food safety and quality: Monitoring microorganisms and genetically modified organisms. *Trends in Food Science & Technology, 18*, 306–319., ,), for detection of ingredients in food products (Lopez-Calleja et al., 2013López-Calleja, I., de la Cruz, S., Pegels, N., González, I., García, T., & Martín, R. (2013). Development of a real time PCR assay for detection of allergenic trace amounts of peanut (*Arachis hypogaea*) in processed foods. *Food Control, 30*, 480–490.10.1016/j.foodcont.2012.09.017, ,) and different meats and fish species identification (Chapela et al., 2002Chapela, M. J., Sotelo, C. G., Calo-Mata, P., Perez-Martin, R. I., Rehbein, H.,

Hold, G. L., … Santos, A. T. (2002). Identification of cephalopod species (Ommastrephidae and Loliginidae) in seafood products by forensically informative nucleotide sequencing (FINS). *Journal of Food Science, 67*, 1672–1676.10.1111/jfds.2002.67.issue-5, , ; Mackie et al., 1999Mackie, I. M., Pryde, S. E., Gonzales-Sotelo, C., Medina, I., Perez-Martin, R., Quinteiro, J., … Rehbein, R. (1999). Challenges in the identification of species of canned fish. *Trends in Food Science & Technology, 10*, 9–14., , ; Quinteiro et al., 1998Quinteiro, J., Sotelo, C. G., Rehbein, H., Pryde, S. E., Medina, I., Pérez-Martín, R. I., … Mackie, I. M.(1998). Use of mtDNA direct polymerase chain reaction (PCR) sequencing and PCR–restriction fragment length polymorphism methodologies in species identification of canned tuna. *Journal of Agricultural and Food Chemistry, 46*, 1662–1669.10.1021/jf970552+, , ; Rodríguez-Lázaro, Cook, & Hernández, 2013Rodríguez-Lázaro, D., Cook, N., & Hernández, M. (2013). Real-time PCR in food science: PCR diagnostics. *Current Issues in Molecular Biology, 15*, 39–44., , ; Soares, Amaral, Oliveira, & Mafra, 2013Soares, S., Amaral, J. S., Oliveira, M. B. P. P., & Mafra, I. (2013). A SYBR green real-time PCR assay to detect and quantify pork meat in processed poultry meat products. *Meat Science, 94*, 115–120.10.1016/j.meatsci.2012.12.012 ; Sotelo et al., 2001 Sotelo, C. G., Calo-Mata, P., Chapela, M. J., Pérez-Martín, R. I., Rehbein, H., Hold, G. L., … Santos, A. T. (2001). Identification of flatfish (*Pleuronectiforme*) species using DNA-based techniques. *Journal of Agricultural and Food Chemistry, 49*, 4562–4569.10.1021/jf010452a ; Zhang, 2013 Zhang, C. (2013). Semi-nested multiplex PCR enhanced method sensitivity of species detection in further-processed meats. *Food Control, 31*, 326–330.10.1016/j.foodcont.2012.11.002, ,).

PCR and Multiplex PCR

Conventional PCR can be described as an exponential amplification of a DNA fragment. Its principle is based on the mechanism of DNA replication *in vivo*, i.e. double-stranded DNA (dsDNA) is denatured to single-stranded DNA (ssDNA) and duplicated by a DNA polymerase (Higuchi, Fockler, Dollinger, & Watson, 1993Higuchi, R., Fockler, C., Dollinger, G., & Watson, R. (1993). Kinetic PCR analysis: Real-time monitoring of DNA amplification reactions. *Bio/Technology, 11*, 1026–1030.10.1038/nbt0993-1026). Usually, PCR is performed using one pair of primers to amplify a specific sequence; however, multiplex PCR allows the simultaneous amplification of more than one target sequence in a single reaction by using different sets of primers (Elnifro, Ashshi, Cooper, & Klapper, 2000Elnifro,

E. M., Ashshi, A. M., Cooper, R. J., & Klapper, P. E. (2000). Multiplex PCR: Optimization and application in diagnostic virology. *Clinical Microbiology Reviews*, 13, 559–570.10.1128/CMR.13.4.559-570.2000). This option saves considerable time and effort, and decreases the number of reactions that need to be performed to detect the desired targets in the sample. However, the presence of many PCR primers in a single tube can cause some problems, such as the increased formation of misprimed PCR products, known as "primer dimers," and the amplification of unspecific DNA fragments. Primers should be chosen with similar annealing temperatures for multiplex PCR assays. Also, the lengths of amplified products should be similar, since large differences in the lengths of the target DNAs may favor the amplification of shorter targets over the longer ones, which results in different yields of amplified products. In standard PCR protocols, amplification products are detected by gel electrophoresis using UV luminescence and the ethidium bromide double-stranded DNA intercalating dye or similar non-toxic compounds. These procedures have the big disadvantage of relying on end-point analysis; therefore, amplification products cannot give quantitative information regarding the initial amount of target molecules (Rodríguez-Lázaro & Hernández, 2013Rodríguez-Lázaro, D., & Hernández, M. (2013). Real-time PCR in food science: Introduction. *Current Issues in Molecular Biolog, 15*, 25–38., ,).

qPCR and Multiplex qPCR

Higuchi and collaborators (Higuchi, Dollinger, Walsh, & Griffith, 1992Higuchi, R., Dollinger, G., Walsh, P. S., & Griffith, R. (1992). Simultaneous amplification and detection of specific DNA-sequences. *Bio/Technology, 10*, 413–417.10.1038/nbt0492-413 ; Higuchi et al., 1993Higuchi, R., Fockler, C., Dollinger, G., & Watson, R. (1993). Kinetic PCR analysis: Real-time monitoring of DNA amplification reactions. *Bio/Technology, 11*, 1026–1030.10.1038/nbt0993-1026) constructed a system that detects PCR products as soon as they accumulate. This was the beginning of the real time or qPCR, also known as the second generation of PCR. qPCR allows the monitoring of the amplification process in real time by using fluorescence, and not only at the end of the reaction as occurs in conventional PCR. Its major advantages are that it is a fast and easy-to-perform analysis, the extremely wide dynamic range of quantification (more than eight orders of magnitude), and the significantly higher reliability and sensitivity of the results compared to conventional PCR. Since the first scientific work published in 1996 (Heid, Stevens, Livak, & Williams,

1996Heid, C. A., Stevens, J., Livak, K. J., & Williams, P. M. (1996). Real time quantitative PCR. *Genome Research, 6*, 986–994.10.1101/gr.6.10.986), the number of publications where qPCR is used has increased nearly exponentially. Specific detection and quantification of pathogens in food by qPCR has been evaluated for a wide variety of micro-organisms, with emphasis on the main foodborne pathogens responsible for important medical and economic outbreaks: *Salmonella* spp., *L. monocytogenes*, *E. coli* O157:H7, and *Staphylococcus aureus*, among others (Rodríguez-Lázaro et al., 2013Rodríguez-Lázaro, D., Cook, N., & Hernández, M. (2013). Real-time PCR in food science: PCR diagnostics. *Current Issues in Molecular Biology, 15*, 39–44., ,).

The major advantage of this molecular method with respect to standard methods is the shorter time required to obtain reliable results. Typical culture methods as ISO or FDA Bacteriological Analytical Manual—for *Shigella* spp., *E. coli* O157, and different human pathogenic *Vibrio* spp. (*V. cholerae*, *V. parahemolyticus,*and *V. vulnificus*) detection—need two days to guarantee a negative result. Regarding *Salmonella* spp. and *L. monocytogenes,* three and five days are needed for a definitive negative result. In contrast, when a positive result is found, the duration of the analyses is greatly extended as additional isolation and identification steps must be performed. However, recent studies have demonstrated that times can be significantly shortened to only 24 h by optimizing qPCR methods for most of these bacteria (Chapela et al., 2010Chapela, M. J., Fajardo, P., Garrido, A., Cabado, A. G., Ferreira, M., Lago, J., & Vieites, J. M. (2010). Comparison between a TaqMan polymerase chain reaction assay and a culture method for ctx-positive *Vibrio cholerae* detection. *Journal of Agricultural and Food Chemistry, 58*, 4051–4055.10.1021/jf903658k ; Garrido, Chapela, Ferreira, et al., 2012Garrido, A., Chapela, M. J., Ferreira, M., Atanassova, M., Fajardo, P., Lago, J., ... Cabado, A. G. (2012). Development of a multiplex real-time PCR method for pathogenic *Vibrio parahaemolyticus* detection (*tdh*+ and *trh*+). *Food Control, 24*, 128–135.10.1016/j.foodcont.2011.09.015, , ; Garrido, Chapela, Román, et al., 2012Garrido, A., Chapela, M. J., Román, B., Ferreira, M., Lago, J., Vieites, J. M., & Cabado, A. G. (2012). Development of a multiplex real-time PCR method for simultaneous detection of *Salmonella enterica*, *Shigella flexneri* and *Listeria monocytogenes* in processed food samples. *European Food Research and Technology, 234*, 571–580.10.1007/s00217-012-1665-3, , ; Garrido, Chapela, Román, Fajardo, Lago, et al., 2013Garrido, A., Chapela, M.-J., Román, B., Fajardo, P., Lago, J., Vieites, J. M., & Cabado, A. G. (2013). A new multiplex real-time PCR developed method for *Salmonella* spp. and *Listeria monocytogenes* detection in food and environmental samples. *Food Control, 30*, 76–85.10.1016/j.foodcont.2012.06.029, ,).

Digital PCR

Digital PCR (dPCR) allows the precise quantification of nucleic acids, facilitating the measurement of small percentage differences and quantification of rare variants. It was reported to be more reproducible and less susceptible to inhibition than qPCR. Its main applications are: rare variant measurement, molecular counting, and applications that need higher precision. Due to its growing interest, a "Minimum Information for Publication of Quantitative dPCR Experiments" guide has been recently published (Huggett et al., 2013Huggett, J. F., Foy, C. A., Benes, V., Emslie, K., Garson, J. A., Haynes, R., … Bustin, S. A. (2013). The digital MIQE guidelines: Minimum information for publication of quantitative digital PCR experiments. *Clinical Chemistry, 59*, 892–902.10.1373/clinchem.2013.206375).

Limit of Detection and Limit of Quantification of qPCR

Concerning the limit of detection (LOD) of qPCR, it was evaluated in several multiplex studies targeting different types of bacteria (including Gram positive and Gram negative) from different origins. Obtained results were similar to traditional culture methods or even lower, with values in general below 10 CFU/25 g, with the added advantage that, with multiplex qPCR, several targets were detected simultaneously (Garrido, Chapela, Ferreira, et al., 2012Garrido, A., Chapela, M. J., Ferreira, M., Atanassova, M., Fajardo, P., Lago, J., … Cabado, A. G. (2012). Development of a multiplex real-time PCR method for pathogenic *Vibrio parahaemolyticus* detection (*tdh*+ and *trh*+). *Food Control, 24*, 128–135.10.1016/j.foodcont.2011.09.015, , ; Garrido, Chapela, Román, et al., 2012Garrido, A., Chapela, M. J., Román, B., Ferreira, M., Lago, J., Vieites, J. M., & Cabado, A. G. (2012). Development of a multiplex real-time PCR method for simultaneous detection of *Salmonella enterica, Shigella flexneri* and *Listeria monocytogenes* in processed food samples. *European Food Research and Technology, 234*, 571–580.10.1007/s00217-012-1665-3, , ; Garrido, Chapela, Román, Fajardo, Lago, et al., 2013Garrido, A., Chapela, M.-J., Román, B., Fajardo, P., Lago, J., Vieites, J. M., & Cabado, A. G. (2013). A new multiplex real-time PCR developed method for *Salmonella* spp. and *Listeria monocytogenes* detection in food and environmental samples. *Food Control, 30*, 76–85.10.1016/j.foodcont.2012.06.029, ,). Low levels of target pathogens can also be detected in food matrices contaminated by other dominant micro-organisms. Garrido and collaborators demonstrated in two independent studies that *L. monocytogenes* could be successfully enriched and detected in the presence of high initial numbers of other interfering bacteria, generally above 10^6 CFU/25 g (Garrido, Chapela, Román, Fajardo, Lago, et al., 2013Garrido, A., Chapela, M.-J., Román, B., Fajardo, P., Lago, J., Vieites, J. M., & Cabado, A. G. (2013). A new multiplex real-time PCR developed method for *Salmonella* spp. and *Listeria monocytogenes* detection in food and environmental samples. *Food Control, 30*, 76–85.10.1016/j.foodcont.2012.06.029, ,).

However, in a great number of cases, the detection and quantification limits obtained without enrichment of the food samples prior to performing qPCR were in the range of 102–103 CFU/g (or ml) (Hierro, Esteve-Zarzoso, Gonzalez, Mas, & Guillamon, 2006Hierro, N., Esteve-Zarzoso, B., Gonzalez, A., Mas, A., & Guillamon, J. M. (2006). Real-time quantitative PCR (QPCR) and reverse transcription-QPCR for detection and enumeration of total yeasts in wine. *Applied and Environmental Microbiology, 72*, 7148–7155.10.1128/AEM.00388-06 ; Takahashi et al., 2009Takahashi, H., Kimura, B., Tanaka, Y., Shinozaki, J., Suda, T., & Fujii, T. (2009). Real-time PCR and enrichment culture for sensitive detection and enumeration of *Escherichia coli*. *Journal of Microbiological Methods, 79*, 124–127.10.1016/j.mimet.2009.08.002). Therefore, in order to meet the microbiological criteria required by international legislations for foodstuffs, it is sometimes necessary to associate qPCR with an enrichment step of a few hours.

When qPCR was used, not only for detection, but also as a quantitative tool, linear quantifications were reported over a large range of at least five logs and very good correlations with plate counts were obtained. In some cases, however, discrepancies between microbiological counts and qPCR have been reported with higher bacterial counts with the molecular method (Martínez-Blanch, Sánchez, Garay, & Aznar, 2009Martínez-Blanch, J. F., Sánchez, G., Garay, E., & Aznar, R. (2009). Development of a real-time PCR assay for detection and quantification of enterotoxigenic members of *Bacillus cereus*group in food samples. *International Journal of Food Microbiology, 135*, 15–21.10.1016/j.ijfoodmicro.2009.07.013). Several reasons can explain the differences: (1) the presence of intact DNA from dead cells, (2) the presence of viable but non-culturable forms, which can be quantified by qPCR but not by plate counts, (3) the fact that one CFU on plate might be generated from more than one cell, and (4) the use of PCR primers targeting varying numbers of multicopy genes (e.g. 16S rRNA). An important distinction should be made between LOD and quantification limits (LOQ). This distinction may lead to confusing comparisons between sensitivities of different methods. The LOD is the lowest population of micro-organisms that can be detected by the method. The LOQ is the minimal population that can be accurately quantified. Usually, LOD is lower than LOQ, meaning that when qPCR gives a positive signal, the amount of template is too low and provides a quantification cycle (Cq) that falls out of the linear range of quantification curves. When quantifying pathogens in focd, the lower LOQ in the food matrix should be considered and not the lower LOQ obtained from pure cultures because it takes into account the efficiency of nucleic acid extraction and possible interactions of food components with PCR amplification. In agreement with EN ISO 22174:2005, a standard for PCR application in order to detect foodborne pathogens, and with experts in PCR/qPCR (Hoorfar et al., 2003Hoorfar, J., Cook, N., Malorny, B., Wagner, M., De Medici, D., Abdulmawjood, A., & Fach, P.(2003). Making internal amplification control mandatory for diagnostic PCR. *Journal of Clinical Microbiology, 41*, 5835–5835.10.1128/JCM.41.12.5835.2003 ; Hoorfar, Cook, et al., 2004 Hoorfar, J., Cook, N., Malorny, B., Wagner, M., De Medici, D., Abdulmawjood, A., & Fach, P.(2004). Letter to the editor. *Journal of Applied Microbiology, 96*, 221–222.10.1046/j.1365-2672.2003.02188.x ; Hoorfar, Malorny, et al., 2004Hoorfar, J., Malorny, B., Abdulmawjood, A., Cook, N., Wagner, M., & Fach, P. (2004). Practical considerations in design of internal amplification controls for diagnostic PCR assays. *Journal of Clinical Microbiology, 42*, 1863–1868.10.1128/JCM.42.5.1863-1868.2004 ; ISO, 2005ISO. (2005). *Microbiology of food and animal feeding stuffs—Polymerase chain reaction (PCR) for the detection of food-borne pathogens* (General requirements and definitions). *22174*, 19.), an increasing number of studies have included an internal amplification control to qPCR protocols, to avoid negative deviations (Calvó, Martínez-Planells, Pardos-Bosch, & Garcia-Gil, 2008Calvó, L., Martínez-Planells, A., Pardos-Bosch, J., & Garcia-Gil, L. J. (2008). A new real-time PCR assay for the specific detection of *Salmonella*spp. targeting the *bipA* gene. *Food Analytical Methods, 1*, 236–242.10.1007/s12161-007-9008-x, , ; Garrido, Chapela, Ferreira, et al., 2012Garrido, A., Chapela, M. J., Ferreira, M., Atanassova, M., Fajardo, P., Lago, J., … Cabado, A. G. (2012). Development of a multiplex real-time PCR method for pathogenic *Vibrio parahaemolyticus* detection (*tdh*+ and *trh*+). *Food Control, 24*, 128–

135.10.1016/j.foodcont.2011.09.015, , ; Garrido, Chapela, Román, et al., 2012Garrido, A., Chapela, M. J., Román, B., Ferreira, M., Lago, J., Vieites, J. M., & Cabado, A. G. (2012). Development of a multiplex real-time PCR method for simultaneous detection of *Salmonella enterica*, *Shigella flexneri* and *Listeria monocytogenes* in processed food samples. *European Food Research and Technology, 234*, 571–580.10.1007/s00217-012-1665-3, , ; Garrido, Chapela, Román, Fajardo, Lago, et al., 2013Garrido, A., Chapela, M.-J., Román, B., Fajardo, P., Lago, J., Vieites, J. M., & Cabado, A. G. (2013). A new multiplex real-time PCR developed method for *Salmonella* spp. and *Listeria monocytogenes* detection in food and environmental samples. *Food Control, 30*, 76–85.10.1016/j.foodcont.2012.06.029, , ; Jofré et al., 2005Jofré, A., Martin, B., Garriga, M., Hugas, M., Pla, M., Rodríguez-Lázaro, D., & Aymerich, T. (2005). Simultaneous detection of *Listeria monocytogenes* and *Salmonella* by multiplex PCR in cooked ham. *Food Microbiology, 22*, 109–115.10.1016/j.fm.2004.04.009, , ; Nordstrom, Vickery, Blackstone, Murray, & DePaola, 2007Nordstrom, J. L., Vickery, M. C. L., Blackstone, G. M., Murray, S. L., & DePaola, A. (2007). Development of a multiplex real-time PCR assay with an internal amplification control for the detection of total and pathogenic V*ibrio parahaemolyticus* bacteria in oysters. *Applied and Environmental Microbiology, 73*, 5840–5847.10.1128/AEM.00460-07).

Nowadays, the current trend is moving towards identification of several pathogens in the same reaction tube, by applying multiplex amplification. Some qPCR technologies are particularly adapted to multiplexing. For instance, using TaqMan chemistry, several sequence-specific probes can be labeled with different fluorophores and many targets can be co-amplified and quantified within a single reaction. By contrast, in theory, the widely used SYBR Green chemistry does not allow multiplexing reactions because binding takes place non-specifically in the presence of DNA. However, some authors have circumvented the problem and successfully proposed multiplex qPCR protocols with SYBR Green by performing subsequent melting curve analysis. Distinct melt curves were obtained for each target micro-organism (Wehrle, Didier, Moravek, Dietrich, & Märtlbauer, 2010Wehrle, E., Didier, A., Moravek, M., Dietrich, R., & Märtlbauer, E. (2010). Detection of *Bacillus cereus* with enteropathogenic potential by multiplex real-time PCR based on SYBR green I. *Molecular and Cellular Probes, 24*, 124–130.10.1016/j.mcp.2009.11.004). Therefore, multiplex qPCR is an interesting tool to quickly detect different genera or species which are potentially present in the same food matrices.

qPCR has revolutionized the molecular approaches in food microbiology; specific micro-organisms can be quantified in food, and there is no need for post-amplification treatment of the samples, such as gel electrophoresis, saving time and efforts. Currently, a high number of qPCR chemistries are commercially available. These can be divided into those that are not sequence specific, such as DNA minor groove-binding dyes, and those that are sequence specific and that allow simultaneous detection and confirmation of the target amplicon during the PCR reaction (Rodríguez-Lázaro et al., 2013Rodríguez-Lázaro, D., Cook, N., & Hernández, M. (2013). Real-time PCR in food science: PCR diagnostics. *Current Issues in Molecular Biology, 15*, 39–44., ,).

qPCR Primers and Probes Design

When a qPCR assay is designed, the most important parameters are the amplicon length and the Tm of the primers and probe. The optimal amplicon length should be less than 150 bp; however, amplicons up to 300 bp amplify efficiently. Shorter amplicons amplify more efficiently than longer ones and are more tolerant to suboptimal reaction conditions. See Table 1 for details regarding TaqMan® probes assays that may be extensive to other type of qPCR methods.

Table 1. General guide for qPCR hydrolysis probes design

Primers length	20–24 bp
Primers Tm	60 ± 2°C
Amplicon size	140–300 bp
Probe length	20–24 bp
Probe Tm	Primer Tm + 10°C
Additional considerations	Avoid "G" in 5′ end, secondary and repetitive structures
	Probe placed no more than 10 bp from the forward primer and not less than 2 bp

IMPORTANCE OF SAMPLE PREPARATION METHODS

Enrichment broths

The combination of microbiological enrichment and molecular pathogen detection is a useful link between traditional and molecular microbiology. The advantages of both methods add up, and the disadvantages are partly eliminated. Then, the LOD of qPCR is far exceeded after a successful enrichment. Short enrichments, if performed in an appropriate way, are mostly sufficient to provide enough DNA for detection (Rossmanith & Wagner, 2010Rossmanith, P., & Wagner, M. (2010). The challenge to quantify *Listeria monocytogenes*-a model leading to new aspects in molecular biological food pathogen detection. *Journal of Applied Microbiology, 110*, 605–617.). In general, pathogenic bacteria are present in foods in very low numbers; thus, an enrichment step to increase their number is crucial in order to guarantee the achievement of legal limits (absence in 25 g for most bacterial pathogens). Extensive evaluation of optimal enrichment conditions, as well as incubation times and DNA extraction protocols has been approached for different pathogens (Gattuso et al., 2014Gattuso, A., Gianfranceschi, M. V., Sonnessa, M., Delibato, E., Marchesan, M., Hernandez, M., … Rodriguez-Lazaro, D. (2014). Optimization of a real time PCR based method for the detection of *Listeria monocytogenes* in pork meat. *International Journal of Food Microbiology,184*, 106–108. ; Rodríguez-Lázaro, Gonzalez-García, Gattuso, Gianfranceschi, & Hernandez, 2014Rodríguez-Lázaro, D., Gonzalez-García, P., Gattuso, A., Gianfranceschi, M. V., & Hernandez, M. (2014). Reducing time in the analysis of *Listeria monocytogenes* in meat, dairy and vegetable products. *International Journal of Food Microbiology, 184*, 98–105.).

When facing multipathogen detection, a popular approach is to select different broths for the enrichment of every pathogen; then DNA is extracted and multiplex qPCR is performed (Elizaquível, Gabaldón, & Aznar, 2011Elizaquível, P., Gabaldón, J. A., & Aznar, R.(2011). Quantification of *Salmonella* spp, *Listeria monocytogenes* and *Escherichia coli*O157 H7 in non-spiked food products and evaluation of real-time PCR as a diagnostic tool in routine food analysis. *Food Control, 22*, 158–164.10.1016/j.foodcont.2010.05.018, , ; Jofré et al., 2005Jofré, A., Martin, B., Garriga, M., Hugas, M., Pla, M., Rodríguez-Lázaro, D., & Aymerich, T. (2005). Simultaneous detection of *Listeria monocytogenes* and *Salmonella* by multiplex PCR in cooked ham. *Food Microbiology, 22*, 109–115.10.1016/j.fm.2004.04.009, , ; Ruiz-Rueda, Soler, & Calvo, 2010Ruiz-Rueda, O., Soler, M., & Calvo, L. (2010). Multiplex real-time PCR for the simultaneous detection of *Salmonella* spp. and *Listeria monocytogenes* in food samples. *Food Analytical Methods, 4*, 131–138., ,). An alternative is the use of a common broth for all the target pathogens, but the broth design and its performance must be carefully evaluated. In this sense, one of the first broths designed for several bacteria was the Universal Pre-enrichment Broth (UPB) for the simultaneous recovery of *Salmonella* spp. and *L. monocytogenes* (Bailey & Cox, 1992Bailey, J. S., & Cox, N. A. (1992). Universal preenrichment broth for the simultaneous detection of *Salmonella* and *Listeria* in foods. *Journal of Food Protection, 55*, 256–259.,). This broth, which is still in use (Bhagwat, 2004Bhagwat, A. A. (2004). Rapid detection of *Salmonella* from vegetable rinse-water using real-time PCR. *Food Microbiology, 21*, 73–78.10.1016/S0740-0020(03)00020-0, , ; Hammack, Amaguana, Johnson, & Andrews, 2003Hammack, T. S., Amaguana, R. M., Johnson, M. L., & Andrews, W. H. (2003). Effectiveness of universal pre-enrichment broth for recovery of *Salmonella* from selected dairy foods. *Journal of Aoac International, 86*, 714–718., ,), does not contain any inhibitory substance, allowing a better recovery of the bacteria due to its high buffer concentration. Previous studies have demonstrated that methods which incorporate a non-selective enrichment step gave better results than completely selective media, specially for the recovery of slow growing bacteria such as *L. monocytyogenes* (Rijpens & Herman, 2004Rijpens, N., & Herman, L. (2004). Comparison of selective and nonselective primary enrichments for the detection of *Listeria monocytogenes* in cheese. *International Journal of Food Microbiology, 94*, 15–22.10.1016/j.ijfoodmicro.2003.10.016). Following this concept, Kawasaki et al. developed a similar broth named N17 (Kawasaki et al., 2005Kawasaki, S., Horikoshi, N., Okada, Y., Takeshita, K., Sameshima, T., & Kawamoto, S. (2005). Multiplex PCR for simultaneous detection of *Salmonella spp.*, *Listeria monocytogenes*, and *Escherichia coli* O157:H7 in meat samples. *Journal of Food Protection, 68*, 551–556., ,) which has been applied in different studies for recovery of *Salmonella* spp., *L. monocytogenes,* and *E. coli* O157 (Kamisaki-Horikoshi et al., 2011Kamisaki-Horikoshi, N., Okada, Y., Takeshita, K., Sameshima, T., Kawasaki, S., Kawamoto, S., & Fratamico, P. M. (2011). Evaluation of TA10 broth for recovery of heat- and freeze-injured *Salmonella* from beef. *Journal of Aoac International, 94*, 857–862., , ; Kawasaki et al., 2009Kawasaki, S., Fratamico, P. M.,

Horikoshi, N., Okada, Y., Takeshita, K., Sameshima, T., & Kawamoto, S. (2009). Evaluation of a multiplex PCR system for simultaneous detection of *Salmonella* spp., *Listeria monocytogenes*, and *Escherichia coli* O157:H7 in foods and in food subjected to freezing. *Foodborne Pathogens and Disease, 6*, 81–89.10.1089/fpd.2008.0153 , 2010Kawasaki, S., Fratamico, P. M., Horikoshi, N., Okada, Y., Takeshita, K., Sameshima, T., & Kawamoto, S. (2010). Multiplex real-time polymerase chain reaction assay for simultaneous detection and quantification of *Salmonella* species, *Listeria monocytogenes*, and *Escherichia coli* O157:H7 in ground pork samples. *Foodborne Pathogens and Disease, 7*, 549–554.10.1089/fpd.2009.0465 , 2011Kawasaki, S., Fratamico, P. M., Kamisaki-Horikoshi, N., Okada, Y., Takeshita, K., Sameshima, T., & Kawamoto, S. (2011). Development of the multiplex PCR detection kit for *Salmonella* spp., *Listeria monocytogenes*, and *Escherichia coli* O157:H7. *Jarq-Japan Agricultural Research Quarterly, 45*, 77–81., , ; Mahmuda & Kawamoto, 2007Mahmuda, Y., & Kawamoto, K. S. (2007). Evaluation of multiplex PCR system for simultaneous detection of *Escherichia coli*O157:H7, *Listeria monocytogenes* and *Salmonella enteritidis* in shrimp samples. *Bangladesh Journal of Microbiology, 24*, 42–46.). The N17 broth was modified in later studies by independent laboratories for an optimal performance by eliminating dextrose (Omiccioli, Amagliani, Brandi, & Magnani, 2009Omiccioli, E., Amagliani, G., Brandi, G., & Magnani, M. (2009). A new platform for real-time PCR detection of *Salmonella* spp., *Listeria monocytogenes* and *Escherichia coli* O157 in milk. *Food Microbiology, 26*, 615–622.10.1016/j.fm.2009.04.008), and later on, the final pH of the broth was also modified by adjusting the buffering salts concentrations. This optimized broth was successfully applied for the simultaneous enrichment of *Salmonella* spp., *L. monocytogenes*, *Shigella* spp., and *E. coli* O157 (Garrido, Chapela, Román, et al., 2012Garrido, A., Chapela, M. J., Román, B., Ferreira, M., Lago, J., Vieites, J. M., & Cabado, A. G. (2012). Development of a multiplex real-time PCR method for simultaneous detection of *Salmonella enterica*, *Shigella flexneri* and *Listeria monocytogenes* in processed food samples. *European Food Research and Technology, 234*, 571–580.10.1007/s00217-012-1665-3, , ; Garrido, Chapela, Román, Fajardo, Lago, et al., 2013Garrido, A., Chapela, M.-J., Román, B., Fajardo, P., Lago, J., Vieites, J. M., & Cabado, A. G. (2013). A new multiplex real-time PCR developed method for *Salmonella* spp. and *Listeria monocytogenes* detection in food and environmental samples. *Food Control, 30*, 76–85.10.1016/j.foodcont.2012.06.029, ,). Excellent results were obtained with the modified version of N17 broth with all the bacteria tested, obtaining colony counts higher than 10^6 CFU/ml for all micro-organisms tested. Another non-selective enrichment broth recently developed is the simultaneous enrichment broth (SEB) developed by Kobayashi et al. SEB was applied for the simultaneous recovery of *Salmonella* spp., *L. monocytogenes*, *S. aureus*, *Bacillus cereus*, *Vibrio parahaemolyticus,* and *E. coli* O157:H7; and its performance was compared to tryptic soy broth (TSB) (Kobayashi et al., 2009Kobayashi, H., Kubota, J., Fujihara, K., Honjoh, K., Iio, M., Fujiki, N., … Miyamoto, T. (2009). Simultaneous enrichment of *Salmonella* spp, *Escherichia coli* O157:H7, *Vibrio parahaemolyticus*, *Staphylococcus aureus*, *Bacillus cereus*, and *Listeria monocytogenes* by single broth and screening of the pathogens by multiplex real-time PCR. *Food Science and Technology Research, 15*, 427–438.10.3136/fstr.15.427, ,). Even though satisfactory results were obtained when compared against TSB, further studies need to be performed with natural and spiked food and environmental samples.

A different approach was done by other researchers who developed selective enrichment broths for the simultaneous recovery of different bacteria, like the SEB SEL that included four antimicrobial agents—acriflavine, cycloheximide, fosfomycin, and nalidixic acid. SEL showed comparable results to UPB (Kim & Bhunia, 2008Kim, H. C., & Bhunia, A. K. (2008). SEL, a selective enrichment broth for simultaneous growth of *Salmonella* enterica, *Escherichia coli*O157:H7, and *Listeria monocytogenes*. *Applied and Environmental Microbiology, 74*, 4853–4866.10.1128/AEM.02756-07) and it was successfully applied for the recovery of *Salmonella*, *L. monocytogenes,* and *E. coli*O157 in meat products (Suo, He, Tu, & Shi, 2010Suo, B., He, Y. P., Tu, S. I., & Shi, X. M. (2010). A multiplex real-time polymerase chain reaction for simultaneous detection of *Salmonella spp.*, *Escherichia coli* O157, and *Listeria monocytogenes* in meat products. *Foodborne Pathogens and Disease, 7*, 619–628.10.1089/fpd.2009.0430).

DNA Extraction Protocols

Inhibition of the PCR amplification reaction is one of the most limiting factors of the PCR-based methods and it can cause complete reaction failure, leading to false negative results or reducing sensitivity of detection. Inhibition might be driven by interference with the cell lysis necessary for DNA extraction, by nucleic acid degradation or capture, and by inhibition of the polymerase activity necessary for amplification of the target DNA. Organic and phenolic compounds and polysaccharides are also possible inhibitors (Elizaquivel & Aznar, 2008Elizaquivel, P., & Aznar, R. (2008).

Comparison of four commercial DNA extraction kits for PCR detection of *Listeria monocytogenes*, *Salmonella*, *Escherichia coli* O157:H7, and *Staphylococcus aureus* in fresh, minimally processed vegetables. *Journal of Food Protection, 71*, 2110–2114., ,). In this sense, a critical factor when facing multipathogen detection is the selection of an appropriate DNA extraction protocol. In addition, the type of bacteria of interest must be kept in mind as it is more difficult to extract DNA from Gram positive bacteria, than from Gram negative. For Gram negative bacteria, a thermal cell lysis in the presence of a chelating resin—e.g. 6–10% (w/v) Chelex-100 suspension—is usually enough to obtain good-quality DNA (Garrido, Chapela, Román, et al., 2012Garrido, A., Chapela, M. J., Román, B., Ferreira, M., Lago, J., Vieites, J. M., & Cabado, A. G. (2012). Development of a multiplex real-time PCR method for simultaneous detection of *Salmonella enterica*, *Shigella flexneri* and *Listeria monocytogenes* in processed food samples. *European Food Research and Technology, 234*, 571–580.10.1007/s00217-012-1665-3, , ; Kim & Cho, 2010Kim, H.-J., & Cho, J.-C. (2010). Simple and rapid detection of *Listeria monocytogenes* in fruit juice by real-time PCR without enrichment culture. *Food Control, 21*, 1419–1423.10.1016/j.foodcont.2010.04.006, , ; Malorny, Huehn, Dieckmann, Kramer, & Helmuth, 2009Malorny, B., Huehn, S., Dieckmann, R., Krämer, N., & Helmuth, R. (2009). Polymerase chain reaction for the rapid detection and serovar identification of *Salmonella* in food and feeding stuff. *Food Analytical Methods, 2*, 81–95.10.1007/s12161-008-9057-9, , ; Rodriguez-Lazaro, Jofre, Aymerich, Hugas, & Pla, 2004Rodriguez-Lazaro, D., Jofre, A., Aymerich, T., Hugas, M., & Pla, M. (2004). Rapid quantitative detection of *Listeria monocytogenes* in meat products by real-time PCR. *Applied and Environmental Microbiology, 70*, 6299–6301.10.1128/AEM.70.10.6299-6301.2004). Reyes-Escogido et al. developed a chelex-based method that replaced the boiling water bath by a microwave heating procedure, as the thermal lysis source, and by adding proteinase K and RNase A obtained large amounts of highly pure DNA from Gram positive and negative bacteria (Reyes-Escogido et al., 2010Reyes-Escogido, L., Balam-Chi, M., Rodríguez-Buenfil, I., Valdés, J., Kameyama, L., & Martínez-Pérez, F. (2010). Purification of bacterial genomic DNA in less than 20 min using chelex-100 microwave: Examples from strains of lactic acid bacteria isolated from soil samples. *Antonie van Leeuwenhoek, 98*, 465–474.10.1007/s10482-010-9462-0); but this method has been scarcely used.

In-house developed protocols have shown appropriate results regarding the quantity and purity of DNA extracted from pure and mixed cultures, as well as from natural food and environmental samples. Kawasaki et al. developed and compared a DNA extraction protocol based on the use of a strong chaotropic agent (guanidine isothiocyanate) combined with a surfactant agent (Tween 20) and an enzymatic lysis step (combination of lysozyme and achromopeptidase enhanced the enzymatic extraction) (Kawasaki et al., 2005Kawasaki, S., Horikoshi, N., Okada, Y., Takeshita, K., Sameshima, T., & Kawamoto, S. (2005). Multiplex PCR for simultaneous detection of *Salmonella spp.*, *Listeria monocytogenes*, and *Escherichia coli* O157:H7 in meat samples. *Journal of Food Protection, 68*, 551–556., ,). This protocol was successfully applied in different studies and compared against other in-house protocols, showing similar or better results, specially focused on Gram positive bacteria (Garrido, Chapela, Román, Fajardo, Lago, et al., 2013Garrido, A., Chapela, M.-J., Román, B., Fajardo, P., Lago, J., Vieites, J. M., & Cabado, A. G. (2013). A new multiplex real-time PCR developed method for *Salmonella* spp. and *Listeria monocytogenes* detection in food and environmental samples. *Food Control, 30*, 76–85.10.1016/j.foodcont.2012.06.029, , ; Garrido, Chapela, Román, Fajardo, Vieites, et al., 2013Garrido, A., Chapela, M.-J., Román, B., Fajardo, P., Vieites, J. M., & Cabado, A. G. (2013). In-house validation of a multiplex real-time PCR method for simultaneous detection of *Salmonella* spp., *Escherichia coli* O157 and *Listeria monocytogenes*. *International Journal of Food Microbiology, 164*, 92–98.10.1016/j.ijfoodmicro.2013.03.024 ; Germini, Masola, Carnevali, & Marchelli, 2009Germini, A., Masola, A., Carnevali, P., & Marchelli, R. (2009). Simultaneous detection of *Escherichia coli* O175:H7, *Salmonella* spp., and *Listeria monocytogenes* by multiplex PCR. *Food Control, 20*, 733–738.10.1016/j.foodcont.2008.09.010, , ; Mahmuda & Kawamoto, 2007Mahmuda, Y., & Kawamoto, K. S. (2007). Evaluation of multiplex PCR system for simultaneous detection of *Escherichia coli*O157:H7, *Listeria monocytogenes* and *Salmonella enteritidis* in shrimp samples. *Bangladesh Journal of Microbiology, 24*, 42–46.). Later studies reported that DNA extraction may be enhanced by the addition of a filtration step through a 11–5 μm sterile micropore filter (Kawasaki et al., 2009Kawasaki, S., Fratamico, P. M., Horikoshi, N., Okada, Y., Takeshita, K., Sameshima, T., & Kawamoto, S. (2009). Evaluation of a multiplex PCR system for simultaneous detection of *Salmonella* spp., *Listeria monocytogenes*, and *Escherichia coli* O157:H7 in foods and in food subjected to freezing. *Foodborne Pathogens and Disease, 6*, 81–89.10.1089/fpd.2008.0153 ; Rodriguez-Lazaro, Jofre, et al., 2004Rodriguez-Lazaro, D., Jofre, A., Aymerich, T., Hugas, M., & Pla, M. (2004). Rapid quantitative detection of *Listeria monocytogenes* in meat products by real-time PCR. *Applied and Environmental Microbiology, 70*, 6299–6301.10.1128/AEM.70.10.6299-6301.2004), being this in agreement with other studies reporting that the presence of food particles should be avoided (Malorny et al., 2009Malorny, B., Huehn, S., Dieckmann, R., Krämer, N., & Helmuth, R. (2009). Polymerase chain reaction for the rapid detection and serovar identification of *Salmonella* in food and feeding stuff. *Food Analytical Methods, 2*, 81–95.10.1007/s12161-008-9057-9, ,) as it may have different effects in the qPCR reaction, like the Cq of the final fluorescence obtained (Wright et al., 2007Wright, A. C., Garrido, V., Debuex, G., Farrell-Evans, M., Mudbidri, A. A., & Otwell, W. S.(2007). Evaluation of postharvest-processed oysters by using PCR-based most-probable-number enumeration of *Vibrio vulnificus*bacteria. *Applied and Environmental Microbiology, 73*, 7477–7481.10.1128/AEM.01118-07).

On the opposite side of in-house protocols, are commercial DNA extraction kits. Different studies have been applied and several of these kits were compared, like PrepMan Ultra® (Applied Biosystems), InstaGene® (BioRad), DNeasy Tissue Kit® (Quiagen), UltraClean Microbial DNA isolation kit® (MoBio), NucleoSpin Tissue Kit® (Machery-Nagel), Nexttec kit for genomic DNA from Bacteria® (Nexttec), QIAmp DNA stool mini kit® (Quiagen), and High Pure PCR template® (Roche), among others (Elizaquivel & Aznar, 2008Elizaquivel, P., & Aznar, R. (2008). Comparison of four commercial DNA extraction kits for PCR detection of *Listeria monocytogenes*, *Salmonella*, *Escherichia coli* O157:H7, and *Staphylococcus aureus* in fresh, minimally processed vegetables. *Journal of Food Protection, 71*, 2110–2114., , ; Flekna, Schneeweiss, Smulders, Wagner, & Hein, 2007Flekna, G., Schneeweiss, W., Smulders, F. J. M., Wagner, M., & Hein, I. (2007). Real-time PCR method with statistical analysis to compare the potential of DNA isolation methods to remove PCR inhibitors from samples for diagnostic PCR. *Molecular and Cellular Probes, 21*, 282–287.10.1016/j.mcp.2007.02.001). Out of these, PrepMan Ultra® and InstaGene® have been extensively applied, especially with Gram negative bacteria. The DNeasy Tissue Kit is one of the most frequently selected kit as it has proven suitable for different types of samples and with different types of micro-organisms (Elizaquivel & Aznar, 2008Elizaquivel, P., & Aznar, R. (2008). Comparison of four commercial DNA extraction kits for PCR detection of *Listeria monocytogenes*, *Salmonella*, *Escherichia coli* O157:H7, and *Staphylococcus aureus* in fresh, minimally processed vegetables. *Journal of Food Protection, 71*, 2110–2114., , ; Elizaquível et al., 2011Elizaquível,

P., Gabaldón, J. A., & Aznar, R.(2011). Quantification of *Salmonella* spp, *Listeria monocytogenes* and *Escherichia coli*O157 H7 in non-spiked food products and evaluation of real-time PCR as a diagnostic tool in routine food analysis. *Food Control, 22*, 158–164.10.1016/j.foodcont.2010.05.018, , ; Jofré et al., 2005Jofré, A., Martin, B., Garriga, M., Hugas, M., Pla, M., Rodríguez-Lázaro, D., & Aymerich, T. (2005). Simultaneous detection of *Listeria monocytogenes* and *Salmonella* by multiplex PCR in cooked ham. *Food Microbiology, 22*, 109–115.10.1016/j.fm.2004.04.009, , ; Kobayashi et al., 2009Kobayashi, H., Kubota, J., Fujihara, K., Honjoh, K., Iio, M., Fujiki, N., ... Miyamoto, T. (2009). Simultaneous enrichment of *Salmonella* spp, *Escherichia coli* O157:H7, *Vibrio parahaemolyticus*, *Staphylococcus aureus*, *Bacillus cereus*, and *Listeria monocytogenes* by single broth and screening of the pathogens by multiplex real-time PCR. *Food Science and Technology Research, 15*, 427–438.10.3136/fstr.15.427, , ; McGuinness et al., 2009McGuinness, S., McCabe, E., O'Regan, E., Dolan, A., Duffy, G., Burgess, C., ... O'Grady, J. (2009). Development and validation of a rapid real-time PCR based method for the specific detection of *Salmonella* on fresh meat. *Meat Science, 83*, 555–562.10.1016/j.meatsci.2009.07.004).

TARGET GENES AND MICRO-ORGANISMS

The bacterial ribosomal operon (16S rRNA, 23S rRNA, and intergenic spacer (IGS) region) has been frequently used as target for PCR amplification. It is ubiquitous, bears both variable and highly conserved sequences, is easily available from public databases for many species, and often results in sensitive detection due to its multicopy nature. Detailed sequence analysis has demonstrated that the 16S rRNA gene is suitable for accurate PCR identification of many pathogens (Chakravorty, Helb, Burday, Connell, & Alland, 2007Chakravorty, S., Helb, D., Burday, M., Connell, N., & Alland, D. (2007). A detailed analysis of 16S ribosomal RNA gene segments for the diagnosis of pathogenic bacteria. *Journal of Microbiological Methods, 69*, 330–339.10.1016/j.mimet.2007.02.005). While its discriminating power might be sufficient for some genera and species, it is not always enough to distinguish closely related species (e.g. for enterococci or *B. cereus* group members). Therefore, other housekeeping genes have been studied, as well as functional genes involved in virulence or metabolism. Greater availability of genome sequences, generalization of high-throughput molecular tools and progress in computational genomics are facilitating the set up of new qPCR protocols based on a large variety of genes, thus increasing test specificity.

Nowadays, bacterial foodborne pathogen detection by PCR and qPCR is based on the detection of one or multiple specific genes; thus, correct selection of the target along with the design/selection of the primers and probes is a key factor for the correct identification of bacterial hazards. For every single pathogenic bacterium, there is a plethora of possible gene targets; thus lack of consensus is a general situation, even though a higher degree of agreement may be observed for the detection of certain microorganisms.

According to the FoodNet, in 2012, 19,531 cases of infection were laboratory confirmed in the US, being the main bacterial pathogens identified *Salmonella* spp. (16.42%), *Campylobacter* spp. (14.30%), *Shigella* spp. (4.50%), Shiga-toxin-producing serotypes (STEC) non-O157 (1.16%), STEC O157 (1.12%), *Vibrio* spp. (0.41%), and *Listeria* (0.25%) (CDC, 2013CDC. (2013). Incidence and trends of infection with pathogens transmitted commonly through food-foodborne diseases active surveillance network, 10 US sites, 1996–2012. *MMWR. Morbidity and mortality weekly report, 62*, 283–287., ,). These data highlight the importance that these pathogens still represent for human health, even some have been studied for long time.

Legislated pathogens according to Commission Regulation (EC) No. 2073/2005 on microbiological criteria for foodstuffs

Salmonella spp.

Nowadays, several alternatives exist for the specific detection of *Salmonella* spp. These include the *bipA*which encodes a protein of the guanosine -5′-triphosphate (GTP)-elongation binding family that displays global modulating properties by regulating a wide variety of downstream processes (Calvó et al., 2008Calvó, L., Martínez-Planells, A., Pardos-Bosch, J., & Garcia-Gil, L. J. (2008). A new real-time PCR assay for the specific detection of *Salmonella*spp. targeting the *bipA* gene. *Food Analytical Methods, 1*, 236–242.10.1007/s12161-007-9008-x, ,), the *fimC*, which is involved in the synthesis of type 1 fimbria, and has been successfully applied for the detection of the species *S. enterica* (Krascsenicsová, Piknová, Kaclíková, & Kuchta, 2008Krascsenicsová, K., Piknová, L., Kaclíková, E., & Kuchta, T. (2008). Detection of *Salmonella enterica* in food using two-step enrichment and real-time polymerase chain reaction. *Letters in Applied Microbiology, 46*, 483–487.10.1111/j.1472-765X.2008.02342.x ; Piknová, Kaclíková,

Pangallo, Polek, & Kuchta, 2005Piknová, L., Kaclíková, E., Pangallo, D., Polek, B., & Kuchta, T. (2005). Quantification of *Salmonella* by 5′-nuclease real-time polymerase chain reaction targeted to fimC gene. *Current Microbiology, 50*, 38–42.10.1007/s00284-004-4374-8), the *ttrRSBCA* locus, which is located near the *Salmonella* pathogenicity island 2 at centisome 30.5. It is required for tetrathionate respiration in *Salmonella* (Hyeon et al., 2010Hyeon, J. Y., Hwang, I. G., Kwak, H. S., Park, C., Choi, I. S., & Seo, K. H. (2010). Evaluation of PCR inhibitory effect of enrichment broths and comparison of DNA extraction methods for detection of *Salmonella* Enteritidis using real-time PCR assay. *Journal of Veterinary Science,11*, 143–149.10.4142/jvs.2010.11.2.143 ; Malorny et al., 2004Malorny, B., Paccassoni, E., Fach, P., Bunge, C., Martin, A., & Helmuth, R. (2004). Diagnostic real-time PCR for detection of *Salmonella* in food. *Applied and Environmental Microbiology,70*, 7046–7052.10.1128/AEM.70.12.7046-7052.2004). Delibato et al., reported a European interlaboratory validation based on the method described by Josefsen et al., (Delibato et al., 2014Delibato, E., Rodriguez-Lázaro, D., Gianfranceschi, M., De Cesare, A., Comin, D., Gattuso, A., ... De Medici, D. (2014). European validation of real-time PCR method for detection of *Salmonella* spp. in pork meat. *International Journal of Food Microbiology, 184*, 134–138. ; Josefsen, Krause, Hansen, & Hoorfar, 2007Josefsen, M. H., Krause, M., Hansen, F., & Hoorfar, J. (2007). Optimization of a 12-h TaqMan PCR-based method for detection of *Salmonella* bacteria in meat. *Applied and Environmental Microbiology, 73*, 3040–3048.10.1128/AEM.02823-06). Another gene used for *Salmonella* detection is the *iagA*, the invasion associated gene (Chua & Bhagwat, 2009Chua, T., & Bhagwat, A. A. (2009). A rapid and simple DNA extraction procedure to detect *Salmonella* spp. and *Listeria monocytogenes*from fresh produce using real-time PCR. *Food Analytical Methods, 2*, 96–101.10.1007/s12161-008-9032-5, ,).

One of the most frequently used genes to perform the detection of *Salmonella* spp. is the *invA*. This gene allows the entrance of the bacterium in the intestinal epithelial cells. Even though previous authors have described the existence of strains lacking this gene, like *S.* Saintpaul, while others stated, in the case of this strain, that they could successfully detect it. It was hypothesized that strains lacking this gene could not penetrate the intestinal cells, thus not being able to cause disease. A similar situation is observed with serovars Litchfield and Senftenberg, although, in general, absence of *invA* seems to be rare (Anderson et al., 2011Anderson, A., Pietsch, K., Zucker, R., Mayr, A., Müller-Hohe, E., Messelhäusser, U., ...

Huber, I.(2011). Validation of a duplex real-time PCR for the detection of *Salmonella* spp. in different food products. *Food Analytical Methods, 4*, 259–267.10.1007/s12161-010-9142-8, , ; Calvó et al., 2008Calvó, L., Martínez-Planells, A., Pardos-Bosch, J., & Garcia-Gil, L. J. (2008). A new real-time PCR assay for the specific detection of *Salmonella*spp. targeting the *bipA* gene. *Food Analytical Methods, 1*, 236–242.10.1007/s12161-007-9008-x, , ; Cheng et al., 2008Cheng, C. M., Lin, W., Van, K. T., Phan, L., Tran, N. N., & Farmer, D. (2008). Rapid detection of *Salmonella* in foods using real-time PCR. *Journal of Food Protection, 71*, 2436–2441., , ; D'Urso et al., 2009D'Urso, O. F., Poltronieri, P., Marsigliante, S., Storelli, C., Hernández, M., & Rodríguez-Lázaro, D. (2009). A filtration-based real-time PCR method for the quantitative detection of viable *Salmonella enterica* and *Listeria monocytogenes* in food samples. *Food Microbiology, 26*, 311–316.10.1016/j.fm.2008.12.006 ; González-Escalona, Brown, & Zhang, 2012González-Escalona, N., Brown, E. W., & Zhang, G. (2012). Development and evaluation of a multiplex real-time PCR (qPCR) assay targeting *ttrRSBCA* locus and *invA* gene for accurate detection of *Salmonella* spp. in fresh produce and eggs. *Food Research International, 48*, 202–208., , ; Jofré et al., 2005Jofré, A., Martin, B., Garriga, M., Hugas, M., Pla, M., Rodríguez-Lázaro, D., & Aymerich, T. (2005). Simultaneous detection of *Listeria monocytogenes* and *Salmonella* by multiplex PCR in cooked ham. *Food Microbiology, 22*, 109–115.10.1016/j.fm.2004.04.009, , ; Kimura et al., 1999Kimura, B., Kawasaki, S., Fujii, T., Kusunoki, J., Itoh, T., & Flood, S. J. A. (1999). Evaluation of TaqMan PCR assay for detecting *Salmonella* in raw meat and shrimp. *Journal of Food Protection, 62*, 329–335., , ; Malorny, Hoorfar, Bunge, & Helmuth, 2003Malorny, B., Hoorfar, J., Bunge, C., & Helmuth, R. (2003). Multicenter validation of the analytical accuracy of *Salmonella* PCR: Towards an international standard. *Applied and Environmental Microbiology, 69*, 290–296.10.1128/AEM.69.1.290-296.2003 ; Malorny et al., 2004Malorny, B., Paccassoni, E., Fach, P., Bunge, C., Martin, A., & Helmuth, R. (2004). Diagnostic real-time PCR for detection of *Salmonella* in food. *Applied and Environmental Microbiology,70*, 7046–7052.10.1128/AEM.70.12.7046-7052.2004 ; Rahn et al., 1992Rahn, K., De Grandis, S. A., Clarke, R. C., McEwen, S. A., Galán, J. E., Ginocchio, C., … Gyles, C. (1992). Amplification of an *invA* gene sequence of *Salmonella typhimurium* by polymerase chain reaction as a specific method of detection of *Salmonella*. *Molecular and Cellular Probes, 6*, 271–279.10.1016/0890-8508(92)90002-F). Even though these slight discrepancies were observed by different authors, independent laboratories have selected this target for interlaboratory validation of different PCR and qPCR methods like the study of Malorny et al. (2003Malorny, B., Hoorfar, J., Bunge, C., & Helmuth, R. (2003). Multicenter validation of the analytical accuracy of *Salmonella* PCR: Towards an international standard. *Applied and Environmental Microbiology, 69*, 290–296.10.1128/AEM.69.1.290-296.2003 , 2004Malorny, B., Paccassoni, E., Fach, P., Bunge, C., Martin, A., & Helmuth, R. (2004). Diagnostic real-time PCR for detection of *Salmonella* in food. *Applied and Environmental Microbiology,70*, 7046–7052.10.1128/AEM.70.12.7046-7052.2004) in Europe, or the one carried out by Cheng et al. in the USA (Cheng et al., 2008Cheng, C. M., Lin, W., Van, K. T., Phan, L., Tran, N. N., & Farmer, D. (2008). Rapid detection of *Salmonella* in foods using real-time PCR. *Journal of Food Protection, 71*, 2436–2441., , ; Cheng, Van, Lin, & Ruby, 2009Cheng, C. M., Van, K. T., Lin, W., & Ruby, R. M.(2009). Interlaboratory validation of a real-time PCR 24-hour rapid method for detection of *Salmonella* in foods. *Journal of Food Protection,72*, 945–951., ,).

Recently, Malorny et al. (2007Malorny, B., Bunge, C., & Helmuth, R. (2007). A real-time PCR for the detection of *Salmonella*Enteritidis in poultry meat and consumption eggs. *Journal of Microbiological Methods, 70*, 245–251.10.1016/j.mimet.2007.04.013) developed a qPCR method for the specific detection of serovar Enteritidis. This is of special relevance, since in the last amendment of a European Regulation 2073/2005 ((EC), 2005(EC), C. R. (2005). Microbiological criteria for foodstuffs. *Official Journal of the European Union,2073/2005.* ; (EU), 2011(EU), C. R. (2011). Amending Annex II to Regulation (EC) No. 2160/2003 of the European parliament and of the council and annex I to Commission Regulation (EC) No. 2073/2005 as regards *Salmonella* in fresh poultry meat. *Official Journal of the European Union,1086/2011.*), it is specifically pointed out that the investigation of *Salmonella* spp. in chicken meat should be restricted to serovars Enteritidis and Typhimurium. The method targeted the Prot6e gene located on the *S.*Enteritidis specific 60-kb virulence plasmid. Even though the gene is promising, further studies should be performed, due to the fact that it fails to detect 4 out of the 79 isolates tested as they lacked the plasmid. Previous studies targeted the *sefA* gene encoding a fimbrial antigen termed SEF14. This gene is exclusively unique to *Salmonella* serogroup D serovars such as Gallinarum, Pullorum, Dublin, Rostock, Semban, and Typhi, and consequently not specific enough to identify only *S.* Enteritidis. Another conventional PCR assay used a chromosomal located fragment, called SdfI which has a high exclusivity but did not detect phage types 6A, 9A, 11, 16,

20, and 27 and was only tested on pure cultures (Malorny et al., 2007Malorny, B., Bunge, C., & Helmuth, R. (2007). A real-time PCR for the detection of *Salmonella*Enteritidis in poultry meat and consumption eggs. *Journal of Microbiological Methods, 70*, 245–251.10.1016/j.mimet.2007.04.013).

Listeria monocytogenes

Concerning *L. monocytogenes,* the most widely used targets for its specific detection are the genes *hlyA*which encodes for the listeriolisin O (Elizaquível et al., 2011Elizaquível, P., Gabaldón, J. A., & Aznar, R.(2011). Quantification of *Salmonella* spp, *Listeria monocytogenes* and *Escherichia coli*O157 H7 in non-spiked food products and evaluation of real-time PCR as a diagnostic tool in routine food analysis. *Food Control, 22*, 158–164.10.1016/j.foodcont.2010.05.018, , ; Kobayashi et al., 2009Kobayashi, H., Kubota, J., Fujihara, K., Honjoh, K., Iio, M., Fujiki, N., … Miyamoto, T. (2009). Simultaneous enrichment of *Salmonella* spp, *Escherichia coli* O157:H7, *Vibrio parahaemolyticus*, *Staphylococcus aureus*, *Bacillus cereus*, and *Listeria monocytogenes* by single broth and screening of the pathogens by multiplex real-time PCR. *Food Science and Technology Research, 15*, 427–438.10.3136/fstr.15.427, , ; Navas et al., 2006Navas, J., Ortiz, S., Lopez, P., Jantzen, M. M., Lopez, V., & Martinez-Suarez, J. V. (2006). Evaluation of effects of primary and secondary enrichment for the detection of listeria monocytogenes by real-time PCR in retail ground chicken meat. *Foodborne Pathogens and Disease, 3*, 347–354.10.1089/fpd.2006.3.347 ; Omiccioli et al., 2009Omiccioli, E., Amagliani, G., Brandi, G., & Magnani, M. (2009). A new platform for real-time PCR detection of *Salmonella* spp., *Listeria monocytogenes* and *Escherichia coli* O157 in milk. *Food Microbiology, 26*, 615–622.10.1016/j.fm.2009.04.008 ; Rodriguez-Lazaro, Hernández, et al., 2004Rodriguez-Lazaro, D., Hernandez, M., Scortti, M., Esteve, T., Vazquez-Boland, J. A., & Pla, M.(2004). Quantitative detection of *Listeria monocytogenes* and *Listeria innocua* by real-time PCR: Assessment of *hly*, *iap*, and *lin02483*targets and amplifluor technology. *Applied and Environmental Microbiology, 70*, 1366–1377.10.1128/AEM.70.3.1366-1377.2004 ; Ruiz-Rueda et al., 2010Ruiz-Rueda, O., Soler, M., & Calvo, L. (2010). Multiplex real-time PCR for the simultaneous detection of *Salmonella* spp. and *Listeria monocytogenes* in food samples. *Food Analytical Methods, 4*, 131–138., , ; Suo et al., 2010Suo, B., He, Y. P., Tu, S. I., & Shi, X. M. (2010). A multiplex real-time polymerase chain reaction for simultaneous detection of *Salmonella spp.*, *Escherichia coli* O157, and *Listeria monocytogenes* in meat products. *Foodborne Pathogens* *and Disease, 7*, 619–628.10.1089/fpd.2009.0430), and the *prfA* which is a central virulence regulator of this bacterium (D'Urso et al., 2009D'Urso, O. F., Poltronieri, P., Marsigliante, S., Storelli, C., Hernández, M., & Rodríguez-Lázaro, D. (2009). A filtration-based real-time PCR method for the quantitative detection of viable *Salmonella enterica* and *Listeria monocytogenes* in food samples. *Food Microbiology, 26*, 311–316.10.1016/j.fm.2008.12.006 ; Jofré et al., 2005Jofré, A., Martin, B., Garriga, M., Hugas, M., Pla, M., Rodríguez-Lázaro, D., & Aymerich, T. (2005). Simultaneous detection of *Listeria monocytogenes* and *Salmonella* by multiplex PCR in cooked ham. *Food Microbiology, 22*, 109–115.10.1016/j.fm.2004.04.009, , ; Rossmanith, Krassnig, Wagner, & Hein, 2006Rossmanith, P., Krassnig, M., Wagner, M., & Hein, I. (2006). Detection of *Listeria monocytogenes* in food using a combined enrichment/real-time PCR method targeting the *prfA* gene. *Research in Microbiology, 157*, 763–771.10.1016/j.resmic.2006.03.003). Additionally, the *iap* gene, coding the protein p60 essential for cell viability, has also been applied (Flekna, Stefanic, et al., 2007Flekna, G., Schneeweiss, W., Smulders, F. J. M., Wagner, M., & Hein, I. (2007). Real-time PCR method with statistical analysis to compare the potential of DNA isolation methods to remove PCR inhibitors from samples for diagnostic PCR. *Molecular and Cellular Probes, 21*, 282–287.10.1016/j.mcp.2007.02.001 ; Mukhopadhyay & Mukhopadhyay, 2007Mukhopadhyay, A., Mukhopadhyay, K., & U.(2007). Novel multiplex PCR approaches for the simultaneous detection of human pathogens: *Escherichia coli* 0157:H7 and *Listeria monocytogenes*. *Journal of Microbiological Methods, 68*, 193–200.10.1016/j.mimet.2006.07.009 ; Wuenscher, Köhler, Bubert, Gerike, & Goebel, 1993Wuenscher, M. D., Köhler, S., Bubert, A., Gerike, U., & Goebel, W. (1993). The *iap* gene of *Listeria monocytogenes* is essential for cell viability, and its gene product, p60, has bacteriolytic activity. *Journal of bacteriology, 175*, 3491–3501., ,).

The most popular targets are *hlyA* and *prfA*. Regarding the former, the qPCR method described by Rodriguez-Lazaro, Jofre, et al. (2004Rodriguez-Lazaro, D., Jofre, A., Aymerich, T., Hugas, M., & Pla, M. (2004). Rapid quantitative detection of *Listeria monocytogenes* in meat products by real-time PCR. *Applied and Environmental Microbiology, 70*, 6299–6301.10.1128/AEM.70.10.6299-6301.2004) was included in a European interlaboratory assay to evaluate different methodologies' performance (Dalmasso et al., 2014Dalmasso, M., Bolocan, A. S., Hernandez, M., Kapetanakou, A. E., Kuchta, T., Manios, S. G., … Jordan, K. (2014). Comparison of polymerase chain reaction methods and plating for analysis of enriched cultures of *Listeria monocytogenes*when using the ISO11290-1 method. *Journal of Microbiological Methods, 98*, 8–14.10.1016/j.mimet.2013.12.018). Later on, the same group of laboratories performed its validation (Gianfranceschi et al., 2014Gianfranceschi, M. V., Rodriguez-Lazaro, D., Hernandez, M., Gonzalez-Garcia, P., Comin, D., Gattuso, A., … De Medici, D. (2014). European validation of a real-time PCR-based method for detection of *Listeria monocytogenes* in soft cheese. *International Journal of Food Microbiology, 184*, 128–133.). Concerning *prfA*, it was selected for an interlaboratory validation study involving several laboratories in Europe (D'Agostino et al., 2004D'Agostino, M., Wagner, M., Vazquez-Boland, J. A., Kuchta, T., Karpiskova, R., Hoorfar, J., … Cook, N. (2004). A validated PCR-based method to detect *Listeria monocytogenes* using raw milk as a food model—Towards an international standard. *Journal of Food Protection, 67*, 1646–1655., ,). In a later different report, a specific fluorescent probe was designed in order to adapt the method to qPCR (Rossmanith et al., 2006Rossmanith, P., Krassnig, M., Wagner, M., & Hein, I. (2006). Detection of *Listeria monocytogenes* in food using a combined enrichment/real-time PCR method targeting the *prfA* gene. *Research in Microbiology, 157*, 763–771.10.1016/j.resmic.2006.03.003) and extensively evaluated with natural samples (Rossmanith, Mester, Wagner, & Schoder, 2010Rossmanith, P., Mester, P., Wagner, M., & Schoder, D. (2010). Demonstration of the effective performance of a combined enrichment/real-time PCR method targeting the *prfA* gene of *Listeria monocytogenes* by testing fresh naturally contaminated acid curd cheese. *Letters in Applied Microbiology, 51*, 480–484.10.1111/j.1472-765X.2010.02925.x). Primers and probe targeting *prfA* were recently adopted in an in-house multiplex qPCR method for the simultaneous detection of *Salmonella* spp., *E. coli* O157, and *L. monocytogenes* returning comparable results to those of the original studies where *L. monocytogenes* was the only target (Garrido, Chapela, Román, Fajardo, Vieites, et al., 2013Garrido, A., Chapela, M.-J., Román, B., Fajardo, P., Vieites, J. M., & Cabado, A. G. (2013). In-house validation of a multiplex real-time PCR method for simultaneous detection of *Salmonella* spp., *Escherichia coli* O157 and *Listeria monocytogenes*. *International Journal of Food Microbiology, 164*, 92–98.10.1016/j.ijfoodmicro.2013.03.024).

Emerging Pathogens

Campylobacter spp.

Campylobacter spp. Detection has been performed in two different ways, either the main pathogenic species (*C. jejuni, C. coli,* and *C. lari*) or the complete genus. Different genes have been described for each species like *hipO* (gene encoding the hippuricase), *mapA* (membrane-associated protein), *rpoB* (gene encoding the β-subunit of the RNA polymerase), and the VS1 sequence of *C. jejuni*'s genome (Debretsion, Habtemariam, Wilson, Nganwa, & Yehualaeshet, 2007Debretsion, A., Habtemariam, T., Wilson, S., Nganwa, D., & Yehualaeshet, T. (2007). Real-time PCR assay for rapid detection and quantification of *Campylobacter jejuni* on chicken rinses from poultry processing plant. *Molecular and Cellular Probes, 21*, 177–181.10.1016/j.mcp.2006.10.006 ; He et al., 2010He, Y., Yao, X., Gunther, N. W., Xie, Y., Tu, S.-I., & Shi, X. (2010). Simultaneous detection and differentiation of *Campylobacter jejuni, C. coli*, and *C. lari* in chickens using a multiplex real-time PCR assay. *Food Analytical Methods, 3*, 321–329.10.1007/s12161-010-9136-6, , ; Mayr et al., 2010Mayr, A. M., Lick, S., Bauer, J., Thaerigen, D., Busch, U., & Huber, I. (2010). Rapid detection and differentiation of *Campylobacter jejuni, Campylobacter coli*, and *Campylobacter lari* in food, using multiplex real-time PCR. *Journal of Food Protection, 73*, 241–250., , ; Rantsiou, Lamberti, & Cocolin, 2010Rantsiou, K., Lamberti, C., & Cocolin, L. (2010). Survey of *Campylobacter jejuni* in retail chicken meat products by application of a quantitative PCR protocol. *International Journal of Food Microbiology, 141*, S75–S79.10.1016/j.ijfoodmicro.2010.02.002 ; Yang, Jiang, Huang, Zhu, & Yin, 2003Yang, C. B., Jiang, Y., Huang, K. H., Zhu, C. Q., & Yin, Y. L. (2003). Application of real-time PCR for quantitative detection of *Campylobacter jejuni*in poultry, milk and environmental water. *FEMS Immunology & Medical Microbiology, 38*, 265–271.10.1016/S0928-8244(03)00168-8).

Regarding *C. lari,* the *gyrA* (a fragment of the gyrase A subunit gene) and *pepT* (gene encoding a peptidase) have been used for the specific detection of this species.

Finally, the *ceuE* (periplasmic substrate binding protein) and the *cdtA* (gene encoding a subunit of cytolethal toxin, an important virulence factor) have been applied (He et al., 2010He, Y., Yao, X., Gunther, N. W., Xie, Y., Tu, S.-I., & Shi, X. (2010). Simultaneous detection and differentiation of *Campylobacter jejuni, C. coli*, and *C. lari* in chickens using a multiplex real-time PCR assay. *Food Analytical Methods, 3*, 321–329.10.1007/s12161-010-9136-6, , ; Mayr et al., 2010Mayr, A. M., Lick, S., Bauer, J., Thaerigen, D., Busch, U., & Huber, I. (2010). Rapid detection and differentiation of *Campylobacter jejuni*, *Campylobacter coli*, and *Campylobacter lari* in food, using multiplex real-time PCR. *Journal of Food Protection, 73*, 241–250., ,). As it can be observed, there is not much agreement on the PCR targets for the detection of these species. Typical differentiation of *C. jejuni*, from *C. coli*, is based on hippurate hydrolysis, but atypical strains exist, thus Toplak et al. combined a *hipO* and *ccoN* (encoding the cytochromo c oxidase of *C. jejuni*), from *C. coli cadF*(fibronectin-binding protein) for the simultaneous detection of both species (Toplak, Kovac, Piskernik, Mozina, & Jersek, 2011Toplak, N., Kovac, M., Piskernik, S., Mozina, S. S., & Jersek, B. (2011). Detection and quantification of *Campylobacter jejuni* and *Campylobacter coli* using real-time multiplex PCR. *Journal of Applied Microbiology, 112*, 752–764., ,).

The second alternative, detection of genus *Campylobacter* spp. without differentiating the species, may be an ideal approach from a risk assessment point of view, specially focused on thermotolerant campylobacters (*C. jejuni*, *C. coli*, and *C. lari*). Many studies have successfully applied different primers targeting 16S rRNA, specific for genus *Campylobacter* and/or thermotolerant campylobacters (Botteldoorn et al., 2008Botteldoorn, N., Van Coillie, E., Piessens, V., Rasschaert, G., Debruyne, L., Heyndrickx, M., … Messens, W. (2008). Quantification of *Campylobacter* spp. in chicken carcass rinse by real-time PCR. *Journal of Applied Microbiology, 105*, 1909–1918.10.1111/jam.2008.105.issue-6 ; Josefsen, Cook, et al., 2004Josefsen, M. H., Cook, N., D'Agostino, M., Hansen, F., Wagner, M., Demnerova, K., … Hoorfar, J. (2004). Validation of a PCR-based method for detection of food-borne thermotolerant campylobacters in a multicenter collaborative trial. *Applied and Environmental Microbiology, 70*, 4379–4383.10.1128/AEM.70.7.4379-4383.2004 ; Josefsen, Jacobsen, & Hoorfar, 2004Josefsen, M. H., Jacobsen, N. R., & Hoorfar, J.(2004). Enrichment followed by quantitative PCR both for rapid detection and as a tool for quantitative risk assessment of food-borne thermotolerant campylobacters. *Applied and Environmental Microbiology, 70*, 3588–3592.10.1128/AEM.70.6.3588-3592.2004 ; Leblanc-Maridor et al., 2011Leblanc-Maridor, M., Garénaux, A., Beaudeau, F., Chidaine, B., Seegers, H., Denis, M., & Belloc, C. (2011). Quantification of *Campylobacter* spp. in pig feces by direct real-time PCR with an internal control of extraction and amplification. *Journal of Microbiological Methods, 85*, 53–61.10.1016/j.mimet.2011.01.013 ; Lubeck,

Cook, Wagner, Fach, & Hoorfar, 2003Lubeck, P. S., Cook, N., Wagner, M., Fach, P., & Hoorfar, J. (2003). Toward an international standard for PCR-based detection of food-borne thermotolerant campylobacters: Validation in a multicenter collaborative trial. *Applied and Environmental Microbiology, 69*, 5670–5672.10.1128/AEM.69.9.5670-5672.2003 ; Lubeck, Wolffs, et al., 2003Lubeck, P. S., Wolffs, P., On, S. L. W., Ahrens, P., Radstrom, P., & Hoorfar, J. (2003). Toward an international standard for PCR-based detection of food-borne thermotolerant campylobacters: Assay development and analytical validation. *Applied and Environmental Microbiology, 69*, 5664–5669.10.1128/AEM.69.9.5664-5669.2003 ; Wolffs, Norling, Hoorfar, Griffiths, & Radstrom, 2005Wolffs, P., Norling, B., Hoorfar, J., Griffiths, M., & Radstrom, P. (2005). Quantification of *Campylobacter* spp. in chicken rinse samples by using flotation prior to real-time PCR. *Applied and Environmental Microbiology, 71*, 5759–5764.10.1128/AEM.71.10.5759-5764.2005).

Due to the fact that the group of thermotolerant campylobacters are the main human hazard, 16S rRNA was selected as the optimal target in an interlaboratorystudyperformedinEurope(Josefsen,Cook,etal.,2004Josefsen, M. H., Cook, N., D'Agostino, M., Hansen, F., Wagner, M., Demnerova, K., ... Hoorfar, J. (2004). Validation of a PCR-based method for detection of food-borne thermotolerant campylobacters in a multicenter collaborative trial. *Applied and Environmental Microbiology, 70*, 4379–4383.10.1128/AEM.70.7.4379-4383.2004 ; Josefsen, Jacobsen, et al., 2004Josefsen, M. H., Jacobsen, N. R., & Hoorfar, J.(2004). Enrichment followed by quantitative PCR both for rapid detection and as a tool for quantitative risk assessment of food-borne thermotolerant campylobacters. *Applied and Environmental Microbiology, 70*, 3588–3592.10.1128/AEM.70.6.3588-3592.2004 ; Lubeck, Cook, et al., 2003Lubeck, P. S., Cook, N., Wagner, M., Fach, P., & Hoorfar, J. (2003). Toward an international standard for PCR-based detection of food-borne thermotolerant campylobacters: Validation in a multicenter collaborative trial. *Applied and Environmental Microbiology, 69*, 5670–5672.10.1128/AEM.69.9.5670-5672.2003 ; Lubeck, Wolffs, et al., 2003Lubeck, P. S., Wolffs, P., On, S. L. W., Ahrens, P., Radstrom, P., & Hoorfar, J. (2003). Toward an international standard for PCR-based detection of food-borne thermotolerant campylobacters: Assay development and analytical validation. *Applied and Environmental Microbiology, 69*, 5664–5669.10.1128/AEM.69.9.5664-5669.2003). A later study developed an appropriate fluorescent probe suitable for its application along with the European-validated primers (Josefsen, Jacobsen, et al., 2004Josefsen, M. H.,

Jacobsen, N. R., & Hoorfar, J.(2004). Enrichment followed by quantitative PCR both for rapid detection and as a tool for quantitative risk assessment of food-borne thermotolerant campylobacters. *Applied and Environmental Microbiology, 70*, 3588–3592.10.1128/AEM.70.6.3588-3592.2004).

Escherichia coli O157 and other STEC

In 2004, Perelle et al. designed a method for the identification of the most frequent clinical serotypes of Shiga-toxin producing *E. coli*, which included O26, O55, O91, O103, O111, O113, O145, and O157:H7. For this purpose, genes *stx1*, *stx2*, *rfbE* (lipopolysaccharide O-antigen synthesis gene for O157), *wbdI* (O111), *wzx*(flippase gene for O26), *wzy* (O-antigen polymerase specific for O113), *wzy* (O91), *wbgN* (glycosyltransferase gene specific for O55), *ihpl* (based on the polymorphism between the genes encoding the putative adhesin from "O-island 29" of STEC O157 and STEC O145 O145), *eae* (adherence intimin protein for O103), and *fliC* H7 (flagellar antigen H7) were successfully applied (DebRoy et al., 2004DebRoy, C., Roberts, E., Kundrat, J., Davis, M. A., Briggs, C. E., & Fratamico, P. M. (2004). Detection of *Escherichia coli* serogroups O26 and O113 by PCR amplification of the wzx and wzy genes. *Applied and Environmental Microbiology, 70*, 1830–1832.10.1128/AEM.70.3.1830-1832.2004 ; Gonzales et al., 2011Gonzales, T. K., Kulow, M., Park, D. J., Kaspar, C. W., Anklam, K. S., Pertzborn, K. M., et al. (2011). A high-throughput open-array qPCR gene panel to identify, virulotype, and subtype O157 and non-O157 enterohemorrhagic *Escherichia coli. Molecular and Cellular Probes, 25*, 222–230.10.1016/j.mcp.2011.08.004 ; Kagkli et al., 2011Kagkli, D. M., Weber, T. P., Van den Bulcke, M., Folloni, S., Tozzoli, R., Morabito, S., … Van den Eede, G. (2011). Application of the modular approach to an in-house validation study of real-time PCR methods for the detection and serogroup determination of verocytotoxigenic *Escherichia coli. Applied and Environmental Microbiology, 77*, 6954–6963.10.1128/AEM.05357-11 ; Perelle, Dilasser, Grout, & Fach, 2003Perelle, S., Dilasser, F., Grout, J., & Fach, P.(2003). Development of a 5′ nuclease PCR assay for detecting Shiga toxin-producing *Escherichia coli* O145 based on the identification of an "O-island 29" homologue. *Journal of Applied Microbiology, 94*, 587–594.10.1046/j.1365-2672.2003.01872.x , 2004Perelle, S., Dilasser, F., Grout, J. L., & Fach, P.(2004). Detection by 5′-nuclease PCR of Shiga-toxin producing *Escherichia coli* O26, O55, O91, O103, O111, O113, O145 and O157:H7, associated with the world's most frequent clinical cases. *Molecular and Cellular Probes,18*, 185–192.10.1016/j.mcp.2003.12.004 ;

Samuel, Hogbin, Wang, & Reeves, 2004Samuel, G., Hogbin, J.-P., Wang, L., & Reeves, P. R. (2004). Relationships of the *Escherichia coli*O157, O111, and O55 O-antigen gene clusters with those of *Salmonella* enterica and *Citrobacter freundii*, which express identical O antigens. *Journal of Bacteriology, 186*, 6536–6543.10.1128/JB.186.19.6536-6543.2004). Out of these, genes *stx1* and *stx2* (Shiga-toxin genes 1 and 2) are considered two major virulence factors for *E. coli*O157 and other STEC (Bonetta et al., 2011Bonetta, S., Borelli, E., Bonetta, S., Conio, O., Palumbo, F., & Carraro, E. (2011). Development of a PCR protocol for the detection of *Escherichia coli* O157:H7 and *Salmonella* spp. in surface water. *Environmental Monitoring and Assessment, 177*, 493–503.10.1007/s10661-010-1650-x ; Prendergast et al., 2011Prendergast, D. M., Lendrum, L., Pearce, R., Ball, C., McLernon, J., O'Grady, D., … Gutierrez, M.(2011). Verocytotoxigenic *Escherichia coli* O157 in beef and sheep abattoirs in Ireland and characterisation of isolates by pulsed-field gel electrophoresis and multi-locus variable number of tandem repeat analysis. *International Journal of Food Microbiology, 144*, 519–527.10.1016/j.ijfoodmicro.2010.11.012) and the *eae* gene has also been applied for serotype O157 detection by other authors (Adiguzel et al., 2012Adiguzel, G., Gulluce, M., Bozoglu, C., Yanmis, D., Gormez, A., Atasever, M., & Adiguzel, A.(2012). Molecular characterization of *Escherichia coli* O157:H7 from retail beef in Erzurum, Turkey. *Journal of Pure and Applied Microbiology, 6*, 1033–1041., ; Franz, Klerks, De Vos, Termorshuizen, & van Bruggen, 2007Franz, E., Klerks, M. A., De Vos, O. J., Termorshuizen, A. J., & van Bruggen, A. H. C.(2007). Prevalence of Shiga toxin-producing *Escherichia coli stx1*, *stx2*, *eaeA*, and *rfbE* genes and survival of *E. coli* O157:H7 in manure from organic and low-input conventional dairy farms. *Applied and Environmental Microbiology, 73*, 2180–2190.10.1128/AEM.01950-06 ; Fratamico & DebRoy, 2010Fratamico, P. M., & DebRoy, C. (2010). Detection of *Escherichia coli* O157:H7 in food using real-time multiplex PCR assays targeting the *stx 1*, *stx 2*, *wzy O157*, and the *fliC h7* or *eae* genes. *Food Analytical Methods, 3*, 330–337.10.1007/s12161-010-9140-x, , ; Kawasaki et al., 2010Kawasaki, S., Fratamico, P. M., Horikoshi, N., Okada, Y., Takeshita, K., Sameshima, T., & Kawamoto, S. (2010). Multiplex real-time polymerase chain reaction assay for simultaneous detection and quantification of *Salmonella* species, *Listeria monocytogenes*, and *Escherichia coli* O157:H7 in ground pork samples. *Foodborne Pathogens and Disease, 7*, 549–554.10.1089/fpd.2009.0465 ; Zhang et al., 2009Zhang, D., Zhang, H., Yang, L., Guo, J., Li, X., & Feng, Y. (2009). Simultaneous detection of *Listeria monocytogenes*, *Staphylococcus* *aureus*, *Salmonella enterica* and *Escherichia coli* O157:H7 in food samples using multiplex PCR method. *Journal of Food Safety, 29*, 348–363.10.1111/jfs.2009.29.issue-3, ,).

Different combinations of several genes have been published and alternative targets have been selected, like *hlyA* (hemolysin) in combination with *fliC*, *stx1*, *stx2*, *eae,* and *rfbE*(Bai, Shi, & Nagaraja, 2010Bai, J., Shi, X., & Nagaraja, T. G. (2010). A multiplex PCR procedure for the detection of six major virulence genes in *Escherichia coli*O157:H7. *Journal of Microbiological Methods,82*, 85–89.10.1016/j.mimet.2010.05.003). Other options included the *uidA* (β-glucuronidase gene) which was only found in O157:H7 and H- but not other O157 (Al-Ajmi et al., 2006Al-Ajmi, D., Padmanabha, J., Denman, S. E., Gilbert, R. A., Al Jassim, R. A. M., & McSweeney, C. S. (2006). Evaluation of a PCR detection method for *Escherichia coli* O157:H7/H-bovine faecal samples. *Letters in Applied Microbiology,42*, 386–391.10.1111/lam.2006.42.issue-4 ; Elizaquível et al., 2011Elizaquível, P., Gabaldón, J. A., & Aznar, R.(2011). Quantification of *Salmonella* spp, *Listeria monocytogenes* and *Escherichia coli*O157 H7 in non-spiked food products and evaluation of real-time PCR as a diagnostic tool in routine food analysis. *Food Control, 22*, 158–164.10.1016/j.foodcont.2010.05.018, ,).

A large outbreak caused by Shiga-toxin 2 (*stx2*)-producing enterohemorrhagic *E. coli* (EHEC) O104:H4 that involved 4,000 cases, including 855 patients with hemolytic-uremic syndrome (HUS) and 53 fatalities, occurred in Germany from May to July 2011. A combination of the previously mentioned genes was specifically optimized and combined for the detection of this specific serotype of STEC including wzy_{O104}, *stx2,* and $fliC_{H4}$ (Zhang et al., 2012Zhang, W., Bielaszewska, M., Bauwens, A., Fruth, A., Mellmann, A., & Karch, H. (2012). Real-time multiplex PCR for detecting Shiga toxin 2-producing *Escherichia coli* O104:H4 in human stools. *Journal of Clinical Microbiology, 50*, 1752–1754.10.1128/JCM.06817-11).

The wide variety of possible targets is mainly associated with high number of possible serotypes included in the group STEC. Due to the fact that the most prevalent serotype associated with human infection was O157, a European validation study, gene *rfbE* was selected for the specific detection of this serotype independently of the "H" antigen. In the validation study, special attention was focused on the specificity of the primers due to the fact that this gene is also present in the genus *Vibrio* spp.

As expected, after careful design and optimization of the method, optimal results were obtained (Abdulmawjood et al., 2003Abdulmawjood, A., Bülte, M., Cook, N., Roth, S., Schönenbrücher, H., & Hoorfar, J. (2003). Toward an international standard for PCR-based detection of *Escherichia coli* O157. *Journal of Microbiological Methods, 55*, 775–786.10.1016/j.mimet.2003.08.012 ; Abdulmawjood et al., 2004Abdulmawjood, A., Bulte, M., Roth, S., Schonenbrucher, H., Cook, N., Heuvelink, A. E., et al. (2004). Development, validation, and standardization of polymerase chain reaction-based detection of *E. coli* O157. *Journal of Aoac International, 87*, 596–603., ,).

Yersinia Enterocolitica

Regarding *Yersinia enterocolitica*, it was observed that few studies performed the detection at the species level 16S rRNA (Lantz et al., 1998Lantz, P. G., Knutsson, R., Blixt, Y., Al-Soud, W., Borch, E., & Rådström, P. (1998). Detection of pathogenic *Yersinia enterocolitica* in enrichment media and pork by a multiplex PCR: A study of sample preparation and PCR-inhibitory components. *International Journal of Food Microbiology, 45*, 93–105.10.1016/S0168-1605(98)00152-4 ; Wolffs, Knutsson, Norling, & Radstrom, 2004Wolffs, P., Knutsson, R., Norling, B., & Radstrom, P. (2004). Rapid quantification of *Yersinia enterocolitica* in pork samples by a novel sample preparation method, flotation, prior to real-time PCR. *Journal of Clinical Microbiology,42*, 1042–1047.10.1128/JCM.42.3.1042-1047.2004). Most studies are focused on the detection of pathogenic strains of this species. There is a high degree of consensus on the application of genes *ail* (attachment invasion locus, which mediates cell invasion) as well as plasmid genes such as *yadA* (whose product is involved in autoagglutination, serum resistance, and adhesion) (Cheyne, Van Dyke, Anderson, & Huck, 2010Cheyne, B., Van Dyke, M., Anderson, W., & Huck, P. (2010). The detection of *Yersinia enterocolitica* in surface water by quantitative PCR amplification of the *ail* and *yadA* genes. *Journal of Water and Health, 8*, 487–499.10.2166/wh.2009.215 ; Jourdan, Johnson, & Wesley, 2000Jourdan, A. D., Johnson, S. C., & Wesley, I. V.(2000). Development of a fluorogenic 5' nuclease PCR assay for detection of the ail gene of pathogenic *Yersinia enterocolitica*. *Applied and Environmental Microbiology, 66*, 3750–3755.10.1128/AEM.66.9.3750-3755.2000 ; Lambertz, Nilsson, Hallanvuo, & Lindblad, 2008Lambertz, S. T., Nilsson, C., Hallanvuo, S., & Lindblad, M. (2008). Real-time PCR method for detection of pathogenic *Yersinia enterocoliticain* food. *Applied and Environmental Microbiology, 74*, 6060–6067.10.1128/

AEM.00405-08 ; Lantz et al., 1998Lantz, P. G., Knutsson, R., Blixt, Y., Al-Soud, W., Borch, E., & Rådström, P. (1998). Detection of pathogenic *Yersinia enterocolitica* in enrichment media and pork by a multiplex PCR: A study of sample preparation and PCR-inhibitory components. *International Journal of Food Microbiology, 45*, 93–105.10.1016/S0168-1605(98)00152-4 ; Najdenski, Heyndrickx, Herman, Werbrouck, & Van Coillie, 2012Najdenski, H., Heyndrickx, M., Herman, L., Werbrouck, H., & Van Coillie, E. (2012). Quantification of *Yersinia enterocolitica* in raw milk using qPCR. *Veterinary Microbiology, 160*, 428–434.10.1016/j.vetmic.2012.06.013 ; Wiemer et al., 2011Wiemer, D., Loderstaedt, U., von Wulffen, H., Priesnitz, S., Fischer, M., Tannich, E., & Hagen, R. M. (2011). Real-time multiplex PCR for simultaneous detection of *Campylobacter jejuni*, *Salmonella*, *Shigella* and *Yersinia* species in fecal samples. *International Journal of Medical Microbiology, 301*, 577–584.10.1016/j.ijmm.2011.06.001).

Additionally, other genes like *inv* (invasive gene, which mediates cell invasion), *ystA* (which is responsible for the production of a heat-stable enterotoxin in virulent *Y. enterocolitica*), *ystB* and *ystC* (which have been observed to encode an enterotoxin present mainly in avirulent *Y. enterocolitica*), *rfbC* (which can be used to identify pathogenic *Y. enterocolitica* O:3 strains), and *virF* or *lcrF* (for *Y. enterocolitica* and *Y. pseudotuberculosis*, respectively, which encodes transcriptional activators of the *yop* regulon) have also been used (Zheng, Sun, Mao, & Jiang, 2008Zheng, H., Sun, Y., Mao, Z., & Jiang, B. (2008). Investigation of virulence genes in clinical isolates of *Yersinia enterocolitica*. *FEMS Immunology & Medical Microbiology, 53*, 368–374.10.1111/fim.2008.53.issue-3).

Shigella spp.

A high degree of homogeneity was observed for the detection of *Shigella* spp., both by PCR and by qPCR. The most commonly used target is the *ipaH* gene, the invasion plasmid antigen H. This gene was also observed in some strains of entero invasive *E. coli* (EIEC) (Gómez-Duarte, Bai, & Newell, 2009Gómez-Duarte, O. G., Bai, J., & Newell, E. (2009). Detection of *Escherichia coli*, *Salmonella spp.*, *Shigella spp.*, *Yersinia enterocolitica*, *Vibrio cholerae*, and *Campylobacter* spp.enteropathogens by 3-reaction multiplex polymerase chain reaction. *Diagnostic Microbiology and Infectious Disease, 63*, 1–9.10.1016/j.diagmicrobio.2008.09.006 ; Hsu et al., 2010Hsu, B. M., Wu, S. F., Huang, S. W., Tseng, Y. J., Ji, D. D., Chen, J. S., & Shih, F. C. (2010). Differentiation and identification of *Shigella*spp. and enteroinvasive *Escherichia coli* in environmental waters by a molecular method and biochemical test. *Water Research, 44*, 949–955.10.1016/j.watres.2009.10.004 ; Li, Zhuang, & Mustapha, 2005Li, Y., Zhuang, S., & Mustapha, A. (2005). Application of a multiplex PCR for the simultaneous detection of *Escherichia coli*O157:H7, *Salmonella* and *Shigella* in raw and ready-to-eat meat products. *Meat Science, 71*, 402–406.10.1016/j.meatsci.2005.04.013 ; Thiem et al., 2004Thiem, V. D., Sethabutr, O., von Seidlein, L., Van Tung, T., Canh, D. G., Chien, B. T., … Trach, D. D.(2004). Detection of *Shigella* by a PCR assay targeting the ipaH gene suggests increased prevalence of shigellosis in Nha Trang, Vietnam. *Journal of Clinical Microbiology, 42*, 2031–2035.10.1128/JCM.42.5.2031-2035.2004 ; Wang, Li, & Mustapha, 2007Wang, L. X., Li, Y., & Mustapha, A. (2007). Rapid and simultaneous quantitation of *Escherichia coli* O157:H7, *Salmonella*, and *Shigella* in ground beef by multiplex real-time PCR and immunomagnetic separation. *Journal of Food Protection, 70*(6), 1366–1372., ,). It was included in an in-house multiplex qPCR validation method which simultaneously detected *Salmonella* spp., *Shigella* spp., and *L. monocytogenes*. Excellent results were obtained with natural and spiked samples regarding relative sensitivity, relative specificity, and relative accuracy for the detection of *Shigella* spp. as well as for the other two bacteria (Garrido, Chapela, Román, et al., 2012Garrido, A., Chapela, M. J., Román, B., Ferreira, M., Lago, J., Vieites, J. M., & Cabado, A. G. (2012). Development of a multiplex real-time PCR method for simultaneous detection of *Salmonella enterica*, *Shigella flexneri* and *Listeria monocytogenes* in processed food samples. *European Food Research and Technology, 234*, 571–580.10.1007/s00217-012-1665-3, ,).

Vibrio cholerae

The most widely used genes for the detection of this species are *toxR* (the transcriptional regulator for the genes encoding the major outer membrane porins OmpU and OmpT) and *ompW* (outermembrane protein) (Baron, Chevalier, & Lesne, 2007Baron, S., Chevalier, S., & Lesne, J. (2007). *Vibrio cholerae* in the environment: A simple method for reliable identification of the species. *Journal of Health Population and Nutrition, 25*, 312–318., , ; Goel, Tamrakar, Nema, Kamboj, & Singh, 2005Goel, A. K., Tamrakar, A. K., Nema, V., Kamboj, D. V., & Singh, L. (2005). Detection of viable toxigenic *Vibrio cholerae* from environmental water sources by direct cell duplex PCR assay. *World Journal of Microbiology & Biotechnology,21*, 973–976., , ; Gubala & Proll, 2006Gubala, A. J., & Proll, D. F. (2006). Molecular-beacon multiplex real-time PCR assay for detection of *Vibrio cholerae*. *Applied and Environmental Microbiology, 72*, 6424–6428.10.1128/AEM.02597-05 ; Nandi et al., 2000Nandi, B., Nandy, R. K., Mukhopadhyay, S., Nair, G. B., Shimada, T., & Ghose, A. C. (2000). Rapid method for species-specific identification of *Vibrio cholerae* using primers targeted to the gene of outer membrane protein OmpW. *Journal of Clinical Microbiology, 38*, 4145–4151., , ; Neogi et al., 2010Neogi, S. B., Chowdhury, N., Asakura, M., Hinenoya, A., Haldar, S., Saidi, S. M., … Yamasaki, S. (2010). A highly sensitive and specific multiplex PCR assay for simultaneous detection of *Vibrio cholerae*, *Vibrio parahaemolyticus* and *Vibrio vulnificus*. *Letters in Applied Microbiology, 51*, 293–300.10.1111/lam.2010.51.issue-3 ; Sheikh, Goodarzi, & Aslani, 2012Sheikh, A. F., Goodarzi, H., & Aslani, S. (2012). Identification of *Vibrio cholerae* pathogenicity island (*ctxA*, *OmpW* and *tcpA*) in non-O139 and non-O1 *V. cholerae* strains isolated from Karun River in Ahvaz, Iran. *African Journal of Microbiology Research, 6*, 1185–1189., ; Shuan Ju Teh et al., 2009Shuan Ju Teh, C., Lin Thong, K., Tein Ngoi, S., Ahmad, N., Balakrish Nair, G., & Ramamurthy, T. (2009). Molecular characterization of serogrouping and virulence genes of Malaysian *Vibrio cholerae* isolated from different sources. *The Journal of General and Applied Microbiology, 55*, 419–425.10.2323/jgam.55.419), even though recently the *lolB* (outermembrane lipoprotein) gene has also been described as an optimal target for *V. cholerae* detection (Chua et al., 2011Chua, A. L., Elina, H. T., Lim, B. H., Yean, C. Y., Ravichandran, M., & Lalitha, P. (2011). Development of a dry reagent-based triplex PCR for the detection of toxigenic and non-toxigenic *Vibrio cholerae*. *Journal of Medical Microbiology, 60*, 481–485.10.1099/jmm.0.027433-0 ; Lalitha et al., 2008Lalitha, P., Sid Suraiya, M. N., Lim, K. L., Lee, S. Y., Nur Haslindawaty, A. R., Chan, Y. Y., … Ravichandran, M. (2008). Analysis of *lolB* gene sequence and its use in the development of a PCR assay for the detection of *Vibrio cholerae*. *Journal of Microbiological Methods, 75*, 142–144.10.1016/j.mimet.2008.05.001). Out of these, the most extensively applied and evaluated gene was the *ompW*. Further studies need to be performed with the other targets.

The most important pathogenic gene associated with *V. cholerae* is *ctxA* (gene coding for the A subunit of the cholera toxin) (Blackstone et al., 2007Blackstone, G. M., Nordstrom, J. L., Bowen, M. D., Meyer, R. F., Imbro, P., & DePaola, A. (2007). Use of a real time PCR assay for detection of the *ctxA* gene of *Vibrio cholerae* in an environmental survey of Mobile Bay. *Journal of Microbiological Methods, 68*, 254–259.10.1016/j.mimet.2006.08.006 ; Chapela et al., 2010Chapela, M. J., Fajardo, P., Garrido, A., Cabado, A. G., Ferreira, M., Lago, J., & Vieites, J. M. (2010). Comparison between a TaqMan polymerase chain reaction assay and a culture method for ctx-positive *Vibrio cholerae* detection. *Journal of Agricultural and Food Chemistry, 58*, 4051–4055.10.1021/jf903658k ; Fedio et al., 2007Fedio, W., Blackstone, G. M., Kikuta-Oshima, L., Wendakoon, C., McGrath, T. H., & DePaola, A.(2007). Rapid detection of the *Vibrio cholerae ctx* gene in food enrichments using real-time polymerase chain reaction. *Journal of Aoac International, 90*, 1278–1283., ,), and also *hly* (coding for a hemolysin) (Huang et al., 2009Huang, J., Zhu, Y., Wen, H., Zhang, J., Huang, S., Niu, J., & Li, Q. (2009). Quadruplex real-time PCR assay for detection and identification of *Vibrio cholerae* O1 and O139 strains and determination of their toxigenic potential. *Applied and Environmental Microbiology, 75*, 6981–6985.10.1128/AEM.00517-09 ; Lyon, 2001Lyon, W. J. (2001). TaqMan PCR for detection of *Vibrio cholerae* O1, O139, non-O1, and non-O139 in pure cultures, raw oysters, and synthetic seawater. *Applied and Environmental Microbiology, 67*, 4685–4693.10.1128/AEM.67.10.4685-4693.2001 ; Saravanan, Sanath Kumar, Karunasagar, & Karunasagar, 2007Saravanan, V., Sanath Kumar, H., Karunasagar, I., & Karunasagar, I. (2007). Putative virulence genes of *Vibrio cholerae* from seafoods and the coastal environment of Southwest India. *International Journal of Food Microbiology, 119*, 329–333.10.1016/j.ijfoodmicro.2007.08.023 ; Shuan Ju Teh et al., 2009Shuan Ju Teh, C., Lin Thong, K., Tein Ngoi, S., Ahmad, N., Balakrish Nair, G., & Ramamurthy, T. (2009). Molecular characterization of serogrouping and virulence genes of Malaysian *Vibrio cholerae* isolated from different sources. *The Journal of General and Applied Microbiology, 55*, 419–425.10.2323/jgam.55.419), *tcpA* (toxin-coregulated pilus A) (Sharma & Chaturvedi, 2006Sharma, A., & Chaturvedi, A. N. (2006). Prevalence of virulence genes (*ctxA*, *stn*, *OmpW*and *tcpA*) among non-O1 *Vibrio cholerae*isolated from fresh water environment. *International Journal of Hygiene and Environmental Health, 209*, 521–526.10.1016/j.ijheh.2006.06.005 ; Sheikh et al., 2012Sheikh, A. F., Goodarzi, H., & Aslani, S. (2012). Identification of *Vibrio cholerae* pathogenicity island (*ctxA*, *OmpW* and *tcpA*) in non-O139 and non-O1 *V. cholerae* strains isolated from Karun River in Ahvaz, Iran. *African Journal of Microbiology Research, 6*, 1185–1189.,), and the *zot* (zonula ocludens toxin) (Ang et al., 2010Ang, G. Y., Yu, C. Y., Balqis, K., Elina, H. T., Azura, H., Hani, M. H., & Yean, C. Y. (2010). Molecular evidence of cholera outbreak caused by a toxigenic *Vibrio cholerae* O1 El Tor variant strain in Kelantan, Malaysia. *Journal of Clinical Microbiology, 48*, 3963–3969.10.1128/JCM.01086-10).

Vibrio parahaemolyticus

Detection of *V. parahaemolyticus* has been accomplished by the detection of the *toxR* gene and the *tlh*(thermolabile hemolysin) (Croci et al., 2007Croci, L., Suffredini, E., Cozzi, L., Toti, L., Ottaviani, D., Pruzzo, C., … Mioni, R. (2007). Comparison of different biochemical and molecular methods for the identification of *Vibrio parahaemolyticus*. *Journal of Applied Microbiology, 102*, 229–237.10.1111/jam.2007.102.issue-1 ; Neogi et al., 2010Neogi, S. B., Chowdhury, N., Asakura, M., Hinenoya, A., Haldar, S., Saidi, S. M., … Yamasaki, S. (2010). A highly sensitive and specific multiplex PCR assay for simultaneous detection of *Vibrio cholerae*, *Vibrio parahaemolyticus* and *Vibrio vulnificus*. *Letters in Applied Microbiology, 51*, 293–300.10.1111/lam.2010.51.issue-3 ; Nordstrom et al., 2007Nordstrom, J. L., Vickery, M. C. L., Blackstone, G. M., Murray, S. L., & DePaola, A. (2007). Development of a multiplex real-time PCR assay with an internal amplification control for the detection of total and pathogenic V*ibrio parahaemolyticus* bacteria in oysters. *Applied and Environmental Microbiology, 73*, 5840–5847.10.1128/AEM.00460-07 ; Parveen et al., 2008Parveen, S., Hettiarachchi, K. A., Bowers, J. C., Jones, J. L., Tamplin, M. L., McKay, R., … DePaola, A. (2008). Seasonal distribution of total and pathogenic *Vibrio parahaemolyticus* in Chesapeake Bay oysters and waters. *International Journal of Food Microbiology, 128*, 354–361.10.1016/j.ijfoodmicro.2008.09.019 ; Rizvi & Bej, 2010Rizvi, A. V., & Bej, A. K. (2010). Multiplexed real-time PCR amplification of *tlh*, *tdh* and *trh* genes in *Vibrio parahaemolyticus* and its rapid detection in shellfish and Gulf of Mexico water. *Antonie Van Leeuwenhoek International Journal of General and Molecular Microbiology,98*, 279–290.10.1007/s10482-010-9436-2 ; Rizvi, Panicker, Myers, & Bej, 2006Rizvi, A. V., Panicker, G., Myers, M. L., & Bej, A. K.(2006). Detection of pandemic *Vibrio parahaemolyticus* O3:K6 serovar in Gulf of Mexico water and shellfish using real-time PCR with Taqman (R) fluorescent probes. *FEMS Microbiology Letters, 262*, 185–192.10.1111/fml.2006.262.issue-2). Virulence of *V. parahaemoyliticus* has historically been associated with the presence of two hemolysins encoded in the *tdh* and *trh* genes (thermostable direct hemolysin and hemolysin related with tdh) (Blackstone et al., 2003Blackstone, G. M., Nordstrom, J. L., Vickery, M. C. L., Bowen, M. D., Meyer, R. F., & DePaola, A.(2003). Detection of pathogenic *Vibrio parahaemolyticus* in oyster enrichments by real time PCR. *Journal of Microbiological Methods,53*, 149–155.10.1016/S0167-7012(03)00020-4 ; Garrido, Chapela, Ferreira, et al., 2012Garrido, A., Chapela, M. J., Ferreira, M., Atanassova, M., Fajardo, P., Lago, J., …

Cabado, A. G. (2012). Development of a multiplex real-time PCR method for pathogenic *Vibrio parahaemolyticus* detection (*tdh*+ and *trh*+). *Food Control, 24*, 128–135.10.1016/j.foodcont.2011.09.015, , ; Nordstrom et al., 2007Nordstrom, J. L., Vickery, M. C. L., Blackstone, G. M., Murray, S. L., & DePaola, A. (2007). Development of a multiplex real-time PCR assay with an internal amplification control for the detection of total and pathogenic V*ibrio parahaemolyticus* bacteria in oysters. *Applied and Environmental Microbiology, 73*, 5840–5847.10.1128/AEM.00460-07). Recently, the type three secretion system has been observed to be related with the pathogenic potential of the species (Hiyoshi, Kodama, Iida, & Honda, 2010Hiyoshi, H , Kodama, T., Iida, T., & Honda, T.(2010). Contribution of *Vibrio parahaemolyticus*virulence factors to cytotoxicity, enterotoxicity, and lethality in mice. *Infection and Immunity,78*, 1772–1780.10.1128/IAI.01051-09 ; Pineyro et al., 2010Pineyro, P., Zhou, X. H., Orfe, L. H., Friel, P. J., Lahmers, K., & Call, D. R. (2010). Development of two animal models to study the function of *Vibrio parahaemolyticus* type III secretion systems. *Infection and Immunity, 78*, 4551–4559.10.1128/IAI.00461-10).

Multiplex qPCR methods have been developed for the simultaneous detection of *tlh*, *tdh,* and *trh* (Nordstrom et al., 2007Nordstrom, J. L., Vickery, M. C. L., Blackstone, G. M., Murray, S. L., & DePaola, A. (2007). Development of a multiplex real-time PCR assay with an internal amplification control for the detection of total and pathogenic V*ibrio parahaemolyticus* bacteria in oysters. *Applied and Environmental Microbiology, 73*, 5840–5847.10.1128/AEM.00460-07 ; Ward & Bej, 2006Ward, L. N., & Bej, A. K. (2006). Detection of *Vibrio parahaemolyticus* in shellfish by Use of multiplexed real-time PCR with TaqMan fluorescent probes. *Applied and Environmental Microbiology, 72*, 2031–2042.10.1128/AEM.72.3.2031-2042.2006). These methods successfully detected *V. parahemolyticus* in food and environmental samples.

Vibrio vulnificus

Uniformity has been observed for the detection of *V. vulnificus* at the species level, where *vvhA* (exotoxin hemolysin/cytolysin) (Campbell & Wright, 2003Campbell, M. S., & Wright, A. C. (2003). Real-time PCR analysis of *Vibrio vulnificus* from oysters. *Applied and Environmental Microbiology, 69*, 7137–7144.10.1128/AEM.69.12.7137-7144.2003 ; Neogi et al., 2010Neogi, S. B., Chowdhury, N., Asakura, M., Hinenoya, A., Haldar, S., Saidi, S. M., … Yamasaki, S. (2010). A highly sensitive and specific multiplex PCR assay for simultaneous detection of *Vibrio cholerae*, *Vibrio parahaemolyticus* and *Vibrio vulnificus*. *Letters in Applied Microbiology, 51*, 293–300.10.1111/lam.2010.51.issue-3 ; Panicker & Bej, 2005Panicker, G., & Bej, A. K. (2005). Real-time PCR detection of *Vibrio vulnificus* in oysters: Comparison of oligonucleotide primers and probes targeting *vvhA*. *Applied and Environmental Microbiology, 71*, 5702–5709.10.1128/AEM.71.10.5702-5709.2005 ; Wright et al., 2007Wright, A. C., Garrido, V., Debuex, G., Farrell-Evans, M., Mudbidri, A. A., & Otwell, W. S.(2007). Evaluation of postharvest-processed oysters by using PCR-based most-probable-number enumeration of *Vibrio vulnificus*bacteria. *Applied and Environmental Microbiology, 73*, 7477–7481.10.1128/AEM.01118-07) is one of the most frequently applied targets. This homogeneity is not observed when trying to assess the pathogenic potential of the bacterium, where several targets have been proposed, such as the *rtxA*, *vcg*C/E (virulence-correlated gene clinical/environmental), and 16S rRNA type A (environmental) and type B (clinical) (Drake, Whitney, Levine, DePaola, & Jaykus, 2010Drake, S. L., Whitney, B., Levine, J. F., DePaola, A., & Jaykus, L.-A. (2010). Correlation of mannitol fermentation with virulence-associated genotypic characteristics in *Vibrio vulnificus* isolates from oysters and water samples in the Gulf of Mexico. *Foodborne Pathogens and Disease, 7*, 97–101.10.1089/fpd.2009.0362 ; Rosche, Binder, & Oliver, 2010Rosche, T. M., Binder, E. A., & Oliver, J. D. (2010). *Vibrio vulnificus* genome suggests two distinct ecotypes. *Environmental Microbiology Reports,2*, 128–132.10.1111/j.1758-2229.2009.00119.x ; Rosche, Yano, & Oliver, 2005Rosche, T. M., Yano, Y., & Oliver, J. D. (2005). A rapid and simple PCR analysis indicates there are two subgroups of *Vibrio vulnificus* which correlate with clinical or environmental isolation. *Microbiology and Immunology, 49*, 381–389.10.1111/mim.2005.49.issue-4). Recently, a polimorfism in the *pilF* gene has been described as a good marker of the pathogenic potential of *V. vulnificus* (Baker-Austin et al., 2012Baker-Austin, C., Lemm, E., Hartnell, R., Lowther, J., Onley, R., Amaro, C., … Lees, D.(2012). *pilF* polymorphism-based real-time PCR to distinguish *Vibrio vulnificus* strains of human health relevance. *Food Microbiology, 30*, 17–23., ; Fajardo et al., 2014Fajardo, P., Atanassova, M., Garrido-Maestu, A., Wortner-Smith, T., Cotterill, J., & Cabado, A. G.(2014). Bacteria isolated from shellfish digestive gland with antipathogenic activity as candidates to increase the efficiency of shellfish depuration process. *Food Control, 46*, 272–281.10.1016/j.foodcont.2014.05.038, , ; Garrido-Maestu, Chapela, Román, Vieites, & Cabado, 2014Garrido-Maestu, A., Chapela, M. J., Román, B., Vieites, J. M., & Cabado, A. G. (2014). Application of a novel pathogenicity marker in a multiplex real-time PCR method to assess total and pathogenic *Vibrio vulnificus* in food and environmental samples. *Food Control, 35*, 274–283.10.1016/j.foodcont.2013.07.007, , ; Roig, Sanjuan, Llorens, & Amaro, 2010Roig, F. J., Sanjuan, E., Llorens, A., & Amaro, C.(2010). *pilF* polymorphism-based PCR to distinguish *Vibrio vulnificus* strains potentially dangerous to public health. *Applied and Environmental Microbiology, 76*, 1328–1333.10.1128/AEM.01042-09).

Multiplex qPCR methods for the simultaneous detection of total and pathogenic *V. vulnificus* are scarce (Garrido-Maestu et al., 2014Garrido-Maestu, A., Chapela, M. J., Román, B., Vieites, J. M., & Cabado, A. G. (2014). Application of a novel pathogenicity marker in a multiplex real-time PCR method to assess total and pathogenic *Vibrio vulnificus* in food and environmental samples. *Food Control, 35*, 274–283.10.1016/j.foodcont.2013.07.007, ,). Most methods combined specific gene targets for the detection at the species level *V. vulnificus*, *V. cholerae,* and *V. parahemolyticus* (Bauer & Rørvik, 2007Bauer, A., & Rørvik, L. M. (2007). A novel multiplex PCR for the identification of *Vibrio parahaemolyticus*, *Vibrio cholerae* and *Vibrio vulnificus*. *Letters in Applied Microbiology, 45*, 371–375.10.1111/lam.2007.45.issue-4 ; Garrido-Maestu, Chapela, Peñaranda, Vieites, & Cabado, 2014Garrido-Maestu, A., Chapela, M.-J., Peñaranda, E., Vieites, J. M., & Cabado, A. G. (2014). In-house validation of novel multiplex real-time PCR gene combination for the simultaneous detection of the main human pathogenic vibrios (*Vibrio cholerae*, *Vibrio parahaemolyticus*, and *Vibrio vulnificus*). *Food Control, 37*, 371–379.10.1016/j.foodcont.2013.09.026, , ; Izumiya et al., 2011Izumiya, H., Matsumoto, K., Yahiro, S., Lee, J., Morita, M., Yamamoto, S., … Ohnishi, M. (2011). Multiplex PCR assay for identification of three major pathogenic *Vibrio* spp., *Vibrio cholerae*, *Vibrio parahaemolyticus*, and *Vibrio vulnificus*. *Molecular and Cellular Probes, 25*, 174–176.10.1016/j.mcp.2011.04.004 ; Neogi et al., 2010Neogi, S. B., Chowdhury, N., Asakura, M., Hinenoya, A., Haldar, S., Saidi, S. M., … Yamasaki, S. (2010). A highly sensitive and specific multiplex PCR assay for simultaneous detection of *Vibrio cholerae*, *Vibrio parahaemolyticus* and *Vibrio vulnificus*. *Letters in Applied Microbiology, 51*, 293–300.10.1111/lam.2010.51.issue-3 ; Panicker, Call, Krug, & Bej, 2004 Panicker, G., Call, D. R., Krug, M. J., & Bej, A. K.(2004). Detection of pathogenic *Vibrio* spp. in shellfish by using multiplex PCR and DNA microarrays. *Applied and Environmental Microbiology, 70*, 7436–7444.10.1128/AEM.70.12.7436-7444.2004).

COMMERCIAL KITS AVAILABLE FOR MULTIPLEX QPCR

At present, a wide variety of commercial qPCR kits are available for detection of single micro-organisms from different companies as iQ-Check from Bio-Rad, DuPont Qualicon™ BAX®, foodproof® Salmonella from Merck-Millipore, AnDiaTec® from Roche or MicroSEQ® from Applied Biosystems. All these brands have validated solutions for *Salmonella* spp., *L. monocytogenes*, *Campylobacter* spp., and *Legionella*, among others. The problem arises when a multiplex qPCR approach is desired, as the commercial kits are scarce. A possible solution may be the selection of open-formula studies that have gone through an in-house validation where parameters like specificity of the primers and probes, LOD, diagnostic specificity, diagnostic sensitivity, and diagnostic accuracy are evaluated against a reference method. In this context, a very common combination of bacterial pathogens found in the literature include simultaneous detection of *Salmonella* spp. and *L. monocytogenes* (Badosa, Chico, Pla, Pares, & Montesinos, 2009Badosa, E., Chico, N., Pla, M., Parés, D., & Montesinos, E. (2009). Evaluation of ISO enrichment real-time PCR methods with internal amplification control for detection of *Listeria monocytogenes* and *Salmonella enterica* in fresh fruit and vegetables. *Letters in Applied Microbiology, 49*, 105–111.10.1111/lam.2009.49.issue-1 ; Chua & Bhagwat, 2009Chua, T., & Bhagwat, A. A. (2009). A rapid and simple DNA extraction procedure to detect *Salmonella* spp. and *Listeria monocytogenes*from fresh produce using real-time PCR. *Food Analytical Methods, 2*, 96–101.10.1007/s12161-008-9032-5, , ; Garrido, Chapela, Román, Fajardo, Lago, et al., 2013Garrido, A., Chapela, M.-J., Román, B., Fajardo, P., Lago, J., Vieites, J. M., & Cabado, A. G. (2013). A new multiplex real-time PCR developed method for *Salmonella* spp. and *Listeria monocytogenes* detection in food and environmental samples. *Food Control, 30*, 76–85.10.1016/j.foodcont.2012.06.029, , ; Jofré et al., 2005Jofré, A., Martin, B., Garriga, M., Hugas, M., Pla, M., Rodríguez-Lázaro, D., & Aymerich, T. (2005). Simultaneous detection of *Listeria monocytogenes* and *Salmonella* by multiplex PCR in cooked ham. *Food Microbiology, 22*, 109–115.10.1016/j.fm.2004.04.009, , ; Singh, Batish, & Grover, 2011Singh, J., Batish, V. K., Grover, S. (2011). Simultaneous detection of *Listeria monocytogenes* and *Salmonella* spp. in dairy products using real time PCR-melt curve analysis. *Journal of Food Science and Technology, 49*, 234–239.). This combination should be more useful when the method also includes of *E. coli* O157 (Elizaquível et al., 2011Elizaquível,

P., Gabaldón, J. A., & Aznar, R.(2011). Quantification of *Salmonella* spp, *Listeria monocytogenes* and *Escherichia coli*O157 H7 in non-spiked food products and evaluation of real-time PCR as a diagnostic tool in routine food analysis. *Food Control, 22*, 158–164.10.1016/j.foodcont.2010.05.018, , ; Garrido, Chapela, Román, Fajardo, Vieites, et al., 2013Garrido, A., Chapela, M.-J., Román, B., Fajardo, P., Vieites, J. M., & Cabado, A. G. (2013). In-house validation of a multiplex real-time PCR method for simultaneous detection of *Salmonella* spp., *Escherichia coli* O157 and *Listeria monocytogenes*. *International Journal of Food Microbiology, 164*, 92–98.10.1016/j.ijfoodmicro.2013.03.024 ; Kawasaki et al., 2010Kawasaki, S., Fratamico, P. M., Horikoshi, N., Okada, Y., Takeshita, K., Sameshima, T., & Kawamoto, S. (2010). Multiplex real-time polymerase chain reaction assay for simultaneous detection and quantification of *Salmonella* species, *Listeria monocytogenes*, and *Escherichia coli* O157:H7 in ground pork samples. *Foodborne Pathogens and Disease, 7*, 549–554.10.1089/fpd.2009.0465 ; Omiccioli et al., 2009Omiccioli, E., Amagliani, G., Brandi, G., & Magnani, M. (2009). A new platform for real-time PCR detection of *Salmonella* spp., *Listeria monocytogenes* and *Escherichia coli* O157 in milk. *Food Microbiology, 26*, 615–622.10.1016/j.fm.2009.04.008) or others food-borne pathogens like *Shigella* spp. (Garrido, Chapela, Román, et al., 2012Garrido, A., Chapela, M. J., Román, B., Ferreira, M., Lago, J., Vieites, J. M., & Cabado, A. G. (2012). Development of a multiplex real-time PCR method for simultaneous detection of *Salmonella enterica*, *Shigella flexneri* and *Listeria monocytogenes* in processed food samples. *European Food Research and Technology, 234*, 571–580.10.1007/s00217-012-1665-3, , ; Wang, Li, & Mustapha, 2007Wang, L. X., Li, Y., & Mustapha, A. (2007). Rapid and simultaneous quantitation of *Escherichia coli* O157:H7, *Salmonella*, and *Shigella* in ground beef by multiplex real-time PCR and immunomagnetic separation. *Journal of Food Protection, 70*(6), 1366–1372., ,).

FUTURE CHALLENGES

An international standard guideline (ISO 16140:2003) was prepared by CEN/TC275 in collaboration with the Technical Committee ISO/TC34 to establish the general principle and technical procedure for the validation of alternative methods in the field of microbiological analysis of food, animal feeding stuffs, and environmental and veterinary samples. Many of the performance criteria are well recognized and accepted by numerous internationally recognized organizations for standardization (ISO, EMEA, Codex Alimentarius, etc.). There is a series of concepts that define an analytical method and that are applied to diagnostic qPCR. The main concepts are: linearity, accuracy, precision, sensitivity, and specificity. Progressive introduction of automated equipment for qPCR has enabled it to become a high-throughput method. The numerous examples cited in this review and their recentness illustrate the current interest for qPCR methods in food microbiology. However, not all fields of application are equally developed yet, in particular, the multiplex strategies as described above. There is an obvious lack of consensus considering the appropriate approach to perform the experiments and interpret the data (Bustin et al., 2009 Bustin, S. A., Benes, V., Garson, J. A., Hellemans, J., Huggett, J., Kubista, M., … Wittwer, C. T.(2009). The MIQE guidelines: Minimum information for publication of quantitative real-time PCR experiments. *Clinical Chemistry, 55*, 611–622.10.1373/clinchem.2008.112797 , 2010Bustin, S. A., Beaulieu, J.-F., Huggett, J., Jaggi, R., Kibenge, F. S. B., Olsvik, P. A., … Toegel, S.(2010). MIQE precis: Practical implementation of minimum standard guidelines for fluorescence-based quantitative real-time PCR experiments. *BMC Molecular Biology, 11*, 74. ; Huggett et al., 2013Huggett, J. F., Foy, C. A., Benes, V., Emslie, K., Garson, J. A., Haynes, R., … Bustin, S. A. (2013). The digital MIQE guidelines: Minimum information for publication of quantitative digital PCR experiments. *Clinical Chemistry, 59*, 892–902.10.1373/clinchem.2013.206375).

CONCLUSIONS

Molecular techniques applied in food microbiology are very promising alternatives to classical cultures, biochemical and serological identification and typification methods. Among them, PCR and real-time PCR have been successfully used, not only in control laboratories, but also in research to gain deeper knowledge in the biology and pathogenesis of bacterial species.

ACKNOWLEDGMENT

We would like to thank Mr Manuel Mazoy, head of the Border Inspection Post (BIP)-Vigo (Spain), for the kind collaboration in data searching.

REFERENCES

1. Abdulmawjood, A., Bülte, M., Cook, N., Roth, S., Schönenbrücher, H., & Hoorfar, J. (2003). Toward an international standard for PCR-based detection of *Escherichia coli* O157. *Journal of Microbiological Methods, 55*, 775–786.10.1016/j.mimet.2003.08.012
2. Abdulmawjood, A., Bulte, M., Roth, S., Schonenbrucher, H., Cook, N., Heuvelink, A. E., et al. (2004). Development, validation, and standardization of polymerase chain reaction-based detection of *E. coli* O157. *Journal of Aoac International, 87*, 596–603.
3. Adiguzel, G., Gulluce, M., Bozoglu, C., Yanmis, D., Gormez, A., Atasever, M., & Adiguzel, A. (2012). Molecular characterization of *Escherichia coli* O157:H7 from retail beef in Erzurum, Turkey. *Journal of Pure and Applied Microbiology, 6*, 1033–1041.
4. Agroalimentaria, E. F. V. p. l. S. (2012). *Informe RASFF*.
5. Al-Ajmi, D., Padmanabha, J., Denman, S. E., Gilbert, R. A., Al Jassim, R. A. M., & McSweeney, C. S. (2006). Evaluation of a PCR detection method for *Escherichia coli* O157:H7/H-bovine faecal samples. *Letters in Applied Microbiology, 42*,386–391.10.1111/lam.2006.42.issue-4
6. Anderson, A., Pietsch, K., Zucker, R., Mayr, A., Müller-Hohe, E., Messelhäusser, U., … Huber, I. (2011). Validation of a duplex real-time PCR for the detection of *Salmonella* spp. in different food products. *Food Analytical Methods, 4*, 259–267.10.1007/s12161-010-9142-8
7. Ang, G. Y., Yu, C. Y., Balqis, K., Elina, H. T., Azura, H., Hani, M. H., & Yean, C. Y. (2010). Molecular evidence of cholera outbreak caused by a toxigenic *Vibrio cholerae* O1 El Tor variant strain in Kelantan, Malaysia. *Journal of Clinical Microbiology, 48*, 3963–3969.10.1128/JCM.01086-10
8. Badosa, E., Chico, N., Pla, M., Parés, D., & Montesinos, E. (2009). Evaluation of ISO enrichment real-time PCR methods with internal amplification control for detection of *Listeria monocytogenes* and *Salmonella enterica* in fresh fruit and vegetables. *Letters in Applied Microbiology, 49*, 105–111.10.1111/lam.2009.49.issue-1
9. Bai, J., Shi, X., & Nagaraja, T. G. (2010). A multiplex PCR procedure for the detection of six major virulence genes in*Escherichia coli* O157:H7. *Journal of Microbiological Methods, 82*, 85–89.10.1016/j.mimet.2010.05.003
10. Bailey, J. S., & Cox, N. A. (1992). Universal preenrichment broth for

the simultaneous detection of *Salmonella* and*Listeria* in foods. *Journal of Food Protection, 55*, 256–259.

11. Baker-Austin, C., Lemm, E., Hartnell, R., Lowther, J., Onley, R., Amaro, C., … Lees, D. (2012). *pilF* polymorphism-based real-time PCR to distinguish *Vibrio vulnificus* strains of human health relevance. *Food Microbiology, 30*, 17–23.
12. Baron, S., Chevalier, S., & Lesne, J. (2007). *Vibrio cholerae* in the environment: A simple method for reliable identification of the species. *Journal of Health Population and Nutrition, 25*, 312–318.
13. Bauer, A., & Rørvik, L. M. (2007). A novel multiplex PCR for the identification of *Vibrio parahaemolyticus*, *Vibrio cholerae* and *Vibrio vulnificus*. *Letters in Applied Microbiology, 45*, 371–375.10.1111/lam.2007.45.issue-4
14. Bhagwat, A. A. (2004). Rapid detection of *Salmonella* from vegetable rinse-water using real-time PCR. *Food Microbiology, 21*, 73–78.10.1016/S0740-0020(03)00020-0
15. Blackstone, G. M., Nordstrom, J. L., Bowen, M. D., Meyer, R. F., Imbro, P., & DePaola, A. (2007). Use of a real time PCR assay for detection of the *ctxA* gene of *Vibrio cholerae* in an environmental survey of Mobile Bay. *Journal of Microbiological Methods, 68*, 254–259.10.1016/j.mimet.2006.08.006
16. Blackstone, G. M., Nordstrom, J. L., Vickery, M. C. L., Bowen, M. D., Meyer, R. F., & DePaola, A. (2003). Detection of pathogenic *Vibrio parahaemolyticus* in oyster enrichments by real time PCR. *Journal of Microbiological Methods, 53*,149–155.10.1016/S0167-7012(03)00020-4
17. Bonetta, S., Borelli, E., Bonetta, S., Conio, O., Palumbo, F., & Carraro, E. (2011). Development of a PCR protocol for the detection of *Escherichia coli* O157:H7 and *Salmonella* spp. in surface water. *Environmental Monitoring and Assessment, 177*, 493–503.10.1007/s10661-010-1650-x
18. Botteldoorn, N., Van Coillie, E., Piessens, V., Rasschaert, G., Debruyne, L., Heyndrickx, M., … Messens, W. (2008).Quantification of *Campylobacter* spp. in chicken carcass rinse by real-time PCR. *Journal of Applied Microbiology, 105*,1909–1918.10.1111/jam.2008.105.issue-6
19. Bustin, S. A., Beaulieu, J.-F., Huggett, J., Jaggi, R., Kibenge, F.

S. B., Olsvik, P. A., … Toegel, S. (2010). MIQE precis: Practical implementation of minimum standard guidelines for fluorescence-based quantitative real-time PCR experiments. *BMC Molecular Biology, 11*, 74.

20. Bustin, S. A., Benes, V., Garson, J. A., Hellemans, J., Huggett, J., Kubista, M., … Wittwer, C. T. (2009). The MIQE guidelines: Minimum information for publication of quantitative real-time PCR experiments. *Clinical Chemistry, 55*,611–622.10.1373/clinchem.2008.112797
21. Calvó, L., Martínez-Planells, A., Pardos-Bosch, J., & Garcia-Gil, L. J. (2008). A new real-time PCR assay for the specific detection of *Salmonella* spp. targeting the *bipA* gene. *Food Analytical Methods, 1*, 236–242.10.1007/s12161-007-9008-x
22. Campbell, M. S., & Wright, A. C. (2003). Real-time PCR analysis of *Vibrio vulnificus* from oysters. *Applied and Environmental Microbiology, 69*, 7137–7144.10.1128/AEM.69.12.7137-7144.2003
23. CDC. (2013). Incidence and trends of infection with pathogens transmitted commonly through food-foodborne diseases active surveillance network, 10 US sites, 1996–2012. *MMWR. Morbidity and mortality weekly report, 62*, 283–287.
24. Chakravorty, S., Helb, D., Burday, M., Connell, N., & Alland, D. (2007). A detailed analysis of 16S ribosomal RNA gene segments for the diagnosis of pathogenic bacteria. *Journal of Microbiological Methods, 69*, 330–339.10.1016/j.mimet.2007.02.005
25. Chapela, M. J., Fajardo, P., Garrido, A., Cabado, A. G., Ferreira, M., Lago, J., & Vieites, J. M. (2010). Comparison between a TaqMan polymerase chain reaction assay and a culture method for ctx-positive *Vibrio cholerae* detection. *Journal of Agricultural and Food Chemistry, 58*, 4051–4055.10.1021/jf903658k
26. Chapela, M. J., Sotelo, C. G., Calo-Mata, P., Perez-Martin, R. I., Rehbein, H., Hold, G. L., Santos, A. T. (2002).Identification of cephalopod species (Ommastrephidae and Loliginidae) in seafood products by forensically informative nucleotide sequencing (FINS). *Journal of Food Science, 67*, 1672–1676.10.1111/jfds.2002.67.issue-5
27. Cheng, C. M., Lin, W., Van, K. T., Phan, L., Tran, N. N., & Farmer, D. (2008). Rapid detection of *Salmonella* in foods using real-time PCR. *Journal of Food Protection, 71*, 2436–2441.
28. Cheng, C. M., Van, K. T., Lin, W., & Ruby, R. M. (2009). Interlaboratory

validation of a real-time PCR 24-hour rapid method for detection of *Salmonella* in foods. *Journal of Food Protection, 72*, 945–951.

29. Cheyne, B., Van Dyke, M., Anderson, W., & Huck, P. (2010). The detection of *Yersinia enterocolitica* in surface water by quantitative PCR amplification of the *ail* and *yadA* genes. *Journal of Water and Health, 8*, 487–499.10.2166/wh.2009.215
30. Chua, A. L., Elina, H. T., Lim, B. H., Yean, C. Y., Ravichandran, M., & Lalitha, P. (2011). Development of a dry reagent-based triplex PCR for the detection of toxigenic and non-toxigenic *Vibrio cholerae*. *Journal of Medical Microbiology,60*, 481–485.10.1099/jmm.0.027433-0
31. Chua, T., & Bhagwat, A. A. (2009). A rapid and simple DNA extraction procedure to detect *Salmonella* spp. and *Listeria monocytogenes* from fresh produce using real-time PCR. *Food Analytical Methods, 2*, 96–101.10.1007/s12161-008-9032-5
32. Commission, E. (2011). *RASFF-Annual Report.*
33. Croci, L., Suffredini, E., Cozzi, L., Toti, L., Ottaviani, D., Pruzzo, C., … Mioni, R. (2007). Comparison of different biochemical and molecular methods for the identification of *Vibrio parahaemolyticus*. *Journal of Applied Microbiology, 102*, 229–237.10.1111/jam.2007.102.issue-1
34. D'Agostino, M., Wagner, M., Vazquez-Boland, J. A., Kuchta, T., Karpiskova, R., Hoorfar, J., … Cook, N. (2004). A validated PCR-based method to detect *Listeria monocytogenes* using raw milk as a food model—Towards an international standard. *Journal of Food Protection, 67*, 1646–1655.
35. D'Urso, O. F., Poltronieri, P., Marsigliante, S., Storelli, C., Hernández, M., & Rodríguez-Lázaro, D. (2009). A filtration-based real-time PCR method for the quantitative detection of viable *Salmonella enterica* and *Listeria monocytogenes*in food samples. *Food Microbiology, 26*, 311–316.10.1016/j.fm.2008.12.006
36. Dalmasso, M., Bolocan, A. S., Hernandez, M., Kapetanakou, A. E., Kuchta, T., Manios, S. G., Jordan, K. (2014).Comparison of polymerase chain reaction methods and plating for analysis of enriched cultures of *Listeria monocytogenes* when using the ISO11290-1 method. *Journal of Microbiological Methods, 98*, 8–14.10.1016/j.mimet.2013.12.018
37. Debretsion, A., Habtemariam, T., Wilson, S., Nganwa, D., & Yehualaeshet, T. (2007). Real-time PCR assay for rapid detection and quantification of *Campylobacter jejuni* on chicken rinses from

poultry processing plant. *Molecular and Cellular Probes, 21*, 177–181.10.1016/j.mcp.2006.10.006

38. DebRoy, C., Roberts, E., Kundrat, J., Davis, M. A., Briggs, C. E., & Fratamico, P. M. (2004). Detection of *Escherichia coli*serogroups O26 and O113 by PCR amplification of the wzx and wzy genes. *Applied and Environmental Microbiology,70*, 1830–1832.10.1128/AEM.70.3.1830-1832.2004

39. Delibato, E., Rodriguez-Lázaro, D., Gianfranceschi, M., De Cesare, A., Comin, D., Gattuso, A., ... De Medici, D. (2014).European validation of real-time PCR method for detection of *Salmonella* spp. in pork meat. *International Journal of Food Microbiology, 184*, 134–138.

40. Drake, S. L., Whitney, B., Levine, J. F., DePaola, A., & Jaykus, L.-A. (2010). Correlation of mannitol fermentation with virulence-associated genotypic characteristics in *Vibrio vulnificus* isolates from oysters and water samples in the Gulf of Mexico. *Foodborne Pathogens and Disease, 7*, 97–101.10.1089/fpd.2009.0362

41. (EC), C. R. (2002). Regulation (EC) No. 178/2002 of the European Parliament and of the Council of 28 January2002.Laying down the general principles and requirements of food law, establishing the European food safety authorityand laying down procedures in matters of food safety. *Official Journal of the European Communities, Official Journal of the European Union, 178/2002.*

42. (EC), C. R. (2005). Microbiological criteria for foodstuffs. *Official Journal of the European Union, 2073/2005.*

43. Elizaquivel, P., & Aznar, R. (2008). Comparison of four commercial DNA extraction kits for PCR detection of *Listeria monocytogenes, Salmonella, Escherichia coli* O157:H7, and *Staphylococcus aureus* in fresh, minimally processed vegetables. *Journal of Food Protection, 71*, 2110–2114.

44. Elizaquível, P., Gabaldón, J. A., & Aznar, R. (2011). Quantification of *Salmonella* spp, *Listeria monocytogenes* and*Escherichia coli* O157 H7 in non-spiked food products and evaluation of real-time PCR as a diagnostic tool in routine food analysis. *Food Control, 22*, 158–164.10.1016/j.foodcont.2010.05.018

45. Elnifro, E. M., Ashshi, A. M., Cooper, R. J., & Klapper, P. E. (2000). Multiplex PCR: Optimization and application in diagnostic virology. *Clinical Microbiology Reviews*, 13, 559–570.10.1128/CMR.13.4.559-570.2000

46. (EU), C. R. (2011). Amending Annex II to Regulation (EC) No. 2160/2003 of the European parliament and of the council and annex I to Commission Regulation (EC) No. 2073/2005 as regards *Salmonella* in fresh poultry meat. *Official Journal of the European Union, 1086/2011*.

47. Fajardo, P., Atanassova, M., Garrido-Maestu, A., Wortner-Smith, T., Cotterill, J., & Cabado, A. G. (2014). Bacteria isolated from shellfish digestive gland with antipathogenic activity as candidates to increase the efficiency of shellfish depuration process. *Food Control, 46*, 272–281.10.1016/j.foodcont.2014.05.038

48. Fedio, W., Blackstone, G. M., Kikuta-Oshima, L., Wendakoon, C., McGrath, T. H., & DePaola, A. (2007). Rapid detection of the *Vibrio cholerae ctx* gene in food enrichments using real-time polymerase chain reaction. *Journal of Aoac International, 90*, 1278–1283.

49. Flekna, G., Schneeweiss, W., Smulders, F. J. M., Wagner, M., & Hein, I. (2007). Real-time PCR method with statistical analysis to compare the potential of DNA isolation methods to remove PCR inhibitors from samples for diagnostic PCR. *Molecular and Cellular Probes, 21*, 282–287.10.1016/j.mcp.2007.02.001

50. Flekna, G., Štefanič, P., Wagner, M., Smulders, F. J. M., Možina, S. S., & Hein, I. (2007). Insufficient differentiation of live and dead *Campylobacter jejuni* and *Listeria monocytogenes* cells by ethidium monoazide (EMA) compromises EMA/real-time PCR. *Research in Microbiology, 158*, 405–412.10.1016/j.resmic.2007.02.008

51. Franz, E., Klerks, M. A., De Vos, O. J., Termorshuizen, A. J., & van Bruggen, A. H. C. (2007). Prevalence of Shiga toxin-producing *Escherichia coli stx1*, *stx2*, *eaeA*, and *rfbE* genes and survival of *E. coli* O157:H7 in manure from organic and low-input conventional dairy farms. *Applied and Environmental Microbiology, 73*, 2180–2190.10.1128/AEM.01950-06

52. Fratamico, P. M., & DebRoy, C. (2010). Detection of *Escherichia coli* O157:H7 in food using real-time multiplex PCR assays targeting the *stx 1*, *stx 2*, *wzy O157*, and the *fliC h7* or *eae* genes. *Food Analytical Methods, 3*, 330–337.10.1007/s12161-010-9140-x

53. Garrido, A., Chapela, M. J., Ferreira, M., Atanassova, M., Fajardo, P., Lago, J., … Cabado, A. G. (2012). Development of a multiplex real-time PCR method for pathogenic *Vibrio parahaemolyticus* detection (*tdh*+ and *trh*+). *Food Control, 24*,128–135.10.1016/j.foodcont.2011.09.015

54. Garrido, A., Chapela, M.-J., Román, B., Fajardo, P., Lago, J., Vieites, J.

M., & Cabado, A. G. (2013). A new multiplex real-time PCR developed method for *Salmonella* spp. and *Listeria monocytogenes* detection in food and environmental samples. *Food Control, 30*, 76–85.10.1016/j.foodcont.2012.06.029

55. Garrido, A., Chapela, M.-J., Román, B., Fajardo, P., Vieites, J. M., & Cabado, A. G. (2013). In-house validation of a multiplex real-time PCR method for simultaneous detection of *Salmonella* spp., *Escherichia coli* O157 and *Listeria monocytogenes*. *International Journal of Food Microbiology, 164*, 92–98.10.1016/j.ijfoodmicro.2013.03.024
56. Garrido, A., Chapela, M. J., Román, B., Ferreira, M., Lago, J., Vieites, J. M., & Cabado, A. G. (2012). Development of a multiplex real-time PCR method for simultaneous detection of *Salmonella enterica*, *Shigella flexneri* and *Listeria monocytogenes* in processed food samples. *European Food Research and Technology, 234*, 571–580.10.1007/s00217-012-1665-3
57. Garrido-Maestu, A., Chapela, M.-J., Peñaranda, E., Vieites, J. M., & Cabado, A. G. (2014). In-house validation of novel multiplex real-time PCR gene combination for the simultaneous detection of the main human pathogenic vibrios (*Vibrio cholerae*, *Vibrio parahaemolyticus*, and *Vibrio vulnificus*). *Food Control, 37*, 371–379.10.1016/j.foodcont.2013.09.026
58. Garrido-Maestu, A., Chapela, M. J., Román, B., Vieites, J. M., & Cabado, A. G. (2014). Application of a novel pathogenicity marker in a multiplex real-time PCR method to assess total and pathogenic *Vibrio vulnificus* in food and environmental samples. *Food Control, 35*, 274–283.10.1016/j.foodcont.2013.07.007
59. Gattuso, A., Gianfranceschi, M. V., Sonnessa, M., Delibato, E., Marchesan, M., Hernandez, M., ... Rodriguez-Lazaro, D.(2014). Optimization of a real time PCR based method for the detection of *Listeria monocytogenes* in pork meat.*International Journal of Food Microbiology,184*, 106–108.
60. Germini, A., Masola, A., Carnevali, P., & Marchelli, R. (2009). Simultaneous detection of *Escherichia coli* O175:H7,*Salmonella* spp., and *Listeria monocytogenes* by multiplex PCR. *Food Control, 20*, 733–738.10.1016/j.foodcont.2008.09.010
61. Gianfranceschi, M. V., Rodriguez-Lazaro, D., Hernandez, M., Gonzalez-Garcia, P., Comin, D., Gattuso, A., ... De Medici,D. (2014). European validation of a real-time PCR-based method for detection of

Listeria monocytogenes in soft cheese. *International Journal of Food Microbiology, 184*, 128–133.

62. Goel, A. K., Tamrakar, A. K., Nema, V., Kamboj, D. V., & Singh, L. (2005). Detection of viable toxigenic *Vibrio cholerae*from environmental water sources by direct cell duplex PCR assay. *World Journal of Microbiology & Biotechnology,21*, 973–976.
63. Gómez-Duarte, O. G., Bai, J., & Newell, E. (2009). Detection of *Escherichia coli, Salmonella spp., Shigella spp., Yersinia enterocolitica, Vibrio cholerae*, and *Campylobacter* spp. enteropathogens by 3-reaction multiplex polymerase chain reaction. *Diagnostic Microbiology and Infectious Disease, 63*, 1–9.10.1016/j.diagmicrobio.2008.09.006
64. Gonzales, T. K., Kulow, M., Park, D. J., Kaspar, C. W., Anklam, K. S., Pertzborn, K. M., et al. (2011). A high-throughput open-array qPCR gene panel to identify, virulotype, and subtype O157 and non-O157 enterohemorrhagic *Escherichia coli. Molecular and Cellular Probes, 25*, 222–230.10.1016/j.mcp.2011.08.004
65. González-Escalona, N., Brown, E. W., & Zhang, G. (2012). Development and evaluation of a multiplex real-time PCR (qPCR) assay targeting *ttrRSBCA* locus and *invA* gene for accurate detection of *Salmonella* spp. in fresh produce and eggs. *Food Research International, 48*, 202–208.
66. Gubala, A. J., & Proll, D. F. (2006). Molecular-beacon multiplex real-time PCR assay for detection of *Vibrio cholerae.Applied and Environmental Microbiology, 72*, 6424–6428.10.1128/AEM.02597-05
67. Hammack, T. S., Amaguana, R. M., Johnson, M. L., & Andrews, W. H. (2003). Effectiveness of universal pre-enrichment broth for recovery of *Salmonella* from selected dairy foods. *Journal of Aoac International, 86*, 714–718.
68. He, Y., Yao, X., Gunther, N. W., Xie, Y., Tu, S.-I., & Shi, X. (2010). Simultaneous detection and differentiation of*Campylobacter jejuni, C. coli*, and *C. lari* in chickens using a multiplex real-time PCR assay. *Food Analytical Methods, 3*,321–329.10.1007/s12161-010-9136-6
69. Heid, C. A., Stevens, J., Livak, K. J., & Williams, P. M. (1996). Real time quantitative PCR. *Genome Research, 6*, 986–994.10.1101/gr.6.10.986
70. Hierro, N., Esteve-Zarzoso, B., Gonzalez, A., Mas, A., & Guillamon, J. M. (2006). Real-time quantitative PCR (QPCR) and reverse

transcription-QPCR for detection and enumeration of total yeasts in wine. *Applied and Environmental Microbiology, 72*, 7148–7155.10.1128/AEM.00388-06

71. Higuchi, R., Dollinger, G., Walsh, P. S., & Griffith, R. (1992). Simultaneous amplification and detection of specific DNA-sequences. *Bio/Technology, 10*, 413–417.10.1038/nbt0492-413
72. Higuchi, R., Fockler, C., Dollinger, G., & Watson, R. (1993). Kinetic PCR analysis: Real-time monitoring of DNA amplification reactions. *Bio/Technology, 11*, 1026–1030.10.1038/nbt0993-1026
73. Hiyoshi, H., Kodama, T., Iida, T., & Honda, T. (2010). Contribution of *Vibrio parahaemolyticus* virulence factors to cytotoxicity, enterotoxicity, and lethality in mice. *Infection and Immunity, 78*, 1772–1780.10.1128/IAI.01051-09
74. Hoorfar, J., Cook, N., Malorny, B., Wagner, M., De Medici, D., Abdulmawjood, A., & Fach, P. (2003). Making internal amplification control mandatory for diagnostic PCR. *Journal of Clinical Microbiology, 41*, 5835–5835.10.1128/JCM.41.12.5835.2003
75. Hoorfar, J., Cook, N., Malorny, B., Wagner, M., De Medici, D., Abdulmawjood, A., & Fach, P. (2004). Letter to the editor. *Journal of Applied Microbiology, 96*, 221–222.10.1046/j.1365-2672.2003.02188.x
76. Hoorfar, J., Malorny, B., Abdulmawjood, A., Cook, N., Wagner, M., & Fach, P. (2004). Practical considerations in design of internal amplification controls for diagnostic PCR assays. *Journal of Clinical Microbiology, 42*, 1863–1868.10.1128/JCM.42.5.1863-1868.2004
77. Hsu, B. M., Wu, S. F., Huang, S. W., Tseng, Y. J., Ji, D. D., Chen, J. S., & Shih, F. C. (2010). Differentiation and identification of *Shigella* spp. and enteroinvasive *Escherichia coli* in environmental waters by a molecular method and biochemical test. *Water Research, 44*, 949–955.10.1016/j.watres.2009.10.004
78. Huang, J., Zhu, Y., Wen, H., Zhang, J., Huang, S., Niu, J., & Li, Q. (2009). Quadruplex real-time PCR assay for detection and identification of *Vibrio cholerae* O1 and O139 strains and determination of their toxigenic potential. *Applied and Environmental Microbiology, 75*, 6981–6985.10.1128/AEM.00517-09
79. Huggett, J. F., Foy, C. A., Benes, V., Emslie, K., Garson, J. A., Haynes, R., … Bustin, S. A. (2013). The digital MIQE guidelines: Minimum information for publication of quantitative digital PCR experiments.

Clinical Chemistry, 59, 892–902.10.1373/clinchem.2013.206375

80. Hyeon, J. Y., Hwang, I. G., Kwak, H. S., Park, C., Choi, I. S., & Seo, K. H. (2010). Evaluation of PCR inhibitory effect of enrichment broths and comparison of DNA extraction methods for detection of *Salmonella* Enteritidis using real-time PCR assay. *Journal of Veterinary Science, 11*, 143–149.10.4142/jvs.2010.11.2.143
81. ISO. (1996). *Microbiology of food and animal feeding stuffs—Horizontal method for the detection and enumeration of Listeria monocytogenes-Part 1: Detection method* (Vol. 11290-1:1996).
82. ISO. (1998). *Microbiology of food and animal feeding stuffs—Horizontal method for the detection and enumeration of Listeria monocytogenes-Part 2: Enumeration method* (Vol. ISO 11290-2:1998).
83. ISO. (2001). *Microbiology of food and animal feeding stuffs—Horizontal method for the detection of Escherichia coliO157* (Vol. 16654:2001).
84. ISO. (2003). *Microbiology of food and animal feeding stuffs—Horizontal method for the detection of Salmonella spp.*(Vol. 6579:2002).
85. ISO. (2005). *Microbiology of food and animal feeding stuffs—Polymerase chain reaction (PCR) for the detection of food-borne pathogens* (General requirements and definitions). *22174*, 19.
86. Izumiya, H., Matsumoto, K., Yahiro, S., Lee, J., Morita, M., Yamamoto, S., … Ohnishi, M. (2011). Multiplex PCR assay for identification of three major pathogenic *Vibrio* spp., *Vibrio cholerae, Vibrio parahaemolyticus*, and *Vibrio vulnificus*.*Molecular and Cellular Probes, 25*, 174–176.10.1016/j.mcp.2011.04.004
87. Jofré, A., Martin, B., Garriga, M., Hugas, M., Pla, M., Rodríguez-Lázaro, D., & Aymerich, T. (2005). Simultaneous detection of *Listeria monocytogenes* and *Salmonella* by multiplex PCR in cooked ham. *Food Microbiology, 22*, 109–115.10.1016/j.fm.2004.04.009
88. Josefsen, M. H., Cook, N., D'Agostino, M., Hansen, F., Wagner, M., Demnerova, K., … Hoorfar, J. (2004). Validation of a PCR-based method for detection of food-borne thermotolerant campylobacters in a multicenter collaborative trial.*Applied and Environmental Microbiology, 70*, 4379–4383.10.1128/AEM.70.7.4379-4383.2004
89. Josefsen, M. H., Jacobsen, N. R., & Hoorfar, J. (2004). Enrichment followed by quantitative PCR both for rapid detection and as a tool for quantitative risk assessment of food-borne thermotolerant

campylobacters. *Applied and Environmental Microbiology, 70*, 3588–3592.10.1128/AEM.70.6.3588-3592.2004

90. Josefsen, M. H., Krause, M., Hansen, F., & Hoorfar, J. (2007). Optimization of a 12-h TaqMan PCR-based method for detection of *Salmonella* bacteria in meat. *Applied and Environmental Microbiology, 73*, 3040–3048.10.1128/AEM.02823-06
91. ourdan, A. D., Johnson, S. C., & Wesley, I. V. (2000). Development of a fluorogenic 5' nuclease PCR assay for detection of the ail gene of pathogenic *Yersinia enterocolitica. Applied and Environmental Microbiology, 66*, 3750–3755.10.1128/AEM.66.9.3750-3755.2000
92. Kagkli, D. M., Weber, T. P., Van den Bulcke, M., Folloni, S., Tozzoli, R., Morabito, S., … Van den Eede, G. (2011).Application of the modular approach to an in-house validation study of real-time PCR methods for the detection and serogroup determination of verocytotoxigenic *Escherichia coli. Applied and Environmental Microbiology, 77*, 6954–6963.10.1128/AEM.05357-11
93. Kamisaki-Horikoshi, N., Okada, Y., Takeshita, K., Sameshima, T., Kawasaki, S., Kawamoto, S., & Fratamico, P. M. (2011).Evaluation of TA10 broth for recovery of heat- and freeze-injured *Salmonella* from beef. *Journal of Aoac International,94*, 857–862.
94. Kawasaki, S., Fratamico, P. M., Horikoshi, N., Okada, Y., Takeshita, K., Sameshima, T., & Kawamoto, S. (2009).Evaluation of a multiplex PCR system for simultaneous detection of *Salmonella* spp., *Listeria monocytogenes*, and*Escherichia coli* O157:H7 in foods and in food subjected to freezing. *Foodborne Pathogens and Disease, 6*, 81–89.10.1089/fpd.2008.0153
95. Kawasaki, S., Fratamico, P. M., Horikoshi, N., Okada, Y., Takeshita, K., Sameshima, T., & Kawamoto, S. (2010). Multiplex real-time polymerase chain reaction assay for simultaneous detection and quantification of *Salmonella* species,*Listeria monocytogenes*, and *Escherichia coli* O157:H7 in ground pork samples. *Foodborne Pathogens and Disease, 7*,549–554.10.1089/fpd.2009.0465
96. Kawasaki, S., Fratamico, P. M., Kamisaki-Horikoshi, N., Okada, Y., Takeshita, K., Sameshima, T., & Kawamoto, S. (2011).Development of the multiplex PCR detection kit for *Salmonella* spp., *Listeria monocytogenes*, and *Escherichia coli*O157:H7. *Jarq-Japan Agricultural Research Quarterly, 45*, 77–81.
97. Kawasaki, S., Horikoshi, N., Okada, Y., Takeshita, K., Sameshima, T.,

& Kawamoto, S. (2005). Multiplex PCR for simultaneous detection of *Salmonella spp.*, *Listeria monocytogenes*, and *Escherichia coli* O157:H7 in meat samples.*Journal of Food Protection, 68*, 551–556.

98. Kim, H. C., & Bhunia, A. K. (2008). SEL, a selective enrichment broth for simultaneous growth of *Salmonella* enterica,*Escherichia coli* O157:H7, and *Listeria monocytogenes*. *Applied and Environmental Microbiology, 74*, 4853–4866.10.1128/AEM.02756-07

99. Kim, H.-J., & Cho, J.-C. (2010). Simple and rapid detection of *Listeria monocytogenes* in fruit juice by real-time PCR without enrichment culture. *Food Control, 21*, 1419–1423.10.1016/j.foodcont.2010.04.006

100. Kimura, B., Kawasaki, S., Fujii, T., Kusunoki, J., Itoh, T., & Flood, S. J. A. (1999). Evaluation of TaqMan PCR assay for detecting *Salmonella* in raw meat and shrimp. *Journal of Food Protection, 62*, 329–335.

101. Kobayashi, H., Kubota, J., Fujihara, K., Honjoh, K., Iio, M., Fujiki, N., … Miyamoto, T. (2009). Simultaneous enrichment of*Salmonella* spp, *Escherichia coli* O157:H7, *Vibrio parahaemolyticus*, *Staphylococcus aureus*, *Bacillus cereus*, and*Listeria monocytogenes* by single broth and screening of the pathogens by multiplex real-time PCR. *Food Science and Technology Research, 15*, 427–438.10.3136/fstr.15.427

102. Krascsenicsová, K., Piknová, L., Kaclíková, E., & Kuchta, T. (2008). Detection of *Salmonella enterica* in food using two-step enrichment and real-time polymerase chain reaction. *Letters in Applied Microbiology, 46*, 483–487.10.1111/j.1472-765X.2008.02342.x

103. Lalitha, P., Sid Suraiya, M. N., Lim, K. L., Lee, S. Y., Nur Haslindawaty, A. R., Chan, Y. Y., … Ravichandran, M. (2008).Analysis of *lolB* gene sequence and its use in the development of a PCR assay for the detection of *Vibrio cholerae*.*Journal of Microbiological Methods, 75*, 142–144.10.1016/j.mimet.2008.05.001

104. Lambertz, S. T., Nilsson, C., Hallanvuo, S., & Lindblad, M. (2008). Real-time PCR method for detection of pathogenic*Yersinia enterocolitica* in food. *Applied and Environmental Microbiology, 74*, 6060–6067.10.1128/AEM.00405-08

105. Lantz, P. G., Knutsson, R., Blixt, Y., Al-Soud, W., Borch, E., & Rådström, P. (1998). Detection of pathogenic *Yersinia enterocolitica* in enrichment media and pork by a multiplex PCR: A study of sample preparation and PCR-inhibitory components. *International Journal of Food Microbiology, 45*, 93–105.10.1016/S0168-1605(98)00152-4

106. Leblanc-Maridor, M., Garénaux, A., Beaudeau, F., Chidaine, B., Seegers, H., Denis, M., & Belloc, C. (2011). Quantification of *Campylobacter* spp. in pig feces by direct real-time PCR with an internal control of extraction and amplification. *Journal of Microbiological Methods, 85*, 53–61.10.1016/j.mimet.2011.01.013

107. Li, Y., Zhuang, S., & Mustapha, A. (2005). Application of a multiplex PCR for the simultaneous detection of *Escherichia coli* O157:H7, *Salmonella* and *Shigella* in raw and ready-to-eat meat products. *Meat Science, 71*, 402–406.10.1016/j.meatsci.2005.04.013

108. López-Calleja, I., de la Cruz, S., Pegels, N., González, I., García, T., & Martín, R. (2013). Development of a real time PCR assay for detection of allergenic trace amounts of peanut (*Arachis hypogaea*) in processed foods. *Food Control, 30*,480–490.10.1016/j.foodcont.2012.09.017

109. Lubeck, P. S., Cook, N., Wagner, M., Fach, P., & Hoorfar, J. (2003). Toward an international standard for PCR-based detection of food-borne thermotolerant campylobacters: Validation in a multicenter collaborative trial. *Applied and Environmental Microbiology, 69*, 5670–5672.10.1128/AEM.69.9.5670-5672.2003

110. Lubeck, P. S., Wolffs, P., On, S. L. W., Ahrens, P., Radstrom, P., & Hoorfar, J. (2003). Toward an international standard for PCR-based detection of food-borne thermotolerant campylobacters: Assay development and analytical validation.*Applied and Environmental Microbiology, 69*, 5664–5669.10.1128/AEM.69.9.5664-5669.2003

111. Lyon, W. J. (2001). TaqMan PCR for detection of *Vibrio cholerae* O1, O139, non-O1, and non-O139 in pure cultures, raw oysters, and synthetic seawater. *Applied and Environmental Microbiology, 67*, 4685–4693.10.1128/AEM.67.10.4685-4693.2001

112. Mackie, I. M., Pryde, S. E., Gonzales-Sotelo, C., Medina, I., Perez-Martin, R., Quinteiro, J., … Rehbein, R. (1999).Challenges in the identification of species of canned fish. *Trends in Food Science & Technology, 10*, 9–14.

113. Mahmuda, Y., & Kawamoto, K. S. (2007). Evaluation of multiplex PCR system for simultaneous detection of *Escherichia coli* O157:H7, *Listeria monocytogenes* and *Salmonella enteritidis* in shrimp samples. *Bangladesh Journal of Microbiology, 24*, 42–46.

114. Malorny, B., Bunge, C., & Helmuth, R. (2007). A real-time PCR for the detection of *Salmonella* Enteritidis in poultry meat and consumption

eggs. *Journal of Microbiological Methods, 70*, 245–251.10.1016/j.mimet.2007.04.013

115. Malorny, B., Hoorfar, J., Bunge, C., & Helmuth, R. (2003). Multicenter validation of the analytical accuracy of*Salmonella* PCR: Towards an international standard. *Applied and Environmental Microbiology, 69*, 290–296.10.1128/AEM.69.1.290-296.2003
116. Malorny, B., Huehn, S., Dieckmann, R., Krämer, N., & Helmuth, R. (2009). Polymerase chain reaction for the rapid detection and serovar identification of *Salmonella* in food and feeding stuff. *Food Analytical Methods, 2*, 81–95.10.1007/s12161-008-9057-9
117. Malorny, B., Paccassoni, E., Fach, P., Bunge, C., Martin, A., & Helmuth, R. (2004). Diagnostic real-time PCR for detection of *Salmonella* in food. *Applied and Environmental Microbiology, 70*, 7046–7052.10.1128/AEM.70.12.7046-7052.2004
118. Martínez-Blanch, J. F., Sánchez, G., Garay, E., & Aznar, R. (2009). Development of a real-time PCR assay for detection and quantification of enterotoxigenic members of *Bacillus cereus* group in food samples. *International Journal of Food Microbiology, 135*, 15–21.10.1016/j.ijfoodmicro.2009.07.013
119. Mayr, A. M., Lick, S., Bauer, J., Thaerigen, D., Busch, U., & Huber, I. (2010). Rapid detection and differentiation of*Campylobacter jejuni, Campylobacter coli*, and *Campylobacter lari* in food, using multiplex real-time PCR. *Journal of Food Protection, 73*, 241–250.
120. McGuinness, S., McCabe, E., O'Regan, E., Dolan, A., Duffy, G., Burgess, C., ... O'Grady, J. (2009). Development and validation of a rapid real-time PCR based method for the specific detection of *Salmonella* on fresh meat. *Meat Science, 83*, 555–562.10.1016/j.meatsci.2009.07.004
121. Mukhopadhyay, A., Mukhopadhyay, K., & U. (2007). Novel multiplex PCR approaches for the simultaneous detection of human pathogens: *Escherichia coli* 0157:H7 and *Listeria monocytogenes*. *Journal of Microbiological Methods, 68*,193–200.10.1016/j.mimet.2006.07.009
122. Najdenski, H., Heyndrickx, M., Herman, L., Werbrouck, H., & Van Coillie, E. (2012). Quantification of *Yersinia enterocolitica* in raw milk using qPCR. *Veterinary Microbiology, 160*, 428–434.10.1016/j.vetmic.2012.06.013
123. Nandi, B., Nandy, R. K., Mukhopadhyay, S., Nair, G. B., Shimada, T.,

& Ghose, A. C. (2000). Rapid method for species-specific identification of *Vibrio cholerae* using primers targeted to the gene of outer membrane protein OmpW.*Journal of Clinical Microbiology, 38*, 4145–4151.

124. Navas, J., Ortiz, S., Lopez, P., Jantzen, M. M., Lopez, V., & Martinez-Suarez, J. V. (2006). Evaluation of effects of primary and secondary enrichment for the detection of listeria monocytogenes by real-time PCR in retail ground chicken meat. *Foodborne Pathogens and Disease, 3*, 347–354.10.1089/fpd.2006.3.347

125. Neogi, S. B., Chowdhury, N., Asakura, M., Hinenoya, A., Haldar, S., Saidi, S. M., … Yamasaki, S. (2010). A highly sensitive and specific multiplex PCR assay for simultaneous detection of *Vibrio cholerae, Vibrio parahaemolyticus* and *Vibrio vulnificus. Letters in Applied Microbiology, 51*, 293–300.10.1111/lam.2010.51.issue-3

126. Nordstrom, J. L., Vickery, M. C. L., Blackstone, G. M., Murray, S. L., & DePaola, A. (2007). Development of a multiplex real-time PCR assay with an internal amplification control for the detection of total and pathogenic V*ibrio parahaemolyticus* bacteria in oysters. *Applied and Environmental Microbiology, 73*, 5840–5847.10.1128/AEM.00460-07

127. Omiccioli, E., Amagliani, G., Brandi, G., & Magnani, M. (2009). A new platform for real-time PCR detection of*Salmonella* spp., *Listeria monocytogenes* and *Escherichia coli* O157 in milk. *Food Microbiology, 26*, 615–622.10.1016/j.fm.2009.04.008

128. Panicker, G., & Bej, A. K. (2005). Real-time PCR detection of *Vibrio vulnificus* in oysters: Comparison of oligonucleotide primers and probes targeting *vvhA. Applied and Environmental Microbiology, 71*, 5702–5709.10.1128/AEM.71.10.5702-5709.2005

129. Panicker, G., Call, D. R., Krug, M. J., & Bej, A. K. (2004). Detection of pathogenic *Vibrio* spp. in shellfish by using multiplex PCR and DNA microarrays. *Applied and Environmental Microbiology, 70*, 7436–7444.10.1128/AEM.70.12.7436-7444.2004

130. Parveen, S., Hettiarachchi, K. A., Bowers, J. C., Jones, J. L., Tamplin, M. L., McKay, R., … DePaola, A. (2008). Seasonal distribution of total and pathogenic *Vibrio parahaemolyticus* in Chesapeake Bay oysters and waters. *International Journal of Food Microbiology, 128*, 354–361.10.1016/j.ijfoodmicro.2008.09.019

131. Perelle, S., Dilasser, F., Grout, J., & Fach, P. (2003). Development of a 5′ nuclease PCR assay for detecting Shiga toxin-producing *Escherichia coli* O145 based on the identification of an "O-island 29" homologue.

Journal of Applied Microbiology, 94, 587–594.10.1046/j.1365-2672.2003.01872.x

132. Perelle, S., Dilasser, F., Grout, J. L., & Fach, P. (2004). Detection by 5′-nuclease PCR of Shiga-toxin producing *Escherichia coli* O26, O55, O91, O103, O111, O113, O145 and O157:H7, associated with the world's most frequent clinical cases.*Molecular and Cellular Probes, 18*, 185–192.10.1016/j.mcp.2003.12.004
133. Piknová, L., Kaclíková, E., Pangallo, D., Polek, B., & Kuchta, T. (2005). Quantification of *Salmonella* by 5′-nuclease real-time polymerase chain reaction targeted to fimC gene. *Current Microbiology, 50*, 38–42.10.1007/s00284-004-4374-8
134. Pineyro, P., Zhou, X. H., Orfe, L. H., Friel, P. J., Lahmers, K., & Call, D. R. (2010). Development of two animal models to study the function of *Vibrio parahaemolyticus* type III secretion systems. *Infection and Immunity, 78*, 4551–4559.10.1128/IAI.00461-10
135. Prendergast, D. M., Lendrum, L., Pearce, R., Ball, C., McLernon, J., O'Grady, D., … Gutierrez, M. (2011).Verocytotoxigenic *Escherichia coli* O157 in beef and sheep abattoirs in Ireland and characterisation of isolates by pulsed-field gel electrophoresis and multi-locus variable number of tandem repeat analysis. *International Journal of Food Microbiology, 144*, 519–527.10.1016/j.ijfoodmicro.2010.11.012
136. Quinteiro, J., Sotelo, C. G., Rehbein, H., Pryde, S. E., Medina, I., Pérez-Martín, R. I., … Mackie, I. M. (1998). Use of mtDNA direct polymerase chain reaction (PCR) sequencing and PCR−restriction fragment length polymorphism methodologies in species identification of canned tuna. *Journal of Agricultural and Food Chemistry, 46*, 1662–1669.10.1021/jf970552+
137. Rahn, K., De Grandis, S. A., Clarke, R. C., McEwen, S. A., Galán, J. E., Ginocchio, C., … Gyles, C. (1992). Amplification of an*invA* gene sequence of *Salmonella typhimurium* by polymerase chain reaction as a specific method of detection of*Salmonella*. *Molecular and Cellular Probes, 6*, 271–279.10.1016/0890-8508(92)90002-F
138. Rantsiou, K., Lamberti, C., & Cocolin, L. (2010). Survey of *Campylobacter jejuni* in retail chicken meat products by application of a quantitative PCR protocol. *International Journal of Food Microbiology, 141*, S75–S79.10.1016/j.ijfoodmicro.2010.02.002
139. Reyes-Escogido, L., Balam-Chi, M., Rodríguez-Buenfil, I., Valdés, J., Kameyama, L., & Martínez-Pérez, F. (2010).Purification of bacterial

genomic DNA in less than 20 min using chelex-100 microwave: Examples from strains of lactic acid bacteria isolated from soil samples. *Antonie van Leeuwenhoek, 98*, 465–474.10.1007/s10482-010-9462-0

140. Rijpens, N., & Herman, L. (2004). Comparison of selective and nonselective primary enrichments for the detection of*Listeria monocytogenes* in cheese. *International Journal of Food Microbiology, 94*, 15–22.10.1016/j.ijfoodmicro.2003.10.016

141. Rizvi, A. V., & Bej, A. K. (2010). Multiplexed real-time PCR amplification of *tlh*, *tdh* and *trh* genes in *Vibrio parahaemolyticus* and its rapid detection in shellfish and Gulf of Mexico water. *Antonie Van Leeuwenhoek International Journal of General and Molecular Microbiology, 98*, 279–290.10.1007/s10482-010-9436-2

142. Rizvi, A. V., Panicker, G., Myers, M. L., & Bej, A. K. (2006). Detection of pandemic *Vibrio parahaemolyticus* O3:K6 serovar in Gulf of Mexico water and shellfish using real-time PCR with Taqman (R) fluorescent probes. *FEMS Microbiology Letters, 262*, 185–192.10.1111/fml.2006.262.issue-2

143. Rodríguez-Lázaro, D., Cook, N., & Hernández, M. (2013). Real-time PCR in food science: PCR diagnostics. *Current Issues in Molecular Biology, 15*, 39–44.

144. Rodríguez-Lázaro, D., Gonzalez-García, P., Gattuso, A., Gianfranceschi, M. V., & Hernandez, M. (2014). Reducing time in the analysis of *Listeria monocytogenes* in meat, dairy and vegetable products. *International Journal of Food Microbiology, 184*, 98–105.

145. Rodríguez-Lázaro, D., & Hernández, M. (2013). Real-time PCR in food science: Introduction. *Current Issues in Molecular Biolog, 15*, 25–38.

146. Rodriguez-Lazaro, D., Hernandez, M., Scortti, M., Esteve, T., Vazquez-Boland, J. A., & Pla, M. (2004). Quantitative detection of *Listeria monocytogenes* and *Listeria innocua* by real-time PCR: Assessment of *hly*, *iap*, and *lin02483*targets and amplifluor technology. *Applied and Environmental Microbiology, 70*, 1366–1377.10.1128/AEM.70.3.1366-1377.2004

147. Rodriguez-Lazaro, D., Jofre, A., Aymerich, T., Hugas, M., & Pla, M. (2004). Rapid quantitative detection of *Listeria monocytogenes* in meat products by real-time PCR. *Applied and Environmental Microbiology, 70*, 6299–6301.10.1128/AEM.70.10.6299-6301.2004

148. Rodriguez-Lazaro, D., Lombard, B., Smith, H., Rzezutka, A., D'Agostino, M., Helmuth, R., ... Cook, N. (2007). Trends in analytical methodology in food safety and quality: Monitoring microorganisms and genetically modified organisms.*Trends in Food Science & Technology, 18*, 306–319.

149. Roig, F. J., Sanjuan, E., Llorens, A., & Amaro, C. (2010). *pilF* polymorphism-based PCR to distinguish *Vibrio vulnificus*strains potentially dangerous to public health. *Applied and Environmental Microbiology, 76*, 1328–1333.10.1128/AEM.01042-09

150. Rosche, T. M., Binder, E. A., & Oliver, J. D. (2010). *Vibrio vulnificus* genome suggests two distinct ecotypes.*Environmental Microbiology Reports, 2*, 128–132.10.1111/j.1758-2229.2009.00119.x

151. Rosche, T. M., Yano, Y., & Oliver, J. D. (2005). A rapid and simple PCR analysis indicates there are two subgroups of*Vibrio vulnificus* which correlate with clinical or environmental isolation. *Microbiology and Immunology, 49*, 381–389.10.1111/mim.2005.49.issue-4

152. Rossmanith, P., Krassnig, M., Wagner, M., & Hein, I. (2006). Detection of *Listeria monocytogenes* in food using a combined enrichment/real-time PCR method targeting the *prfA* gene. *Research in Microbiology, 157*, 763–771.10.1016/j.resmic.2006.03.003

153. Rossmanith, P., & Wagner, M. (2010). The challenge to quantify *Listeria monocytogenes*-a model leading to new aspects in molecular biological food pathogen detection. *Journal of Applied Microbiology, 110*, 605–617.

154. Rossmanith, P., Mester, P., Wagner, M., & Schoder, D. (2010). Demonstration of the effective performance of a combined enrichment/real-time PCR method targeting the *prfA* gene of *Listeria monocytogenes* by testing fresh naturally contaminated acid curd cheese. *Letters in Applied Microbiology, 51*, 480–484.10.1111/j.1472-765X.2010.02925.x

155. Ruiz-Rueda, O., Soler, M., & Calvo, L. (2010). Multiplex real-time PCR for the simultaneous detection of *Salmonella*spp. and *Listeria monocytogenes* in food samples. *Food Analytical Methods, 4*, 131–138.

156. Samuel, G., Hogbin, J.-P., Wang, L., & Reeves, P. R. (2004). Relationships of the *Escherichia coli* O157, O111, and O55 O-antigen gene clusters with those of *Salmonella* enterica and *Citrobacter freundii*, which express identical O antigens.*Journal of Bacteriology, 186*, 6536–6543.10.1128/JB.186.19.6536-6543.2004

157. Saravanan, V., Sanath Kumar, H., Karunasagar, I., & Karunasagar, I. (2007). Putative virulence genes of *Vibrio cholerae*from seafoods and the coastal environment of Southwest India. *International Journal of Food Microbiology, 119*, 329–333.10.1016/j.ijfoodmicro.2007.08.023

158. Sharma, A., & Chaturvedi, A. N. (2006). Prevalence of virulence genes (*ctxA*, *stn*, *OmpW* and *tcpA*) among non-O1*Vibrio cholerae* isolated from fresh water environment. *International Journal of Hygiene and Environmental Health,209*, 521–526.10.1016/j.ijheh.2006.06.005

159. Sheikh, A. F., Goodarzi, H., & Aslani, S. (2012). Identification of *Vibrio cholerae* pathogenicity island (*ctxA*, *OmpW* and*tcpA*) in non-O139 and non-O1 *V. cholerae* strains isolated from Karun River in Ahvaz, Iran. *African Journal of Microbiology Research, 6*, 1185–1189.

160. Shuan Ju Teh, C., Lin Thong, K., Tein Ngoi, S., Ahmad, N., Balakrish Nair, G., & Ramamurthy, T. (2009). Molecular characterization of serogrouping and virulence genes of Malaysian *Vibrio cholerae* isolated from different sources.*The Journal of General and Applied Microbiology, 55*, 419–425.10.2323/jgam.55.419

161. Singh, J., Batish, V. K., Grover, S. (2011). Simultaneous detection of *Listeria monocytogenes* and *Salmonella* spp. in dairy products using real time PCR-melt curve analysis. *Journal of Food Science and Technology, 49*, 234–239.

162. Soares, S., Amaral, J. S., Oliveira, M. B. P. P., & Mafra, I. (2013). A SYBR green real-time PCR assay to detect and quantify pork meat in processed poultry meat products. *Meat Science, 94*, 115–120.10.1016/j.meatsci.2012.12.012

163. Sotelo, C. G., Calo-Mata, P., Chapela, M. J., Pérez-Martín, R. I., Rehbein, H., Hold, G. L., … Santos, A. T. (2001).Identification of flatfish (*Pleuronectiforme*) species using DNA-based techniques. *Journal of Agricultural and Food Chemistry, 49*, 4562–4569.10.1021/jf010452a

164. Suo, B., He, Y. P., Tu, S. I., & Shi, X. M. (2010). A multiplex real-time polymerase chain reaction for simultaneous detection of *Salmonella spp., Escherichia coli* O157, and *Listeria monocytogenes* in meat products. *Foodborne Pathogens and Disease, 7*, 619–628.10.1089/fpd.2009.0430

165. Takahashi, H., Kimura, B., Tanaka, Y., Shinozaki, J., Suda, T., & Fujii, T. (2009). Real-time PCR and enrichment culture for sensitive detection and enumeration of *Escherichia coli. Journal of Microbiological*

Methods, 79, 124–127.10.1016/j.mimet.2009.08.002

166. Thiem, V. D., Sethabutr, O., von Seidlein, L., Van Tung, T., Canh, D. G., Chien, B. T., … Trach, D. D. (2004). Detection of*Shigella* by a PCR assay targeting the ipaH gene suggests increased prevalence of shigellosis in Nha Trang, Vietnam.*Journal of Clinical Microbiology, 42*, 2031–2035.10.1128/JCM.42.5.2031-2035.2004
167. Toplak, N., Kovac, M., Piskernik, S., Mozina, S. S., & Jersek, B. (2011). Detection and quantification of *Campylobacter jejuni* and *Campylobacter coli* using real-time multiplex PCR. *Journal of Applied Microbiology, 112*, 752–764.
168. Wang, L. X., Li, Y., & Mustapha, A. (2007). Rapid and simultaneous quantitation of *Escherichia coli* O157:H7,*Salmonella*, and *Shigella* in ground beef by multiplex real-time PCR and immunomagnetic separation. *Journal of Food Protection, 70*(6), 1366–1372.
169. Ward, L. N., & Bej, A. K. (2006). Detection of *Vibrio parahaemolyticus* in shellfish by Use of multiplexed real-time PCR with TaqMan fluorescent probes. *Applied and Environmental Microbiology, 72*, 2031–2042.10.1128/AEM.72.3.2031-2042.2006
170. Wehrle, E., Didier, A., Moravek, M., Dietrich, R., & Märtlbauer, E. (2010). Detection of *Bacillus cereus* with enteropathogenic potential by multiplex real-time PCR based on SYBR green I. *Molecular and Cellular Probes, 24*,124–130.10.1016/j.mcp.2009.11.004
171. Wernars, K., Delfgou, E., Soentoro, P. S., & Notermans, S. (1991). Successful approach for detection of low numbers of enterotoxigenic *Escherichia coli* in minced meat by using the polymerase chain-reaction. *Applied and Environmental Microbiology, 57*, 1914–1919.
172. Wernars, K., Heuvelman, C. J., Chakraborty, T., & Notermans, S. H. W. (1991). Use of the polymerase chain reaction for direct detection of *Listeria monocytogenes* in soft cheese. *Journal of Applied Bacteriology, 70*, 121–126.10.1111/jam.1991.70.issue-2
173. Wiemer, D., Loderstaedt, U., von Wulffen, H., Priesnitz, S., Fischer, M., Tannich, E., & Hagen, R. M. (2011). Real-time multiplex PCR for simultaneous detection of *Campylobacter jejuni*, *Salmonella*, *Shigella* and *Yersinia* species in fecal samples. *International Journal of Medical Microbiology, 301*, 577–584.10.1016/j.ijmm.2011.06.001
174. Wolffs, P., Knutsson, R., Norling, B., & Radstrom, P. (2004). Rapid quantification of *Yersinia enterocolitica* in pork samples by a novel

sample preparation method, flotation, prior to real-time PCR. *Journal of Clinical Microbiology, 42*,1042–1047.10.1128/JCM.42.3.1042-1047.2004

175. Wolffs, P., Norling, B., Hoorfar, J., Griffiths, M., & Radstrom, P. (2005). Quantification of *Campylobacter* spp. in chicken rinse samples by using flotation prior to real-time PCR. *Applied and Environmental Microbiology, 71*, 5759–5764.10.1128/AEM.71.10.5759-5764.2005
176. Wright, A. C., Garrido, V., Debuex, G., Farrell-Evans, M., Mudbidri, A. A., & Otwell, W. S. (2007). Evaluation of postharvest-processed oysters by using PCR-based most-probable-number enumeration of *Vibrio vulnificus* bacteria.*Applied and Environmental Microbiology, 73*, 7477–7481.10.1128/AEM.01118-07
177. Wuenscher, M. D., Köhler, S., Bubert, A., Gerike, U., & Goebel, W. (1993). The *iap* gene of *Listeria monocytogenes* is essential for cell viability, and its gene product, p60, has bacteriolytic activity. *Journal of bacteriology, 175*, 3491–3501.
178. Yang, C. B., Jiang, Y., Huang, K. H., Zhu, C. Q., & Yin, Y. L. (2003). Application of real-time PCR for quantitative detection of *Campylobacter jejuni* in poultry, milk and environmental water. *FEMS Immunology & Medical Microbiology, 38*,265–271.10.1016/S0928-8244(03)00168-8
179. Zhang, C. (2013). Semi-nested multiplex PCR enhanced method sensitivity of species detection in further-processed meats. *Food Control, 31*, 326–330.10.1016/j.foodcont.2012.11.002
180. Zhang, D., Zhang, H., Yang, L., Guo, J., Li, X., & Feng, Y. (2009). Simultaneous detection of *Listeria monocytogenes*,*Staphylococcus aureus*, *Salmonella enterica* and *Escherichia coli* O157:H7 in food samples using multiplex PCR method. *Journal of Food Safety, 29*, 348–363.10.1111/jfs.2009.29.issue-3
181. Zhang, W., Bielaszewska, M., Bauwens, A., Fruth, A., Mellmann, A., & Karch, H. (2012). Real-time multiplex PCR for detecting Shiga toxin 2-producing *Escherichia coli* O104:H4 in human stools. *Journal of Clinical Microbiology, 50*,1752–1754.10.1128/JCM.06817-11
182. Zheng, H., Sun, Y., Mao, Z., & Jiang, B. (2008). Investigation of virulence genes in clinical isolates of *Yersinia enterocolitica*. *FEMS Immunology & Medical Microbiology, 53*, 368–374.10.1111/fim.2008.53.issue-3

CHAPTER 9

DEVELOPMENT OF A PEROXIDE BIODETECTOR FOR A DIRECT DETECTION OF BIOFILMS PRODUCED BY CATALASE-POSITIVE BACTERIA ON FOOD-CONTACT SURFACES

Carolina Ripolles-Avila, Abel G. Ríos-Castillo & José J. Rodríguez-Jerez

Hygiene and Food Inspection Unit, Department of Food and Animal Science, Universitat Autònoma de Barcelona, Barcelona, Spain

---

CITATION: Ripolles-Avila, C., A. G. Ríos-Castillo, and J. J. Rodríguez-Jerez. "Development of a peroxide biodetector for a direct detection of biofilms produced by catalase-positive bacteria on food-contact surfaces". Journal of Food (2018). https://doi.org/10.1080/19476337.2017.1418434.

ABSTRACT

Detection of biofilms on food-contact surfaces is essential for establishing sanitization procedures and avoiding microbial contamination of food products. The objective of our study was to develop a peroxide biodetector, to detect and reveal the presence of biofilms. A positive reaction was observed when biodetector bubbles were formed on stainless steel and polypropylene surfaces. A range of microorganisms able to form biofilms was evaluated. Catalase-positive bacteria: *Staphylococcus aureus, Escherichia coli, Listeria monocytogenes, Pseudomonas aeruginosa, Salmonella enterica ser*. Typhimurium and *Cronobacter sakazakii* were employed as well as the catalase-negative bacterium: *Lactobacillus brevis*. With the exception of *C. sakazakii*, all catalase-positive foodborne pathogens forming the biofilms showed a positive detection by the biodetector, being 10^4 CFU cm^{-2} the minimum microbial load detected. The strongest positive reaction was for *P. aeruginosa*. Results demonstrated the potential of this biodetector to detect biofilms, particularly when used as a tool in the food industry.

Keywords: Biofilms, detection, biodetector, food hygiene, rapid method

INTRODUCTION

Contaminated food-contact surfaces may allow bacteria to attach themselves and subsequently develop biofilms, which become a potential source of cross-contamination of food products (Hood & Zottola, 1995Hood, S., & Zottola, E. A. (1995). Biofilms in food processing. *Food Control*, 6, 9–18., , ; Lee Wong, 1998Lee Wong, A. C. (1998). Biofilms in food processing environments. *Journal of Dairy Science*, 81, 2765–2770. Myszka & Czaczyk, 2011Myszka, K., & Czaczyk, K. (2011). Bacterial biofilms on food contact surfaces - a review. *Polish Journal of Food and Nutrition Sciences*, 61, 173–180.,). A biofilm is a community of microorganisms attached to a surface that produces an amorphous three-dimensional matrix of extracellular polymeric substances (EPS). They simultaneously serve as a sticking vehicle to attach different microorganisms (LeChevallier, Cawthon, & Lee, 1988LeChevallier, M. W., Cawthon, C. D., & Lee, R. G.(1988). Inactivation of biofilm bacteria. *Applied Environmental Microbiology*, 54, 2492–2499., , ; Stepanović, Cirković, Ranin, & Svabić-Vlahović, 2004Stepanović, S., Cirković, I., Ranin, L., & Svabić-Vlahović, M. (2004). Biofilm formation by *Salmonella* spp. and *Listeria monocytogenes* on plastic surface. *Letters in Applied Microbiology*, 38, 428–432. Whittenbury, 1964Whittenbury, R. (1964). Hydrogen peroxide formation and catalase activity in the lactic acid bacteria. *Journal of General Microbiology*, 35, 13–26., ,). An exceptional biofilm characteristic is that once it is formed on food or food-processing facility surfaces, its complete removal using common cleaning and sanitation procedures is very difficult (Baumann, Martin, & Feng, 2009Baumann, A. R., Martin, S. E., & Feng, H. (2009). Removal of *Listeria monocytogenes* biofilms from stainless steel by use of ultrasound and ozone. *Journal of Food Protection*, 72, 1306–1309. Torlak & Sert, 2013Torlak, E., & Sert, D. (2013). Combined effect of benzalkonium chloride and ultrasound against *Listeria monocytogenes* biofilm on plastic surface. *Letters in Applied Microbiology*, 57, 220–226.). This fact poses a serious concern to food safety as pathogenic microorganisms can be involved in the development of biofilms in food and on food-contact surfaces, and therefore biofilm cells can be released resulting in a cross-contamination to food products (Shi & Zhu, 2009Shi, X., & Zhu, X. (2009). Biofilm formation and food safety in food industries. *Trends in Food Science and Technology*, 20, 407–413., , ; Simões, Simões, & Vieira, 2010Simões, M., Simões, L. C., & Vieira, M. J. (2010). A review of current and emergent biofilm control strategies. *LWT Food Science and Technology*, 43, 573–583., ,).

Listeria monocytogenes and other pathogens readily produce biofilms, which helps them to survive for prolonged periods in food production plants (Swaminathan & Gerner-Smidt, 2007Swaminathan, B., & Gerner-Smidt, P. (2007). The epidemiology of human listeriosis. *Microbes and Infection*, 9, 1236–1243. Tompkin, 2002Tompkin, R. B. (2002). Control of *Listeria monocytogenes* in the food-processing environment. *Journal of Food Protection*, 65, 709–725.). To prevent biofilm formation, sanitization programmes that target mature biofilms for maximal bacterial inactivation should be established (Yang, Kendall, Medeiros, & Sofos, 2009Yang, H., Kendall, P. A., Medeiros, L. C., & Sofos, J. N. (2009). Efficacy of sanitizing agents against *Listeria monocytogenes* biofilms on high-density polyethylene cutting board surfaces. *Journal of Food Protection*, 72, 990–998.). Current available methods and procedures are insufficient to achieve an efficient control of the different kinds of microorganisms found not only in the food itself, but also in the food production process, the processing equipment and related environments (Carlton, Noordman, Biswas, de Meester, & Loessner, 2005Carlton, R. M., Noordman, W. H., Biswas, B., de Meester, E. D., & Loessner, M. J. (2005). Bacteriophage P100 for control of *Listeria monocytogenes* in foods: Genome sequence, bioinformatic analyses, oral toxicity study, and application. *Regulatory Toxicology and Pharmacology*, 43, 301–312.). In fact, reported outbreaks of *E. coli* O157:H7 and *Salmonella* spp. have been related to cross-contamination and subsequent biofilm formation on lettuce, spinach, and cantaloupe melons (Annous, Fratamico, & Smith, 2009Annous, B. A., Fratamico, P. M., & Smith, J. L.(2009). Quorum sensing in biofilms: Why bacteria behave the way they do. *Journal of Food Science*, 74, 24–36.,), indicating that there is a need for better methods to control contamination.

In general, the research regarding biofilms has been centred on monospecies and pure cultures. However, biofilms are formed by multiple microbial species in natural environments (Manuzon & Wang, 2007Manuzon, M. Y., & Wang, H. H. (2007). Mixed-species biofilms. In H. P. Blaschek, H. H. Wang, & M. E. Agle (Eds.), *Biofilms in food environment* (pp. 105–125). Ames, IA: Blackwell Publishing Ltd. ; Montañez-Izquierdo, Salas-Vázquez, & Rodríguez-Jerez, 2012Montañez-Izquierdo, V. Y., Salas-Vázquez, D. I., & Rodríguez-Jerez, J. J. (2012). Use of epifluorescence microscopy to assess the effectiveness of phage P100 in controlling *Listeria monocytogenes* biofilms on stainless steel surfaces. *Food Control*, 23, 470–477., , ; Moons, Michiels, & Aertsen, 2009Moons, P., Michiels, C. W., & Aertsen, A. (2009). Bacterial interactions in biofilms. *Critical Reviews in Microbiology*, 35, 157–168.[Taylor & Francis Online], , ; Percival, Walker, & Hunter, 2000Percival, S. L., Walker, J. T., & Hunter, P. R. (2000). *Microbiological aspects on biofilms and drinking water*. Boca Raton, FL: CRC Press.,). Different studies have outlined the significance of biofilms formed by mixed species for the food industry (Jahid & Ha, 2014Jahid, I. K., & Ha, S. (2014). The paradox of mixed-species biofilms in the context of food safety. *Comprehensive Reviews of Food Science and Food Safety*, 13, 990–1011., , ; Percival et al., 2000Percival, S. L., Walker, J. T., & Hunter, P. R. (2000). *Microbiological aspects on biofilms and drinking water*. Boca Raton, FL: CRC Press.,). The importance for this industry is attributed to the existence of interspecies interactions that might enhance not only the survival of pathogens to sanitizers, disinfectants, UV light, or desiccation, but also its persistence on food-contact surfaces (Lee et al., 2014Lee, K. W., Periasamy, S., Mukherjee, M., Xie, C., Kjelleberg, S., & Rice, S. A. (2014). Biofilm development and enhanced stress resistance of a model, mixed-species community biofilm. *ISME Journal*, 8, 894–907.). In a study conducted by Jahid, Han, Srey, and Ha (2014bJahid, I. K., Han, N., Srey, S., & Ha, S. D. (2014b). Competitive interactions inside mixed-species biofilms of *Salmonella Typhimurium* and cultivable indigenous microorganisms on lettuce enhance microbial resistance of their sessile cells to ultraviolet C (UV-C) irradiation. *Food Research International*, 55, 445–454., ,), it was demonstrated that biofilms formed by *S.* Typhimurium and cultivable microflora from lettuce became more resistant to UV-C treatments than biofilms formed with mono-cultures. In another study, it was reported how mixed biofilms of *L. monocytogenes, Pseudomonas fluorescens, Serratia proteamaculans,* and *Shewanella baltica* formed on stainless steel surfaces produced more EPS and were more resistant to desiccation (Alavi & Hansen, 2013Alavi, H. E. D., & Hansen, L. T. (2013). Kinetics of biofilm formation and desiccation survival of *Listeria monocytogenes* in single and dual species biofilms with *Pseudomonas fluorescens, Serratia proteamaculans* or *Shewanella baltica* on food-grade stainless steel surfaces. *Biofouling*, 29, 1253–1268.[Taylor & Francis Online], ,).

Until now, identifying the type of microorganisms involved as contaminating agents and establishing the source and extent of contamination by biofilms have been conducted using enumeration techniques. The different methods employed range from the most conventional techniques to the most recent developments. Some traditional cultivation techniques such as swabbing, rinsing, and agar contact methods have been used to quantify the microorganisms on external and internal surfaces (Kumar & Anand, 1998Kumar, C. G., & Anand, S. (1998). Significance of microbial biofilms in food industry: A review. *International Journal of Food Microbiology*, 42, 9–27.). These methods can present some standardization problems when collecting samples from different surfaces in the food industry. For this reason, a surface sensor named SCH (Hygiene Control Sensor, Premiumlab, Sant Boi del Llobregat, Spain) was developed to monitor the biological contamination of surfaces and offer an alternative to traditional methods when analysing product safety and quality. However, the detection of pathogens on food-contact surfaces is mainly founded on traditional microbiological culture methods that can take considerable time to complete (Branda, Vik, Friedman, & Kolter, 2005Branda, S. S., Vik, A., Friedman, L., & Kolter, R.(2005). Biofilms: The matrix revisited. *Current Trends in Microbiology*, 13, 20–26.). This is problematic because reducing the time required for the detection and confirmation of microbial contamination is highly important for the food industry.

Methods based on microscopy, particularly direct epifluorescence microscopy (DEM) and scanning electron microscopy (SEM) of surfaces, have attracted significant attention in the study of biofilms (Notermans, Nauta, Jansen, Jouve, & Mead, 1998Notermans, S., Nauta, M. J., Jansen, J.,

Jouve, J. L., & Mead, G. C. (1998). A risk assessment approach to evaluating food safety based on product surveillance. *Food Control*, 9, 217–223., ,). DEM with digital image analysis is a rapid method used to enumerate microorganisms. This technique is able to distinguish between viable and nonviable cells (Holah, Betts, & Thorpe, 1989Holah, J. T., Betts, R. P., & Thorpe, R. H. (1989). The use of epifluorescence microscopy to determine surface hygiene. *International Biodeterioration & Biodegradation*, 25, 147–153., , ; Lunau, Lemke, Walther, Martens-Habbena, & Simon, 2005Lunau, M., Lemke, A., Walther, K., Martens-Habbena, W., & Simon, M. (2005). An improved method for counting bacteria from sediments and turbid environments by epifluorescence microscopy. *Environmental Microbiology*, 7, 961–968. Pettipher & Rodrigues, 1982Pettipher, G. L., & Rodrigues, U. M. (1982). Rapid enumeration of microorganisms in foods by the direct epifluorescent filter technique. *Applied and Environmental Microbiology*, 44, 809–813., ,). Moreover, DEM approaches provide the opportunity to visualize bacteria on the substrate in the state of adherence (Hannig et al., 2007Hannig, C., Hannig, M., Rehmer, O., Braun, G., Hellwig, E., & Al-Ahmad, A. (2007). Fluorescence microscopic visualization and quantification of initial bacterial colonization on enamel in situ. *Archives Oral Biology*, 52, 1048–1056.). A strong linear correlation between the counts by DEM and conventional culture has been observed (Montañez-Izquierdo et al., 2012Montañez-Izquierdo, V. Y., Salas-Vázquez, D. I., & Rodríguez-Jerez, J. J. (2012). Use of epifluorescence microscopy to assess the effectiveness of phage P100 in controlling *Listeria monocytogenes* biofilms on stainless steel surfaces. *Food Control*, 23, 470–477., ,). However, when microorganisms are exposed to adverse conditions, the correlation for both methods decreases significantly. This reduction is attributed to the presence of a bacterial stress response that cannot be detected by cultivation (Fuster-Valls, Hernández-Herrero, Marín-de-Matero, & Rodríguez-Jerez, 2008Fuster-Valls, N., Hernández-Herrero, M., Marín-de-Matero, M., & Rodríguez-Jerez, J. J. (2008). Effect of different environmental conditions on the bacteria survival on stainless steel surfaces. *Food Control*, 19, 308–314., ,).

For the food industry, it is vital to obtain rapid results of the different microbiological analysis of food-contact surfaces in order to know, not only their hygienic state, but also to be able to take decisions. The development of a product that can detect the presence of biofilms in a short time is a powerful tool to ensure the safety of foods.

The objective of this study was to consider the potential for developing a biodetector for a direct biofilm detection on surfaces of catalase-positive bacteria. With this purpose, an evaluation of the environmental conditions to ensure the development of biofilms on stainless steel and polypropylene surfaces was performed. Furthermore, different formulas to obtain an efficient biofilm biodetector were evaluated.

MATERIALS AND METHODS

Test Surfaces

Two materials were assayed: different-sized 1 mm thick stainless steel 304 grade 2B coupons (first area: diameter disc of 2 cm, second area: 5 × 5 cm, third area: 20 × 20 cm) and different-sized 0.4 mm thick polypropylene coupons (first area: 5 × 5 cm, second area: 20 × 20 cm). In both cases and prior to being used, the surfaces were cleaned and disinfected in accordance with ISO 13697:2001 (Anonymous, 2002Anonymous. (2002). ISO-13697:2002. Chemical disinfectants and antiseptics. Quantitative non-porous surface test for the evaluation of bactericidal and/or fungicidal activity of chemical disinfectants used in food, industrial, domestic and institutional.). First, a wash was performed with a surfactant (Dipol Bac Stop, D.I.S. DINO S.L., Madrid, Spain). Second, a solution of 2-propanol in water (70%) was applied to eliminate the residue of the surfactant (Panreac Química, Barcelona, Spain). The stainless steel surfaces were further sterilized by autoclave at 121°C for 15 min. The polypropylene coupons were sterilized by placing them in a laminar flux sterile cabinet (Telstar, Terrassa, Spain) and exposing them to the antibacterial action and evaporation of 2-propanol for 30 min (Anonymous, 2007Anonymous. (2007). ISO-22196:2007. Plastics - Measurement of antibacterial activity on plastics surfaces.).

Test organisms

Frozen-dried cultures of different strains of microorganisms were employed in this study: *Staphylococcus aureus* ATCC 6538, *Escherichia coli* CECT 515, *Listeria monocytogenes* CCUG 15526, *Pseudomonas aeruginosa* ATCC 15442, *Salmonella enterica ser.* Typhimurium CCUG 29478, *Cronobacter sakazakii* ATCC BAA-894 and *Lactobacillus brevis* CECT 216.

Bacterial Cultures for the tests

Bacterial cultures were recovered using buffered peptone water (BPW, *Buffered Peptone Water*; bioMérieux®, Marcy l'Etoile, France) and incubated at 37°C for 18–24 h. After the incubation period, the different bacterial cultures were streaked on TSA (*Tryptic Soy Agar*; Biokar Diagnostics, Beauvais, France) and incubated at 30°C for 24 h. Only *L. brevis* was rehydrated in MRS broth (Man-Rogosa-Sharpe Broth, Oxoid, UK), incubated at 30°C for 48 h and finally streaked on MRS agar (30°C for 24 h). From each strain, isolated colonies were selected and prepared as stock cultures in TSA or MRS agar slants. These slants were stored as working cultures at 4°C for a maximum period of 30 days.

Each stock culture in slants was subcultured prior to preparing the bacterial inoculums. This procedure was made (i) in 9 ml of BPW in the preliminary step before the biofilm formation, and (ii) in 9 ml of TSB (*Tryptic Soy Broth*; Biokar Diagnostics, Beauvais, France) in the final step during biofilm formation. The incubation periods were selected taking into account the optimal temperature and times for each microorganism. Incubations were performed at 37°C for 18–20 h in order to reach the stationary phase. However, for *L. monocytogenes* the incubation period was set at 20–22 h. In the case of *L. brevis*, 5 ml of MRS broth was used to prepare the inoculum, which was incubated at 30°C for 22–24 h. From each bacterial culture obtained, decimal dilutions were conducted using TSS (*Tryptone Saline Solution*) (1 g of tryptone [Becton Dickinson and Co.], 8.5 g of sodium chloride [Panreac Química]) and 1000 ml of deionized water (pH 7.0 ± 0.2)).

Biofilm Formation

To ensure biofilm formation, sterile coupons of stainless steel and polypropylene were placed in sterile Petri dishes. Each coupon was inoculated with 100 µl of TSS containing a concentration of 10^4 CFU ml^{-1} cells of corresponding bacterial inoculums described above. The surfaces inoculated were placed in Petri dishes and then placed into a humidified chamber and incubated at room temperature (22–24°C) for 48 or 72 h (Fuster-Valls et al., 2008Fuster-Valls, N., Hernández-Herrero, M., Marín-de-Matero, M., & Rodríguez-Jerez, J. J. (2008). Effect of different environmental conditions on the bacteria survival on stainless steel surfaces. *Food Control*, 19, 308–314., ,) under five different sets of conditions to form biofilms (1) Static condition. Without renewal growth culture medium. The bacterial inoculums on the surfaces were not manipulated until the end of the incubation period. (2) With two washes and the addition of culture medium. The washing steps were performed with 5 ml of physiologic saline solution (PSS: 8.5 g of sodium chloride in 1000 ml of deionized water; pH

7.0 ± 0.2), followed by an inoculation of 100 μl of sterile TSS or MRS broth. This step was performed at 24 and 48 h. (3) Three washes and the addition of culture medium. The washings were performed with 5 ml of PSS, followed by an inoculation of 100 μl of sterile TSS or MRS broth. This step was performed at 3, 24, and 48 h. (4) With addition of 100 μl of sterile TSS or MRS broth in order to just renew the nutrients. This procedure was performed without the PSS wash and was performed after 24 h. (5) With addition of 50 μl of sterile TSS or MRS broth with the same objective mentioned above, the renewal of nutrients. This procedure was performed without the PSS wash and after 24 h of the incubation.

Furthermore, with the aim of ensuring a positive control for every condition, surfaces were inoculated with 400 μl of a solution with 10^4 CFU ml^{-1} and 400 μl of sterile TSS in order to have biofilms that occupied a wide surface area. Surfaces were subsequently placed in a saturated humidified environment for 72 h. After 24 and 48 h, surfaces were washed with 5 ml of PSS followed by an inoculation of 800 μl of sterile TSS or MRS broth.

Biodetector

Different formulas for the development of the biodetector were evaluated: (i) Solution with hydrogen peroxide; (ii) Solution with hydrogen peroxide and crystal violet dye; (iii) Solution with hydrogen peroxide, phosphonates (<5%), surfactans (5–15%), bleaching agents (5–15%), and an orange colorant.

Biofilm Detection using Epifluorescent Microscopy and the Biodetector

After the incubation time, biofilms that formed on surfaces were washed with 5 ml of PSS and analysed by both, direct epifluorescent microscopy and the biodetector.

Epifluorescent Microscopy

After surfaces were washed, vital staining was employed with 20 μl of the kit LIVE/DEAD® BacLight™(L13152; Molecular Probes Inc., Oregon, USA). This kit is composed of two fluorescent dyes of nucleic acids, SYTO®9 and propidium iodide (PI). SYTO®9, which penetrate cells with either intact or injured membranes. In contrast, PI penetrates only injured membrane cells and reduces the dye SYTO®9. Therefore, with the simultaneous application of these two dyes in appropriate proportions viable cells with intact membranes show up in fluorescent green and dead or injured cells show up in fluorescent red.

After applying the stain, the samples were incubated at room temperature for 15 min in the dark according to the manufacturer's instructions. Readings were performed with an epifluorescent microscope Olympus BX51/BX52 (Olympus, Tokyo, Japan) equipped with a mercury lamp of 100 W (USH-103OL, Olympus), a double pass filter (U-M51004 F/R – V2, Olympus), and a digital camera DP50-CU (Olympus). Stained samples were observed with 20× and 40× objectives.

Detecting Biofilms

After the corresponding incubation periods to form biofilms of the different bacterial strains, the different hydrogen peroxide formulas were evaluated. The evaluation was performed by spraying 100 μl of the biodetector on the surface to be analysed and the result was observed after 10 min. A reaction was considered positive when a formation of bubbles was observed from the biofilms formed on stainless steel or polypropylene surfaces. The bubbles are caused by an enzymatic reaction that releases oxygen when catalase enzyme from microorganisms is present.

Statistical Data Analysis

All tests were performed in three separate experiments by triplicate (n = 9). Images obtained by DEM were analysed by using the analySIS Auto 3.2 software (Soft Imaging System GMBH, Münster, Germany), and the obtained counts were converted into decimal logarithmic values to nearly match the assumption of a normal distribution. The SAS® v9.1.3.4. statistical software package (SAS Inst. Inc., Cary, N.C., U.S.A.) was employed to evaluate differences between the conditions for biofilm formation by analysing the variance (ANOVA). A *p*value of <0.05 was considered to be significant.

RESULTS AND DISCUSSION

Detection of Biofilms using DEM and the Different Biodetector Formulas

From the data obtained, it was observed that biofilms of *E. coli, P. aeruginosa,* and *S.* Typhimurium were intensely formed after three washes (Figures 1 and 2), indicating that the bacterial density was clearly greater in this condition. Nevertheless, the count of microorganisms obtained by the image analysis system of biofilms formed with three washes compared from those formed with two washes did not show significant differences ($P > 0.05$). *L. monocytogenes* did not develop biofilms after the first washing step and attachment was observed in the static conditions, showing that this pathogen might need more time for its initial adhesion to the stainless steel. Our results are in accordance with those described by Chae and Schraft (2000Chae, M. S., & Schraft, H. (2000). Comparative evaluation of adhesion and biofilm formation of different *Listeria monocytogenes* strains. *International Journal of Food Microbiology*, 62, 103–111.), showing that a minimum incubation period of 3 h was needed for the adhesion of this pathogen to glass, suggesting that there is also an optimum incubation period for its adhesion to stainless steel and polypropylene. The rest of the bacteria, *S. aureus, L. brevis,* and *C. sakazakii* showed a low attachment capacity compared with *L. monocytogenes* in this study. These microorganisms probably needed more incubation periods, the presence of more active microorganisms to develop the polymeric structure and other environmental factors that have an influence on the development of biofilms (Garrett, Bhakoo, & Zhang, 2008Garrett, T. R., Bhakoo, M., & Zhang, Z. (2008). Bacterial adhesion and biofilms on surfaces. *Progress in Natural Science*, 18(9), 1049–1056., , ; Van Houdt & Michiels, 2010Van Houdt, R., & Michiels, C. W. (2010). Biofilm formation and the food industry, a focus on the bacterial outer surface. *Journal of Applied Microbiology*, 109, 1117–1131.).

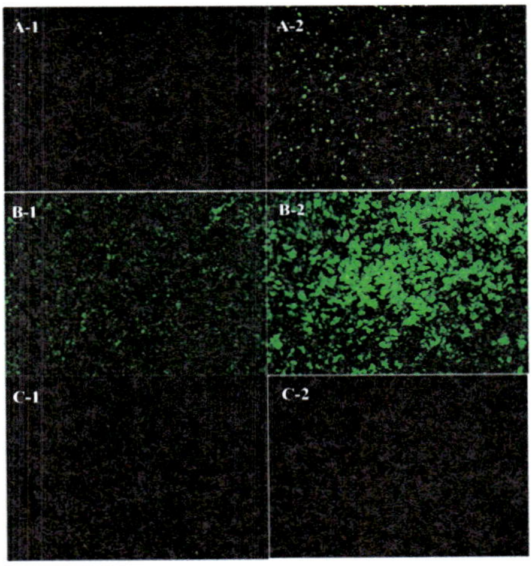

Figure 1. Biofilms of: (A) *L. brevis*, (B) *S. Typhimurium*, and (C) *C. sakazakii*.

Biofilms formed on stainless steel, without washes (A-1, B-1, and C-1), and with three washes (A-2, B-2, and C-2). Epifluorescent images obtained with LIVE/DEAD® stain. Viable cells stained in green colour and not viable cells in red. (All figures at 40× magnification).

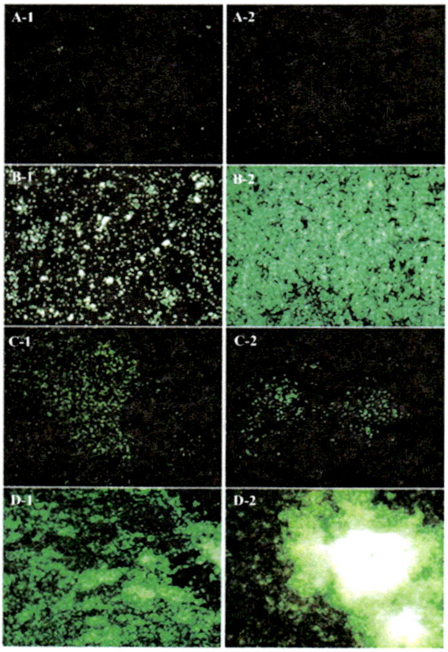

Figure 2. Biofilms of: (A) *S. aureus*, *E. coli* (B), *L. monocytogenes* (C), and *P. aeruginosa* (D). Biofilms formed on stainless steel, without washes (A-1, B-1, C-1, and D-1), and with three washes (A-2, B-2, C-3, and D-2). Epifluorescent images obtained with LIVE/DEAD® stain. Viable cells stained in green colour and not viable cells in red. (All figures at 40× magnification).

When surfaces were washed twice, the results were very similar to those obtained with three washes for *E. coli, P. aeruginosa,* and *S.* Typhimurium. These microorganisms developed more intense biofilms with a higher bacterial density and a higher microbial count. Significant differences ($P < 0.05$) were found on the counts when compared to those that were incubated in static conditions without the renewal of nutrients. Previous reports have shown the importance of a constant flow of nutrients for the development of biofilms that is vital in the early stages of their formation (Myszka & Czaczyk, 2011Myszka, K., & Czaczyk, K. (2011). Bacterial biofilms on food contact surfaces - a review. *Polish Journal of Food and*

Nutrition Sciences, 61, 173–180., ; Simões et al., 2010Simões, M., Simões, L. C., & Vieira, M. J. (2010). A review of current and emergent biofilm control strategies. *LWT Food Science and Technology*, 43, 573–583., ,) Furthermore, under these conditions increased biofilm formation for *S. aureus, L. brevis,* and *C. sakazakii*was observed. Our results reveal that even under adverse conditions, a catalase-positive microorganism such as *S. aureus* represent a resistant bacterium to common hygienic procedures (Ríos-Castillo, González-Rivas, & Rodríguez-Jerez, 2017Ríos-Castillo, A. G., González-Rivas, F., & Rodríguez-Jerez, J. J. (2017). Bactericidal efficacy of hydrogen peroxide-based disinfectants against gram-positive and gram-negative bacteria on stainless steel surfaces. *Journal of Food Science*, 82, 2351–2356.) and is able to form biofilms (Di Ciccio et al., 2015Di Ciccio, P., Vergara, A. R., Festino, A., Paludi, D., Zanardi, E., Ghidini, S., & Ianieri, A. (2015). Biofilm formation by *Staphylococcus aureus* on food contact surfaces: Relationship with temperature and cell surface hydrophobicity. *Food Control*, 50, 930–936., , ; Fuster-Valls et al., 2008Fuster-Valls, N., Hernández-Herrero, M., Marín-de-Matero, M., & Rodríguez-Jerez, J. J. (2008). Effect of different environmental conditions on the bacteria survival on stainless steel surfaces. *Food Control*, 19, 308–314., ,). The difference of biofilm formation obtained in this experiment in comparison with the previous experiment can be justified because in the previous experiment the first wash was performed after 24 h, whereas in this experimental design the first wash was performed after an incubation period of 3 days. Therefore, it might be concluded that a long incubation period before the first wash is necessary to increase the biofilm capacity of these microorganisms. Similar results were found in a study conducted by Wang et al. (2017Wang, H., Qi, J., Dong, Y., Li, Y., Xu, X., & Zhou, G.(2017). Characterization of attachment and biofilm formation by meat-borne *Enterobacteriaceae* strains associated with spoilage. *LWT – Food Science and Technology*, 86, 399–407., ,) on where it was demonstrated that the increase in biofilm formation was time-dependent before 5-days incubation period, obtaining nearly 10.5 log CFU cm^{-2} cells of *Enterobacter cloacae* on stainless steel when incubated under long-term (7 days) growth scenario.

L. monocytogenes developed a high-density biofilm when the static condition was applied and behaved differently to the rest of microorganisms. However, biofilms obtained for *E. coli, P. aeruginosa,* and *S.* Typhimurium were denser after two washes, the first one starting after 24 h of incubation.

The strongest biofilm formers obtained were then employed for the evaluation of the biofilm detection capacity of the different biodetectors formulated. When the solution with just hydrogen peroxide was applied on *E. coli, P. aeruginosa,* and *S. Typhimurium* biofilms, bubble formations were detected, and it was considered as a positive reaction to the presence of biofilms. The election of hydrogen peroxide as an indicator of the presence of biofilms in this study was justified by the positive activity of this compound in different types of bacteria commonly found in food-processing environments reported by Elkins, Hassett, Stewart, Schweizer, and MCDermott (1999Elkins, J. G., Hassett, D. J., Stewart, P. S., Schweizer, H. P., & MCDermott, T. R. (1999). Protective role of catalase in *Pseudomonas aeruginosa* biofilm resistance to hydrogen peroxide. *Applied and Environmental Microbiology*, 65(10), 4594–4600., ,); Christensen, Trønnes, Vollan, Smidsrød, and Bakke (1990Christensen, B. E., Trønnes, H. N., Vollan, K., Smidsrød, O., & Bakke, R. (1990). Biofilm removal by low concentration of hydrogen peroxide. *Biofouling*, 2, 165–175.[Taylor & Francis Online],); and Ríos-Castillo et al. (2017Ríos-Castillo, A. G., González-Rivas, F., & Rodríguez-Jerez, J. J. (2017). Bactericidal efficacy of hydrogen peroxide-based disinfectants against gram-positive and gram-negative bacteria on stainless steel surfaces. *Journal of Food Science*, 82, 2351–2356.). When the solution of hydrogen peroxide that contained crystal violet stain was applied to the surfaces (Figure 3), the biofilm area stained blue and an intense bubble formation was observed, indicating that biofilms are more visual when colorants are included in the formula. Finally, when the solution with the hydrogen peroxide and the colorant was applied in a gel form, an even stronger positive reaction was detected, with an abundant bubble formation on the surfaces that contained the biofilms from *E. coli* and *P. aeruginosa* (Figure 4). Surfaces with biofilms from *S. aureus, S.* Typhimurium, and *L. monocytogenes* showed a weaker reaction. However, surfaces with biofilm from *C. sakazakii* did not show any positive reaction. This could be explained by the fact that, according to previous studies, *C. sakazakii* cells imbedded within biofilms have a high resistance to stress conditions (Yan & Gurtler, 2014Yan, X., & Gurtler, J. B. (2014). *Cronobacter(Enterobacter) sakazakii*. In C. A. Batt & M. L.Tortorello (Eds.), *Encyclopedia of food microbiology*, (2nd ed. pp. 528–532). Oxford, UK: Academic Press. doi:10.1016/B978-0-12-384730-0.00382-7,) and is affected by atmospheric relative humidity (Beuchat et al., 2009Beuchat, L. R., Kim, H., Gurtler, J. B., Lin, L. C., Ryu, J. H., & Richards, G. M. (2009). *Cronobacter sakazakii* in foods and factors affecting its survival, growth, and inactivation. *International Journal of Food Microbiology*, 136(2), 204–213.). Therefore, may be for this bacterium, a larger amount of hydrogen peroxide is necessary to show a positive reaction. Surfaces with biofilms of *L. brevis* did not show any positive reaction because this microorganism is catalase negative (Whittenbury, 1964Whittenbury, R. (1964). Hydrogen peroxide formation and catalase activity in the lactic acid bacteria. *Journal of General Microbiology*, 35, 13–26., ,) and no positive action was expected. In food industry environment, the non-detection of negative catalase microorganisms would not be a problem because biofilms are formed by mixed species (Colagiorgi et al., 2017Colagiorgi, A., Bruini, I., Di Ciccio, P. A., Zanardi, E., Ghidini, S., & Ianieri, A. (2017). *Listeria monocytogenes* biofilms in the wonderland of food industry. *Pathogens*, 6, 41–50., , ; Elias & Banin, 2012Elias, S., & Banin, E. (2012). Multi-species biofilms: Living with friendly neighbors. *FEMS Microbiology Reviews*, 36, 990–1004. Phillips, 2016Phillips, C. A. (2016). Bacterial biofilms in food processing environments: A review of recent developments in chemical and bioloical control. *International Journal of Food Science and Technology*, 51, 1731–1743., ,) and therefore the use of the biodetector would show microbial activity. Biofilms formed by mixed species constitutes a general adaptation of foodborne pathogens and microbiota associated to the food-processing environment for extended survival in their food niche (Jahid & Ha, 2014Jahid, I. K., & Ha, S. (2014). The paradox of mixed-species biofilms in the context of food safety. *Comprehensive Reviews of Food Science and Food Safety*, 13, 990–1011., ,).

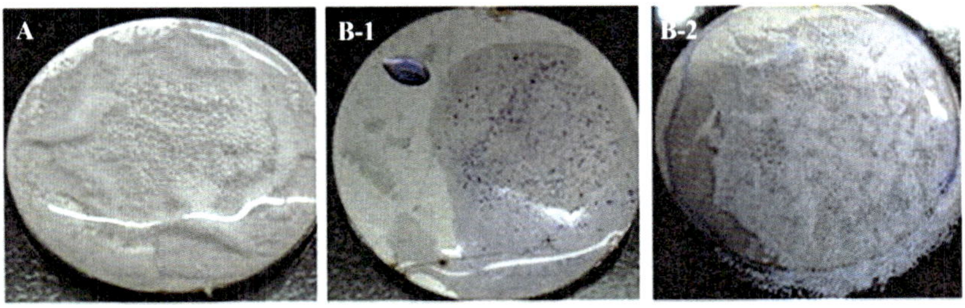

Figure 3. Detection of *P. aeruginosa* biofilms by (A) solution with hydrogen peroxide, (B-1) crystal violet, and (B-2) solution with hydrogen peroxide and crystal violet.

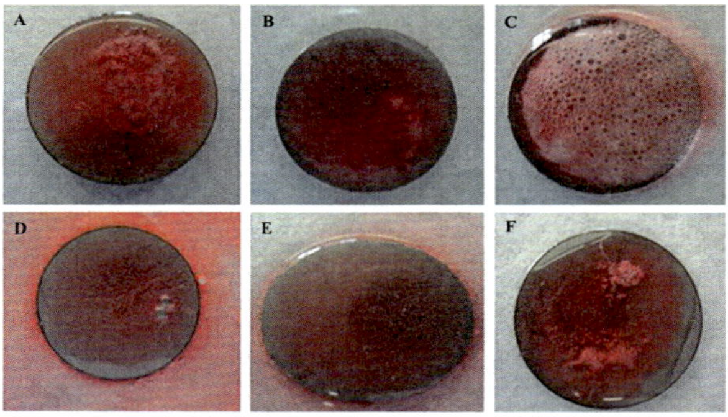

Figure 4. Reactions after the application of the solution with hydrogen peroxide, phosphonates, surfactants, and bleaching agents on stainless steel surfaces with biofilms of (A) *E. coli*, (B) *S. aureus*, (C) *P. aeruginosa*, (D) *S.* Typhimurium, (E) *C. sakazakii*, and (F) *L. monocytogenes.*

As the formula that showed the most visual results was the latest, it was further selected for the development of the industrial product. By adapting the classic hydrogen peroxide detection method, a novel gel has been formulated for its direct use in the food industry. It shows bacterial presence on food-contact surfaces, demonstrating its capacity for being an innovative and effective tool for hygiene monitoring. In order to screen its detection power, a spray was applied to surfaces in a flat position on were biofilms were formed. As shown in Figure 5, positive reactions were very clear and there was an intense bubble formation. When the surface was placed in an upright position, a bubble formation was observed after the biodetector was sprayed onto it (Figure 6). They were laid flat to be able to mark the positive area clearly. Furthermore, for practical use in the food industry as an indicator of biofilm presence, the biodetector was developed to be easy to clean after its use. For this reason, the biodetector was formulated for being highly water soluble, which aids in rinsing and does not leave stain or residues on surfaces.

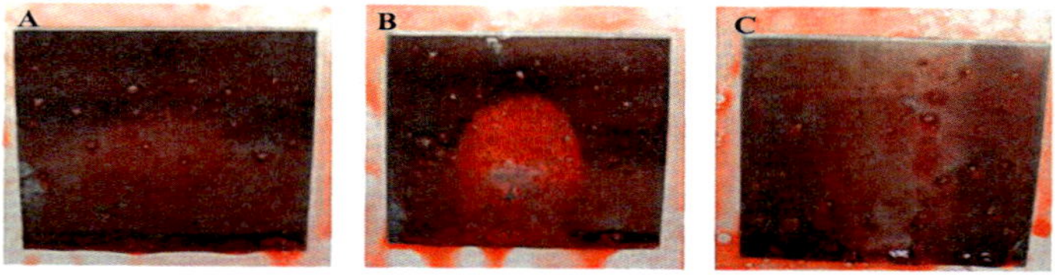

Figure 5. Biofilms on stainless steel coupons detected after the spray application of the solution with hydrogen peroxide, phosphonates, surfactants, and

bleaching agents of (A) *E. coli*, (B) *P. aeruginosa*, and (C) *L. brevis*, placed in a flat position.

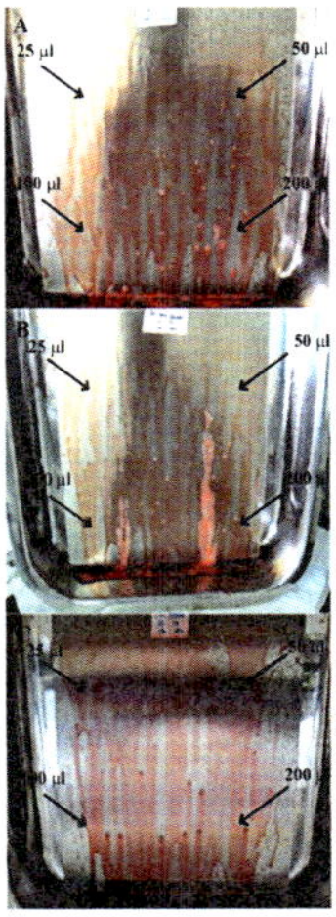

Figure 6. Reactions after the spray application of the solution with hydrogen peroxide, phosphonates, surfactants, and bleaching agents on stainless steel surfaces for the detection of biofilms from (A) *E. coli*, (B) *P. aeruginosa*, and (C) *L. brevis*, formed with different inoculum volumes (25 μl, 50 μl, 100 μl, and 200 μl).

The inoculum volume not only affected the biofilm formation, but also caused a positive reaction after the biodetector was applied. On comparing the positive reaction of the biofilms formed by *E. coli* and *P. aeruginosa* and considering the inoculum volumes, 25 μl, 50 μl, 100 μl, and 200 μl (Table 1), 25 μl showed no positive reaction for any microorganism. For *L. brevis* a negative reaction was detected in all cases, as was expected.

Table 1. Counts of microorganisms by microscopic epifluorescent assay on the biofilms formed with inoculum of 25 μl, 50 μl, 100 μl, and 200 μl. Results expressed as $\log_{10}$ CFU cm^{-2}.

Volume of inoculum	*E. coli*	*P. aeruginosa*	*L. brevis*
25 μl	5.66	5.82	5.01
50 μl	5.96	6.21	5.67
100 μl	6.55	6.30	5.19
200 μl	6.94	6.48	5.23

Biodetector Final Formula

From the three formulas tested, the one that gave the best results, as the contrast was clearer was the solution that contained hydrogen peroxide, phosphonates, surfactans, and bleaching agents. When this formula was used on polypropylene surfaces with *P. aeruginosa* biofilms, a strong positive reaction was observed after 5 s. This reaction was maintained for more than 5 min. Not only did the orange colour contrast well with the white bubbles, but it also did not stain the material employed (Figure 7). When the use of this solution was compared with a solution of hydrogen peroxide, the solution was shown to have a poor contrast. The use of the developed biofinder, however, maintained the foam formed for a long time and ensured better detection (Figure 8).

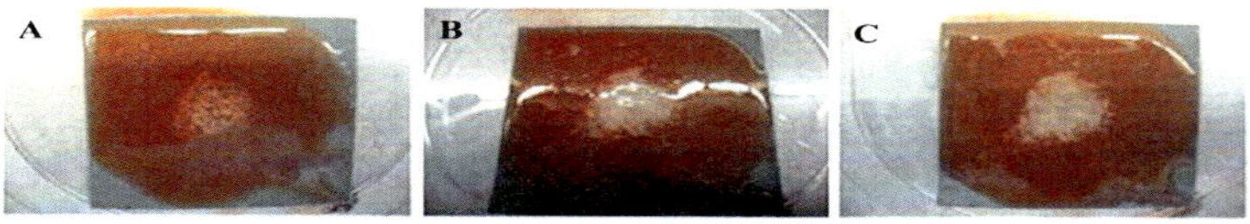

Figure 7. Reaction from *P. aeruginosa* biofilms on polypropylene surfaces using the solution with hydrogen peroxide, phosphonates, surfactants, and bleaching agents after (A) zero, (B) 5 s, and (C) 5 min.

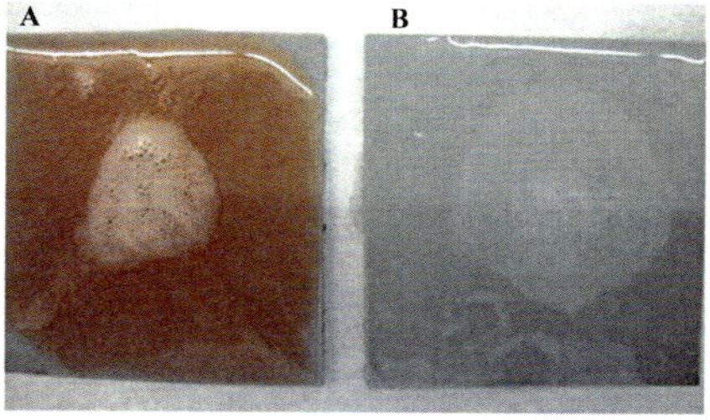

Figure 8. Reaction from *P. aeruginosa* biofilms on polypropylene surfaces using (A) solution with hydrogen peroxide, phosphonates, surfactants, and bleaching agents, and (B) solution with hydrogen peroxide.

Processed products can be contaminated through biofilms from food-processing environments, increasing disease transmission and reducing the shelf life of food products (Chmielewski & Frank, 2003Chmielewski, R. A. N., & Frank, J. F. (2003). Biofilm formation and control in food processing facilities. *Comprehensive Reviews on Food Science and Food Safety*, 2, 22–32.,). In fact, the significance of biofilms in food-processing environments is still not well understood owing to a lack of methodology that allows the direct observation of biofilms in these environments. The development of a system that enables a rapid and visual biofilm detection on food-contact surfaces is highly interesting for controlling the hygienic state of these environments in food industries.

As it has been outlined before, biofilms in natural environments are defined as microbial communities formed of multiple species, including pathogens. In fact, some studies carried on the microbial composition of biofilms on food-processing surfaces have revealed that those which are present on processing line surfaces are established by various microorganisms (Bagge-Ravn et al., 2003Bagge-Ravn, D., Ng, Y., Hjelm, M., Christiansen, J. N., Johansen, C., & Gram, L. (2003). The microbial ecology of processing equipment in different fish industries. Analysis of the microflora during processing and following cleaning and disinfection. *International Journal of Food Microbiology*, 87, 239–250. Gunduz & Tuncel, 2006Gunduz, G. T., & Tuncel, G. (2006). Biofilm formation in an ice cream plant. *Antonie Van Leeuwenhoek*, 89, 329–336. Guobjornsdottir, Einarsson, & Thorkelsson, 2005Guobjornsdottir, B., Einarsson, H., & Thorkelsson, G. (2005). Microbial adhesion to processing lines for fish fillets and cooked shrimp: Influence of stainless steel surface finish and presence of gram-negative bacteria on the attachment of *Listeria monocytogenes*. *Food Technology and Biotechnology*, 43, 55–61.,). If food-contact surfaces are contaminated with biofilms, numerous microbial species will be involved in their formation, including both negative and positive catalase producers (Sharma & Anand, 2002Sharma, M., & Anand, S. K. (2002). Biofilms evaluation as an essential component of HACCP for food/dairy processing industry – A case. *Food Control*, 13, 469–477., ,). For that reason, the application of the biodetector on those surfaces will show an intense bubble formation, and therefore a positive result revealing microbial activity, even when negative catalase producers are forming the biofilm, as catalase producers are also present. Furthermore, the biodetector can be used not only to reveal microbial activity, but also as a tool for the food-processing line workers responsible for making decisions in regard to cleaning procedures.

DETECTION LIMIT

The minimum microbial load for a positive reaction was 10^4 CFU cm^{-2}, a concentration obtained when biofilms were formed after 24–48 h on surfaces with humid environmental conditions. Traditional swabbing or contact plate does not easily detect this number of microorganisms and in many cases, very low recovery is obtained (Nivens, Co, & Franklin, 2009Nivens, D. E., Co, B. M., & Franklin, M. J. (2009). Sampling and quantification of biofilms in food processing and other environments. In P. M.Fratamico, B. A. Annous, & N. W. Guenther(Eds.), *Biofilms in the food and beverage industries* (pp. 539–568). Abington Hall, UK: Woodhead Publishing. doi:10. 1533/9781845697167.5.539,). Those false negative results can be explained by the fact that, when using these techniques, the EPS of the biofilm is not being broken and thus the real number of bacteria is not sampled (Jun et al., 2010Jun, W., Kim, M. S., Cho, B. K., Millner, P. D., Chao, K., & Chan, D. E. (2010). Microbial biofilm detection on food contact surfaces by macro-scale fluorescence imaging. *Journal of Food Engineering*, 99, 314–322., ,). Although there is also the ATP system to measure the microbial presence on surfaces, it is not a very good tool to sample biofilms, particularly when biofilms are mature as the transference of the ATP inside the biofilm is very low. In these situations, low ATP numbers would be obtained, indicating low microbial load, when in fact there might be a mature biofilm composed by a high concentration of microorganisms. Consequently, using the biodetector will help to control the hygienic state of the food-processing surfaces and making decisions in regard to cleaning and disinfection procedures within seconds, preventing cross-uncontrolled contamination from surfaces to foods.

CONCLUSIONS

Many attempts have been made to better understand the main causes of cross-contamination to food products. Food-contact surfaces are known to be the principal site of contamination and controlling them is a key to food hygiene. The results obtained in this study indicate that with the use of the solution that contained hydrogen peroxide, phosphonates, surfactans, and bleaching agents, biofilms of specific microorganisms on different surfaces can be visually detected. The data demonstrated that the strongest biofilm identified was *P. aeruginosa*, followed by *E. coli, S.* Thyphimurium, *L. monocytogenes,* and *S. aureus*. Biofilms formed by *L. brevis* used as a negative control were not detected when the biodetector was applied because this microorganism does not have catalase activity. Based on the results obtained and the literature survey, it can be concluded that the developed biodetector as a biofilm detection system is a potentially powerful tool to prevent both the transmission of foodborne diseases and the spoilage of food originating from food-contact surfaces.

ACKNOWLEDGMENTS

The authors thank Dolors Busquets enormously for her technical assistance. This study was supported by Research Project grants RTA2014-00045-C03-03 (INIA) from the Spanish Ministry of Economy and Competitiveness. The authors alone are responsible for the content and writing of the paper. Authors acknowledge to Mrs. Sarah Davis for English grammar revision.

REFERENCES

1. Alavi, H. E. D., & Hansen, L. T. (2013). Kinetics of biofilm formation and desiccation survival of *Listeria monocytogenes*in single and dual species biofilms with *Pseudomonas fluorescens, Serratia proteamaculans* or *Shewanella baltica* on food-grade stainless steel surfaces. *Biofouling*, 29, 1253–1268.
2. Annous, B. A., Fratamico, P. M., & Smith, J. L. (2009). Quorum sensing in biofilms: Why bacteria behave the way they do. *Journal of Food Science*, 74, 24–36.
3. Anonymous. (2002). ISO-13697:2002. Chemical disinfectants and antiseptics. Quantitative non-porous surface test for the evaluation of bactericidal and/or fungicidal activity of chemical disinfectants used in food, industrial, domestic and institutional.
4. Anonymous. (2007). ISO-22196:2007. Plastics - Measurement of antibacterial activity on plastics surfaces.
5. Bagge-Ravn, D., Ng, Y., Hjelm, M., Christiansen, J. N., Johansen, C., & Gram, L. (2003). The microbial ecology of processing equipment in different fish industries. Analysis of the microflora during processing and following cleaning and disinfection. *International Journal of Food Microbiology*, 87, 239–250.
6. Baumann, A. R., Martin, S. E., & Feng, H. (2009). Removal of *Listeria monocytogenes* biofilms from stainless steel by use of ultrasound and ozone. *Journal of Food Protection*, 72, 1306–1309.
7. Beuchat, L. R., Kim, H., Gurtler, J. B., Lin, L. C., Ryu, J. H., & Richards, G. M. (2009). *Cronobacter sakazakii* in foods and factors affecting its survival, growth, and inactivation. *International Journal of Food Microbiology*, 136(2), 204–213.
8. Branda, S. S., Vik, A., Friedman, L., & Kolter, R. (2005). Biofilms: The matrix revisited. *Current Trends in Microbiology*, 13, 20–26.
9. Carlton, R. M., Noordman, W. H., Biswas, B., de Meester, E. D., & Loessner, M. J. (2005). Bacteriophage P100 for control of *Listeria monocytogenes* in foods: Genome sequence, bioinformatic analyses, oral toxicity study, and application. *Regulatory Toxicology and Pharmacology*, 43, 301–312.
10. Chae, M. S., & Schraft, H. (2000). Comparative evaluation of adhesion and biofilm formation of different *Listeria monocytogenes* strains. *International Journal of Food Microbiology*, 62, 103–111.

11. Chmielewski, R. A. N., & Frank, J. F. (2003). Biofilm formation and control in food processing facilities. *Comprehensive Reviews on Food Science and Food Safety*, 2, 22–32.
12. Christensen, B. E., Trønnes, H. N., Vollan, K., Smidsrød, O., & Bakke, R. (1990). Biofilm removal by low concentration of hydrogen peroxide. *Biofouling*, 2, 165–175.
13. Colagiorgi, A., Bruini, I., Di Ciccio, P. A., Zanardi, E., Ghidini, S., & Ianieri, A. (2017). *Listeria monocytogenes* biofilms in the wonderland of food industry. *Pathogens*, 6, 41–50.
14. Di Ciccio, P., Vergara, A. R., Festino, A., Paludi, D., Zanardi, E., Ghidini, S., & Ianieri, A. (2015). Biofilm formation by *Staphylococcus aureus* on food contact surfaces: Relationship with temperature and cell surface hydrophobicity. *Food Control*, 50, 930–936.
15. Elias, S., & Banin, E. (2012). Multi-species biofilms: Living with friendly neighbors. *FEMS Microbiology Reviews*, 36, 990–1004.
16. Elkins, J. G., Hassett, D. J., Stewart, P. S., Schweizer, H. P., & MCDermott, T. R. (1999). Protective role of catalase in *Pseudomonas aeruginosa* biofilm resistance to hydrogen peroxide. *Applied and Environmental Microbiology*, 65(10), 4594–4600.
17. Fuster-Valls, N., Hernández-Herrero, M., Marín-de-Matero, M., & Rodríguez-Jerez, J. J. (2008). Effect of different environmental conditions on the bacteria survival on stainless steel surfaces. *Food Control*, 19, 308–314.
18. Garrett, T. R., Bhakoo, M., & Zhang, Z. (2008). Bacterial adhesion and biofilms on surfaces. *Progress in Natural Science*, 18(9), 1049–1056.
19. Gunduz, G. T., & Tuncel, G. (2006). Biofilm formation in an ice cream plant. *Antonie Van Leeuwenhoek*, 89, 329–336.
20. Guobjornsdottir, B., Einarsson, H., & Thorkelsson, G. (2005). Microbial adhesion to processing lines for fish fillets and cooked shrimp: Influence of stainless steel surface finish and presence of gram-negative bacteria on the attachment of *Listeria monocytogenes*. *Food Technology and Biotechnology*, 43, 55–61.
21. Hannig, C., Hannig, M., Rehmer, O., Braun, G., Hellwig, E., & Al-Ahmad, A. (2007). Fluorescence microscopic visualization and quantification of initial bacterial colonization on enamel in situ. *Archives Oral Biology*, 52, 1048–1056.
22. Holah, J. T., Betts, R. P., & Thorpe, R. H. (1989). The use of

epifluorescence microscopy to determine surface hygiene. *International Biodeterioration & Biodegradation*, 25, 147–153.

23. Hood, S., & Zottola, E. A. (1995). Biofilms in food processing. *Food Control*, 6, 9–18.
24. Jahid, I. K., & Ha, S. (2014). The paradox of mixed-species biofilms in the context of food safety. *Comprehensive Reviews of Food Science and Food Safety*, 13, 990–1011.
25. Jahid, I. K., Han, N., Srey, S., & Ha, S. D. (2014b). Competitive interactions inside mixed-species biofilms of *Salmonella Typhimurium* and cultivable indigenous microorganisms on lettuce enhance microbial resistance of their sessile cells to ultraviolet C (UV-C) irradiation. *Food Research International*, 55, 445–454.
26. Jun, W., Kim, M. S., Cho, B. K., Millner, P. D., Chao, K., & Chan, D. E. (2010). Microbial biofilm detection on food contact surfaces by macro-scale fluorescence imaging. *Journal of Food Engineering*, 99, 314–322.
27. Kumar, C. G., & Anand, S. (1998). Significance of microbial biofilms in food industry: A review. *International Journal of Food Microbiology*, 42, 9–27.
28. LeChevallier, M. W., Cawthon, C. D., & Lee, R. G. (1988). Inactivation of biofilm bacteria. *Applied Environmental Microbiology*, 54, 2492–2499.
29. Lee, K. W., Periasamy, S., Mukherjee, M., Xie, C., Kjelleberg, S., & Rice, S. A. (2014). Biofilm development and enhanced stress resistance of a model, mixed-species community biofilm. *ISME Journal*, 8, 894–907.
30. Lee Wong, A. C. (1998). Biofilms in food processing environments. *Journal of Dairy Science*, 81, 2765–2770.
31. Lunau, M., Lemke, A., Walther, K., Martens-Habbena, W., & Simon, M. (2005). An improved method for counting bacteria from sediments and turbid environments by epifluorescence microscopy. *Environmental Microbiology*, 7, 961–968.
32. Manuzon, M. Y., & Wang, H. H. (2007). Mixed-species biofilms. In H. P. Blaschek, H. H. Wang, & M. E. Agle (Eds.), *Biofilms in food environment* (pp. 105–125). Ames, IA: Blackwell Publishing Ltd.
33. Montañez-Izquierdo, V. Y., Salas-Vázquez, D. I., & Rodríguez-Jerez, J. J. (2012). Use of epifluorescence microscopy to assess the

effectiveness of phage P100 in controlling *Listeria monocytogenes* biofilms on stainless steel surfaces. *Food Control*, 23, 470–477.

34. Moons, P., Michiels, C. W., & Aertsen, A. (2009). Bacterial interactions in biofilms. *Critical Reviews in Microbiology*, 35, 157–168.

35. Myszka, K., & Czaczyk, K. (2011). Bacterial biofilms on food contact surfaces - a review. *Polish Journal of Food and Nutrition Sciences*, 61, 173–180.

36. Nivens, D. E., Co, B. M., & Franklin, M. J. (2009). Sampling and quantification of biofilms in food processing and other environments. In P. M. Fratamico, B. A. Annous, & N. W. Guenther (Eds.), *Biofilms in the food and beverage industries*(pp. 539–568). Abington Hall, UK: Woodhead Publishing. doi:10.1533/9781845697167.5.539

37. Notermans, S., Nauta, M. J., Jansen, J., Jouve, J. L., & Mead, G. C. (1998). A risk assessment approach to evaluating food safety based on product surveillance. *Food Control*, 9, 217–223.

38. Percival, S. L., Walker, J. T., & Hunter, P. R. (2000). *Microbiological aspects on biofilms and drinking water*. Boca Raton, FL: CRC Press.

39. Pettipher, G. L., & Rodrigues, U. M. (1982). Rapid enumeration of microorganisms in foods by the direct epifluorescent filter technique. *Applied and Environmental Microbiology*, 44, 809–813.

40. Phillips, C. A. (2016). Bacterial biofilms in food processing environments: A review of recent developments in chemical and bioloical control. *International Journal of Food Science and Technology*, 51, 1731–1743.

41. Ríos-Castillo, A. G., González-Rivas, F., & Rodríguez-Jerez, J. J. (2017). Bactericidal efficacy of hydrogen peroxide-based disinfectants against gram-positive and gram-negative bacteria on stainless steel surfaces. *Journal of Food Science*, 82, 2351–2356.

42. Sharma, M., & Anand, S. K. (2002). Biofilms evaluation as an essential component of HACCP for food/dairy processing industry – A case. *Food Control*, 13, 469–477.

43. Shi, X., & Zhu, X. (2009). Biofilm formation and food safety in food industries. *Trends in Food Science and Technology*, 20, 407–413.

44. Simões, M., Simões, L. C., & Vieira, M. J. (2010). A review of current and emergent biofilm control strategies. *LWT Food Science and Technology*, 43, 573–583.

45. Stepanović, S., Cirković, I., Ranin, L., & Svabić-Vlahović, M. (2004).

Biofilm formation by *Salmonella* spp. and *Listeria monocytogenes* on plastic surface. *Letters in Applied Microbiology*, 38, 428–432.

46. Swaminathan, B., & Gerner-Smidt, P. (2007). The epidemiology of human listeriosis. *Microbes and Infection*, 9, 1236–1243.
47. Tompkin, R. B. (2002). Control of *Listeria monocytogenes* in the food-processing environment. *Journal of Food Protection*, 65, 709–725.
48. Torlak, E., & Sert, D. (2013). Combined effect of benzalkonium chloride and ultrasound against *Listeria monocytogenes* biofilm on plastic surface. *Letters in Applied Microbiology*, 57, 220–226.
49. Van Houdt, R., & Michiels, C. W. (2010). Biofilm formation and the food industry, a focus on the bacterial outer surface. *Journal of Applied Microbiology*, 109, 1117–1131.
50. Wang, H., Qi, J., Dong, Y., Li, Y., Xu, X., & Zhou, G. (2017). Characterization of attachment and biofilm formation by meat-borne *Enterobacteriaceae* strains associated with spoilage. *LWT – Food Science and Technology*, 86, 399–407.
51. Whittenbury, R. (1964). Hydrogen peroxide formation and catalase activity in the lactic acid bacteria. *Journal of General Microbiology*, 35, 13–26.
52. Yan, X., & Gurtler, J. B. (2014). *Cronobacter* (*Enterobacter*) *sakazakii*. In C. A. Batt & M. L. Tortorello (Eds.), *Encyclopedia of food microbiology*, (2nd ed. pp. 528–532). Oxford, UK: Academic Press. doi:10.1016/B978-0-12-384730-0.00382-7
53. Yang, H., Kendall, P. A., Medeiros, L. C., & Sofos, J. N. (2009). Efficacy of sanitizing agents against *Listeria monocytogenes* biofilms on high-density polyethylene cutting board surfaces. *Journal of Food Protection*, 72, 990–998

CHAPTER 10

DEVELOPMENT OF COLLOIDAL GOLD-BASED IMMUNOCHROMATOGRAPHIC ASSAY FOR RAPID DETECTION OF ALICYCLOBACILLUS CIDO-TERRESTRIS IN APPLE JUICE

Yingsha Liu, Jianke Li, Lin Zhang, Xiaoxia Wu, Feng Wang and Bingbing Song

College of Food Engineering and Nutritional Science, Shaanxi Normal University, Xi' an, P. R. China

CITATION: Liu, Y., et al., "Development of colloidal gold-based immunochromatographic assay for rapid detection of Alicyclobacillus acidoterrestris in apple juice", Journal of Food (2017), https://doi.org/10.1080/19476337.2016.1266032..

ABSTRACT

The colloidal gold quick test card which we developed was used for rapid detection of *Alicyclobacillus acidoterrestris* (*A. acidoterrestris*) in apple juice concentrate. Two antibodies against *A. acidoterrestris*, obtained from Sprague Dawley (SD) rats and Japan White Rabbits, were adopted to construct this quick test card. The results showed that the colloidal gold quick test card had better specificity, and was more rapid and convenient which deserved notice especially (the test could be completed within 5–10 min) and concurred with ELISA method (we previously report) and K medium methods (traditional classical method). This new developed colloidal gold quick test card (we named it TAB quick test card) could be more convenient for apple juice safety testing.

INTRODUCTION

Alicyclobacillus acidoterrestris is a thermophilic, acidophilic, and spore-forming bacterium (TAB) in many fruit juices (Cerny, Hennlich, & Poralla, 1984Cerny, G., Hennlich, W., & Poralla, K. (1984). Spoilage of fruit juice by bacilli: Isolation and characterization of the spoiling microorganisms. *Zeitschrift Fur Lebensmittel-Untersuchung Und -Forschung*, 179(3), 224–227. doi:10.1007/BF01041898, ,), such as apple, pear, orange, citrus, and so on. It can survive through pasteurization and cause juice spoilage (Breveglieri, Masiero, Spisani, & De Taddeo, 2011Breveglieri, G., Masiero, L., Spisani, S., & De Taddeo, H. (2011). Detection methods for the identification of Alicyclobacillus species involved in food spoilage. *Minerva Biotecnologica*, 23(1), 23–32., ; Huang, Yuan, Guo, Gekas, & Yue, 2014Huang, X.C., Yuan, Y.H., Guo, C.F., Gekas, V., & Yue, T.L. (2014). *Alicylobacillus* in the fruit juice industry: Spoilage, detection, and prevention/control. *Food Reviews International*, *31*(13), 91–124.). The traditional method for the detection of *A. acidoterrestris* is plate cultivation, which requires 4–5 days to identify the target bacteria (Pinhatti, Variane, Eguchi, & Manfio, 1997Pinhatti, M.E.M.C., Variane, S., Eguchi, S.Y., & Manfio, G.P. (1997). Detection of *acidothermic bacilli* in industrialized fruit juices. *Fruit Process*, 7, 350–353.). In order to monitor and control these spoilage organisms timely and effectively, a rapid detection technology is needed. Numerous studies for rapid detection of *A. acidoterrestris* have been reported previously. For example, the reverse-transcription polymerase chain reaction (RT-PCR) detection by Yamazaki, Teduka, Inoue, and Shinano (1996Yamazaki, K., Teduka, H., Inoue, N.,

& Shinano, H. (1996). Specific primers for detection of *Alicyclobacillus acidoterrestris* by RT-PCR. *Letters in Applied Microbiology*, 23(5), 350–354. doi:10.1111/j.1472-765X.1996.tb00206.x), a real-time PCR rapid detection method by Luo, Yousef, and Wang (2004Luo, H., Yousef, A.E., & Wang, H.H. (2004). A real-time polymerase chain reaction-based method for rapid and specific detection of spoilage *Alicyclobacillus* spp. in apple juice. *Letters in Applied Microbiology*, 39(4), 376–382. doi:10.1111/lam.2004.39. issue-4), a quantitatively competitive PCR (QC-PCR) method by Li, Feng, and Qiu (2006Li, J.K., Feng, Z.P., & Qiu, N.X. (2006). Optimization and establishment of quantitatively competitive PCR system for the detection of. *Alicyclobacillus Acidoterrestris. Scientia Agricultura Sinica*, 39(2), 375–380.), and a Fourier transform infrared (FT-IR) spectroscopy detection by Al-Qadiri, Lin, Cavinato, and Rasco (2006Al-Qadiri, H.M., Lin, M.S., Cavinato, A.G., & Rasco, B.A. (2006). Fourier transform infrared spectroscopy, detection and identification of Escherichia coli O157: H7and *Alicyclobacillus*strains in apple juice. *International Journal of Food Microbiology*, 111(1), 73–80. doi:10.1016/j.ijfoodmicro.2006.05.004). However, those methods are difficult to use in factory because it requires skilled personnel and relatively expensive equipments (Li, Huang, Xia, & Liu, 2014Li, J.K., Huang, R.R., Xia, K., & Liu, L. (2014). Double antibodies sandwich enzyme-linked immunosorbent assay for the detection of *Alicyclobacillus acidoterrestris* in apple juice concentrate. *Food Control*, 40, 172–176. doi:10.1016/j.foodcont.2013.11.037, ,).

We reported a novel method of assay indirect enzyme-linked immunosorbent (ID-ELISA) and staphylococcal protein A enzyme-linked immunosorbent assay (SPA-ELISA) to rapidly detect *A. acidoterrestris* in apple juice concentrate (AJC) (Li, Wang, & Xia, 2011Li, J.K., Wang, F., & Xia, K. (2011). An indirect and accurate ELISA for detection of *Alicyclobacillus acidoterrestris* in apple juice concentrate. *Scientia Agricultura Sinica*, 44(22), 4669–4677. ; Li, Xia, & Yu, 2013Li, J.K., Xia, K., & Yu, C.Z. (2013). Detection of *Alicyclobacillus acidoterrestris* in apple juice concentrate by enzyme-linked immunosorbent assay. *Food Control*, 30(1), 251–254. doi:10.1016/j.foodcont.2012.07.009, ,) for the first time. And recently an improved double-antibody sandwich ELISA (DAS-ELISA) (Li et al., 2014Li, J.K., Huang, R.R., Xia, K., & Liu, L. (2014). Double antibodies sandwich enzyme-linked immunosorbent assay for the detection of *Alicyclobacillus acidoterrestris* in apple juice concentrate. *Food Control*, 40, 172–176. doi:10.1016/j.foodcont.2013.11.037, ,) has been proved to be convenient and suitable for application.

In this study, a new colloidal gold quick card (TAB quick test card) was developed with the aid of colloidal gold immunochromatographic assay. This test card showed more rapidness and convenience for rapid detection of *A. acidoterrestris* in apple juice.

MATERIALS AND METHODS

Bacterial Strains and Materials

Alicyclobacillus acidoterrestris DSM 3922 was purchased from DSMZ (Brunswick, Germany). *Escherichia coli* ATCC 8739, *Bacillus subtilis* ATCC 6633, *Bacillus cereus* ATCC 11778, *Staphylococcus aureus* ATCC 6538 and *Shigella* ATCC 12022 were obtained from the Laboratory of Microbiology, College of LifeScience, Shaanxi Normal University. Japanese White Rabbits and Sprague Dawley (SD) rats were obtained from Laboratory Animal Research Center, College of Medicine, Xi'an Jiaotong University. Juice concentrate (pH 3.2–4.6) was obtained from Shaanxi Hengxing Fruit Juice Co., Ltd. Freund's complete and incomplete adjuvant was purchased from Sigma Chemical (St. Louis, MO, USA). ELISA Well Strips and Plate Frames were bought from JET BIOFIL (Canada). Colloidal gold and Goat anti-rabbit were purchased from Shanghai Jiening Biotechnology Co., Ltd (Shanghai, CN). Nitrocellulose membranes, sample pads, absorbent pads and conjugate pads were purchased from Millipore Corporation. Polyvinylpyrrolidone (PVP) was purchased from JingBo, Ltd (Shaanxi, CN). All inorganic chemicals and organic solvents used in this study were of reagent grade or chemically pure, which were purchased from DingGuo Biotech (Beijing, CN) or Sigma (USA).

The formula of K medium (pH 4.0): Tween 80(1.0 g), glucose (1.0 g), yeast extract (2.5 g), peptone (5.0 g), agar (15 g) and distilled water (1000 mL).

The formula of 402 medium(pH 4.0): $(NH_4)_2SO_4$(0.2 g), $CaCl_2 \cdot 2H_2O$ (0.25 g), $MgSO_4$(0.5 g), yeast extract (2.0 g), glucose (5.0 g),K H_2PO_4 (5.0 g)and distilled water (1000 mL).

Antigen Preparation

A. acidoterrestris were cultured on K medium at pH 3.7 at43 C for 36 h and then they were harvested with sterile saline (0.9% NaCl), and the cells were washed three times by centrifugation (3500 g, 10 min), and keep resuspension in sterile saline. Bacterial density of the final suspension was 2×10^9 CFU/mL, which was determined by spectrophotometer (OD/600 nm) (Wang, Li, Liu, & Jiang, 2009Wang, F., Li, J.K., Liu, H.X., & Jiang, K. (2009). Preparation and purification of polyclonal antibody against *Alicyclobacillus Acidoterrestrisfrom* apple juice concentrate. *Food and Fermentation Industries*, 5(4), 33–37. ; Xia, Li, & Yu, 2012Xia, K., Li, J.K., & Yu, Z. (2012). Preparation and purification of rat polyclonal antibody against. *Aliyclobacillua Acidoterrestris. Science and Technology of Food Industry*, 33(15), 162–168.).

Antibody Preparation and Purification

The preparation of Japanese White Rabbits and SD rats polyclonal antibodies against *A. acidoterrestris* was reported by Wang et al. (2009Wang, F., Li, J.K., Liu, H.X., & Jiang, K. (2009). Preparation and purification of polyclonal antibody against *Alicyclobacillus Acidoterrestris from* apple juice concentrate. *Food and Fermentation Industries*, 5(4), 33–37.) and Xia et al. (2012Xia, K., Li, J.K., & Yu, Z. (2012). Preparation and purification of rat polyclonal antibody against. *Aliyclobacillua Acidoterrestris. Science and Technology of Food Industry*, 33(15), 162–168.). The polyclonal antibody was precipitated by caprylic acid and ammonium sulfate and then purified by Hitrap Protein A chromatography column (Wang et al., 2009Wang, F., Li, J.K., Liu, H.X., & Jiang, K. (2009). Preparation and purification of polyclonal antibody against *Alicyclobacillus Acidoterrestris from* apple juice concentrate. *Food and Fermentation Industries*, 5(4), 33–37. ; Xia et al., 2012Xia, K., Li, J.K., & Yu, Z. (2012). Preparation and purification of rat polyclonal antibody against. *Aliyclobacillua Acidoterrestris. Science and Technology of Food Industry*, 33(15), 162–168.).

Preparation of Colloidal Gold Probe

The colloidal gold probe was prepared according to the standard method with modifications (Ju et al., 2010Ju, Y., Hao, H.J., Xiong, G.H., Geng, H.R., Zheng, Y.L., Wang, J. … Jiang, Y.Q. (2010). Development of colloidal gold-based immunochromatographic assay for rapid detection of *Streptococcus suis serotype* 2. *Veterinary Immunology and Immunopathology*, 133(2–4), 207–211. doi:10.1016/j.vetimm.2009.08.010). The optimal pH and concentration of capture antibody were added to the colloidal gold solution (10 mL), shaking for 30 min. Then 200 μL 10% BSA (Bull Serum Albumin) were added for 15 min after, the mixtures were washed three times by centrifugation(10 × 10^3 g,10 min). The precipitate was collected and suspended with 1 mL of the suspension buffer (50 mM pH 8.1Tris-HCl, 0.1% PVP, 0.2% tween-20, 0.2% BSA). The colloidal gold was stored in a dark bottle at 4 C.

Preparation of the Immunochromatographic Strips

The immunochromatographic strip (ICS) included four parts (named as TAB quick test card): a sample pad, a conjugate pad, an immobilized nitrocellulose membrane, and an absorbent pad. The sample pads were treated with the treatment liquid (50 mM pH 8.1Tris-HCl, 0.3% tween) for 30 min. The conjugate pads were sprayed with colloidal gold-labeled rabbit antibody against *A. acidoterrestris*. The T line was coated with rat antibody and the C line was coated with goat anti-rabbit IgG. When the samples were slowly added to the sample pads, it was moved through immobilized nitrocellulose membrane under capillarity field effect. After 5–10 min, the positive effect displayed two red lines, while the negative effect displayed one red line in the control lines. The immunochromatographic strips lost effectiveness if there was nothing in the control lines.

Specificity and Sensitivity of the TAB Quick Test Card

In order to determine the specificity of the TAB quick test card, we used the test card to test *Escherichia coli* ATCC 8739, *Bacillus subtilis* ATCC 6633, *Bacillus cereus* ATCC 11778, *Staphylococcus aureus* ATCC 6538 and *Shigella* ATCC 12022.

The five representative food-borne bacteria were cultured on medium at 37 C for 12 h, then they were harvested with sterile saline (0.9% NaCl), and the cells were washed three times by centrifugation (3500 g, 10 min), and kept in resuspension in sterile saline.

The bacterial suspensions (*A. acidoterrestris*) were diluted by pH 7.4 PBS buffer (0.2 g KH_2PO_4, 2.9 g Na_2HPO_4, 0.2 g KCl, 8.0 g NaCl, 1000 mL distilled water) from concentrations of 1 × 10^{10} CFU/mL, 1 × 10^9CFU/mL, 1 × 10^8 CFU/mL, 1 × 10^7 CFU/mL, 1 × 10^6 CFU/mL to 0 CFU/mL (Y = 0.0844X+0.1126, R^2 = 0.9992. Y: absorbance, X: concentrations(billion/mL).

Each suspension with *A. acidoterrestris* of varying levels was detected by the TAB quick test card. The sensitivity of the TAB quick test card was the positive reaction with the lowest concentration.

AJC Sample Preparation for the TAB Quick test Card Assay

AJC sample preparation was performed using procedures based on the methods of Li et al. (2011Li, J.K., Wang, F., & Xia, K. (2011). An indirect and accurate ELISA for detection of *Alicyclobacillus acidoterrestris* in apple juice concentrate. *Scientia Agricultura Sinica*, 44(22), 4669–4677.). About 10 mL AJC samples was heat-treated at 80 C for 10 min and then deionized water was added. The diluted sample was filter sterilized by discmembrane filters (0.45 μm pore size; 50 mm in diameter). After filtering, the membranes were transported into an enrichment broth of 402 medium (DingGuo Biotech, pH 4.0, 200 mL) and incubated at 45 C for 12 h with shaking for pre-enrichment. After incubation, the broth was centrifuged at 10,000 g for 10 min to collect precipitation and then diluted the precipitation with pH 7.4 PBS buffer (0.2 g KH_2PO_4, 2.9 g Na_2HPO_4, 0.2 g KCl, 8.0 g NaCl, 1000 mL distilled water) (Li et al., 2014Li, J.K., Huang, R.R., Xia, K., & Liu, L. (2014). Double antibodies sandwich enzyme-linked immunosorbent assay for the detection of *Alicyclobacillus acidoterrestris* in apple juice concentrate. *Food Control*, 40, 172–176. doi:10.1016/j.foodcont.2013.11.037, ,) that achieved concentration above the sensitivity of the TAB quick test card.

Analysis of Consistency among Quick test card, ELISA kit and K medium method

Adding 150 μL of samples on the TAB quick test cards, after 5–10 min, we observed the phenomena and compared with the results of DAS-ELISA kit (Li et al., 2014Li, J.K., Huang, R.R., Xia, K., & Liu, L. (2014). Double antibodies sandwich enzyme-linked immunosorbent assay for the detection of *Alicyclobacillus acidoterrestris* in apple juice concentrate. *Food Control*, 40, 172–176. doi:10.1016/j.foodcont.2013.11.037) and K medium method (The bacteria were cultured on K medium(plate) at pH 3.7 at 43 C for 36 h).

RESULTS AND DISCUSSION

Polyclonal Antibodies

The titers of the Japanese White Rabbits polyclonal antibodies against *A. acidoterrestris* were reached at 1/25,600 and the titers of the SD rats polyclonal antibodies against *A. acidoterrestris* were reached at 1/12,800.

Specificity of the TAB Quick Test Card

Escherichia coli ATCC 8739, *Bacillus subtilis* ATCC 6633, *Bacillus cereus* ATCC 11778, *Staphylococcus aureus*ATCC 6538 and *Shigella* ATCC 12022 were cultured on beef-protein medium at 37 C for 12 h. *A. acidoterrestris* was cultured on K medium at pH 3.7 at 43 C for 12 h. Different suspensions were collected to evaluate the specificity of the strip. Only *A. acidoterrestris* showed positive effect and other samples showed negative effect. As shown in Figures 1 and 2, the TAB quick test card exhibited better specificity.

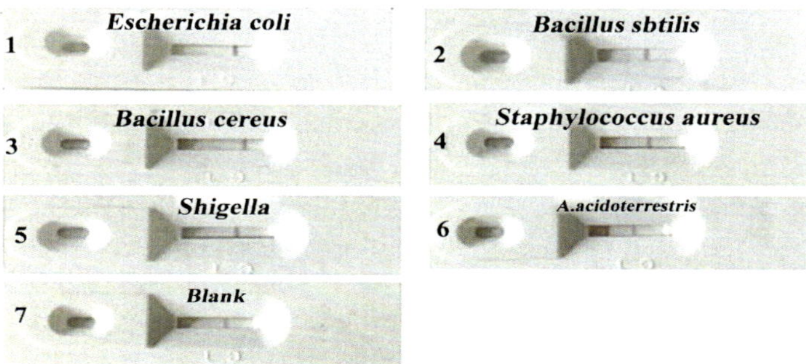

Figure 1. Specificity of the TAB quick test card.

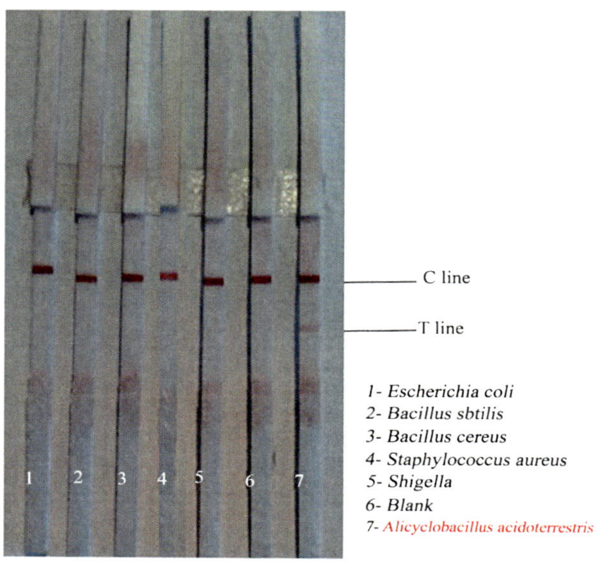

Figure 2. Specificity of the TAB quick test card.

Sensitivity of the TAB quick test card

Different concentrations of *A. acidoterrestris* were detected by the TAB quick test card. When the concentration was 1×10^6 CFU/mL, just the control line showed red, while the concentration exceeded 1×10^7 CFU/mL(45 C for 12 h with shaking for pre-enrichment), the test line and control line both displayed red. The phenomena are shown in Figures 3 and 4, and the sensitivity of the TAB quick test cards is 1×10^7CFU/mL.

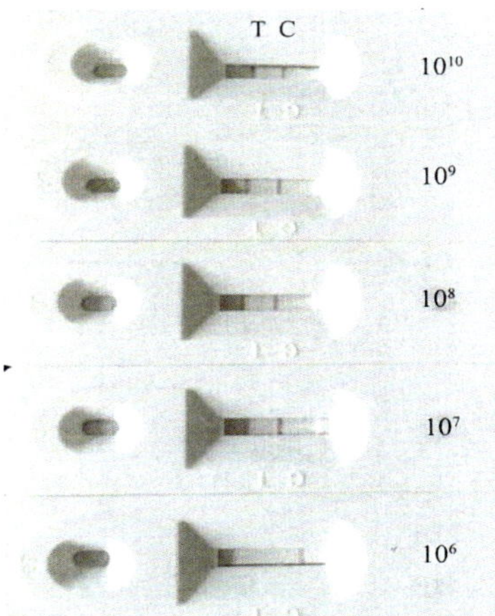

Figure 3. The phenomena of TAB quick test card with different concentrations of *A. acidoterrestris.*

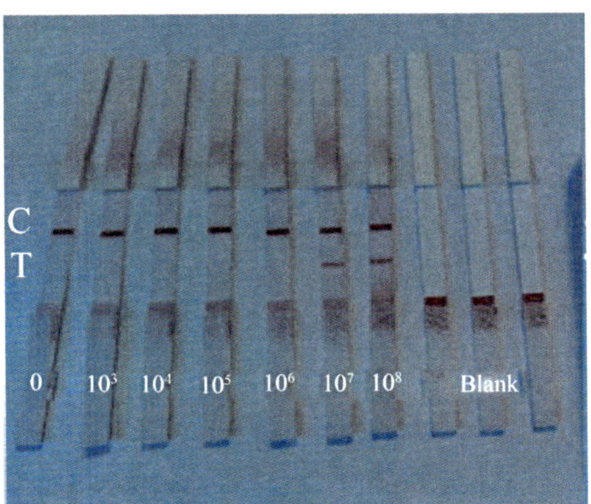

Figure 4. The phenomena of TAB quick test card with different concentrations of *A. acidoterrestris.*

Analysis of Consistency among Quick test card, ELISA kit and K medium method

Twelve samples of AJC were assayed, respectively, by TAB quick test card, ELISA kit and K medium method. For the quick test card assay, samples with the test line and control line both displayed red, which indicated a positive effect. For the DAS-ELISA assay, samples with OD 450 nm more than 0.038 were considered as a positive effect. In the terms of the K medium, 10 mL samples with more than 1 CFU were considered as a positive effect. The results are shown in Table 1. The TAB quick test card assay was well in concordance with ELISA kit and K medium method.

Table 1. Detection of *Alicyclobacillus acidoterrestris* by quick test card, ELISA kit and K medium.

Sample	**1**	**2**	**3**	**4**	**5**	**6**
Quick test card	–	–	–	–	–	+
ELISA kit	0.034 ± 0.011 –	0.028 ± 0.014 –	0.026 ± 0.017 –	0.037 ± 0.010 –	0.037 ± 0.015 –	0.117 ± 0.012 +
K medium method (plate culture method)	<1 –	<1 –	<1 –	<1 –	<1 –	2 +
Sample	**7**	**8**	**9**	**10**	**11**	**12**
Quick test card	–	+	–	–	–	+
ELISA kit	0.024 ± 0.011 –	0.098 ± 0.014 +	0.038 ± 0.011 –	0.034 ± 0.016 –	0.036 ± 0.012 –	0.102 ± 0.013 +
K medium method (plate culture method)	<1 –	2	<1 –	<1 –	<1 –	3 +

The traditional method and the main current way for the detection of *A. acidoterrestris* is plate cultivation (K medium), but it took long times (Chang & Kang, 2004Chang, S.S., & Kang, D.H. (2004). Alicyclobacillus spp. in the fruit juice industry: History, characteristics, and current isolation/detection procedures. *Critical Reviews in Microbiology*, 30(2), 55–74. doi:10.1080/10408410490435089). In our previous study, we reported an ID-

ELISA, SPA-ELISA and DAS-ELISA method to detect the *A. acidoterrestris*, which is suitable and time-saving in ordinary laboratory (Li et al., 2014Li, J.K., Huang, R.R., Xia, K., & Liu, L. (2014). Double antibodies sandwich enzyme-linked immunosorbent assay for the detection of *Alicyclobacillus acidoterrestris* in apple juice concentrate. *Food Control*, 40, 172–176. doi:10.1016/j.foodcont.2013.11.037 2011Li, J.K., Wang, F., & Xia, K. (2011). An indirect and accurate ELISA for detection of *Alicyclobacillus acidoterrestris* in apple juice concentrate. *Scientia Agricultura Sinica*, 44(22), 4669–4677. , 2013Li, J.K., Xia, K., & Yu, C.Z. (2013). Detection of *Alicyclobacillus acidoterrestris* in apple juice concentrate by enzyme-linked immunosorbent assay. *Food Control*, 30(1), 251–254. doi:10.1016/j.foodcont.2012.07.009, ,). The ICS assay method (TAB quick test card) proved to be more rapid and convenient. It took 5–10 min, excluding sample pretreatment and enrichment time. The TAB quick test card (sealed package) could be stored at room temperature in dry condition for 1 year.

CONCLUSION

A high sensitivity and more convenience TAB quick test card based on colloidal gold-based immunochromatographic assay was developed to detect *A. acidoterrestris* in juice. The results can be obtained within 5–10 minutes excluding pretreatment and enrichment (see Section 2.7 for the details). Our results showed that the quick test card had no cross-reaction with the five representative food-borne bacteria and showed high sensitivity and excellent agreement with K medium method and DAS-ELISA kit. TAB quick test card showed a great potential in fruit juice detection.

REFERENCES

1. Al-Qadiri, H.M., Lin, M.S., Cavinato, A.G., & Rasco, B.A. (2006). Fourier transform infrared spectroscopy, detection and identification of Escherichia coli O157: H7and *Alicyclobacillus* strains in apple juice. *International Journal of Food Microbiology*, 111(1), 73–80. doi:10.1016/j.ijfoodmicro.2006.05.004
2. Breveglieri, G., Masiero, L., Spisani, S., & De Taddeo, H. (2011). Detection methods for the identification of Alicyclobacillus species involved in food spoilage. *Minerva Biotecnologica*, 23(1), 23–32.
3. Cerny, G., Hennlich, W., & Poralla, K. (1984). Spoilage of fruit juice by bacilli: Isolation and characterization of the spoiling microorganisms. *Zeitschrift Fur Lebensmittel-Untersuchung Und -Forschung*, 179(3), 224–227. doi:10.1007/BF01041898
4. Chang, S.S., & Kang, D.H. (2004). Alicyclobacillus spp. in the fruit juice industry: History, characteristics, and current isolation/detection procedures. *Critical Reviews in Microbiology*, 30(2), 55–74. doi:10.1080/10408410490435089
5. Huang, X.C., Yuan, Y.H., Guo, C.F., Gekas, V., & Yue, T.L. (2014). *Alicylobacillus* in the fruit juice industry: Spoilage,detection, and prevention/control. *Food Reviews International*, *31*(13), 91–124.
6. Ju, Y., Hao, H.J., Xiong, G.H., Geng, H.R., Zheng, Y.L., Wang, J. … Jiang, Y.Q. (2010). Development of colloidal gold-based immunochromatographic assay for rapid detection of *Streptococcus suis serotype* 2. *Veterinary Immunology and Immunopathology*, 133(2–4), 207–211. doi:10.1016/j.vetimm.2009.08.010
7. Li, J.K., Feng, Z.P., & Qiu, N.X. (2006). Optimization and establishment of quantitatively competitive PCR system for the detection of. *Alicyclobacillus Acidoterrestris. Scientia Agricultura Sinica*, 39(2), 375–380.
8. Li, J.K., Huang, R.R., Xia, K., & Liu, L. (2014). Double antibodies sandwich enzyme-linked immunosorbent assay for the detection of *Alicyclobacillus acidoterrestris* in apple juice concentrate. *Food Control*, 40, 172–176. doi:10.1016/j.foodcont.2013.11.037
9. Li, J.K., Wang, F., & Xia, K. (2011). An indirect and accurate ELISA for detection of *Alicyclobacillus acidoterrestris* in apple juice concentrate. *Scientia Agricultura Sinica*, 44(22), 4669–4677.
10. Li, J.K., Xia, K., & Yu, C.Z. (2013). Detection of *Alicyclobacillus*

acidoterrestris in apple juice concentrate by enzyme-linked immunosorbent assay. *Food Control*, 30(1), 251–254. doi:10.1016/j.foodcont.2012.07.009

11. Luo, H., Yousef, A.E., & Wang, H.H. (2004). A real-time polymerase chain reaction-based method for rapid and specific detection of spoilage *Alicyclobacillus* spp. in apple juice. *Letters in Applied Microbiology*, 39(4), 376–382. doi:10.1111/lam.2004.39.issue-4
12. Pinhatti, M.E.M.C., Variane, S., Eguchi, S.Y., & Manfio, G.P. (1997). Detection of *acidothermic bacilli* in industrialized fruit juices. *Fruit Process*, 7, 350–353.
13. Wang, F., Li, J.K., Liu, H.X., & Jiang, K. (2009). Preparation and purification of polyclonal antibody against *Alicyclobacillus Acidoterrestrisfrom* apple juice concentrate. *Food and Fermentation Industries*, 5(4), 33–37.
14. Xia, K., Li, J.K., & Yu, Z. (2012). Preparation and purification of rat polyclonal antibody against. *Aliyclobacillua Acidoterrestris. Science and Technology of Food Industry*, 33(15), 162–168.
15. Yamazaki, K., Teduka, H., Inoue, N., & Shinano, H. (1996). Specific primers for detection of *Alicyclobacillus acidoterrestris* by RT-PCR. *Letters in Applied Microbiology*, 23(5), 350–354. doi:10.1111/j.1472-765X.1996.tb00206.x

CHAPTER 11

THE DETECTION OF FOODBORNE BACTERIA ON BEEF: THE APPLICATION OF THE ELECTRONIC NOSE

Soad A Abdallah[1], Laila A Al-Shatti[1], Ali F Alhajraf[2], Noura Al-Hammad[1] and Bashayer Al-Awadi[1]

[1]General Science Unit, College of Nursing, Public Authority for Applied Education & Training (PAAET), Al-Shuwaikh B, 70466 Kuwait City, Kuwait. On tenured leave from Women's College for Arts, Science & Education, Botany Department, Ain Shams University. Cairo, Egypt.
[2]Biomedical Sciences Department, College of Nursing, PAAET, Kuwait City, Kuwait

CITATION: Abdallah, et al., "The detection of foodborne bacteria on beef: the application of the electronic nose", SpringerPlus (2013), https://doi.org/10.1186/2193-1801-2-687.

ABSTRACT

This study aims to investigate the application of a fast electronic nose system (Cyranose 320) for detecting foodborne bacteria. The system proved to be very efficient in detecting microbes in beef and sausage samples. In the first part of the study, the total viable counts (TVC) from fresh and frozen beef samples were determined using the standard microbiological method and by the application of the electronic nose. The second part applied the electronic nose to beef before and after contamination with different bacterial pathogens separately: E. coli O157: H7, Salmonellatyphimurium 857, Staphylococcus aureus 29213 and Pseudomonas aeruginosa 27853. The results revealed that the Cyranose 320 can detect the TVC in different beef and sausage samples and quantify the volatile organic compounds produced at concentrations from 50 ppb to > 350 ppb. The concentrations of gases collected from the samples before and after separate contamination with these pathogenic bacteria were highly significantly correlated ($P < 0.005$). From this study one can conclude that the electronic nose system is a rapid way for detecting volatile organic compounds produced by foodborne bacteria that contaminate beef.

Keywords: Food; Pathogens; Rapid detection; Electronic nose

BACKGROUND

Foodborne pathogens such as bacteria or toxins, viruses or parasites may lead to human disease when contaminated food is eaten. The source of contamination may vary but harmful bacteria are mostly responsible for causing gastrointestinal infections (Scallan et al. 2011). The sources could be the animal, the environment or contamination during food processing (McNamara 1998; Slutsker et al. 1998). The main source of meat contamination is animal feces especially during processing at the slaughterhouse (Kudva et al. 1998). Food animals and poultry are the most important reservoirs for many foodborne pathogens (Biswas et al. 2008). Foodborne illnesses associated with meat are caused mostly by certain types of bacteria namely Bacillus cereus, Campylobacter jejuni, Clostridium botulinum, Clostridium perfringens, Escherichia coli O157:H7, Listeria monocytogenes, Salmonella sp., Staphylococcus aureus, Pseudomonas aeruginosa and Yersinia enterocolitica (Atlas 1998; FSIS 1998; Beran et al. 1991; Doyle et al. 1997).

Existing methods for preventing microbial diseases depend on controlling different types of pathogenic bacteria by food safety management and through medical and environmental observation (Meng and Doyle 1998). Classical methods for the cultivation, isolation and identification of bacteria usually comprise morphological assessment in addition to determining the ability of microorganisms to grow in different media under different growth conditions (Ivnitski et al. 2000).

The polymerase chain reaction (PCR) is mainly used in research in food microbiology for the identification of bacterial genus and species due to its high efficacy and accuracy. Although PCR is an advanced technology, inhibitors that occur in foods or the culture media could affect the reaction. However, this technology cannot differentiate between active cells and inactive dead cells so could lead to false results and thus fail to reflect bacterial numbers accurately (Mandal et al. 2011). Instruments used in the food industry can help to detect pathogens. One example, the electronic nose (e-nose), is an electronic instrument that is able to imitate the human ability to detect odor. The Cyranose 320, used in this work contains a group of up to 32 sensors, each one made of a complex of conductive carbon black blended with a non- conductive polymer with each sensor being able to differentiate between different types of gas, whether pure or mixed. Using the e-nose allows an immediate decision regarding the quality of the sample without leaving the sampling location (Hobbs 2003; Mandal et al. 2011). Winquist et al. (1993) showed that the electronic nose could be used within the medical environment and in the food industry where it could discriminate between types of ground beef and assess quality during storage.

Many studies (Falasconi et al. 2012; Sayeed and Shameen 2011; Berna 2010; Dutta et al. 2002; Rossi et al. 1995; Holmberg et al. 1998; Gardner et al. 1998) have used a particular e-nose instrument - First Generation E-Nose, Second Generation E-Nose and Third generation E-Nose especially Cyranose 320- to classify bacteria and to discriminate between different species. This instrument has also been used to detect the presence of E. coli in samples (Powell et al. 2002) and to detect the production of volatile compounds relating to chicken storage time and temperature (Boothe and Arnold 2002). The aim of this study is to compare detection methods for foodborne bacteria such as Salmonella sp., Staphylococcus aureus, Pseudomonas aeruginosa and E. coli O157: H7 from different types of beef using the electronic nose and the standard methods.

RESULTS

The present study is divided into two parts: firstly, a cross validation test for determining the total viable count (TVC) on samples of fresh beef (cut or minced), frozen beef (cut or minced) and sausages (fresh or frozen) on five consecutive days using both the routine method (TVC) and the e-nose and secondly, using the e-nose to detect gases from a range of bacteria contaminating similar beef and sausage samples.

Table 1 shows the total viable counts (TVC): for fresh cut beef samples, it increased up to day 2 then decreased by day 4 then increased to the highest level on day 5, ranging from 1.22×10^3 to 2.42×10^5 CFU/ml. while fresh beef minced samples gave lower levels of TVC which decreased by day 2 then increased on days 3, 4 and 5, ranging from 4.44×10^2 to 6.10×10^4 CFU/ml. For fresh sausage samples the TVC was zero on day 1 and then increased to reach 2.43×10^5 CFU/ml on day 5. The frozen cut beef showed the highest TVC levels on day 1, decreased by days 2 and 3 then increased on day 4 before decreasing sharply on day 5, ranging from 1.11×10^2 to 6.62×10^4 CFU/ml. Frozen minced beef started with the highest TVC level on day one (1.37×10^5 CFU/ml) which then fluctuated during the next 4 days to reach its lowest level on day 5 (1.11×10^2 CFU/ml). The TVC of the frozen sausage samples fluctuated, reaching its highest value on day 4 (3.71×10^5 CFU/ml) then decreasing to its lowest value on day 5 (2.22×10^2 CFU/ml).

According to a statistical analysis using the paired ttest, there was no significant difference between the TVC values from fresh cut and fresh minced beef samples ($P > 0.05$). The same results were obtained between fresh cut beef and fresh sausage, and between fresh minced beef and fresh sausage ($P > 0.05$) (Table 1). Correlation matrix analysis is about observing the interaction of various beef samples' TVC. For instance, correlation is computed into what is known as the correlation coefficient, which ranges between −1 and +1.

Perfect positive correlation (a correlation coefficient of +1) implies that as TVC of one beef sample moves, either up or down, the other sample's TVS will move in the same direction. Alternatively, perfect negative correlation means that if TVC of a beef sample's moves in either direction the TVC of the other sample that is perfectly negatively correlated will move in the opposite direction. If the correlation is 0, the sample's TVCs are said to have no correlation. The correlation matrix using the TVC data from the different types of beef examined in the present study (Table 2), showed that there was a very strong negative correlation between frozen and fresh cut beef (−0.966), a positive moderate correlation between fresh cut and fresh minced beef (0.681) and between fresh cut and fresh sausage samples (0.560), and between fresh sausage and fresh minced beef (0.621). However, negative weak correlations occurred between frozen sausage and fresh sausage samples (−0.205), between frozen minced beef and fresh sausage (−0.341) and between frozen minced with fresh sausage samples (−0.330). According to the P value of all correlations were not significant (P > 0.05) except the correlation between the fresh and frozen cut samples were highly significant (P < 0.01).

Table 1. Total viable count (TVC) CFU/ml of beef and sausage samples

Samples	Fresh cut	Fresh minced	Fresh sausage	Frozen cut	Frozen minced	Frozen sausage
Day 1	$1.33 \times 10^{[illegible]}$	$1.00 \times 10^{[illegible]}$	$1.00 \times 10^{[illegible]}$	6.62×10^{4}	$1.37 \times 10^{[illegible]}$	2.24×10^{4}
Day 2	$1.70 \times 10^{[illegible]}$	$4.44 \times 10^{[illegible]}$	$5.11 \times 10^{[illegible]}$	$5.33 \times 10^{[illegible]}$	$6.67 \times 10^{[illegible]}$	$1.95 \times 10^{[illegible]}$
Day 3	$1.53 \times 10^{[illegible]}$	5.41×10^{4}	$1.11 \times 10^{[illegible]}$	$4.33 \times 10^{[illegible]}$	$3.67 \times 10^{[illegible]}$	$8.33 \times 10^{[illegible]}$
Day 4	$1.22 \times 10^{[illegible]}$	$1.03 \times 10^{[illegible]}$	$5.67 \times 10^{[illegible]}$	6.39×10^{4}	$1.56 \times 10^{[illegible]}$	$3.71 \times 10^{[illegible]}$
Day 5	$2.42 \times 10^{[illegible]}$	$6.10 \times 10^{[illegible]}$	$2.43 \times 10^{[illegible]}$	$1.11 \times 10^{[illegible]}$	$1.11 \times 10^{[illegible]}$	$2.22 \times 10^{[illegible]}$

On the basis of these findings, the frozen minced beef and frozen sausage samples were excluded from the e-nose application because frozen samples gave less TVC when using the routine method (TVC).

Table 2. Correlation matrix between TVC values from all beef and sausage samples

Samples	Fresh minced	Fresh sausage	Fresh cut	Frozen sausage	Frozen minced
Fresh sausage	0.621				
Fresh cut	0.681	0.560			
Frozen sausage	−0.564	−0.205	−0.470		
Frozen minced	−0.453	−0.341	−0.587	−0.330	
Frozen cut	−0.640	−0.344	−0.966*	0.440	0.631

*indicates its P value <0.01.

From the e-nose data, gases produced by the viable bacteria gave higher concentration values after 2 h than after 4 h (Figures 1 and 2). Fresh sausage samples produced the highest concentration of gases, exceeding 350 ppb on day 2 (Figure 1), decreasing to < 200 ppb on day 3, while frozen cut beef samples also showed a high gas concentration, exceeding 350 ppb on day 1, followed by fresh cut beef and fresh minced beef.

After 4 h, all samples showed lower concentrations of gases than after 2 h (Figure 2). Fresh sausage showed the same pattern as after 2 h: it showed the highest concentration of gas collected by the e-nose especially on day 2 when it exceeded 250 ppb, decreasing on days 3 and 4 to fall below 250 ppb. The concentration of gas detected in the fresh minced beef reached 150 ppb on day 2 then decreased to below 100 ppb on day 3 and above 100 ppb on day 4. Fresh cut beef samples showed the lowest gas concentration at below 50 ppb on day 2 (Figure 2) these changes in the gas collected might be attributed to the beef sample type.

After applying the paired t-test to compare the gas concentrations on day 1 with day 5, no significant differences were detected ($P > 0.05$) for all types of beef sample (fresh cut, fresh minced and frozen cut). In contrast, a highly significant difference was found between values for fresh sausage on day 1 and day 5, whether after 2 or 4 h ($P = 0.000$ and 0.009 respectively) (Table 3). The second part of this study was to apply the e-nose to fresh beef samples (cut, minced and sausage) before and after contamination with the different types of bacterial pathogens E. coli O157: H7, Salmonella typhimurium 857, Staphylococcus aureus 29213 and Pseudomonas aeruginosa 27853, after 2 and 4 h.

Figure 3 shows the concentrations of gas collected after 2 and 4 h using the e-nose from the uncontaminated fresh cut samples and those contaminated separately by the above-mentioned pathogens. The data shows that the concentration of gases collected from the fresh cut beef after 4 h was more than that collected after 2 h for all contaminated samples. While the uncontaminated control sample showed the same concentration of gases after 2 and 4 h (>100 ppb), Pseudomonas aeruginosa produced the highest concentration of gases collected after 4 h reaching 200 ppb, followed by E. coli and Salmonella typhimurium which produced gas concentrations above 150 ppb.

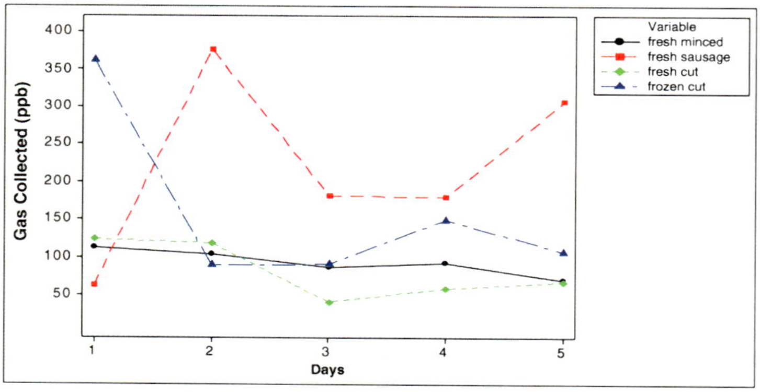

Figure 1. Time series plot of gas concentrations detected by e-nose over 2 h.

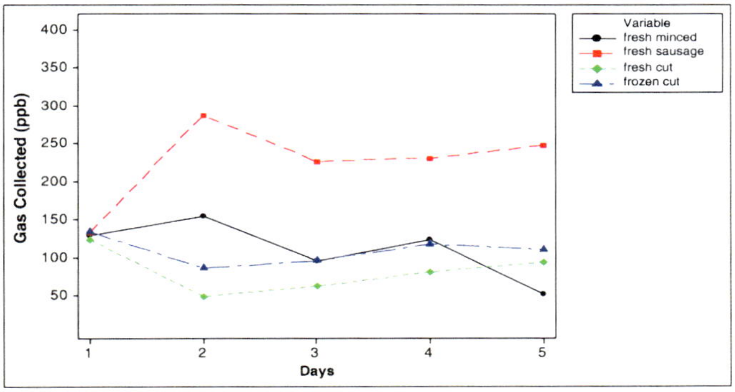

Figure 2. Time series plot of gas concentrations detected by e-nose over 4 h.

For fresh minced beef, the data revealed that the concentration of gases collected from samples exposed to pathogenic bacteria showed differences in value after 2 h. E. coli showed the highest concentration of gases collected, exceeding 200 ppb, followed by samples contaminated with Salmonella typhimurium (> 150 ppb), Pseudomonas aeruginosa and staphylococcus aureus (Figure 4). The uncontaminated fresh minced beef samples produced a gas concentration of 100 ppb after 2 h

For the fresh sausage sample, the concentration of gases produced reached 350 ppb from Pseudomonas aeruginosa after 4 h and above 300 ppb after 2 h. E. coli produced concentrations above 250 ppb after 2 h but below 250 after 4 h. However, Salmonella typhimurium produced gas concentrations of 250 ppb after 2 h which decreased to 150 ppb after 4 h. Staphylococcus aureus produced concentrations above 200 ppb after 4 h and below 200 ppb after 2 h. The lowest concentration of gases collected by the e-nose came from the uncontaminated fresh sausage (Figure 5). Using the paired t-test, comparing the concentration of gases collected after 2 h on days 1 and 5 showed a significant difference for fresh minced beef contaminated with E. coli ($P < 0.05$) and highly significant differences for fresh sausage contaminated by E. coli ($P < 0.005$). Another significant difference occurred between the concentrations of gases produced on days 1 and 5 by Salmonella typhimurium in the fresh sausage samples ($P < 0.05$).

Table 3 Comparison of gas concentrations (ppb) on day 1 and day 5 for fresh cut and minced beef, fresh sausage and frozen cut beef after 2 and 4 h

Samples	Two hours			Four hours		
	Mean	SD	P value	Mean	SD	P value
Fresh cut	56.0	22.7	0.997	30.2	20.18	0.986
Fresh minced	42.4	40.8	0.957	76.8	28.4	0.994
Fresh sausage	−244	20.91	**0.000****	−114	65.8	**0.009****
Frozen cut	253	47	1.000	23.8	22.9	0.960

**indicates P <0.01.

Highly significant differences were found between the concentrations of gases collected from fresh sausage exposed to Staphylococcus aureus and Pseudomonas aeruginosa ($P < 0.005$). Significant differences also occurred between the concentrations of gases collected from fresh cut beef samples exposed to Staphylococcus aureus and Pseudomonas aeruginosa ($P < 0.05$) (Table 4). When the paired t-test was also applied to the gas concentrations after four hours collected on days 1 and 5 (Table 5), it showed highly significant differences between those from minced beef exposed to four different types of pathogenic bacteria separately ($P \leq 0.005$).

Another highly significant difference was found with fresh cut beef and fresh sausage exposed to Pseudomonas aeruginosa ($P < 0.005$). A significant difference was also revealed between the amounts of gases collected from fresh cut beef and fresh sausage after contamination with E. coli ($P < 0.05$). Regarding the potentiometric analysis of the sodium nitrate (NaNO3) content in meat, the results were 19.90, 23.28, and 67.50 ppm in fresh cut beef, fresh minced beef and fresh sausage, respectively.

DISCUSSION

Quality assurance methods in the food industry need to be specific and selective for microbiological examinations. Microbiological processes occurring during the storage of raw meat lead to wastage which threatens meat industry economics and causes a challenging problem form eat businesses. Quantitative assessments of beef contamination can be made by a routine total viable count (TVC) and other tests which measure the number of bacteria (TVC) in a sample that can survive in the conditions on the surface of raw meat or in processed meat, can be harvested by the sampling procedure used and can grow in the presence of air on an agar plate. These bacteria originate both from animals and from the slaughterhouse or meat processing environment. As the TVC includes organisms contaminating the meat, it will also give an indication of its keeping quality. The disadvantages of these routine methods are that they are laborious, costly and time-consuming but these can be overcome by a rapid method that gives instant or real-time results. Rapid methods not only provide the early detection and enumeration of microorganisms but can also characterize these isolates (Naravaneni and Jamil 2005). To determine the extent of spoilage in raw meat, the standard method used to analyze the total viable count of bacteria requires a 1–2 day incubation period to form a bacterial colony on agar plates. Although the literature shows that bacterial growth in meat samples has been widely studied, there is still a need for research on methods to correlate the number of bacteria with shelf-life determination (Panigrahi et al. 2006).

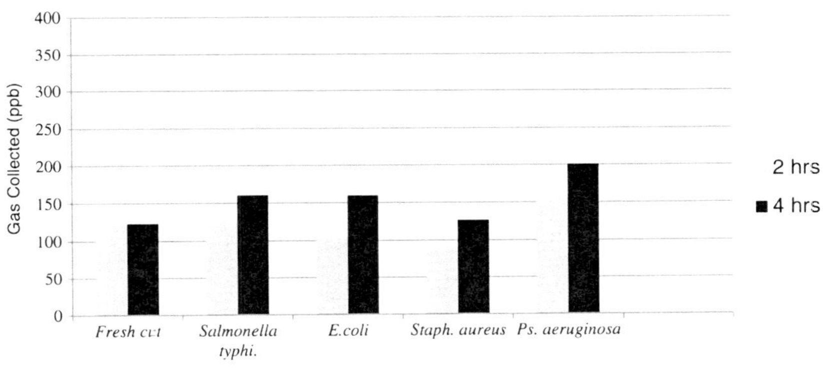

Figure 3 Concentration of gases (ppb) collected by e-nose from fresh cut beef before and after bacterial contamination.

In the present study, the results support using the electronic nose for bacteriological examinations instead of the alternative routine methods. An electronic nose and other measurements were used on beef samples including minced meat and sausages over a period of five days. To use the electronic nose system, the routine bacteriological method was applied as a reference method. The results showed that the highest level of bacterial counts in freshly cut meat occurred on the fifth day whereas in frozen samples, it was the fourth day. The e-nose system can easily distinguish between spoiled and fresh meat.

The positive result obtained indicated that this system can be effectively used to rapidly detect contamination. It is important to note that the purpose of testing against the process criteria that have been set out for raw and certain processed meat is not to assess their fitness for human consumption but to provide an indication of performance and control of the slaughtering, dressing and production processes at the time of sampling and so must be used accordingly. If the criteria are not met, corrective action to improve future production must be initiated but there is no requirement to remove products from the market.

The experiments using the Cyranose e-nose have shown that viable bacteria can be detected in the contaminated samples. From our results, the e-nose has the potential to be used as a tool for the rapid detection of contamination, agreeing with the findings of Ding et al. (2010). The Cyranose 320 equipment has some advantages such as its ability to work with turbid media, high sensitivity and small size (Ivnitski et al. 2000). This electronic nose also has the advantages of short preparation time, being inexpensive to run, safe to use and efficient at volatile detection. Moreover, these e-noses are more sensitive to small bacterial cell numbers, have specificity towards different species of foodborne bacteria in addition to providing results in or near actual time without pre-enrichment. The most important challenge for screening a food sample is achieving the requirements for high sensitivity and a rapid time for effective analysis.

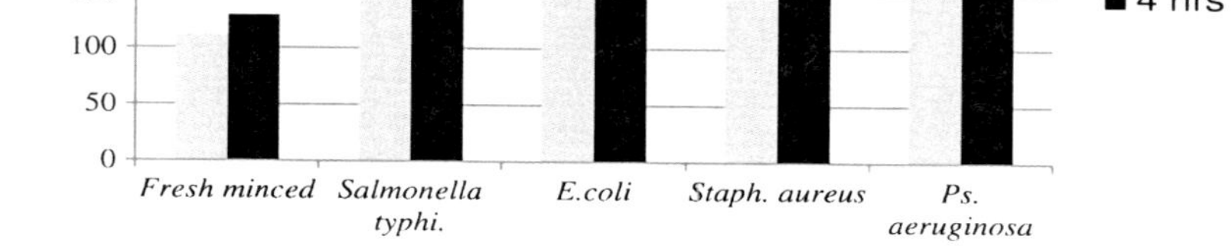

Figure 4 Concentration of gases (ppb) collected by e-nose from fresh minced beef before and after bacterial contamination

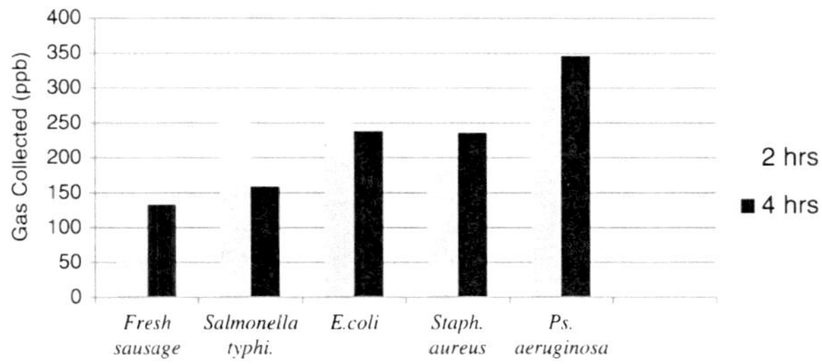

Figure 5 Concentration of gases (ppb) collected by e-nose from fresh sausage before and after bacterial contamination

To assure human health, pathogenic microorganisms in meat products such as Salmonella need to be detected early. The pathogenic strains used in this study were detected by an electronic nose system designed to detect spoiled and unspoiled meat based on contamination. This agrees with previous studies which stated that the electronic nose could detect pathogenic organisms in beef (Balasubramanian et al. 2005; Kress-Rogers 1996; Vernat-Rossi et al. 1996). A study by Siegmund and Pfannhauser (1999) revealed that the electronic nose can detect Salmonella typhimurium at a contamination level of 0.7–2.6 log10CFU/g. As mentioned in the results section, the e-nose detects the volatile organic compounds produced from sausage samples.

Spices, which play a significant role in processed meat products, can be contaminated with a microbial load depending on the processing method, particle size, the variety and moisture content (Akgul 1993). The microbial load of the spice mixture increased that of the sausage samples so that the concentration of gas was higher than that from the beef samples. An instrument to detect sensory properties' reproducibly is needed and could be an alternative to a sensory panel in the food industry (Haugen and Kvaal 1998). Sodium nitrate can be used to extend food storage time without changing the food's color, taste, odor and nutritional value and is used as a preservative in meat curing preparations and meat products at not more than 500 ppm (Branen et al. 1990; Dich et al. 1996; Hsu et al. 2009; Food and Drug Administration 2013). The concentration of gas collected by the e-nose is inversely proportional to the nitrate content of the meat due to its antimicrobial activity, this in harmony with previous studies (Sundberg et al. 2013).

Unfortunately, in the present study, nitrate readings were totally different. This might be attributed to several different causes. The unprocessed meat showed a trend for higher microbial activity over4h compared with 2 h possibly because of the lack of antimicrobial additives but the processed meat showed the same trend. However, the differences between each bacterial species after 2 and 4 h decreased in the processed meat due to the presence of nitrate additives reducing microbial activity. Exceptionally, in S. typhimurium and E. coli on processed meats, the microbial activity decreased after 4 h compared with 2 h proving that $NaNO_3$ is an effective microbial inhibitor towards S. typhimurium and E. coli.

Table 4. Comparison of mean gas concentrations (ppb) on days 1 and 5 from fresh cut and minced beef and fresh sausage samples after 2 h

Samples	*E. coli*			*Salmonella typhimurium*			*Staphylococcus aureus*			*Pseudomonas aeruginosa*		
	Mean	**SD**	**P value**	**Mean**	**SD**	**P value**	**Mean**	**SD**	**P value**	**Mean**	**SD**	**P value**
Fresh cut	18.6	23.5	0.151	−2.2	24.3	0.850	32.8	22.9	**0.033***	−34.4	20.96	**0.021***
Fresh minced	−106	65.0	**0.022***	−61.2	50.9	0.055	−33.4	29.9	0.067	−54.8	46.8	0.059
Fresh sausage	−205	67.9	**0.003****	−195.2	136.5	**0.033***	−127.8	20.86	**0.000****	−255.2	26.2	**0.000****

**indicates P < 0.01, *P < 0.05.

Table 5. Comparison of mean gas concentrations (ppb) on days 1 and 5 from fresh cut and minced beef and fresh sausage samples after 4 h

Samples	*E coli*			*Salmonella*			*Staphylococcus aureus*			*Pseudomonas aeruginosa*		
	Mean	**SD**	**P value**	**Mean**	**SD**	**P value**	**Mean**	**SD**	**P value**	**Mean**	**SD**	**P value**
Fresh cut	−36.2	24.2	**0.029***	−37.0	33.2	0.067	−3.4	25.4	0.780	−77.40	20.48	**0.001****
Fresh minced	−138	55.2	**0.005****	−122.4	44.5	**0.004****	−127.2	47.6	**0.004****	−131.2	33.6	**0.001****
Fresh sausage	−105	78.1	**0.039***	−26.0	85.9	0.536	−103.2	87.0	0.057	−213.2	67.1	**0.002****

**indicates P < 0.01, *P < 0.05.

CONCLUSIONS

The current study suggests that the electronic nose can be used to detect meat contamination with the application of statistical analysis. It can easily run tests to check food products for spoilage, so it would be fast and effective in the food industry. However, there still is a need to ensure its reliability by conducting more research that aims to improve and assure the validity of the results. Gas sensing technologies can be used efficiently to help control food quality. However, this method does not replace reference methods such as using sensory panels because it needs standardization to become the equivalent of other validated reference methods. If the sample handling problems and other instrument performance issues of such a technique can be solved, the application of this instrument in the food industry would be very promising for understanding and assuring food contamination.

METHODS

Sample Collection

Six different types of beef samples were collected from a local market: cut fresh beef, minced fresh beef, cut frozen beef, minced frozen beef, processed fresh sausage and processed frozen sausage. All beef and sausage samples were cut into pieces weighing 10 g ±1, placed immediately in plastic bags and kept in are refrigerator at 4°C.

Microbiological enumeration for contaminated beef samples

A 10 g sample was taken from the refrigerator, placed in 100 ml peptone water (Oxoid) (Thermo Scientific, UK) and left for two minutes (Sneath et al. 1986). Dilutions were prepared using the same diluents. Three dilutions were selected (10^3, 10^5 and 10^8) and plated on Plate Count Agar (Oxoid). The plates were incubated at 30°C for 24-48 hrs. By enumerating the colonies present, the total viable counts (TVC) were obtained and calculated aslog10 colony forming units (CFU)/ml of the sample. This procedure was repeated every day for 10 days. The microbiological enumeration of the six types of beef and sausage samples under investigation was performed separately as mentioned above.

Sample Preparation for the e-nose

A 20 g beef sample was taken from the refrigerator and put into a 100 ml glass bottle and sealed with Parafilm® and held at room temperature (22°C ± 2°C) for two hours and four hours. After 2 h the tube that is connected to the Cyranose 320 was penetrated the parafilm® and dipped in the bottle without touching the liquid media. The Cyranose start button is pressed and its monitor started to detect the gas collected, within few seconds the reading is settled and taken. The instrument is re-set to take the second reading and so on. The same procedure is repeated after 4 h. Duplicate samples from each beef type were used. Measurements by the e-nose were performed after 2 and 4 h each day on 5 consecutive days for each duplicate and each sample (Nychas et al. 2008).

Contaminated Sample Preparation

E. coli O157: H7, Salmonellatyphimurium 857, Pseudomonas aeruginosa 27853 and Staphylococcus aureus 29213 (Obtained from Faculty of Medicine, Kuwait University, Kuwait) were grown separately overnight at 37°C in a 125 ml flask to be used to inoculate the beef samples (Feng and Weagant 2002). The optical density of this culture was adjusted to give a concentration of about 100 colony forming unit per ml (CFU/ml). Beef and sausage samples (10 g) were placed in 250 ml glass bottles with the 100 ml bacterial cultures and sealed with Parafilm® to make it air-tight and incubated at room temperature 22°C. After intervals of 2 and 4 hours the gas produced by the bacteria was collected for measurement by the e-nose.

Electronic Nose Application

The portable electronic nose, the Cyranose-320 (Model XP-329 III R, New Cosmos Electric Co. Ltd. Japan), can be used for detecting odors in a variety of industrial environments. By imprinting an odor print on its 32-sensor nose chip, arranged as an array, the e-nose can be used for non-invasive medical diagnostics, identification of hazardous materials, detection of food spoilage and many other applications. The sensors within the electronic nose are composed of a complex material consisting of conductive carbon black mixed with a non-conductive polymer. When the sensors are exposed to a gas, the polymer absorbs the gas and swells, during which the distance between the conductive carbon particles increases and thus also increases the resistance of the sensor material. This change in resistance is transmitted to a computer with the pattern of change in the sensor array being used to detect the gas

Potentiometric Nitrate Analysis

Fifty grams of meat sample were soaked separately in 250 ml peptone water (Oxoid) broth for 4 h. The nitrate content of the solution was measured by using a pHmeter (Hanna HI 2550 pH/ORP, Woonsocket, RI, USA) with a nitrate combination electrode (Hanna HI 4113).

Statistical Analysis

Statistical analysis was performed using the Minitab statistical package version 16 (Minitab Inc. 2010, State College, PA, USA). The paired t-test, time series and correlation matrix functions were applied to reveal any significant correlations and differences when using the Cyranose 320 using a P value < 0.005 (Daniel 2005).

REFERENCES

1. Akgul A (1993) Spice Science and Technology. Turkish Association Food Technologists, Publ. No. 15, Ankara, Turkey Atlas RM (1998) The medical threat of biological weapons. Crit Rev Microbiol 24:157–168
2. Balasubramanian S, Panigrahi S, Logue CM, Marchello M, Sherwood JS (2005) Identification of Salmonella-inoculated beef using a portable electronic nose system. J Rapid Meth Autom Microbiol 13:71–95
3. Beran GW, Shoeman HP, Anderson KF (1991) Food safety: an overview of problems. Dairy Food Environ Sanit 11:189–194
4. Berna A (2010) Metal oxide sensors for electronic noses and their application to food analysis. Sensors 10(4):3882–3910
5. Biswas AK, Kondaiah N, Bheilegaonkar KN, Anjaneyulu ASR, Mendiratta SK, Jana C, Singh H, Kuma RR (2008) Microbial profiles of frozen trimmings and silver sides prepared at Indian buffalo meatpacking plants. Meat Sci 80:418–422
6. Boothe DDH, Arnold JW (2002) Electronic nose analysis of volatile compounds from poultry meat samples, fresh and after refrigerated storage. J Sci Food Agric 82:315–322
7. Branen AL, Davidson PM, Salminen S (1990) Food Additives. Marcel Dekker, New York, USA
8. Daniel WW (2005) Biostatistics: A foundation for analysis in the health sciences, 8th edition. John Wiley & Sons Inc., New York, USA
9. Dich J, Järvinen R, Knekt P, Penttilä P (1996) Dietary intakes of nitrate, nitrite and NDMA in the Finnish Mobile Clinic Health Examination Survey. Food Addit Contam 13:541–552
10. Ding NY, Lan YB, Zheng XZ (2010) Rapid detection of E. coli on goat meat by electronic nose. Adv Nat Sci 3:185–191
11. Doyle MP, Beuchat LR, Montville TJ (1997) Food Microbiology: Fundamentals and Frontiers. ASM, Washington DC, USA, pp 127–390
12. Dutta R, Hines EL, Gardner JW, Boilot P (2002) Bacteria classification using Cyranose 320 electric nose. Biomed Eng Online 1:4. doi:10.1186/1475-925X-1-4
13. Falasconi M, Concina I, Gobbi E, Sberveglieri V, Pulvirenti A, Sberveglieri G (2012) Electronic nose for microbiological quality control of food products. Int J Electrochem. Article ID 715763,

doi:10.1155/2012/715763

14. FDA (Food and Drug Administration) (2013). http://www.accessdata.fda.gov Feng P, Weagant SD (2002) Diarrheagenic Escherichia coli, Chapter 4A, rev. Sept. 2002. In: FDA Bacteriological analytical manual, 8th edn, Rev A. AOAC International, Gaithersburg, MD, USA
15. FSIS (Food Safety and Inspection Service) (1998) Pathogen reduction and HACCP systems, Final rule, USDA. http://www.fsis.usda.gov/Oa/background/ bkbeyond.htm Accessed 12 Oct 2013
16. Gardner JW, Craven M, Dow C, Hines EL (1998) The prediction of bacteria type and culture growth phase by an electronic nose with multi-layer perceptron network. Meas Sci Technol 9:120–127
17. Haugen JE, Kvaal K (1998) Electronic nose and artificial network. Meat Sci 49 (Supp 1):S273–S286 Hobbs S (2003) A tour of Cyranose 320. http://cyranosciences.com/technology/ tour.html
18. Holmberg M, Gustafsson F, Hörnsten EG, Winquist F, Nilsson LE, Ljung L, Lundström I (1998) Bacteria classification based on feature extraction from sensor data. Biotechnol Tech 12:319–324
19. Hsu J, Arcot J, Lee NA (2009) Nitrate and nitrite quantification from cured meat and vegetables and their estimated dietary intake in Australians. Food Chem 115:334–339
20. Ivnitski D, Abdel-Hamid I, Atanasov P, Wilkins E, Stricker S (2000) Application of electrochemical biosensors for detection of food pathogenic bacteria. Electroanal 12:317–325
21. Kress-Rogers E (1996) Handbook of Biosensors and Electronic Noses: Medicine, Food and the Environment. CRC Press, Boca Raton, FL, USA
22. Kudva T, Blanch K, Hovde CJ (1998) Analysis of Escherichia coli O157:H7 survival in ovine or bovine manure and manure slurry. Appl Environ Microb 64:3166–3174
23. Mandal PK, Biswas AK, Choi K, Pal UK (2011) Methods for rapid detection of foodborne pathogens: an overview. Am J Food Technol 6:87–102
24. McNamara AM (1998) Foodborne pathogens. J Urban Health 75:503–505 Meng J, Doyle MP (1998) Emerging and evolving foodborne pathogens. B I Pasteur 96:151–164
25. Naravaneni R, Jamil K (2005) Rapid detection of food-borne pathogens by using molecular techniques. J Med Microbiol 54:51–54

26. Nychas GJE, Skandamis PN, Tassou CC, Koutsoumanis KP (2008) Meat spoilage during distribution. Meat Sci 78:77–89
27. Panigrahi S, Balasubramanian S, Gu H, Logue C, Marchello M (2006) Neuralnetwork-integrated electronic nose system for identification of spoiled beef. LWT-Food Sci Technol 39:135–145
28. Powell TB, Heinemann P, Demirci A (2002) Electronic nose detection of E. coli on the surface of apples. In: 2002 ASAE Annual International Meeting, Chicago, USA, Paper No 03–3025
29. Rossi V, Talon R, Berdagué J (1995) Rapid discrimination of Micrococcaceae species using semiconductor gas sensors. J Microbiol Meth 24:183–190
30. Sayeed A, Shameen MS (2011) Electronic nose. Adv in Medi Info 1(1):06–09 Scallan E, Griffin PM, Angulo FJ, Tauxe RV, Hoekstra RM (2011) Foodborne illness acquired in the United States–unspecified agents. Emerg Infect Dis 17:16–27
31. Siegmund B, Pfannhauser W (1999) Changes of the volatile fraction of cooked chicken meat during chill storing: results obtained by the electronic nose in comparison to GC-MS and GC Olfactometry. Z Lebensm Unters F A 208:336–341
32. Slutsker L, Altekruse SF, Swerdlow DL (1998) Foodborne diseases: emerging pathogens and trends. Infect Dis Clin N Am 12:199–216
33. Sneath PHA, Mair NS, Sharpe ME, Holt JG (1986) Bergey's Manual of Systematic Bacteriology, vol. 2. Williams and Wilkins Co., Baltimore, MD, USA Sundberg C, Yu D, Whittle I, Kauppi S, Smars S, Insam H, Romantschuk M, Jönsson H (2013) Effects of pH and microbial composition on odour in food waste composition. Waste Manag 33(1):204–211
34. Vernat-Rossi V, Garcia C, Talon R, Denoyer C, Berdagué JL (1996) Rapid discrimination of meat products and bacterial strains using semiconductor gas sensors. Sensor Actuat B: Chemical 37:43–48
35. Winquist F, Hörnsten EG, Sungren H, Lundström I (1993) Performance of an electronic nose for quality estimation of ground meat. Meas Sci Technol 4:1493–1500

CHAPTER 12

A SIMPLE, PORTABLE, ELECTROCHEMICAL BIOSENSOR TO SCREEN SHELLFISH FOR VIBRIO PARAHAEMOLYTICUS

Noordiana Nordin[1,2], Nor Azah Yusof[1,3], Jaafar Abdullah[1,3], Son Radu[2] and Roozbeh Hushiarian[4,5]

[1] Institute of Advanced Technology, Universiti Putra Malaysia (UPM), 43400 Serdang, Selangor, Malaysia.
[2] Food Safety Research Centre, Faculty of Food Science and Technology, Universiti Putra Malaysia, 43400 Serdang, Selangor, Malaysia.
[3] Department of Chemistry, Faculty of Science, Universiti Putra Malaysia (UPM), 43400 Serdang, Selangor, Malaysia.
[4] La Trobe Institute for Molecular Science, La Trobe University, Melbourne, VIC 3086, Australia.
[5] Department of Life Sciences, Imperial College London, London SW7 2AZ, UK.

CITATION: Nordin, N., et al., "A simple, portable, electrochemical biosensor to screen shellfish for Vibrio parahaemolyticus", AMB Express (2017), https://doi.org/10.1186/s13568-017-0339-8.

ABSTRACT

An earlier electrochemical mechanism of DNA detection was adapted and specified for the detection of Vibrio parahaemolyticus in real samples. The reader, based on a screen printed carbon electrode, was modified with polylactidestabilized gold nanoparticles and methylene blue was employed as the redox indicator. Detection was assessed using a microprocessor to measure current response under controlled potential. The fabricated sensor was able to specifically distinguish complementary, non-complementary and mismatched oligonucleotides. DNA was measured in the range of $2.0 \times 10^{-8} - 2.0 \times 10^{-13}$ M with a detection limit of 2.16 pM. The relative standard deviation for 6 replications of differential pulse voltammetry (DPV) measurement of 0.2 μM complementary DNA was 4.33%. Additionally, cross-reactivity studies against various other food-borne pathogens showed a reliably sensitive detection of the target pathogen. Successful identification of Vibrio parahaemolyticus (spiked and unspiked) in fresh cockles, combined with its simplicity and portability demonstrate the potential of the device as a practical screening tool.

Keywords: Vibrio parahaemolyticus, Food-borne pathogens, Electrochemical DNA sensor, Shellfish, Portable biosensor

INTRODUCTION

Vibrio parahaemolyticus (V. parahaemolyticus) a gramnegative, halophilic bacterium is not only the leading cause of seafood-associated bacterial gastroenteritis in the United States (DePaola et al. 2000) but it is also one of the most important food-borne pathogens in Asia, causing around half of the foodborne outbreaks in Southeast Asian countries (Martinez-Urtaza et al. 2004). Additionally, it should be noted that the number of V. parahaemolyticus infections has increased and their reach widened globally during recent years (Nair et al. 2007; Powell et al. 2013). Scientists are currently investigating the conditions that might be fostering this spread and increase (Kaneko and Colwell 1975; Martinez-Urtaza et al. 2016) so that it might be halted, but in the meantime, early detection is important to seafood consumers in Europe, Asia and the US (Terzi Gulel and Martinez-Urtaza 2016). Vibrio parahaemolyticus is the most prevalent of more than 30 Vibrio species reported and is among the 12 which are pathogenic (Skovgaard 2012). With an incubation period of about 15 h (ranging from 4 to 96 h), a dose of about $2 \times 10^5 - 3 \times 10^7$ cfu is sufficient to lead to acute gastroenteritis

(Costa Sobrinho et al. 2014; Ottaviani et al. 2012; Shimohata and Takahashi 2010; Vengadesh et al. 2014) and may be life-threatening for people with weak immune disorders, although the infection is often self-limited (Varnam and Evans 1991).

Because V. parahaemolyticus is usually transmitted along the food supply chain through seafood (Caburlotto et al. 2016; Wong et al. 1999), it has the potential to further increase as the popularity of seafood as a source of healthy protein extends throughout the world (Zhang and Orth 2013). About 90% of global aquaculture products come from sources in the Asian region, particularly China, from where they are exported in massive quantities to overcome a scarcity in other countries (Liao and Chao 2009).

Common techniques used for the detection of V. parahaemolyticus include cultural (Shen et al. 2011), biochemical (Rosec et al. 2012), serological (Bisha et al. 2012), and immunological methods (Maniyankode et al. 2013). Requiring numerous analytical steps, all of these methods take up to a few days to provide a confirmed result. Apart from being laborious, the sensitivity of these methods needs to be improved as interference from other bacteria in the seafood samples, especially other Vibrio spp, can sometimes lead to a false result (Di Pinto et al. 2011). Thus, there is a real need to develop rapid methods and strategies for on-site V. parahaemolyticus monitoring.

Biosensing strategies are showing great promise with such features as being time-saving, cost-effective, practical, and able to perform real-time analysis (Fernandes et al. 2015; Hushiarian et al. 2015b; Lu et al. 2013; Tian et al. 2016; Zhang et al. 2013). In the last decade, electrochemical DNA biosensors have revolutionized modern analysis for detecting contaminants in a range of foods and environments (Celik et al. 2013; Dong et al. 2012; Dutse et al. 2013; Hushiarian et al. 2015a; Singh et al. 2013; Yin et al. 2013). Numerous electrochemical biosensors, based on screen-printing and with DNA immobilized on their surfaces, have been reported in the scientific literature (Alocilja et al. 2013; Das et al. 2014; Ding et al. 2012; Pal and Alocilja 2010; Paniel and Baudart 2013) and have been successfully employed for the fabrication of electrodes for mass production of disposable, low-cost devices (Caramit et al. 2015; Monteiro et al. 2015). A number of attempts to develop portable biosensors for monitoring foodborne pathogens have been reported (Ferguson et al. 2011; Lee et al. 2016; Qin et al. 2016), but there would appear to be none which take this approach to detection of V. parahaemolyticus. Polylactide (PLA)— stabilized gold nanoparticles

(AuNPs) have been widely used in a variety of analytical sensing applications (Han et al. 2014; Nordin et al. 2016; Song et al. 2006; Wu et al. 2011) and here gold nanoparticles (AuNPs) stabilized by the nanofiber were used to modify the electrode surface to increase the active surface area of the working electrode (WE) (Ding et al. 2012; Wu et al. 2011).

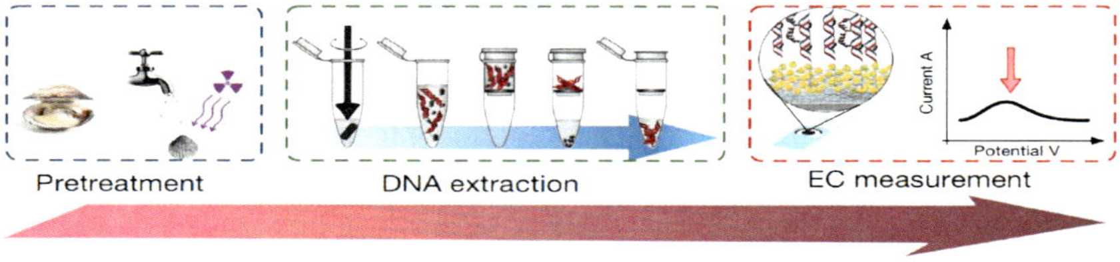

Figure 1. Schematic diagram depicting the steps in the process from pretreatment to electrochemical analysis

MATERIALS AND METHODS

Material and Reagent

Gold (III) chloride trihydrate (Sigma-Aldrich) and poly (lactic acid) resin, commercial grade 4042D (NatureWorks) were of analytical grade and were used without further purification. Gold nanoparticles (AuNPs) and polylactic acid-stabilized gold nanoparticles (PLA-AuNPs) were synthesized and characterized, as described in detail in our previous report which formed the foundation for this improvement (Nordin et al. 2016). Electrochemical measurements were performed using a portable singletechnique customized potentiostat (DropSens, Spain) for use with electrochemical sensor consisting of a screen printed electrode. The device contained a microprocessor, which controls the potential applied to the sensor and measures the current response. The instrument automatically converts current into a value through a calibration equation and is internally recorded and displayed on the LCD simultaneously. Conveniently, the device can be powered by a lithium ion battery and connected directly to a personal computer for data transfer. The sequences of ssDNA probe and complementary DNA were selected by exploring the National Center for Biotechnology Information (NCBI) database. Synthetic oligonucleotides (20- mer ssDNA probe, 20-mer complementary DNA, 20-mer mismatched DNA and 21-mer non-complementary DNA) were purchased (as lyophilized powder) from First BASE Laboratories, Malaysia with the following sequences: thiolated ssDNA probe: 5′-/5ThioMC6-D/CGGATTATGC AGAAGCACTG-3′, complementary DNA: 5′-CAGTGCT TCTGCATAATCCG-3′, one-base mismatched DNA: 5′- CAGTGCTTCTGCṪTAATCCG-3′, three-base mismatched DNA: 5′-CAGTGCTTCTĊṪṪTAATCCG-3′ and non-complementary DNA: 5′-CGCACAAGGCTCGACG GCTGA-3′. The stock solutions of all oligonucleotide (100 μmol l^{-1}) were prepared with sterile Tris–EDTA (TE) solution (10 mM Tris–HCl, 1 mM EDTA, pH 7.5) divided into analytical portions and kept at −4 °C. The appropriate dilutions were made as needed.

Preparation of Bacteria Cell Lysates

Vibrio parahaemolyticus ATCC 17802 as reference strains and nine other bacterial strains of common foodborne pathogen (V. parahaemolyticus, C. jejuni, L. monocytogenes, S. Typhimurium, S. enteritidis, K. pneumonia, E. coli O157:H7, B. cereus and V. alginolyticus) employed for electrochemical DNA biosensor validation were acquired from the Microbial Food Safety and Quality Laboratory, Universiti Putra Malaysia (UPM). We inoculated isolates into a growth broth with 20% glycerol and stored them at −60 °C. We then prepared fresh working culture as needed. We isolated genomic DNA from bacteria by a modified boiled lysis method (Ivanov and Bachvarov 1987) and determined its purity and quantity using an Eppendorf BioPhotometer D30 (Germany). We denatured the DNA in a Thermal Cycler at 92 °C for 2 min and rapidly cooled it in iced water prior to application in the biosensor. The DNA concentration and purity was determined using the biophotometer.

Preparation of Cockle Samples

2 kg of cockles (Anadara granosa) freshly delivered that day were obtained from the wet market in Serdang, Selangor, Malaysia and quickly brought to the laboratory in an iced cooler box. For the study, the cockles were divided into two groups, namely spiked and unspiked group with the assumption that cockles are harvested uniformly from the beginning of harvesting until being placed in cold storage at the market. Half of the cockles in both groups were pre-treated by being stored at −20°C for 24 h, followed by exposure to UV light at 20 °C for 4 h prior to DNA extraction. This pretreatment was considered as a preventive measure to limit or at least minimize naturally accumulated V. parahaemolyticus in the cockles. However, a higher pasteurization regime of 70 °C was not applied as the aim of the controlled

condition in this study was to mimic the actual situation of fresh cockles. Meanwhile, the other half of the samples in both groups were directly sent for analysis as soon as the samples arrived in the laboratory.

Treatment of Cockle Samples

Each cockle was washed in distilled water and scrubbed free of dirt before the tissue was removed from the shell using a sterile forcep in a laminar flow cabinet. About 10 g of cockle tissue sample was homogenized with homogenizer in 90 ml of sterile TSB (tryptic soy broth, 3% NaCl purchased from Merck, Malaysia) for 60 s. A known amount of V. parahaemolyticus inoculum was added to 9 ml of homogenized sample broth for the spiked samples while the unspiked samples were used as a negative control. Genomic DNA of the fresh cockles could be extracted from spiked and unspiked samples where the DNA concentration and purity could be determined using a biophotometer.

Portable Biosensor's Measurement Procedure

The capacity and capability of the developed portable DNA biosensor was investigated by measurement in 0.1 M PBS (phosphate buffer saline pH 7 from Merck Malaysia) after the electrode was immersed in 20 $\mu mol\ l^{-1}$ of MB for 30 min. For this work we used a three electrode carbon screen-printed system (Dropsens, DRP-550) consisting of a carbon working electrode (diameter 4 mm), a platinum counter electrode and a quasi-silver reference electrode. We immersed the electrode in 20 μM MB methylene blue from Merck Malaysia) for 30 min, washed it with 0.5 M PBS/20 mM NaCl (pH 4.5) and rinsed it with deionized water prior to measurement. The same procedure was applied for all interactions including probe DNA, complementary DNA, mismatched DNA and non-complementary DNA samples. We took the DPV measurements of the MB electrochemical reduction in the potential range from −0.5 to 0.25 V at the step potential of 0.005 V and the modulation amplitude of 0.05 V with the scan rate of 7.73 mVs−1 in 0.1 M PBS (pH 7) containing no indicator. We subsequently studied the sensitivity and reproducibility of the customized portable DNA biosensor. Validation studies of the portable DNA biosensor using bacteria cell lysates and fresh cockles were further conducted. All reported results were the measurement of the mean value from three replicates. We investigated hybridization between probe and synthetic oligonucleotides by DPV using a μAutolab III (Eco-chemie, Netherland) voltammetric analyser together with

General Purpose Electrochemical System (GPES 4.9) software. We found significant characteristics of PLA-AuNPs as modifier from preliminary study, which demonstrated good sensitivity, stability, reproducibility, and repeatability. We used SPCE modified with PLA-AuNPs, denoted as SPCE\ PLA-AuNPs for this study. A drop casting method was used for the DNA immobilization and hybridization. We immobilized a 25 μl of thiolated ssDNA probe (1.2 μM) on the SPCE\ PLA-AuNPs and air dried it for 24 h at room temperature. Then we pipetted 25 μl of different concentrations of complementary DNA on to the SPCE\PLA-AuNPs\ ssDNA for 40 min at room temperature. We studied the surface morphology of SPCE prior to modification using scanning electron microscopy (SEM, JEOL JSM 6400). We conducted further studies of hybridization time and temperature using the optimum concentration of complementary DNA.

RESULTS

The selectivity of the optimized DNA biosensor was assessed by measuring its responses towards different gene sequences related to V. parahaemolyticus. The signals measured were as shown in Table 1. After the hybridization of the target DNA, the sensor showed the lowest oxidation signals with peak currents of 1.00 μA. The oxidation signal was about 2.70 times lower than that of the bare electrode. The hybridization with the noncomplementary sequence also show that the peak current was much higher than that obtained from the hybridization of the target DNA.

The percentage of selectivity rate was then calculated based on the following equation: Selectivity rate (%) = $(A_t / A_0) \times 100$, where A0 is the mean MB peak current obtained (n = 3) without hybridization and At is the mean MB peak current obtained (n = 3) with different types of hybridization i.e. non-complementary DNA, 3-base mismatched DNA, 1-base mismatched DNA and complementary DNA. When the ssDNA molecule was used as the capture probe, the hybridization reaction was recorded through the decreases in current signals after the duplex formation on the electrode surface.

Reproducibility

To investigate the reproducibility and precision of the optimized DNA biosensor, we used a freshly prepared biosensor to detect 0.2 μM target DNA.

Table 1. Selectivity of MB peak current in 0.1 mol l−1 PBS (pH 7) at scan rates of 0.1 V s−1 after 30 min incubation in 20 μM MB

Sample	i_{pa} (× 10^{-6} A)	Selectivity rate (%)
Bare electrode	3.63	–
Non-complementary DNA	2.73	75.21
3-Base mismatched DNA	1.83	50.41
1-Base mismatched DNA	1.40	38.57
Complementary DNA	1.00	27.55

Sensitivity

The developed electrochemical DNA biosensor was then studied, using the immobilized ssDNA to hybridize with various concentrations of the target DNA of V. parahaemolyticus as shown in Fig. 2. This device was able to detect target DNA in concentrations ranging from 2×10^{-7} to 2×10^{-2} μM with a linear regression coefficient of 0.989 (Fig. 3).

Crossreactivity Study

Figure 4 shows the results of the cross-reactivity study conducted with the portable DNA biosensor in the presence of various foodborne pathogens which mimicked the environment of this food sample. From these results, it can be seen that the intensity of the oxidation current decreased in the order of V. parahaemolyticus

Evaluation of fresh cockles

Finally, the method trueness of the DNA-based sensing system was evaluated by monitoring real samples (fresh cockles) with the referenced polymerase chain reaction (PCR). It successfully distinguished the available V. parahaemolyticus in the treated group (spiked and unspiked) samples. These results were in positive correlation with PCR results previously discussed.

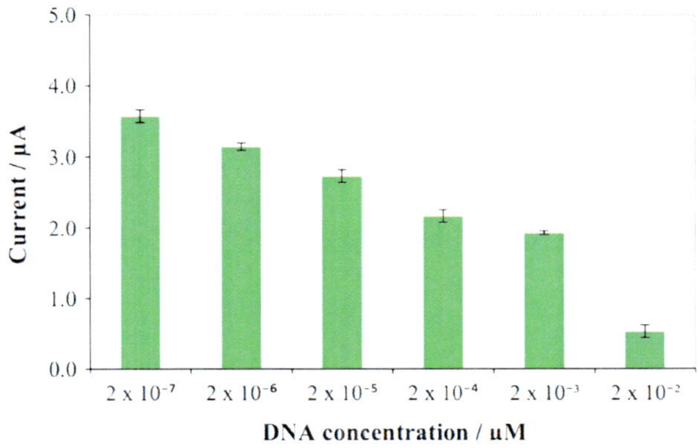

Figure 2. Histogram of effect of different DNA concentration on peak current

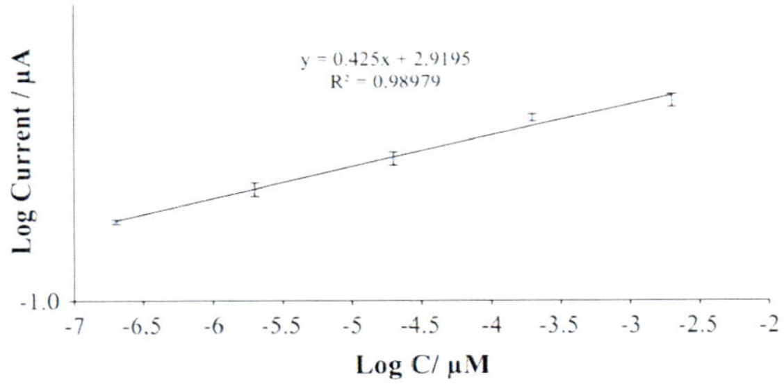

Figure 3. The plot of reduction peak current of MB against log concentration of DNA

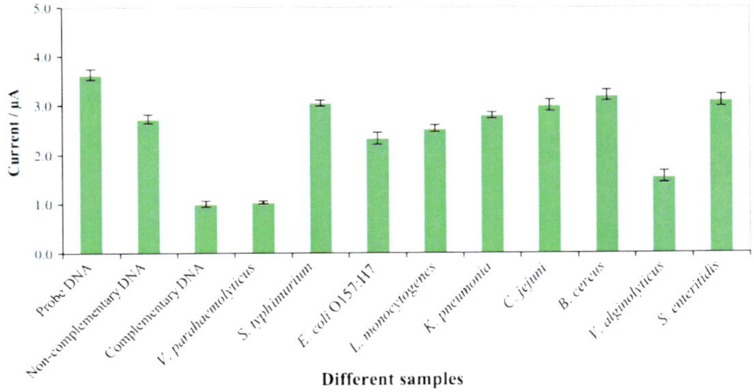

Figure 4. Cross reactivity study of portable DNA biosensor against various foodborne pathogens

When unspiked sample of cockles were hybridized at the SPCE/PLA-AuNPs/ssDNA probe, no significant variation was observed among the samples. These results indicate that peak currents were less varied within a sample of cockles regardless of whether they were treated or not.

The optimized biosensor was validated using treated cockle samples (spiked and unspiked with V. parahaemolyticus culture cell) and untreated cockle samples (spiked and unspiked with V. parahaemolyticus culture cell). The lowest peak current was observed when the target DNA from spiked (treated and untreated cockle samples) was detected. Conversely, unspiked (treated and untreated cockle samples) produced higher current, which was similar to that of the non-complementary DNA signal (Fig. 5).

With this in mind, we used gel electrophoresis to validate the presence of V. parahaemolyticus in fresh cockle samples. From PCR analysis, V. parahaemolyticus were found in spiked (treated and untreated cockle samples) as shown by Lane 7, 8, 9, 10, 11, and 12, whereas unspiked (treated and untreated cockle samples) at Lane 1, 2, 3, 4, 5 and 6 were devoid of V. parahaemolyticus occurrence (Fig. 6). Confirmations of V. parahaemolyticus present in the spiked (treated and untreated cockle samples) by PCR (The appeared band in agarose gel electrophoresis at 368 bp) correlate with the quantity of colony as discussed previously. However, the PCR results showed no detection of V. parahaemolyticus in unspiked (treated and untreated cockle samples) even though the amount of colony indicated presence of V. parahemolyticus in the samples.

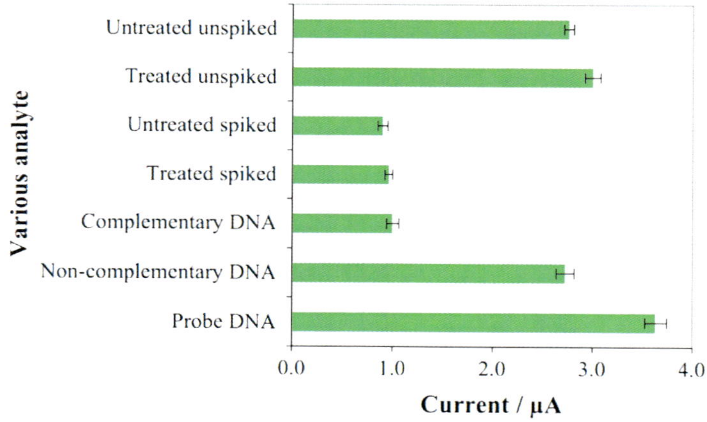

Figure 5. Detection of V. parahaemolyticus in fresh cockle samples using portable DNA biosensor

DISCUSSION

The Vibrio parahaemolyticus pathogen is an important seafood-borne enteropathogen that continues to cause acute human gastroenteritis throughout the world (Zhang et al. 2016) and its rapid detection is of great significance for food safety and disease diagnosis (Cheng et al. 2016).

Although most conventional DNA-based biosensors have the advantage of high selectivity due to the unique structure of the probe DNA, the complicated preparation process and the low electrochemical signal intensity discourage many attempts at practical application (Wang et al. 2016). Additionally, many efforts suffer from one or more other drawbacks including long analytical time, high analytical costs and expensive instrumentation (Sha et al. 2016).

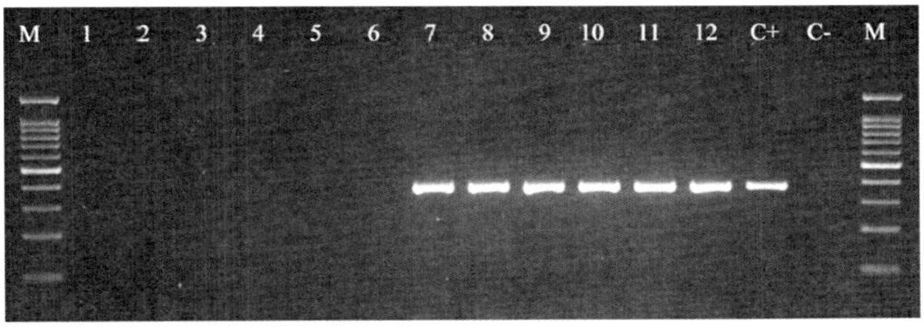

Figure 6. Polymerase chain reaction results of fresh cockle samples (Lane M: 2 μl of 100 bp DNA ladder, Lane 1–3 untreated unspiked samples, Lane 4–6 treated unspiked samples, Lane 7–9 untreated spiked samples, Lane 10–12 treated spiked samples, Lane C+ positive control (V. parahaemolyticus toxR gene), Lane C− negative control (sterile distilled water)

Recent examples which show promise include a highly selective and sensitive SERS-based aptasensor which used Au@Ag core–shell nanoparticles as the active substrate (Duan et al. 2016) which needs to be more sensitive and a label-free electrochemiluminescence (ECL) immunosensor based on multi-functionalized graphene oxide but which needs to be tested in the field (Sha et al. 2016). In the work presented here, the above-mentioned disadvantages were anticipated and overcome.

A previously constructed electrochemical mechanism (Nordin et al. 2016) was modified and a simplified approach for fabrication of the DNA biosensor was based on a screen-printed carbon electrode reader with polylactide—stabilised gold nanoparticles and methylene blue as the redox indicator. Results for the analytical parameters of this sensor (selectivity, reproducibility, sensitivity and cross-reactivity) compare favourably with results from a range of other DNA sensors as recently reviewed by Wang et al. (2016). The percentage of selectivity rate of the optimized sensor was significantly high. After the hybridization of the target DNA, it showed the lowest oxidation signals with peak currents of 1.00 μA. The oxidation signal was about 2.70 times lower than that of the bare electrode. The hybridization with the non-complementary sequence also showed that the peak current was much higher than that obtained from the hybridization of the target DNA. When the ssDNA molecule was used as the capture probe, the hybridization reaction was recorded through the decreases in current signals after the duplex formation on the electrode surface.

For 1-base mismatched and 3-base mismatched DNA, the oxidation current increased by 40 and 38%, respectively, in comparison to that of complementary DNA which suggests that the hybridization was weak. The results indicate limited interaction occurred between MB and guanine bases as only a small amount of MB was available on the surface with hybridized dsDNA. Its reproducibility capacity plays an extremely important role in practical application for a biosensor. With this device, the relative standard deviation (RSD) toward the target DNA concentration of 4.33% (n = 6) was an excellent result in terms of the reproducibility and precision of the optimized DNA biosensor and compares well with other attempts such as that of Sha et al. (2016) who reported RSD measurements of 7.8%.

Additionally, the device demonstrated a low detection limit, calculated to be 2.16×10^{-6} μM (3 σ/m, and a wide linear range for the target DNA sequence being analyzed. The optimized electrochemical DNA biosensor was able to reach a lower quantitation limit (10 σ/m) of 7.2×10^{-6} μM for DPV than had previously been reported (Sha et al. 2016; Zhao et al. 2007). In the presence of various foodborne pathogens, our portable sensor was successfully able to detect V. parahaemolyticus. After hybridization of the SPCE/ PLA-AuNPs/ssDNA with target DNA, the peak current decreased greatly, which suggests that dsDNA was formed at the modified electrode surface. No significant decrement in peak current was observed after the probe DNA was hybridized with non-V. parahaemolyticus target DNA, indicating that it was poorly hybridized. Pleasingly, the optimized DNA biosensor was clearly highly specific because non-V. parahaemolyticus isolates of crude DNA fragments did not show a significant enhancement in peak currents compared with V. parahaemolyticus isolates.

Finally, when the method trueness of the DNA-based sensing system was evaluated by monitoring real samples (fresh cockles) with the referenced polymerase chain reaction (PCR), it was able to successfully distinguish the available V. parahaemolyticus in the treated group (spiked and unspiked) samples. When unspiked sample of cockles were hybridized at the SPCE/PLA-AuNPs/ ssDNA probe, no significant variation was observed among the samples indicating that peak currents were less varied within a sample of cockles regardless of whether they were treated or not. The results suggest that the optimized DNA biosensor is highly selective towards V. parahaemolyticus compared with non-V. parahaemolyticus target DNA. When the optimized biosensor was ultimately validated, the lowest peak current was observed when the target DNA from spiked (treated and untreated cockle samples) was detected while conversely, unspiked (treated and untreated cockle samples) produced higher current, which was similar to that of the non-complementary DNA signal, probably due to the inevitable non-specific binding formed during the hybridization steps. When gel electrophoresis was subsequently tried to validate the presence of V. parahaemolyticus in fresh cockle samples, the PCR analysis found V. parahaemolyticus in spiked (treated and untreated cockle samples) whereas unspiked (treated and untreated cockle samples) were devoid of V. parahaemolyticus occurrence, correlating with the quantity of colony as discussed previously.

Interestingly, the PCR results showed no detection of V. parahaemolyticus in unspiked (treated and untreated cockle samples) even though the amount of colony indicated the presence of V. parahemolyticus in the samples. He et al. (2011) suggest that this might be due to low purity with high contamination of extracted DNA such as protein (He et al. 2011). In summary, the optimized portable electrochemical DNA biosensor described here was able to detect the presence of V. parahaemolyticus in fresh cockle samples even when there were other bacteria present. Thus, not only is this simple device able to discriminate between contaminated and uncontaminated samples, it is also superior to other current detection methods because it dramatically reduces the number of analyses which need to be performed. Its reliable sensitivity together with its portability make it potentially suitable for providing convenient on-site monitoring and analysis of V. parahaemolyticus in real samples.

REFERENCES

1. Alocilja EC, Zhang D, Shi C (2013) AuNP-DNA biosensor for rapid detection of Salmonella enterica serovar enteritidis. ACS Sym Ser 1143:43–53. doi:10.1021/bk-2013-1143.ch003
2. Bisha B, Simonson J, Janes M, Bauman K, Goodridge LD (2012) A review of the current status of cultural and rapid detection of Vibrio parahaemolyticus. Int J Food Sci Tech 47:885–899. doi:10.1111/j.1365-2621.2012.02950.x
3. Caburlotto G, Suffredini E, Toson M, Fasolato L, Antonetti P, Zambon M, Manfrin A (2016) Occurrence and molecular characterisation of Vibrio parahaemolyticus in crustaceans commercialised in Venice area, Italy. Int J Food Microbiol 220:39–49. doi:10.1016/j.ijfoodmicro.2015.12.007
4. Caramit RP, Lucca BG, Souza Ferreira V, Abarza Munoz RA, Richter EM, da Silva RAB (2015) On-site determination of carbendazim, cathecol and hydroquinone in tap water using a homemade batch injection analysis cell for screen printed electrodes. Electroanal 27:271–275. doi:10.1002/elan.201400454
5. Celik G, Eren E, Uygun A (2013) Electrochemical polymerization of 2-thiophen3-yl-malonic acid for biosensor application. J Appl Polym Sci 128:636– 641. doi:10.1002/app.38205
6. Cheng K, Pan D, Teng J, Yao L, Ye Y, Xue F, Xia F, Chen W (2016) Colorimetric integrated pCR protocol for rapid detection of Vibrio parahaemolyticus. Sensors 16(10):1600
7. Costa Sobrinho PdS, Destro MT, Franco BDGM, Landgraf M (2014) A quantitative risk assessment model for Vibrio parahaemolyticus in raw oysters in Sao Paulo State, Brazil. Int J Food Microbiol 180:69–77. doi:10.1016/j.
8. ijfoodmicro.2014.04.008
9. Das R, Sharma MK, Rao VK, Bhattacharya BK, Garg I, Venkatesh V, Upadhyay S (2014) An electrochemical genosensor for Salmonella typhi on gold nanoparticles-mercaptosilane modified screen printed electrode. J Biotechnol 188:9–16. doi:10.1016/j.jbiotec.2014.08.002
10. DePaola A, Kaysner CA, Bowers J, Cook DW (2000) Environmental investigations of Vibrio parahaemolyticus in oysters after outbreaks in Washington, Texas, and New York (1997 and 1998). Appl Environ Microbiol 66(11):4649–4654

11. Di Pinto A, Terio V, Novello L, Tantillo G (2011) Comparison between thiosulphate-citrate-bile salt sucrose (TCBS) agar and CHROMagar Vibrio for isolating Vibrio parahaemolyticus. Food Control 22:124–127. doi:10.1016/j. foodcont.2010.06.013

12. Ding Y, Zhao H, Teng Y, Lan M (2012) Disposable screen-printed biosensor array for simultaneous detection of Microcystis and Escherichia coli oligonucleotides. Adv Sci Lett 6:34–39. doi:10.1166/asl.2012.2016

13. Dong YP, Huang L, Lei ZP, Chu XF, Zhang QF (2012) Electrochemical biosensor of glucose based on the immobilization of glucose oxidase on a carbon microsphere modified glassy carbon electrode. Sens Lett 10:1024–1030. doi:10.1166/sl.2012.2339

14. Duan N, Yan Y, Wu S, Wang Z (2016) Vibrio parahaemolyticus detection aptasensor using surface-enhanced Raman scattering. Food Control 63:122–127

15. Dutse SW, Yusof NA, Ahmad H, Hussein MZ, Hushiarian R (2013) DNA-based biosensor for detection of Ganoderma boninense, an oil palm pathogen utilizing newly synthesized ruthenium complex [Ru (phen) 2 (qtpy)] 2 based on a PEDOT-PSS/Ag nanoparticles modified electrode. Int J Electrochem Sci 8(9):11048–11057

16. Ferguson BS, Buchsbaum SF, Wu TT, Hsieh K, Xiao Y, Sun R, Soh HT (2011) Genetic analysis of H1N1 influenza virus from throat swab samples in a microfluidic system for point-of-care diagnostics. J Am Chem Soc 133:9129–9135. doi:10.1021/ja203981w

17. Fernandes AM, Abdalhai MH, Ji J, Xi BW, Xie J, Sun J, Noeline R, Lee BH, Sun X (2015) Development of highly sensitive electrochemical genosensor based on multiwalled carbon nanotubes-chitosan-bismuth and lead

18. sulfide nanoparticles for the detection of pathogenic Aeromonas. Biosens Bioelectron 63:399–406. doi:10.1016/j.bios.2014.07.054

19. Han HS, You J-M, Jeong H, Jeon S (2014) Electrochemical sensing of H2O2 by the modified electrode with Pd nanoparticles on multi-walled carbon nanotubes-g-poly(lactic acid). J Nanosci Nanotechnol 14:4050–4057. doi:10.1166/jnn.2014.8258

20. He P, Tucker EZ, Gorman CB, He L (2011) Chemical amplification for in-gel DNA detection. Anal Method 3:2463. doi:10.1039/c1ay05514c

21. Hushiarian R, Yusof NA, Abdullah AH, Ahmad SAA, Dutse SW

(2015a) Facilitating the indirect detection of genomic DNA in an electrochemical DNA biosensor using magnetic nanoparticles and DNA ligase. Anal Chem Res 6:17–25. doi:10.1016/j.ancr.2015.10.004

22. Hushiarian R, Yusof NA, Houshiarian N, Abdullah AH, Ahmad SAA (2015b) Computer modeling to optimize the sensitivity of an optical DNA nanosensor. Sens Actuators B Chem 207:716–723. doi:10.1016/j.snb.2014.09.120
23. Ivanov IG, Bachvarov DR (1987) Determination of plasmid copy number by the "boiling" method. Anal Biochem 165:137–141. doi:10.1016/0003-2697(87)90211-9
24. Kaneko T, Colwell R (1975) Incidence of Vibrio parahaemolyticus in Chesapeake Bay. Appl Microbiol 30(2):251–257
25. Lee JR, Choi J, Shultz TO, Wang SX (2016) Small molecule detection in saliva facilitates portable tests of marijuana abuse. Anal Chem 88:7457–7461. doi:10.1021/acs.analchem.6b01688
26. Liao IC, Chao NH (2009) Aquaculture and food crisis: opportunities and constraints. Asia Pac J Clin Nutr. 18:564–569
27. Lu L, Chee G, Yamada K, Jun S (2013) Electrochemical impedance spectroscopic technique with a functionalized microwire sensor for rapid detection of foodborne pathogens. Biosens Bioelectron 42:492–495. doi:10.1016/j.bios.2012.10.060
28. Maniyankode RA, Kingston JJ, Murali HS, Batra HV (2013) Specific identification of Vibrio parahaemolyticus employing monoclonal antibody based immunoassay. Int J Pharma Bio Sci 4:B156–B164
29. Martinez-Urtaza J, Lozano-Leon A, DePaola A, Ishibashi M, Shimada K, Nishibuchi M, Liebana E (2004) Characterization of pathogenic Vibrio parahaemolyticus isolates from clinical sources in Spain and comparison with Asian and North American pandemic isolates. J Clin Microbiol 42(10):4672–4678
30. Martinez-Urtaza J, Powell A, Jansa J, Rey JLC, Montero OP, Campello MG, López MJZ, Pousa A, Valles MJF, Trinanes J (2016) Epidemiological investigation of a foodborne outbreak in Spain associated with US West Coast genotypes of Vibrio parahaemolyticus. SpringerPlus 5(1):1 Monteiro T, Rodrigues PR, Gonçalves AL, Moura JJG, Jubete E, Añorga L, Piknova B, Schechter AN, Silveira CM, Almeida MG (2015) Construction of effective disposable biosensors for point of care testing of nitrite. Talanta 142:246–251. doi:10.1016/j.

talanta.2015.04.057

31. Nair GB, Ramamurthy T, Bhattacharya SK, Dutta B, Takeda Y, Sack DA (2007) Global dissemination of Vibrio parahaemolyticus serotype O3: K6 and its serovariants. Clin Microbiol Rev 20(1):39–48

32. Nordin N, Yusof NA, Abdullah J, Radu S, Hushiarian R (2016) Sensitive detection of multiple pathogens using a single DNA probe. Biosens Bioelectron 86:398–405. doi:10.1016/j.bios.2016.06.077

33. Ottaviani D, Leoni F, Serra R, Serracca L, Decastelli L, Rocchegiani E, Masini L, Canonico C, Talevi G, Carraturo A (2012) Nontoxigenic Vibrio parahaemolyticus strains causing acute gastroenteritis. J Clin Microbiol 50:4141–4143. doi:10.1128/JCM.01993-12

34. Pal S, Alocilja EC (2010) Electrically active magnetic nanoparticles as novel concentrator and electrochemical redox transducer in Bacillus anthracis DNA detection. Biosens Bioelectron 26:1624–1630. doi:10.1016/j. bios.2010.08.035

35. Paniel N, Baudart J (2013) Colorimetric and electrochemical genosensors for the detection of Escherichia coli DNA without amplification in seawater. Talanta 115:133–142. doi:10.1016/j. talanta.2013.04.050

36. Powell A, Baker-Austin C, Wagley S, Bayley A, Hartnell R (2013) Isolation of pandemic Vibrio parahaemolyticus from UK water and shellfish produce. Microb Ecol 65(4):924–927

37. Qin C, Gao Y, Wen W, Zhang X, Wang S (2016) Visual multiple recognition of protein biomarkers based on an array of aptamer modified gold nanoparticles in biocomputing to strip biosensor logic operations. Biosens Bioelectron 79:522–530. doi:10.1016/j.bios.2015.12.096

38. Rosec JP, Causse V, Cruz B, Rauzier J, Carnat L (2012) The international standard ISO/TS 21872-1 to study the occurence of total and pathogenic Vibrio parahaemolyticus and Vibrio cholerae in seafood: ITS improvement by use of a chromogenic medium and PCR. Int J Food Microbiol 157:189–194. doi:10.1016/j.ijfoodmicro.2012.04.026

39. Sha Y, Zhang X, Li W, Wu W, Wang S, Guo Z, Zhou J, Su X (2016) A label-free multi-functionalized graphene oxide based electrochemiluminscence immunosensor for ultrasensitive and rapid detection of Vibrio parahaemolyticus in seawater and seafood. Talanta 147:220–225

40. Shen XS, Qian BL, Wu WH, Cai YQ, Liu CC (2011) Elimination of

Vibrio parahaemolyticus contamination in shucked oysters (Crassostrea plicatula) to with natural antimicrobial agents TeaPolyphenols. Adv Mater Res 320:427–433. doi:10.4028/www.scientific.net/AMR.320.427

41. Shimohata T, Takahashi A (2010) Diarrhea induced by infection of Vibrio parahaemolyticus. J Med Invest. 57:179–182
42. Singh S, Jain DVS, Singla ML (2013) One step electrochemical synthesis of gold-nanoparticles–polypyrrole composite for application in catechin electrochemical biosensor. Anal Method 5:1024. doi:10.1039/ c2ay26201k
43. Skovgaard N (2012) Risk assessment of Vibrio parahaemolyticus in seafood. Interpretative summary and technical report. Int J Food Microbiol 154:215–216. doi:10.1016/j.ijfoodmicro.2011.12.021
44. Song YS, Muthuraman G, Chen YZ, Lin CC, Zen JM (2006) Screen printed carbon electrode modified with poly(L-lactide) stabilized gold nanoparticles for sensitive as(III) detection. Electroanal 18:1763–1770. doi:10.1002/ elan.200603634
45. Terzi Gulel G, Martinez-Urtaza J (2016) Molecular characterizations of Vibrio parahaemolyticus in seafood from the Black Sea, Turkey. Lett Appl Microbiol 62(6):494–500
46. Tian F, Lyu J, Shi J, Tan F, Yang M (2016) A polymeric microfluidic device integrated with nanoporous alumina membranes for simultaneous detection of multiple foodborne pathogens. Sens Actuators B Chem 225:312–318. doi:10.1016/j.snb.2015.11.059
47. Varnam AH, Evans MG (1991) Foodborne pathogens: an illustrated text, 1st edn. Mosby Year Book, London, pp 157–184
48. Vengadesh L, Kok-Gan C, Learn-Han L (2014) Vibrio parahaemolyticus: a review on the pathogenesis, prevalence, and advance molecular identification techniques. Front Microbiol 5:1–13. doi:10.3389/fmicb.2014.00705
49. Wang Q, Gao F, Ni J, Liao X, Zhang X, Lin Z (2016) Facile construction of a highly sensitive DNA biosensor by in situ assembly of electro-active tags on hairpin-structured probe fragment. Sci Rep 6:22441
50. Wong HC, Chen M-C, Liu S-H, Liu D-P (1999) Incidence of highly genetically diversified Vibrio parahaemolyticus in seafood imported from Asian countries. Int J Food Microbiol 52:181–188. doi:10.1016/S0168-1605(99)00143-9

51. Wu X, Jiang H, Zheng J, Wang X, Gu Z, Chen C (2011) Highly sensitive recognition of cancer cells by electrochemical biosensor based on the interface of gold nanoparticles/polylactide nanocomposites. J Electroanal Chem 656:174–178. doi:10.1016/j.jelechem.2010.11.035
52. Yin H, Xu Z, Wang M, Zhang X, Ai S (2013) An electrochemical biosensor for assay of DNA methyltransferase activity and screening of inhibitor. Electrochim Acta 89:530–536. doi:10.1016/j.electacta.2012.11.093
53. Zhang L, Orth K (2013) Virulence determinants for Vibrio parahaemolyticus infection. Curr Opin Microbiol 16:70–77
54. Zhang J, Li Z, Zhang H, Wang J, Liu Y, Chen G (2013) Rapid detection of several foodborne pathogens by F0F1-ATPase molecular motor biosensor. J Microbiol Methods 93:37–41. doi:10.1016/j.mimet.2013.01.011
55. Zhang Q, Dong X, Chen B, Zhang Y, Zu Y, Li W (2016) Zebrafish as a useful model for zoonotic Vibrio parahaemolyticus pathogenicity in fish and human. Dev Comp Immunol 55:159–168
56. Zhao G, Xing F, Deng S (2007) A disposable amperometric enzyme immunosensor for rapid detection of Vibrio parahaemolyticus in food based on agarose/nano-Au membrane and screen-printed electrode. Electrochem Commun 9(6):1263–1268

CHAPTER 13

DIRECT IDENTIFICATION OF MONILINIA BROWN ROT FUNGI ON INFECTED FRUITS BY MATRIX-ASSISTED LASER DESORPTION/IONIZATION (MALDI) MASS SPECTROMETRY

Florian Matthias Freimoser[1], Maja HilberBodmer[1], René Brunisholz[2] and David Drissner[3]

[1] Agroscope, Institute for Plant Production Sciences IPS, Schloss 1, 8820 Wädenswil, Switzerland.
[2] Functional Genomics Center Zurich, ETH Zurich and University of Zurich, Winterthurerstr. 190, 8057 Zurich, Switzer land.
[3] Agroscope, Institute for Food Sciences IFS, Schloss 1, 8820 Wädenswil, Switzerland.

CITATION: Freimoser, F. M., et al., "Direct identification of Monilinia brown rot fungi on infected fruits by matrix-assisted laser desorption/ionization (MALDI) mass spectrometry", Chemical and Biological Technologies in Agriculture (2016), https://doi.org/10.1186/s40538-016-0058-4.

ABSTRACT

Background

Brown rot of stone and pome fruit is a serious fungal disease that is mainly caused by four species in the genus *Monilinia*. Of these four species, *Monilinia fructicola* is the most devastating pathogen and of particular concern because it undergoes sexual recombination and has recently been introduced to Europe. So far, *Monilinia* diagnosis required a multiplex PCR analysis and gel electrophoresis. In contrast, intact-protein biotyping by mass spectrometry is considerably faster and cheaper. However, it usually requires an in vitro cultivation step prior to the MALDI analysis. It was thus attempted to establish a method for the identification of *Monilinia* species by MALDI biotyping with fungal material derived directly from infected fruits; without an in vitro cultivation step.

Results

To simplify and render MALDI biotyping of fungi more reliable, an improved protocol for the preparation of crude protein extracts and for collecting MALDI-TOF MS data for biotyping was developed. We generated reference spectra for all four *Monilinia* brown rot fungi and were able to reliably identify *Monilinia* species based on fungal material that was collected directly from infected fruits. This method allowed the correct, fast and economic identification of *M. fructicola* and *M. laxa*, while *M. fructigena* and *M. polystroma* could not be distinguished reliably.

Conclusions

MALDI biotyping may be used as an economical tool for the routine diagnosis of *Monilinia*brown rot fungi on infected fruits.

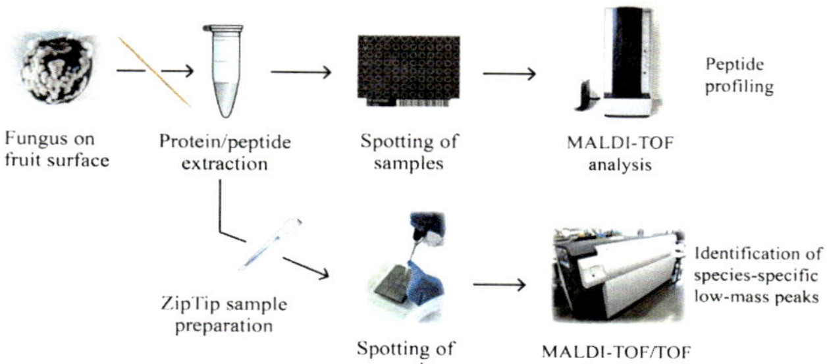

Graphical Abstract

MALDI-TOF and MALDI-TOF/TOF MS workflows for the biotyping and analysis of *Monilinia* species. Optimization of protein extraction, own MSP generation and data analysis permitted the fast and reliable identification of *M. fructicola*, *M. laxa* and *M. fructigena/M. polystroma*

Keywords: Biotyping, Intact-protein profiling, MALDI-TOF/TOF MS, *Monilinia*

BACKGROUND

Brown rot of stone and pome fruit is a fungal disease that causes economic losses and limits fruit production worldwide [1]. The causative agents of this disease are mainly four species in the genus *Monilinia*. Although the four *Monilinia* brown rot fungi exhibit host preference, all are able to infect and cause severe damage in various members of the *Rosaceae* family. The most common of these pathogens worldwide is *Monilinia laxa* (Aderh. & Ruhland) Honey 1945 (Ascomycetes, Helotiales) [2], which also causes twig and branch blight and is one of the most important postharvest pathogens of stone fruits [3]. *Monilinia fructigena* Honey ex Whetzel 1945 (Ascomycetes, Helotiales) occurs worldwide and causes damage in pome fruit [3, 4]. *Monilinia polystroma* van Leeuwen [5] (Ascomycetes, Helotiales), distinguished by molecular and morphological means from *M. fructigena* [5, 6], is mainly found in Asia and Europe [5, 7]. Brown rot caused by *M. fructicola* (G. Winter) Honey 1928 (Ascomycetes, Helotiales) is one of the most devastating stone fruit diseases in Asia, Australasia, North and South America [3, 8]. This species regularly undergoes sexual recombination, which may promote the development and establishment of fungicide-resistant strains [9, 10, 11]. In Europe, this disease was described for the first time in 2001, in southern France, and since then the pathogen has spread [12, 13, 14, 15]. The European Plant Protection Organization recommends *M. fructicola* for regulation as a quarantine pest [16], even though it is widespread now. Local survey and management schemes are thus advised, because phytosanitary measures and chemical control can prevent the establishment of this pathogen [14]. To expedite the identification of *M. fructicola*, we strove for a fast and reliable matrix-assisted laser desorption/ ionization time-of-flight mass spectrometry (MALDI-TOF MS) method that does not require prior in vitro cultivation of the pathogen.

Monilinia brown rot fungi exhibit slight morphological variation, but their identification and delineation usually includes a polymerase chain reaction (PCR)-based analysis [17]. Although this analysis can be performed with fungal material obtained directly from infected fruits (i.e. does not require a cultivation step), it involves DNA extraction, a lengthy PCR protocol and performing gel electrophoresis. *Monilinia polystroma* differs only by five base transitions in the ITS region from *M. fructigena* [6] and thus requires multiplex PCR and the resolution of fragment size differences of 23 bp [7]. In contrast to the DNA-based identification, microbial biotyping builds upon the observation that MALDI-TOF MS of entire microbial cells or of crude extracts reveals species-specific profiles of molecular masses [18]. Most of the mass peaks detected by this method, also called MALDI biotyping, are attributed to basic, abundant and conserved proteins, particularly ribosomal proteins [19, 20] and to a certain degree to cell wall-associated proteins [21]. Sample preparation for MALDI biotyping is considerably faster and cheaper than for the DNA-based methods and the measurement itself is performed within a few minutes. However, MALDI biotyping requires pre-existing reference spectra for the species to be identified. For clinically relevant microorganisms, comprehensive databases of reference spectra exist, but organisms from agricultural ecosystems are often not included in commercially available databases. Furthermore, the method performs reliably for many bacteria, while the MALDI-TOF MS-based identification of fungi is often more difficult and requires optimization of extraction and data acquisition protocols. Finally, a major limitation for the rapid identification of bacteria and fungi is the in vitro cultivation step, which is usually required prior to the MALDI-TOF analysis.

Recently, the potential of different electrophoresis methods and of MS for the identification of *Monilinia* species based on proteins was assessed [22]. The authors successfully used MALDI-TOF MS to identify the four closely related *Monilinia* brown rot species when they were grown in vitro. However, identification of fungal material obtained directly from infected fruits was not possible, because the host affected the result of the MALDI-TOF MS analysis and prevented a reliable identification at the species level. To circumvent this unreliability, a combination of MALDI-TOF MS analysis and capillary isoelectric focusing (CIEF) or capillary zone electrophoresis (CZE) was suggested [22]. It was the goal of the project presented here to build upon this work and to establish the identification of *Monilinia* species by MALDI biotyping with fungal material derived directly from infected fruits; without an in vitro cultivation step. We prepared crude, acid protein extracts from minute sample amounts with glass beads, generated reference spectra for several isolates of all four *Monilinia* brown rot fungi, and were thus able to identify *Monilinia* species with material collected directly from infected fruits. This method allowed the correct, fast and economic identification of *M. fructicola* and *M. laxa*, while *M. fructigena* and *M. polystroma* could not be distinguished reliably. MALDI biotyping may thus be used as an economical tool for the routine diagnostic of *Monilinia* brown rot fungi on infected fruits.

METHODS

Fungal Isolates and Cultivation

All *Monilinia* isolates were obtained from infected apricots or peaches as described previously [11] and grown on Difco™ potato dextrose agar (PDA; Becton, Dickinson and Company, Le Pont de Claix, France) at 20 °C in the dark. The *Monilinia* isolates used for this study are listed in Table 1.

Table 1. *Monilinia* isolates used for generating the reference main spectra (MSP) for this study

Species	Isolate	Origin	Year	Host
M. fructicola	11SD14	Switzerland	2011	Apricot
	M16	USA	2003	Peach
	AS5	France	2003	Apricot
	IT3	Italy	2008	Peach
	SD14	Switzerland	2010	Apricot
M. laxa	E134.5	Switzerland	2009	Apricot
	10S1	Switzerland	2010	Apricot
	10M14	Switzerland	2010	Apricot
	E139.5	Switzerland	2009	Apricot
	E57.7	Switzerland	2009	Apricot
M. fructigena	113.A.15.2	Switzerland	2010	Apricot
	180	Switzerland	2008	Apricot
	E71.1	Switzerland	2009	Apricot
	113A8	Switzerland	2010	Apricot

M. polystroma	Q1	Switzerland	2009	Apricot
	P16	Switzerland	2009	Apricot
	C6	Switzerland	2009	Apricot
	G4	Switzerland	2010	Apricot

Strains were isolated from infected apricots or peach in the years 2003 or 2008–2011 and maintained on PDA medium. The strains were previously identified and characterized molecularly (Hilber-Bodmer et al. [7], Jänsch et al. [11])

Fruit Infection Experiments

Ripe, harvested cherries, prunes or apples were wiped with 70 % ethanol and dried. *Monilinia*mycelium and conidia were harvested with a sterile toothpick from an agar plate (1–2 weeks old). To start the infection process, healthy fruits were pierced with the toothpick. Fruits infected in this manner were either stored in disposable, transparent plastic containers (cherries and prunes) or in transparent plastic bags (apples), kept at 20 °C, and inspected daily. Fruits with clear signs of infection and developing mycelium were removed and the fungus causing the infection was identified by DNA-based methods or by MALDI biotyping as described below.

DNA-based Identification

Mycelium was harvested with a wooden toothpick and DNA was extracted with 50 µl Kawasaki buffer as described [11, 23]. The four *Monilinia* species were identified by a multiplex PCR assay as described by Coté et al. [17]. The PCR conditions were the following: initial denaturation at 95 °C for 15 min, 28 cycles of 94 °C for 1 min, 60 °C for 1 min, 72 °C for 1 min; final extension at 72 °C for 10 min. Alternatively, the fungal ITS region was amplified with primers ITS1f [24] and ITS4 [25] and PCR products were directly used for sequencing.

Protein extraction for MALDI-TOF MS-based identification

For protein extraction, a small amount of mycelium (approx. 1–5 × 1–5 mm), either from a PDA plate or from an infected fruit, was collected with a toothpick and transferred to 1 ml 70 % ethanol. After vigorous shaking, the fungal material was pelleted, the supernatant discarded and the mycelium dried in a concentrator (room temperature, 60 min; Eppendorf, Hamburg, Germany). The pellet was resuspended in 25 µl formic acid (70 %, Carl

Roth GmbH, Karlsruhe, Germany), glass beads (0.5–0.75 mm, approx. 50 µl; Retsch GmbH, Haan, Germany) were added and samples were shaken (Eppendorf Thermomixer, 1400 rpm, room temperature, 15 min; Eppendorf). Following the addition of 25 µl acetonitrile (Carl Roth GmbH), samples were mixed, centrifuged and 1 µl of the supernatant was applied to MALDI-TOF steel target plates (Bruker Daltonics GmbH, Bremen, Germany). After air drying, the spots were overlaid with 1 µl matrix (-cyano-4-hydroxycinnamic acid, HCCA, 10 mg/ml in acetonitrile–water–trifluoroacetic acid (TFA) (50:47.5:2.5); HCCA from Sigma-Aldrich Chemie GmbH, Steinheim, Germany, or Bruker Daltonics GmbH; TFA from Merck KGaA, Darmstadt, Germany). Alternatively, the fungal extract was stored at −20°C for later analysis.

MALDI-TOF MS Measurement

MALDI-TOF MS spectra were acquired with a microflex LT MALDI-TOF mass spectrometer (Bruker Daltonics GmbH). Data collection parameters were defined using flexControl software (Version 3.4; Bruker Daltonics GmbH). Laser frequency of 60 Hz in the positive linear mode with mass acquisition ranging from 2 to 20 kDa. Final spectra consisted of 500 shots per spot (random walk, 50 shots per raster spot). No peak exclusion criterion was defined, but the resolution was required to be higher than 100 and the measurement was quit if 20 subsequently failed judgements were encountered. Data collection was terminated as soon as 10 peaks with an intensity of ≥2000 were detected. The laser intensity was adjusted to obtain spectra with maximal absolute peak intensities ranging from about 5×10^3–10^4 arbitrary units. Extracts that did not reveal any peak with these settings were discarded. The spectra were evaluated using the accompanying MALDI BioTyper OC software (Version 3.1) with standard settings, but without intensity correction (Bruker Daltonics GmbH). According to the BioTyper software, a score of ≥2.0 represents an identification at the species level and values between ≥1.7 and <2 at the genus level. Measurements with a score <1.7 were counted as "not identified".

Creation of MALDI-TOF MS reference spectra

To generate MALDI-TOF main spectra (MSP), eight replicates of a fungal protein extract were spotted and each spot was measured three times. The 24 MALDI-TOF MS spectra were visually inspected and the 20 best spectra were used to calculate an MSP with the default parameter set of the BioTyper MSP Creation Standard Method (Max. mass error of each single spectrum 2000; desired mass error for the MSP 200; desired peak frequency 25 %; max. desired peak number for the MSP 70). The MSP of the 18 *Monilinia* isolates used for this study (see Table 1) were clustered based on Pearson correlation coefficients and UPGMA network creation (BioNumerics 7.5, Applied Maths, Sint-Martens-Latem, Belgium) (see also Fig. 1a).

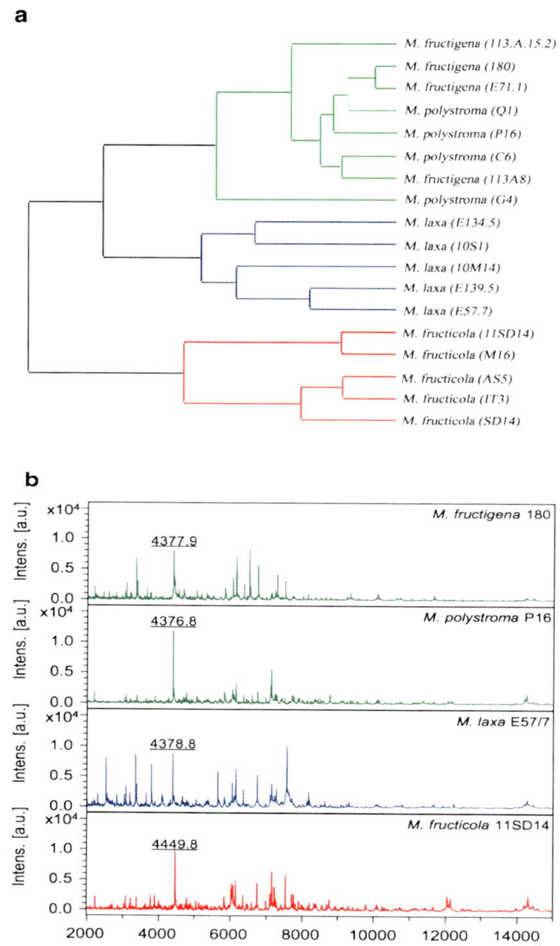

Figure 1. MALDI-TOF MS analysis of the four *Monilinia* brown rot fungi. **a** The MSP of the 18 *Monilinia* isolates (see Table 1) were clustered based on Pearson correlation coefficients and UPGMA network creation (BioNumerics

7.5, Applied Maths, Sint-Martens-Latem, Belgium). **b** Representative MALDI-TOF MS spectra for *M. fructigena* and *M. polystroma* (*green*), *M. laxa* (*blue*) and *M. fructicola* (*red*)

ZipTip preparation of protein extracts for subsequent MALDI-TOF/TOF MS analysis

For the preparation of protein extracts prior to MALDI-TOF/TOF experiments, proteins were desalted and concentrated with C18-ZipTips (Millipore, Bedford, USA). Wetting solution contained 100 % acetonitrile, equilibration solution and wash solution contained 0.1 % TFA in water, and elution solution contained 0.1 % TFA in acetonitrile–water (50:50). The original extract was spiked with 50 µl 0.1 % TFA in water, concentrated to a volume of 30 µl using a SpeedVac (Eppendorf) and subsequently applied to the ziptip sample preparation according to the manufacturer's protocol (Millipore, USA).

MALDI-TOF/TOF MS for in-depth Characterization of Distinct Mass Peaks

MALDI-TOF/TOF experiments were performed on an ultrafleXtreme mass spectrometer (Bruker Daltonics GmbH). The instrument was equipped with a modified neodymium-doped yttrium aluminium garnet (Nd:YAG) laser (smartbeam II) operating at a wavelength of 355 nm and a frequency of 1 kHz. Mass spectra of intact proteins were recorded at matrix-optimized laser fluency in positive ion reflectron mode within a *m/z* range of 400–5000, applying an acceleration voltage of 25 kV. Spectra were generated by accumulating data from approximately 10,000 laser shots. The mass spectra were externally calibrated using the peptide standard mixture II from Bruker Daltonics GmbH. Monoisotopic masses *m/z* = 4373.3 and *m/z* = 4442.0 were selected for fragmentation.

MALDI-TOF/TOF tandem mass spectrometry was performed using the LIFT technology [26] over a *m/z* range of 20–4500 in the positive ion mode. The MALDI target plate was prepared by pipetting 0.7 µl of eluted ziptip protein samples onto the target spots, which were overlaid with 0.7 µl matrix [HCCA (Bruker Daltonics GmbH), 2 mg/ml in acid (TFA) (50:49.9:0.1)]. All mass spectra were acquired and processed using flexControl (version 3.4, build 135) acetonitrile–water–trifluoroacetic and flexAnalysis (version 3.4, build 76), respectively (Bruker Daltonics GmbH).

RESULTS

MALDI-TOF MS analysis of *Monilinia* brown rot fungi cultures

In order to compare different *Monilinia* brown rot fungi by MALDI biotyping, crude formic acid/acetonitrile protein extracts of *M. fructicola* (5 isolates), *M. fructigena* (4 isolates), *M. laxa* (5 isolates) and *M. polystroma* (4 isolates) were prepared and used to generate reference main spectra (MSP). The 18 MSP were used to generate a MSP dendrogram, which revealed well-separated clusters for *M. fructicola* and *M. laxa*, and an unresolved branch containing the *M. polystroma* and *M. fructigena* spectra (Fig. 1a).

The individual MALDI-TOF MS spectra of the four *Monilinia* species revealed reproducible and distinct mass peaks (Fig. 1b). Most noteworthy was a strong signal with a molecular mass between 4370 and 4450 Da that was particularly prominent in *M. fructicola*, *M. fructigena* and *M. polystroma* (Fig. 1B). Interestingly, this mass peak appeared at approx. 4450 Da for *M. fructicola*, while for *M. fructigena* and *M. polystroma* the detected mass was approx. 80 Da lower. The MALDI-TOF MS spectrum for *M. laxa,* in contrast to the other three *Monilinia*species, featured several highly intense masses between 2 and 8 kDa. In this species, the peak at 4380 Da was therefore less remarkable.

The newly generated MSP for four *Monilinia* brown rot fungi, based on material from in vitro cultures, thus clearly delineated three groups: *M. fructicola*, *M. laxa* and *M. fructigena/M. polystroma*. Since the latter two species were not well-separated, these two species were treated as an unresolvable complex in all following analyses. MALDI biotyping experiments where a *M. fructigena* isolate was identified as *M. polystroma*, or vice versa, were thus not counted as wrong identifications (Table 2).

Table 2. Summary of all MALDI-TOF MS results generated for this study

	Species	Extracts	Correct ID (score)		Wrong ID	No ID
			>2.0	1.7–2.0		
Agar	*M. fructicola*	30	30	0	0	0
	M. laxa	22	19	1	1	1
	M. fructigena/M. polystroma	36	30	3	0	3
Apple	*M. fructicola*	25	23	0	0	2
	M. laxa	25	24	0	0	1
	M. fructigena/M. polystroma	80	75	2	0	3
Prune	*M. fructicola*	7	5	1	0	1
	M. laxa	7	3	1	1	2
	M. fructigena/M. polystroma	33	24	2	3	4
Cherry	*M. fructicola*	31	31	0	0	0
	M. laxa	4	4	0	0	0
	M. fructigena/M. polystroma	21	21	0	0	0

Monilinia protein extracts were obtained from different sources (fungus grown on PDA plates or infected apples, prunes or cherries). Samples which did not yield a mass spectrum (less than 5 %) were excluded from this table. All species were identified with a microflex LT MALDI-TOF mass spectrometer (Bruker Daltonics GmbH) using the BioTyper software (Version 3.1, Bruker Daltonics GmbH) and a custom-made main spectra (MSP) reference database

The MSP reference database consisting of the MALDI-TOF MS spectra of 18 *Monilinia* isolates was then used to biotype 88 fresh extracts obtained from in vitro cultures of *M. fructicola* (30 extracts), *M. laxa* (22 extracts) and *M. fructigena/M. polystroma* (36 extracts) (Table 2). None of the extracts was identified as another genus than *Monilinia*. In total, 4 protein extractions failed (i.e. no identification), one identification was wrong and the remaining 83 identifications were correct (4 with a score between 1.7 and 2.0, 79 with a score >2.0). This analysis of closely related *Monilinia* brown rot fungi thus confirmed that MALDI-TOF MS is capable of discriminating *M. fructicola*, *M. laxa* and *M. fructigena/M. polystroma* in vitro cultures and may thus serve as a fast, reliable and economic alternative to the DNA-based identification of these plant pathogens.

Identification of *Monilinia* Species on Infected Fruits

The MALDI-TOF MS-based identification of pathogens requires specific mass peak patterns and, preferably, protein extracts that are free from contaminations of host tissue. Consequently, in most cases an in vitro cultivation step is necessary in order to generate a pure culture from which appropriate extracts can be obtained. In order to facilitate and speed up the MALDI-TOF MS-based identification of *Monilinia* brown rot species, it was attempted to use fungal material obtained directly from infected fruit for preparing the protein extracts.

Ripe cherries, prunes and apples were infected with different *Monilinia* isolates (Table 2) and incubated until signs of colonization were apparent (1–2 weeks, depending on the isolate and fruit). All isolates were able to colonize the three types of fruit and formed mycelium on the fruit surface (Fig. 2a). The negative controls (uninfected fruits) never developed any sign of infection by *Monilinia* species. 130 extracts were prepared from infected apples, 47 extracts from infected prune and 56 extracts from infected cherries (Table 2). Overall, the MALDI-TOF MS spectra acquired directly from infected fruits resembled the spectra obtained from pure cultures of the four *Monilinia* brown rot fungi (Fig. 2b) and led to the correct identification (Table 2). Six and seven extractions of fungal material from apple or prune, respectively, failed (i.e. no identification), while all extractions of fungal material from cherry were successful. The only wrong identifications were observed for fungal extracts from prune (4 cases). 124 of the fungal extracts from apple were correctly identified (two with a score between 1.7 and 2.0, the remaining 122 with a score >2.0). Of the fungal extracts from prune, 36 lead to the correct identification (4 with a score between 1.7 and 2.0, 32 with a score >2.0), and all the 56 fungal extracts from cherry were identified with a high score (>2.0). These results confirmed that, using the extraction protocol presented here, the fruit host did not significantly alter MALDI-TOF MS spectra of the four *Monilinia* species. This approach thus established the reliable identification of *M. fructicola*, *M. laxa* and *M. fructigena*/*M. polystroma* by MALDI-TOF MS by using fungal material obtained directly from an infected host (Fig. 3).

a

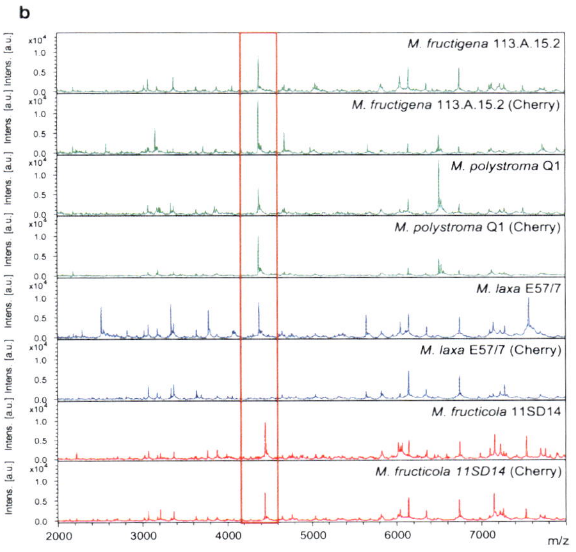

Figure 2. *Monilinia* mycelium obtained from infected cherries can be directly used for MALDI biotyping. **a** Freshly harvested, ripe cherries were infected and colonized by all four species of *Monilinia*. **b** MALDI-TOF MS spectra for *M. fructigena* and *M. polystroma*(*green*), *M. laxa* (*blue*) and *M. fructicola* (*red*). For each species, a representative spectrum obtained from an in vitro culture and from infected cherries is shown

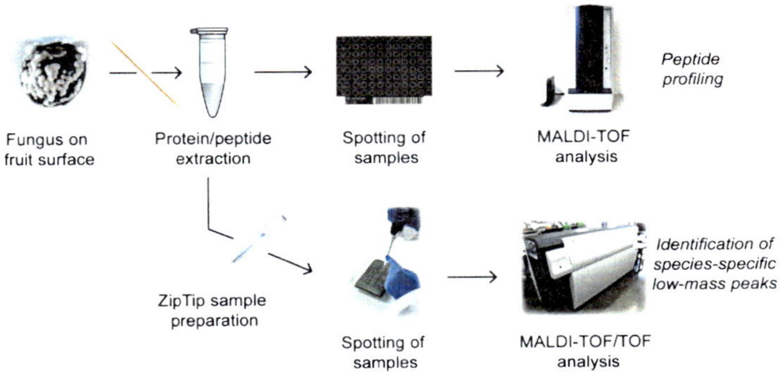

Figure 3. MALDI-TOF and MALDI-TOF/TOF MS workflows for the biotyping and analysis of *Monilinia* species. Optimization of protein extraction, own MSP generation and data analysis permitted the fast and reliable identification of *M. fructicola*, *M. laxa* and *M. fructigena*/*M. polystroma*

Characterization of a potential *Monilinia* low-mass peak biomarker by MALDI-TOF/TOF MS

In the future, closely related species that are difficult to distinguish may be set apart employing species-specific biomarker peptides. The initial MALDI-TOF MS analyses of *Monilinia* species revealed prominent, isolated molecular average masses at approx. 4370 or 4450 Da, respectively, that seemed important for the separation of *M. fructicola* from the other species and offered the basis for a prompt identification of *Monilinia* species in general. To evaluate their potential as biomarkers, we have attempted to structurally characterize these mass peaks further by applying ziptip sample cleaning-up of fungal protein extracts (*M. polystroma* and *M. fructicola*) and subsequent MALDI-TOF/TOF MS analysis (Fig. 3). MALDI-TOF MS operating in the reflectron mode, instead of the linear mode, improved the resolution and revealed monoisotopic masses of 4373.3 and 4442.0 Da for *M. polystroma* and *M. fructicola,*respectively (Fig. 4a). Further structural characterization by the LIFT MALDI-TOF/TOF method resulted in characteristic fragmentation patterns (Fig. 4b), indicating that these peptides of the two *Monilinia* species differ spectrally only slightly from each other and may thus be closely related. These experiments documented that isolated low-mass peaks characteristic of *Monilinia* brown rot fungi can be directly subjected to MALDI-TOF/TOF MS analysis, without prior separation by chromatography, which may be used for developing more sophisticated biomarker analyses in the future.

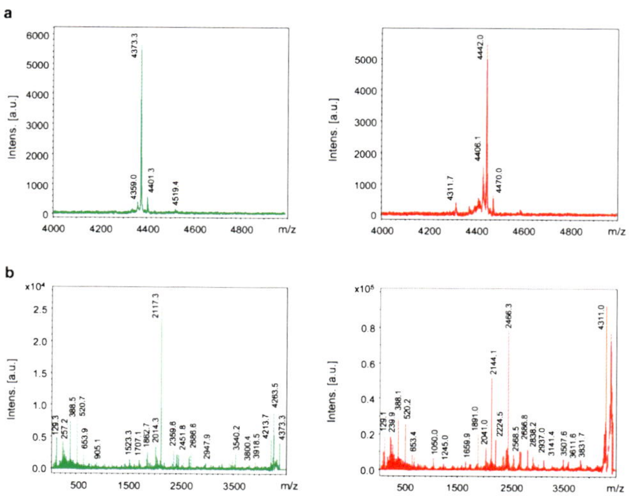

Figure 4. MALDI-TOF/TOF MS of species-specific *Monilinia* peaks. **a**

Species-specific monoisotopic masses for *M. polystroma* (*green*) and *M. fructicola* (*red*). **b** Fragment ions revealed for the masses of 4373.3 and 4442.0 Da, for *M. polystroma* (*green*) and *M. fructicola* (*red*), respectively, after LIFT fragmentation

DISCUSSION

Microbial biotyping of intact cells by mass spectrometry has revolutionized the identification of microorganisms, in particular in the clinical setting, but also holds great promise as a new tool for basic research in microbiology and ecology [27]. Comparison of a newly acquired MALDI-TOF MS spectrum with reference spectra is able to identify an unknown organism with a very short time-to-result (minutes vs. hours or days for classical phenotypic or genotypic identification). The rapid processing and measurement, as well as the negligible requirement of consumables, make MALDI biotyping the method of choice for the routine identification of microorganisms, provided access to a mass spectrometer is guaranteed. In particular in clinical applications, MALDI biotyping is thus widely used and comprehensive databases of reference spectra exist for medically relevant microorganisms. However, while the method performs reliably for many bacteria, the MALDI-TOF MS-based identification of fungi is often more difficult and requires optimization of extraction and data acquisition protocols. Another major limitation for the rapid identification of bacteria and fungi is the cultivation step, which is usually required prior to MALDI-TOF MS analysis.

Until now, MALDI-TOF MS-based identification of fungi has been mainly used in medicine [28, 29, 30, 31, 32, 33, 34], but specific protocols to identify fungal plant pathogens have been developed and are being improved continuously [35]. For example, *Monilinia* brown rot fungi have been subjected to different electrophoresis methods as well as to MALDI-TOF MS analysis [22]. Similarly, downy and powdery mildew, caused by *Bremia lactucae* and *Oidium neolycopersici*, respectively, were identified by MALDI-TOF MS in vitro as well as on infected plant leaves [36]. Furthermore, MALDI-TOF MS, in combination with a DNA-based phylogenetic analysis, was used to compare and differentiate closely related, plant pathogenic *Alternaria* species [37] as well as a collection of 45 Brazilian isolates of the genus *Clonostachys*[38].

In the work presented here, MALDI biotyping alone was able to directly identify *M. fructicola*, *M. laxa* and *M. fructigena/M. polystroma* infections on apples, cherries and prunes, without a prior cultivation in vitro. The same approach also used to identify *Monilinia* brown rot from naturally infected apricots and prunes collected in Switzerland. As for the experimentally infected apples, cherries and prunes, *Monilinia* species were correctly identified (as confirmed by PCR) solely with material obtained from fruit (e.g. without prior cultivation) (data not shown). This indicates that the MALDI-based identification of natural *Monilinia* brown rot infections is possible as long as a particular lesion is caused by one fungal species. However, natural apricot and prune infections were almost exclusively due to *M. laxa* and *M. polystroma*, respectively, which prevented assessing the MALDI-based distinction between the four *Monilinia* brown rot species. Therefore, we only collected and show quantitative data obtained from experimentally infected fruits. The results obtained by these studies are in contrast to earlier reports, where a combination of MALDI-TOF MS and electrophoresis methods was suggested for the identification of *Monilinia* species on fruit samples, because the MALDI-TOF results were affected by the host [22]. It cannot be excluded that other hosts than apple, cherry or prune would affect the MALDI biotyping results and prevent a correct identification. However, the three hosts used in this study include different fruit types that minimally affected the performance of the MALDI biotyping. We thus expect that a correct MALDI biotyping with host-derived material is possible as long as the pathogen forms aerial mycelium and conidia on the host surface. We can only speculate on the factors that allowed the reliable identification of *Monilinia* brown rot fungi on host tissue by MALDI biotyping in this study. First of all, several isolates (18 altogether) were used to generate the MSP; the reference spectra for species identification. Importantly, the extraction procedure was identically performed for the processing of the samples used for the generation of MSP, for the fungal material cultivated in vitro, as well as for the fungal material from infected hosts. The extraction protocol itself has also been optimized by reducing the amount of fungal material that was extracted, the extraction of conidia and mycelium, and the addition of glass beads during the formic acid/acetonitrile extraction step. Finally, the parameters for the actual MALDI-TOF MS measurement, data processing and data analysis were adjusted (laser intensity, peak exclusion criterion, number of shots per spot, intensity correction for identification). In combination, this optimization led to a time-saving procedure and the reliable identification of *M. fructicola*, *M. laxa* and *M. fructigena/M. polystroma*. The fact that *M. fructigena* and *M. polystroma* could not be distinguished by the workflow described in this study underlines the close relatedness of these two species, which are set apart only by five base transitions in their ITS region [6]. It is thus laborious to distinguish these two species, even by molecular methods [7, 39]. The low intensity of most signals for *M. polystroma*, as well as the intense peak with a mass of 4377 Da that was shared with *M. fructigena* (4378 Da), may explain the difficulty in separating these two species by MALDI biotyping (Fig. 1a).

By concentrating and purifying fungal extracts and subsequent MALDI-TOF/TOF MS analysis we have shown that prominent peptide peaks can be structurally characterized without preceding liquid chromatography. This workflow may generally serve as a suitable tool to identify species-specific low-mass peaks of fungal species and therefore contribute to the rapid and efficient identification of species that cannot be differentiated easily so far. The full identification of these peptides will not only allow identifying their biological function, but also to establish a rapid identification of fungi by screening for specific masses in selected reaction monitoring (SRM) experiments.

CONCLUSION

The work described here established a fast and reliable approach for the MALDI-TOF-based identification of four *Monilinia* brown rot fungi directly on infected fruits. This method not only represents an improvement of currently existing diagnostic protocols, but may also facilitate and enable basic research and field studies on the ecology of these important plant pathogens.

AUTHORS' CONTRIBUTIONS

FMF and MHB performed the lab work and all MALDI biotyping experiments. DD and RB performed MS/MS experiments, analysed the MS/MS data and wrote the corresponding sections of the manuscript. FMF and DD planned the study and FMF wrote the manuscript. All authors read and approved the final manuscript.

ACKNOWLEDGEMENTS

We thank Daniel Prata for his help with experimental work. Maria-Theresia Gekenidis and Stefanie Pfrunder are acknowledged for their support with MALDI-TOF MS measurements and data analysis. We recognize Eduard Holliger for his helpful and constructive feedback on the manuscript.

REFERENCES

1. Hrustic J, Mihajlovic M, Grahovac M, Delibasic G, Bulajic A, Krstic B, Tanovic B. Genus *Monilinia* on pome and stone fruit species. Pestic Phytomed (Belgrade). 2012;27(4):283–97.
2. Rungjindamai N, Jeffries P, Xu X-M. Epidemiology and management of brown rot on stone fruit caused by *Monilinia laxa*. Eur J Plant Pathol. 2014;140(1):1–17.
3. Mari M, Martini C, Guidarelli M, Neri F. Postharvest biocontrol of *Monilinia laxa*, *Monilinia fructicola* and *Monilinia fructigena* on stone fruit by two *Aureobasidium pullulans* strains. Biol Control. 2012;60(2):132–40.
4. van Leeuwen GCM, Stein A, Holb IJ, Jeger MJ. Yield loss in apple caused by *Monilinia fructigena* (Aderh. & Ruhl.) Honey, and spatio-temporal dynamics of disease development. Eur J Plant Pathol. 2000;106(6):519–28.
5. van Leeuwen GCM, Baayen RP, Holb IJ, Jeger MJ. Distinction of the Asiatic brown rot fungus *Monilinia polystroma* sp. nov. from *M. fructigena*. Mycol Res. 2002;106(04):444–51.
6. Fulton CE, van Leeuwen GCM, Brown AE. Genetic variation among and within *Monilinia*specis causing brown rot of stone and pome fruits. Eur J Plant Pathol. 1999;105(5):495–500.
7. Hilber-Bodmer M, Knorst V, Smits THM, Patocchi A. First report of asian brown rot caused by *Monilia polystroma* on apricot in Switzerland. Plant Dis. 2012;96(1):146.
8. Hong C, Holtz BA, Morgan DP, Michailides TJ. Significance of thinned fruit as a source of the secondary inoculum of *Monilinia fructicola* in California nectarine orchards. Plant Dis. 1997;81(5):519–24.
9. De Cal A, Eguen B, Melgarejo P. Vegetative compatibility groups and sexual reproduction among Spanish *Monilinia fructicola* isolates obtained from peach and nectarine orchards, but not *Monilinia laxa*. Fungal Biol. 2014;118(5–6):484–94.
10. Chen Y, Zhou MG. Sexual recombination of carbendazim resistance in *Fusarium graminearum* under field conditions. Pest Manag Sci. 2009;65(4):398–403.
11. Jänsch M, Frey JE, Hilber-Bodmer M, Broggini GAL, Weger J, Schnabel G, Patocchi A. SSR marker analysis of *Monilinia fructicola* from Swiss

apricots suggests introduction of the pathogen from neighbouring countries and the United States. Plant Pathol. 2012;61(2):247–54.

12. Lane CR. A synoptic key for differentiation of *Monilinia fructicola*, *M. fructigena* and *M. laxa*, based on examination of cultural charcters. EPPO Bull. 2002;32(3):489–93.
13. OEPP/EPPO. *Monilinia fructicola*. EPPO Bull. 2009;39(3):337–43.
14. EFSA Panel on Plant Health (PLH). Pest risk assessment of *Monilinia fructicola* for the EU territory and identification and evaluation of risk management options. EFSA J. 2011;9(4):2119.
15. Hilber-Bodmer M, Bünter M, Patocchi A. First report of asian brown rot caused by *Monilia fructicola* on apricot in a Swiss orchard. Plant Dis. 2010;94(5):643.
16. EPPO A2. List of pests recommended for regulation as quarantine pests [http://www.eppo.int/QUARANTINE/listA2.htm].
17. Côté M-J, Tardif M-C, Meldrum AJ. Identification of *Monilinia fructigena*, *M. fructicola*, *M. laxa*, and *Monilia polystroma* on inoculated and naturally infected fruit using multiplex PCR. J Plant Dis. 2004;88(11):1219–25.
18. Anhalt JP, Fenselau C. Identification of bacteria using mass spectrometry. Analytical Chem. 1975;47(2):219–25.
19. Fenselau C, Demirev PA. Characterization of intact microorganisms by MALDI mass spectrometry. Mass Spectrom Rev. 2001;20(4):157–71.
20. Sauer S, Kliem M. Mass spectrometry tools for the classification and identification of bacteria. Nat Rev Microbiol. 2010;8(1):74–82.
21. Evason DJ, Claydon MA, Gordon DB. Exploring the limits of bacterial identification by intact cell-mass spectrometry. J Am Soc Mass Spectrom. 2001;12(1):49–54.
22. Horka M, Kubesova A, Salplachta J, Zapletalova E, Horky J, Slais K. Capillary and gel electromigration techniques and MALDI-TOF MS—suitable tools for identification of filamentous fungi. Anal Chim Acta. 2012;716:155–62.
23. Kawasaki ES. Sample preparation from blood, cells, and other fluids. In: Innis MA, Gelfand DH, Sninsky JJ, White TJ, editors. PCR Protocols: a guide to methods and applications. San Diego: Academic Press; 1990. p. 146–52.
24. Gardes M, Bruns TD. ITS primers with enhanced specificity for

basidiomycetes–application to the identification of mycorrhizae and rusts. Mol Ecol. 1993;2(2):113–8

25. White TJ, Bruns TD, Lees S, Taylor JW. Amplification and direct sequencing of fungal ribosomal RNA genes for phylogenetics. In: Innis MA, Gelfand DH, Sninsky JJ, White TJ, editors. PCR Protocols: a guide to methods and applications. San Diego: Academic Press; 1990. p. 315–22.
26. Neubert H, Halket JM. Fernandez Ocana M, Patel RK: MALDI post-source decay and LIFT-TOF/TOF investigation of alpha-cyano-4-hydroxycinnamic acid cluster interferences. J Am Soc Mass Spectrom. 2004;15(3):336–43
27. Ferreira L, Sanchez-Juanes F, Garcia-Fraile P, Rivas R, Mateos PF, Martinez-Molina E, Gonzalez-Buitrago JM, Velazquez E. MALDI-TOF mass spectrometry is a fast and reliable platform for identification and ecological studies of species from family *Rhizobiaceae*. PLoS One. 2011;6(5):e20223.
28. De Carolis E, Posteraro B, Lass-Florl C, Vella A, Florio AR, Torelli R, Girmenia C, Colozza C, Tortorano AM, Sanguinetti M, et al. Species identification of *Aspergillus*, *Fusarium* and *Mucorales* with direct surface analysis by matrix-assisted laser desorption ionization time-of-flight mass spectrometry. Clin Microbiol Infect. 2012;18(5):475–84.
29. Nenoff P, Erhard M, Simon JC, Muylowa GK, Herrmann J, Rataj W, Graser Y. MALDI-TOF mass spectrometry - a rapid method for the identification of dermatophyte species. Med Mycol. 2013;51(1):17–24.
30. Ling H, Yuan Z, Shen J, Wang Z, Xu Y. Accuracy of matrix-assisted laser desorption ionization-time of flight mass spectrometry for identification of clinical pathogenic fungi: a meta-analysis. J Clin Microbiol. 2014;52(7):2573–82.
31. De Carolis E, Vella A, Vaccaro L, Torelli R, Posteraro P, Ricciardi W, Sanguinetti M, Posteraro B. Development and validation of an in-house database for matrix-assisted laser desorption ionization-time of flight mass spectrometry-based yeast identification using a fast protein extraction procedure. J Clin Microbiol. 2014;52(5):1453–8.
32. Posteraro B, De Carolis E, Vella A, Sanguinetti M. MALDI-TOF mass spectrometry in the clinical mycology laboratory: identification of fungi and beyond. Expert Rev Proteomics. 2013;10(2):151–64.
33. Bader O. MALDI-TOF-MS-based species identification and typing

approaches in medical mycology. Proteomics. 2013;13(5):788–99.

34. Del Chierico F, Masotti A, Onori M, Fiscarelli E, Mancinelli L, Ricciotti G, Alghisi F, Dimiziani L, Manetti C, Urbani A, et al. MALDI-TOF MS proteomic phenotyping of filamentous and other fungi from clinical origin. J Proteomics. 2012;75(11):3314–30.

35. Chalupova J, Raus M, Sedlarova M, Sebela M. Identification of fungal microorganisms by MALDI-TOF mass spectrometry. Biotechnol Adv. 2014;32(1):230–41.

36. Chalupova J, Sedlarova M, Helmel M, Rehulka P, Marchetti-Deschmann M, Allmaier G, Sebela M. MALDI-based intact spore mass spectrometry of downy and powdery mildews. J Mass Spectrom. 2012;47(8):978–86.

37. Brun S, Madrid H, Van Gerrits Den Ende B, Andersen B, Marinach-Patrice C, De Hoog GS. Multilocus phylogeny and MALDI-TOF analysis of the plant pathogenic species *Alternaria dauci* and relatives. Fungal Biol. 2013;117(1):32–40.

38. Abreu LM, Moreira GM, Ferreira D, Rodrigues-Filho E, Pfenning LH. Diversity of *Clonostachys* species assessed by molecular phylogenetics and MALDI-TOF mass spectrometry. Fungal Biol. 2014;118(12):1004–12.

39. Chalkley D: Invasive fungi. Asiatic brown fruit rot - *Monilinia polystroma*. In.: Systematic Mycology and Microbiology Laboratory, ARS, USDA.

CHAPTER 14

DIFFERENTIATION BETWEEN TWO STRAINS OF MICROALGA PARACHLORELLA KESSLERI USING MODERN SPECTRO-SCOPIC METHOD

Marwa M Saleh[1,3], Dmitry N Matorin[2], Bolatkhan K Zayadan[1], Daria A Todorenko[2], Evgenii P Lukashov[2] and Mona M Gaballah

[1] Department of Biotechnology, Al-Farabi Kazakh National University, 71, Al-Farabi, ave, Almaty 050040, Kazakhstan.
[2] Faculty of Biology, Moscow State University, Vorobyevi Gory 119992, Moscow, Russia.
[3] Botany Department, Faculty of Science, Suez Canal University, Ismaillia 41522, Egypt.

CITATION: Saleh, M. M., et al., "Differentiation between two strains of microalga Parachlorella kessleri using modern spectroscopic method". Botanical Studies (2014), https://doi.org/10.1186/s40529-014-0053-7

ABSTRACT

Background

The differentiation between wild type of Parachlorella kessleri and its mutant strains PC Mut2, PC Mut4 by using the Multi-functional Plant Efficiency Analyzer (M-PEA-2) was studied. Mutant algal cells of P. kessleri have been obtained by UV-C during 3 and 10 min respectively.

Results

Light-induced kinetics of prompt fluorescence (OJIP transients), delayed fluorescence and modulated reflection at 820 nm (redox transitions of P700 in PSI) showed disturbance of electron transport flow in photosystem II (PSII) and an increase fraction of non-reducing centers of secondary quinone acceptors of electron (QB). In addition, the amplitudes of the fast and slow peak in the kinetics of the delayed light emission and non-photochemical fluorescence quenching (NPQ) were significantly reduced in mutant cells, indicating low level of the membrane energization of photosynthetic membranes. Changes of photosynthetic reactions of mutants may lead to an increase of the carotenoids content, which protect cells against the light stress.

Conclusion

It is suggested to use parameters of induction curves of prompt and delayed fluorescence to characterize mutant algal cells in biotechnological studies.

Keywords: Parachlorella kessleri; Chlorophyll a fluorescence; Delayed fluorescence; JIP-test; Photosynthesis

BACKGROUND

The solar radiation sustains life on Earth. However significant increase of UV-C radiation may lead to inhibition of many biological processes. The major cellular targets of UV-C are various biomolecules, which directly absorb this radiation or are indirectly affected by various UV-C induced photochemical reactions. The biological and, ultimately, ecological consequences are numerous (Charles 2008). It is known that UV-C induces some physiological effects such as declining photosynthetic rates, which can be related not only to damaged biomolecules, but also to ultrastructural changes in organelles or membranes (Holzinger and Lütz 2006).

Typical alterations include swollen mitochondrial cristae, disrupted thylakoids or detached phycobilisomes in chloroplasts, bent-shaped dictyosomes, and damaged plasmalemma. An intact ultrastructure of the algal cell is a prerequisite for the optimum functioning of all physiological processes (Charles 2008). Parachlorella kessleri was described as a new genus, derived from the true spherical Chlorella. The wild strain SAG 211-11 h was formerly referred to as Chlorella vulgaris but later reassigned as Chlorella kessleri (Fott and Nováková 1969). Today this species is referred to as P. kessleri (Krienitz et al. 2004). Fluorescence methods are used in biotechnological experiments for monitoring photosynthesis processes, which give detailed information of primary defects of cell metabolism, mainly at the membrane level (Schreiber 2004; Antal et al. 2009; Matorin et al. 2013; Schansker et al. 2014).

The basis of fluorescence methods lies in the ability of chlorophyll located in photosynthesis membranes to serve as a natural detector for the algae cells, emitting quanta fluorescence. Importantly, such methods allow to gain information of alga state in real time. Measurement of the ratio between the fluorescence intensity under the photosynthesis saturating illumination (FM) and under low light intensity, which induces no changes in the state of the photosynthetic apparatus (FO) makes it possible to determine the maximum efficiency of the PSII processes, which is equal to FV/FM= (FM - FO)/FM. The FV/FM ratio presents dimensionless characteristics of the efficiency of photosynthesis, which is independent of the species of organisms. Measurements of fluorescence induction curves are carried out during several seconds by using PAM or PEA instrument.

Measurements of fluorescence induction curves with high resolution (starting from 20 μs) have been recently used in studies of photosynthetic reactions in higher plants and algae cultures (Schreiber 2004; Strasser et al. 2005; Matorin and Rubin 2012). The M-PEA-2 instrument allows to measure separate reactions in PSI and PSII simultaneously (Strasser et al. 2010; Oukarroum et al. 2013; Bulychev et al. 2013). Moreover, it monitors induction changes of delayed fluorescence, which indicate the level of the membrane energization (Goltsev et al. 2009). In the present paper we investigated processes in PSI and PSII of wild type of P. kessleri and two mutant strains by using a newly designed Multi-functional Plant Efficiency Analyzer (M-PEA-2, Hansatech).

This instrument gives a comprehensive picture of primary photosynthetic events by simultaneous recording the prompt and delayed fluorescence kinetics and reflectance at 820 nm at high time resolution. Here we introduce results of the first attempt of assessment of photosynthetic characteristics of the mutant strains of algal cells by using M-PEA-2.

METHODS

Strains and Growth Conditions

The strain of Parachlorella kessleri microalga was obtained from Al-Farabi Kazakh National University, Biotechnology Department culture collection. P. kessleri wild type strain was grown in tris-acetate-phosphate (TAP) medium (pH 6^{-7}) (Gorman 1965), in 250 ml Erlenmeyer flask at 28°C under continuous illumination of photosynthetic photon flux density (PPFD) 120 µmol photons m^{-2} s^{-1} and constant shaking

UV Irradiation and Mutagenesis

According to Harris (1989), 5 ml of the liquid culture with a density of 1×106 /ml algal cells were placed in 9 cm Petri dish, where the culture formed a thin layer, covering the bottom. The dish was placed on shaker at 20 rpm and exposed to UV-C lamp (254 nm and 40 erg/mm^2) at distance 15 cm during 3 and 10 min respectively. After UV irradiation, the cells were inoculated in solid TAP medium and incubated in darkness for 24 h to prevent photoreactivation. After the 24 h, some dishes were incubated photoautotrophically (photon flux density 120 µmol photons m^{-2} s^{-1}), others were incubated heterotrophically (in dark) during 15 days.

Selection of the mutant strains After the incubation period, two mutant strains were selected based on their phenotypic characteristics, using colonies whose color and size differed from the wild type colonies. These mutant strains, PC Mut2 and PC Mut4, were obtained from P. kessleri after irradiation time of 3 and 10 min respectively. These mutants were transferred from the solid to liquid medium and kept under phototrophic growth conditions (photon flux density 120 µmol photons m^{-2} s^{-1}).

Analysis of Chlorophyll a and b Content

Spectrophotometry method was used according to Merchant et al. (2007). The calculation of the concentration of the pigments was determined by the optical density of pigment solutions at appropriate wavelength.

Chlorophyll Fluorescence and Absorbance Measurements

Light-induced kinetics of prompt Chl fluorescence (PF), delayed Chl fluorescence (DF) and modulated reflectance at 820 nm (MR) were recorded by using Multi-functional Plant Efficiency Analyzer (M-PEA-2, Hansatech Instrument Ltd., King's Lynn, Norfolk, UK). Prior to the measurements, the cells were concentrated on a membrane filter. Then a thin layer of cells was fixed by a special clip on the surface of the measuring chamber and incubated in the darkness for 10 min. It has previously been verified in algal suspension by using Aqua-Pen fluorometer (Photon System Instruments, Czech Republic) that concentration algal cells on the filter surface doesn't affect the physiological state of cells (Matorin et al. 2013).

The design of the measuring chamber, timing protocol of a single measurement, and data acquisition procedure for M-PEA-2 can be found in Strasser et al. (2010). PPFD and duration of actinic illumination were set to 1300 μmol photons m^{-2} s^{-1} and 3 s, respectively. In the M-PEA-2, the DF induction curves are constructed from the DF intensities at certain delay time-point in the kinetics of DF decay recorded during the dark interruptions of the actinic light, which induces OJIP (intermediate levels of the light-induced fluorescence curve) rise. In our experiments, the delay time point in the DF decay curve was set to 50 μs. It is assumed that submillisecond DF components are generated from PSII reaction centers in the state $Z^+ P_{680}Q_A^-$ (Goltsev et al. 2009), where Z is a tyrosine Z residue, P680 is a primary electron donor, and QA is a primary quinone acceptor in PSII.

Fluorescence was measured by using Water-PAM fluorometer (Heiz Walz, Effeltrich, Germany). Ratio FV/FM is the maximum quantum yield of PSII photochemistry was measured in dark-adapted sample, where FV = FM-FO. Measurements in the light were performed with progressive increase of light intensity from 0 to 800 μmol photons m^{-2} s^{-1}. At the end of each illumination session, F_M' and fluorescence yield in light F(t) were measured using saturating flash (0.8 s, 3000 μmol photons m^{-2} s^{-1}) (Schreiber 2004; Herlory et al. 2007). These parameters were used to calculate non-photochemical quenching of chlorophyll fluorescence NPQ = (FM - F_M')/F_M', the quantum yield of photochemical conversion of light energy in PSII, as Y = (F_M'- F(t))/F_M', and the relative yield of noncyclic electron transport at given light intensity rETR = Y × E_i × 0.5, where E_i is illumination intensity (μmol photons m^{-2} s^{-1}) (Lippemeier et al. 1999). Parameters are designated according to generally accepted nomenclature (Schreiber 2004).

Light absorption spectrum of samples was measured by using HITACHI-557 spectrophotometer. Fluorescence spectra, excitation spectra, and fluorescence decay times in algae were measured by using Fluorolog-3 (Horiba Jobin Yvon) spectrofluorometer with the TCSPC (time correlated single photon counting) option (O'Connor and Philips 1984). Fluorescence was measured at 685 nm under excitation with pulse LED at 390 nm. Fluorescence decay kinetics was approximated with the sum of two exponential curves. All measurements were repeated no less than five times. The figures present data of three replicate experiments at least.

RESULTS AND DISCUSSION

In this study mutations were induced by UV-C radiation. UV-C mutagenesis offers many advantages such as less pollution, simple operation, and sterile cultivation conditions. After UV-C irradiation, some colonies appeared with phenotypic characteristics that differed in color and size from the control colonies. The colonies were selected based on their phenotypic appearance, transferred into liquid medium and then subcloned several times to insure that the mutation was stable and no changes in their phenotypic characteristics occurred.

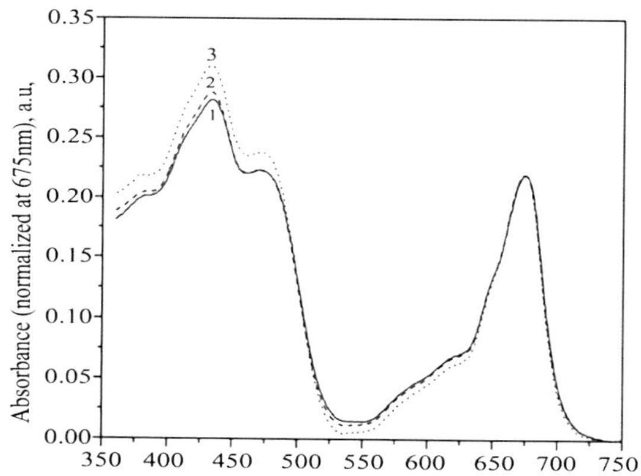

Figure 1. Absorption spectra of suspensions. wildtype (curve 1) and mutant strains PCMut2 (curve 2) and PCMut4 (curve 4). The spectra are normalized to absorption in the red maximum of chlorophylla a at 675 nm.

The major cellular targets of UV-C are different biomolecules, which directly absorb this radiation, or which are indirectly affected by various UV-C induced photochemical reactions (Charles 2008). In the present paper two mutant strains of P. kessleri after irradiation during 3 and 10 min respectively were selected and designated as PC Mut2 and PC Mut4. It was shown microscopically that mutant cells are smaller, than wild type cells. All strains were grown in the liquid TAP medium under phototrophic conditions.

The algal cultures of the wild and mutant strains were different in color. Absorption spectra of wild type and mutant strains of cells are in Figure 1. The spectra have been normalized to the maximum chlorophyll absorption in red region at 675 nm. Relative increase of carotenoid absorption at the Soret band was noted in mutant strains of P. kessleri. The highest increase was observed in PC Mut4. This was confirmed by measurements of fluorescence excitation spectra (Figure 2B) The se spectra showed an increase in the region of carotenoid absorption. Fluorescence spectra did not differ significantly (Figure 2A), which may be related to the fact that carotenoids in algae do not fluorescence, but efficiently transfer excitation energy of chlorophyll a, which is fluorescence emitter.

It is known that the constant fluorescence level F_O highly correlates with the total content of pigments in the photosynthetic apparatus of algal cells, that harvest light energy, and F_O level correlates with cell number (Matorin et al. 2004). The ratio F_V/F_M reflects the maximum quantum yield in PSII. This value is related to the processes of water splitting and oxygen evolution in PSII. Changes of fluorescence parameters (F_O, F_V/F_M) are shown in Figure 3

Figure 3 shows fast increase of algal cells of wild type (curve 1) after the lag-phase (1st day), as measured by F_O. PC Mut2 (curve 2) and PC Mut4 (curve 3) mutants demonstrate a lower growth rate during investigation period. Photosynthetic activity (F_V/F_M ratio) of wild type cells slightly increased during growth, reaching the maximum value at the 3rd day. After prolonged culturing (6th day) photosynthetic activity, judged by FV/FM ratio, was decreased, possibly due to exhaustion of mineral nutrients in the growth medium. Mutants PC Mut2 and PC Mut4 had a lower FV/FM values.

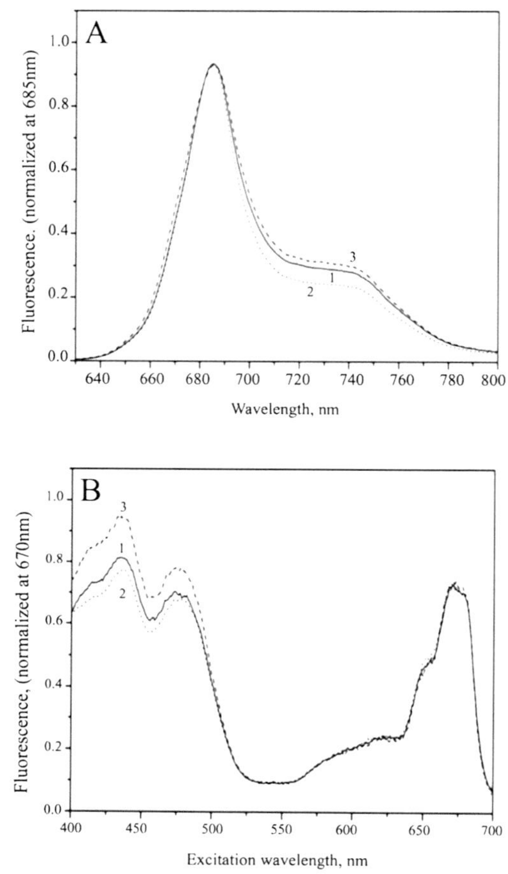

Figure 2. Fluorescence spectra. (A) under excitation at 430 nm (B) at 720 nm in wildtype (curve 1) and mutant strains Parachlorella kessleri cells (PCMut2-curve 2) and (PCMut4 -curve 3). Fluorescence spectra are normalized to the maximum fluorescence values at 690 nm. Fluorescence excitation spectra are normalized to corresponding values of the red absorption maximum at 680 nm.

For more detailed investigation of photosynthetic activity of wild type and two mutant strains P. kessleri cells, fluorescence parameters were measured by using M-PEA2 (Figure 4). M-PEA-2 provides extended possibilities to evaluate the functional state of PSII and PSI by measuring light-induced kinetics of the PF (OJIP transients), DF, and MR with a high resolution (Strasser et al. 2010; Oukarroum et al. 2013). Although there is no full consensus on the origin of the complex OJIP rise kinetics, it is commonly accepted that the underlying mechanisms may involve successive decrease in photochemical and non-photochemical fluorescence quenching due to the light-induced PSII closure and PQs reduction (Antal et al. 2009; Strasser et al. 2010). PSII closure refers to reduction of the primary quinone acceptor, Q_A, in PSII.

Figure 4 presents kinetic curves of fluorescence induction after switching the light on, normalized against the O level. In the control cells, the shape of the fluorescence curve corresponded to that described in the literature (Lazár 2009; Strasser et al. 2010). Usually, several components are observed in the kinetics of fluorescence induction in response to intense illumination, namely, the O-J-I-P transitions. The initial level O corresponds to chlorophyll fluorescence intensity with open RCs of PSII (FO) when all QA are oxidized. The time period required to reach this level is up to 20 µs. The O-J phase is caused by light induced reduction of QA, while the subsequent phases reflect mainly its further accumulation in reduced state, caused by a decrease in its reoxidation rate as a result of reduction of QB acceptors and the pool of quinone.

The shape of the O-J-I-P curve was changed in the mutants (Figure 4). The mutant PC Mut4 showed marked decrease in the contribution of the photochemical phase J-I-P, which indicates interruption of the electron flow from PSII to plastoquinone pool. To quantitatively analyze the characteristics of the primary processes of photosynthesis on the basis of O-J-I-P kinetic curve parameters, the so called JIP test (Strasser et al. 2010) was used. The JIP test uses the following parameters of the fluorescence induction kinetic curve: (a) fluorescence intensity at 20 µs (F_O), 300 µs (F300 µs), 2 ms (F_J), 30 ms (F_I), 6 s (F6s), and FP (F_M, the maximum fluorescence yield); (b) time to reach the maximum fluorescence (t F_M), and (c) area under the kinetic curve below the FM level.

These characteristics were used to calculate the following parameters presented in (Table 1): (1) maximum efficiency of PSII ($F_V/F_M = (F_M - F_O)/F_M$); (2) relative amplitude of the O–J phase ($V_J = (F_J - F_O)/F_V$, which reflects the fraction of non QB-reducing RC of PSII, which lack the contact between the two consecutive PSII acceptors, QA and QB; (3) relative amplitude of the J–I phase ($VI = (F_I - F_J)/F_V$): (4) M_O parameter ($M_O = 4\times(F300\mu s - F_O)/F_V$), which reflects the initial slope of the induction curve; M_O value is proportional to the rate of Q_A reduction under conditions when Q_B and the pool of plastoquinones are mainly in the oxidized state; (5) S_M = (Area)/F_V, normalized value of F_V, the area between the OJIP curve and F_M value, which reflects the total number of PSII turnover during the OJIP phase of growth of the fluorescence yield; (6) ABS/RC = M_O ($1/F_M$) ($1/V_J$), average value of absorbed photon streams in RC of PSII (observed size of the active PSII antenna); (7) capacity for pH induced nonphotochemical fluorescence quenching (q_E = (F_M – F6s)/ F_V); and (8), capability of the quinone pool to quench fluorescence qPQ = (F_M – FI)/FV. Many parameters of the JIP-test are mutually dependent; that is, change in some of them leads to changes in the others. Independent parameters (V_J and M_O) derived from the analysis of the O–J phase provide the information on Q_A reduction.

From the VI and SM parameters, it is possible to derive information on further accumulation of reduced which occurs due to reduction of QB and the quinone pool.

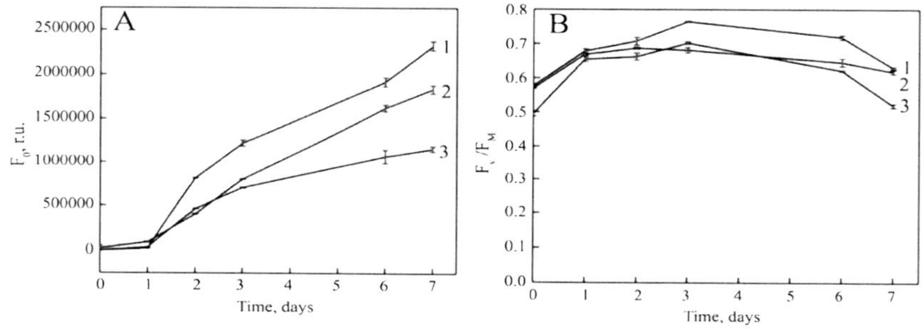

Figure 3. Changes in fluorescence parameters. (A) F_O (B) F_V/F_M in suspension wildtype (curve 1) and mutant strains of Parachlorella kessleri cells PCMut2 (curve 2) and PCMut4 (curve 3) during growth. The graphs are plotted as means ± SEM.

It is assumed that the O–J phase of fluorescence induction reflects the accumulation of both reducing and non-reducing PSII (Antal et al. 2009; Strasser et al. 2010). The hypothesis is supported by the fact that DCMU (3-(3,4-dichlorophenyl)-1,1-dimethylurea), which inhibits electron transport between Q_A and QB, leads to fast growth of fluorescence to the maximum level within 2 ms, corresponding to the appearance of the J peak in the control. Analysis demonstrated that, in cells of the strains obtained, the number of non-Q_B-reducing centers of PSII, which are incapable of the quinone pool reduction, was increased. This was supported by a number of other parameters (Table 1). Accordingly, values of the parameter M_O, which reflects the initial slope of the induction curve, increased. Total flow of photons, absorbed by the pigments of the PSII antenna and normalized against the RC number (ABS/RC), increased in the cultures of mutant strains

P. kessleri, as compared to the control values. At the same time, the relative amplitude of the J–I phase (V_I) was practically unchanged.

Prompt fluorescence induction curves also demonstrated the suppression of fluorescence decay after reaching the maximum due to ΔpH dependent nonphotochemical quenching (qE = (F_M – F6s)/F_V)). The value of this parameter decreased in mutant strains, which evidences a decrease in the membrane energization. The ability of the quinone pool to quench fluorescence qPQ = (F_M – F_I)/F_V in algae of mutant strains was also reduced.

Figure 5 shows the kinetics of fluorescence decay in the wild type and mutant strains of P. kessleri cells. Biexponential fitting of decay kinetics was applied to distinguish two components, one of them is related to energy trapping (fast component) and the other to charge stabilization and recombination in PSII reaction centers (slow ccomponent) (Volgusheva et al. 2007.).

In our experiments we found a fast component with half-time about 1 ns and a slower component with halftime more than 2 ns (Table 2). In the wild type and PC Mut2, the ratio of these components was 40 and 60%, respectively. In PC Mut4, the half-time of the slow component increased from 2.2 to 2.5 ns, and its relative magnitude also increased to 72%.

Slower fluorescence decay and an increase in the contribution of the slow component, generally, reflect a higher reduction of the PSII QA acceptor. It is clearly seen in the presence of DCMU, an inhibitor of electron transport. The fact of slower fluorescence decay in PC Mut4 agrees well with the increase in the concentration of non-QBreducing centers of PSII in cells of this mutant strain. Reflectance changes at 820 nm (MR) are assigned to the redox transitions of P_{700}, a primary electron donor in PSI, and plastocyanin, an electron donor for P_{700}^+. For the simplicity, further in the text we refer MR signal to P700. In the dark adapted state, P_{700} is in the neutral state and the reflectance at 820 nm is high. After the onset of illumination MR value rapidly declines, reaching a dip at about 10 ms, followed by the slow recovery to the initial level. These changes reflect the rapid oxidation of P_{700} by the acceptor side of PSI, and the subsequent reduction of P_{700}^+ by electrons arriving from PSII (Figure 4B) shows MR transients normalized to the initial value at 20 μs (MR/MR_0) recorded simultaneously with the OJIP curves described above.

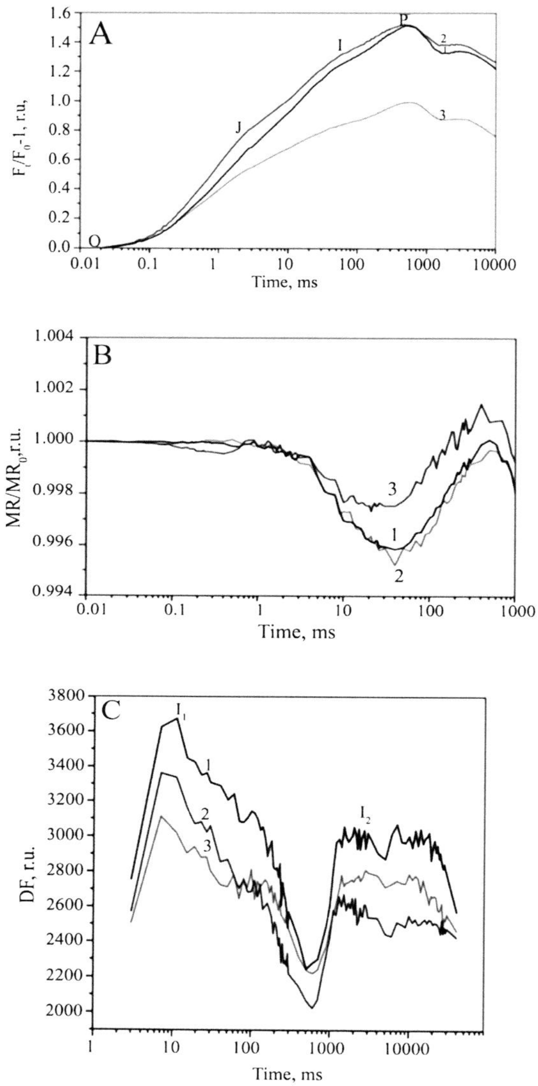

Figure 4. Light-induced kinetics. (A) of PF (OJIP), (B) MR, (C) DF. After turning on the light in the culture of wildtype and mutant strains of Parachlorella kessleri cells. Light intensity was 1300 μE/(m2 s). Kinetic curves were simultaneously recorded with M-PEA instrument. OJIP and MR curves are normalized to the initial values at 20 ns and represented as F(t)/FM – 1 and MR/MR0 ratios, respectively. Each curve represents an average of 3 replicate experiments.

In PC Mut4 mutant strain, P700 oxidation was reduced, indicating an inhibition of electron transport via PSI. Both DF and PF emission originates from radiative deactivation of the singlet excited state of antennae Chl in PSII (Goltsev et al. 2009). In the case of PF, excited states are created mainly through the absorption of photons by photosynthetic pigments, whereas DF emission is coupled to excited states formed as a result of the backward reactions of electron transport. In order to separate a low yield DF emission from high yield PF signal, DF is recorded in the dark after the excitation pulse offset. DF intensity is proportional to the rates of the corresponding recombination reactions which are strongly affected by the energization of a thylakoid membrane (Wraight and Crofts 1971). Thus, changes of DF yield during the induction period depend on the formation of electrical and proton differences across the membrane. Fast and slow phases of the DF induction kinetics recorded with M-PEA in plant leaves have recently been analyzed (Goltsev et al. 2009; Oukarroum et al. 2013). It is assumed that the fast component (I1) (~10-100 ms) reflects transient formation of the electrical gradient across the membrane, likely, due to generation of charge pairs in PSI (P700 + [Fe-S]−). The presence of the second DF peak, I2, in the seconds range, is associated with the light induced formation of the proton transmembrane gradient (ΔpH) that increases the constant of rate of emission transitions in the RC of PSII.

Peaks on DF curve at 20–50 ms, and 1 s were reduced in mutant strains, indicating a decrease in the formation of electrical and proton potential differences across the thylakoid membrane. However, the fast component of DF was reduced to a greater extent in PC Mut4 strain. The slow phase of the DF induction kinetics (ΔpH) was reduced more significantly in PC Mut2, compared to the wild type. The last fact agrees with the reduced non-photochemical quenching $qE = (F_M - F6s)/FV$ in PC Mut2 (Table 1).

These data were confirmed by measurement of fluorescence parameters in algae with a Water-PAM pulse fluorometer. The relative rate of the noncyclic electron transport (ETR), calculated as $rETR = Y \times Ei \times 0.5$, and nonphotochemical fluorescence quenching $NPQ = (F_M - F_M')/ FM'$ by electrochemical gradient of protons were reduced in mutants (Table 3). Thus, simultaneous registration of the kinetics of prompt, delayed fluorescence and redox state of P_{700} made it possible to monitor the individual reactions of accumulation of reduced carriers between the photosystems and kinetics of electrochemical proton gradient on a thylakoid membrane of mutant strains P. kessleri.

It is known that the constant fluorescence level FO highly correlates with the total content of pigments in the photosynthetic apparatus of algal cells that harvest light energy (Matorin et al. 2004). The chlorophyll a and b content in the wild type were 53.3 ± 5.7, 30.2 ± 2.4 mg/l respectively while in the PCMut2 were 48.3 ± 2.3, 25.2 ± 2.3 mg/l respectively and in the PCMut4 were 33.3 ± 3.2, 10.2 ± 3.6 mg/l respectively. Fluorescence parameters are used long ago in hydrobiology and oceanology for an assessment of the concentration of microalgae in water and their photosynthetic activity. There is a large number various the submersible fluorometers of different firms for these purposes. Researchers have therefore been trying to find a method of determining the chlorophyll a concentration from in situ fluorescence measurements.

Table 1 Parameters of OJIP kinetics of fluorescence induction curve of wild type and three mutant strains Parachlorella kessleri cells

JIP-test parameters		Wild type *PC**	*PC Mut2**	*PC Mut4**
F_V/F_M	Maximum quantum yield of charge separation in PS2	0.610 ± 0.01	0.607 ± 0.01	0.506 ± 0.01
V_J	Relative amplitude of the O-J phase	0.38 ± 0.1	0.47 ± 0.01	0.47 ± 0.003
V_I	Relative amplitude of the J-I phase	0.96 ± 0.02	0.96 ± 0.01	0.94 ± 0.004
Mo	Initial slope of the O-J fluorescence phase	0.60 ± 0.04	0.73 ± 0.02	0.87 ± 0.03
S_M	Area between the fluorescence kinetic curve (O-J-I-P) and the level of Fm normalized against the Fv value	40.97 ± 4.24	38.72 ± 2.32	30.06 ± 1.01
ABS/RC	Average value of absorbed photon flows in PS2 RC (of apparent size of the active antenna in PS2)	2.57 ± 0.2	2.85 ± 0.1	3.51 ± 0.2
qE	Capacity for pH-induced non-photochemical fluorescence quenching	0.14 ± 0.01	0.12 ± 0.01	0.10 ± 0.01
qPQ	Capacity of the quinone pool for fluorescence quenching	0.24 ± 0.04	0.20 ± 0.01	0.19 ± 0.02

*Figures are expressed as mean ± SD.
The kinetic parameters of fluorescence induction were recorded with M-PEA-2 instrument under actinic illumination of 1300 μmol photons $m^{-2} s^{-1}$.

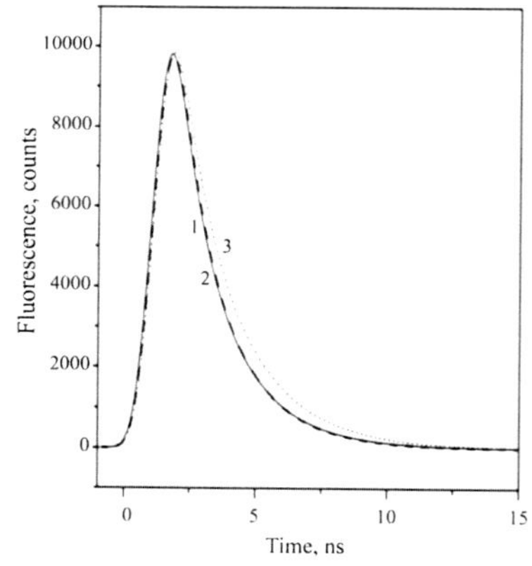

Figure 5. Kinetics of fluorescence decay. In wildtype (curve 1) and mutant strains of Parachlorella kessleri cells (PCMut2- curve 2) and (PCMut4 -curve

3). The measurements were performed with a Fluorolog-3 spectrofluorometer at 685 nm, using TCSPC method with sample activation with light at 1 Mhz frequency with a LED 390 nm by pulses with half-widths of 0.9 ns.

These would cover not only fluorescence induced naturally by sunlight but also that induced by artificial light sources. Measurements of the latter are either contact measurements carried out in situ with submersible fluorometers or remote methods using lidars (Pogosyan and Matorin 2005).

CONCLUSION

Microalgae, which use sun energy for synthesis of organic matter, are the main energy source in aquatic systems and serve as the food for other organisms. The linear chain of photosynthetic electron transport comprises two sequentially operating photosystems (PSII and PSI) interconnected through intermediary quinone pool of electron carriers. According to the Z-scheme of electrontransport chain (ETC.), activities of each photosystem affect the redox state of the other system. Such a relation between PSII and PSI is evident from chlorophyll fluorescence, whose level depends on the redox state of the quinone acceptor QA, as well as from the redox state of chlorophyll P_{700}, the primary electron donor of PSI. The photoreaction of PSII reduces Q_A, thus raising chlorophyll fluorescence and the extent of P_{700} reduction, while the PSI reaction converts P_{700} into its oxidized form (P_{700}^+), which promotes Q_A oxidation and fluorescence lowering. Simultaneous chlorophyll fluorescence and 820 nm transmission measurements by using M-PEA-2 allow evaluating the functional state of PSII and PSI. Analysis of fluorescence induction curves demonstrated inhibition of electron transport in PSII and an increase fraction of non-QB-reducing centers. This manifested itself also in an increase of the slow component in decay curve of fluorescence with half-time 2 ns in PC Mut4. In addition, analysis of induction curves of the delayed fluorescence and a decrease of NPQ revealed a decrease in the energization of photosynthetic membranes in mutant cells. Our study showed that the most sensitive parameters are induction curves of prompt and delayed fluorescence in mutant algal cells. Selected parameters are indeed viable tools for diagnostic alga state in biotechnological studies during obtaining mutant forms.

Table 2. Changes of amplitude and duration of fast (A1, τ1) and slow (A2, τ2) components of fluorescence decay kinetics of wild type and mutant strains Parachlorella kessleri cells

Samples	A1 (%)	A2 (%)	τ_1 (ns)*	τ_2 (ns)*
Wild type	40	60	1.0 ± 0.13	2.24 ± 0.03
PC Mut 2	37	63	0.97 ± 0.14	2.18 ± 0.03
PC Mut 4	28	72	1.09 ± 0.23	2.47 ± 0.04

*Figures are expressed as mean ± SD.

Table 3. Changes of fluorescence parameters of wild type and mutant strains of Parachlorella kessleri cells

Fluorescence parameters	Wild type*	*PC Mut2**	*PC Mut4**
F_V/F_M	0.58 ± 0.01	0.5 ± 0.01	0.45 ± 0.02
NPQ	0.179 ± 0.02	0.133 ± 0.003	0.083 ± 0.004
$rETR_{max.}$ (r.u.)	15.5 ± 1.3	12.06 ± 0.4	8.3 ± 1.4

*Figures are expressed as mean ± SD.
F_V/F_M – samples in darkness, $rETR_{max}$ – maximum relative rate of electron transport and $NPQ = (F_M - F_M')/F_M'$ – non-photochemical fluorescence quenching at illumination 800 μmol photons m^{-2} s^{-1}.

REFERENCES

1. Antal TK, Matorin DN, Ilyash LV, Volgusheva AA, Osipov VA, Konyuhov IV, Krendeleva TE, Rubin AB (2009) Probing of photosynthetic reactions in four phytoplanktonic algae with a PEA fluorometer. Photosynth Res 102:67–76
2. Bulychev AA, Osipov VA, Matorin DN, Vredenberg WJ (2013) Effects of far-red light on fluorescence induction in infiltrated pea leaves under diminished ΔpH and Δφ components of the proton motive force. J Bioenerg Biomembr 45:37–45
3. Charles DA (2008) Algal Chemical Ecology.ch. 13, U. Springer-Verlag Berlin Heidelberg, Karsten Fott B, Nováková M (1969) A monograph of the genus Chlorella. The fresh-water species. In: Fott B (ed) Studies in Phycology. Academia, Prague, pp 10–74
4. Goltsev V, Zaharieva I, Chernev P, Strasser RJ (2009) Delayed fluorescence in photosynthesis. Photosynth Res 101:217–232
5. Gorman DS, Levine RP (1965) Proc Natl Acad Sci U S A 54:1665–1669 Harris EH (1989) The Chlamydomonas Sourcebook: A Comprehensive Guide to Biology and Laboratory Use. Academic Press, San Diego, 780
6. Herlory O, Richard P, Blanchard GF (2007) Methodology of light response curves: Application of chlorophyll fluorescence to microphytobenthic biofilms. Mar Biol 153:91–101
7. Holzinger A, Lütz C (2006) Algae and UV irradiation: effects on ultrastructure and related metabolic functions. Micron 35:190–207
8. Krienitz L, Hegewald EH, Hepperle D, Huss VAR, Rohr T, Wolf M (2004) Phylogenetic relationship of Chlorella and Parachlorella gen. nov. (Chlorophyta, Trebouxiophyceae). Phycologia 43:529–42
9. Lazár D (2009) Modelling of light-induced chlorophyll a fluorescence rise (O-J-I-P transient) and changes in 820 nm-transmittance signal of photosynthesis. Photosynthetica 47:483–498
10. Lippemeier S, Harting P, Colijn F (1999) Direct impact of silicate on the photosynthetic performance of the diatom Thalassiosira weissflogii assessed byon and off-line PAM fluorescence measurements. J Plankton Res 21:269–283
11. Matorin DN, Rubin AB (2012) Chlorophylla fluorescence in higher plants and algae Moscow. IKI-RKhD, Izhevsk, p 256

12. Matorin DN, Antal TK, Ostrowska M, Rubin AB, Ficek D (2004) Chlorophyll fluorometry as a method for studying light absorption by photosynthetic pigments in marine algae. Oceanologia 46:519–531
13. Matorin DN, Todorenko DA, Seifullina NK, Zayadan BK, Rubin AB (2013) Effect of silver nanoparticles on the parameters of chlorophyll fluorescence and P700 reaction in the green alga Chlamydomonas reinhardtii. Microbiology 82:862–867 Merchant SS, Simon EP, Olivier V (2007) The Chlamydomonas Genome Reveals the Evolution of Key Animal and Plant Functions. Sci 318:245–251
14. O'Connor DV, Phillips D (1984) Time correlated single photon counting. Academic Press, London Oukarroum A, Goltsev V, Strasser RJ (2013) Temperature effects on pea plants probed by simultaneous measurements of the kinetics of prompt fluorescence, delayed fluorescence and modulated 820 nm reflection. PLoS One 8:1–10
15. Pogosyan SI, Matorin DN (2005) Variability in the state of the photosynthetic system of the Black Sea phytoplankton. Oceanology 45(1):139–148
16. Schansker G, To'th SZ, Holzwarth AR, Garab G (2014) Chlorophyll a fluorescence: beyong the limits of the Qa model. Photosynth Res 120:43–58
17. Schreiber U (2004) Pulse-Amplitude (PAM) fluorometry and saturation pulse method: an overview. In: Papageorgiou G, Govindjee (eds) Chlorophyll fluorescence: A signature of Photosynthesis and Respiration, vol . Springer, Dordrecht, pp 79–319
18. Strasser RJ, Tsimilli-Michael M, Srivastava A (2005) Analysis of chlorophyll a fluorescence transient. In: Papageorgiou GC and Govindjee (eds) Chlorophyll fluorescence: a signature of photosynthesis. Kluwer Academic Publishers, Dordrecht/Boston/London, pp 321–362
19. Strasser RJ, Tsimilli-Michael M, Sheng Q, Goltsev V (2010) Simultaneous in vivo recording of prompt and delayed fluorescence and 820-nm reflection changes during drying and after rehydration of the resurrection plant Haberlea rhodopensis. Biochim Biophys Acta 1797:1313–1326
20. Volgusheva AA, Zagidullin VE, Antal TK, Korvatovsky BN, Krendeleva TE, Paschenko VZ, Rubin AB (2007) Examination of chlorophyll fluorescence decay kinetics in sulfur deprived algae Chlamydomonas reinhardtii. BiochimBiophys Acta 1767:559–564

21. Wraight CA, Crofts AT (1971) Delayed fluorescence and the high-energy state of chloroplasts. Eur J Biochem 19:386–397

INDEX

D

E

F

G

H

I

T

U

V

W

Y

Printed and bound by PG in the USA